Exploring the Way Life Works

THE SCIENCE of BIOLOGY

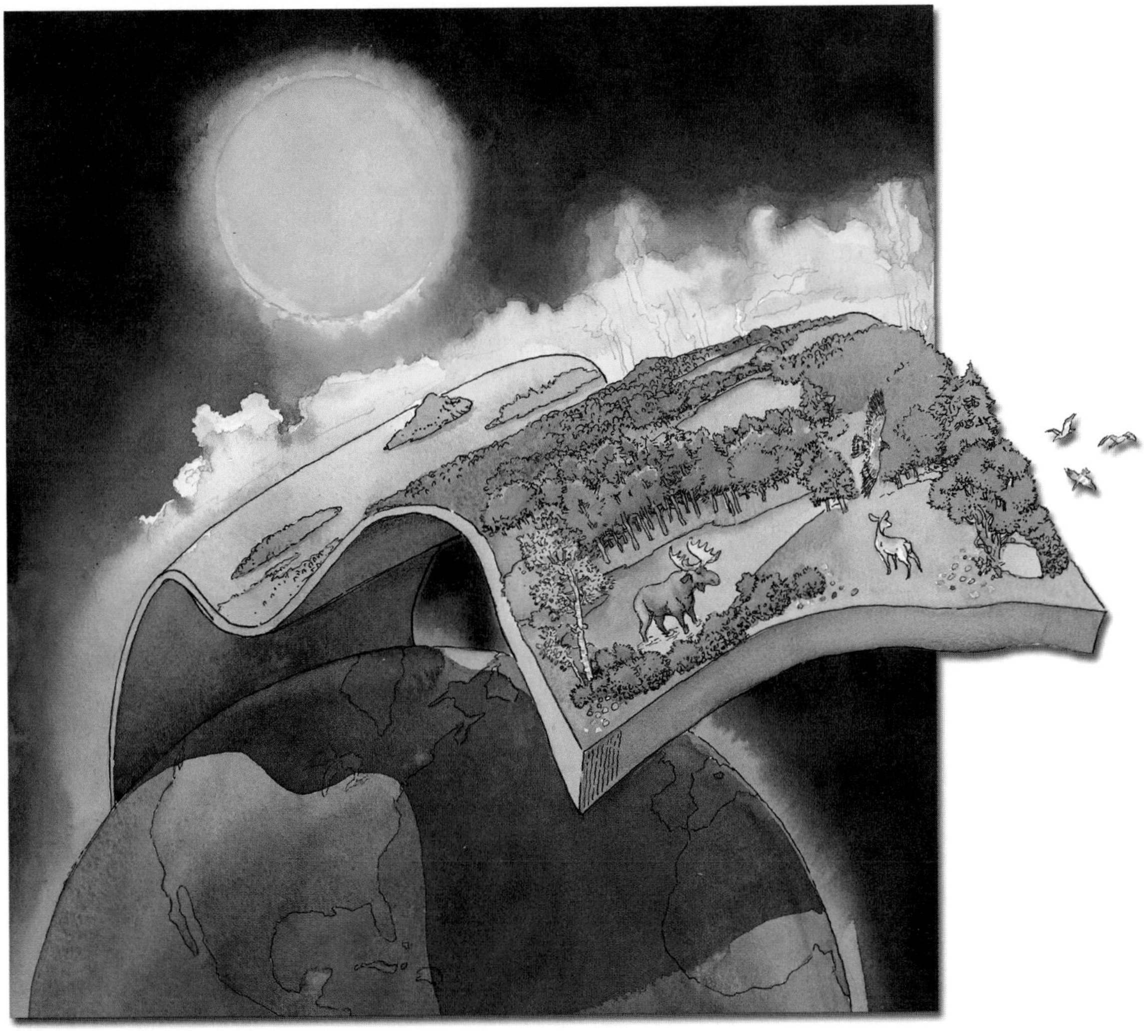

World Headquarters
Jones and Bartlett Publishers
40 Tall Pine Drive
Sudbury, MA 01776
978-443-5000
info@jbpub.com
www.jbpub.com

Jones and Bartlett Publishers Canada
2406 Nikanna Road
Mississauga, ON L5C 2W6
CANADA

Jones and Bartlett Publishers International
Barb House, Barb Mews
London W6 7PA
UK

To Bernard D. Davis,

Judith Church, Robin Hoy,

Jay Hoagland, and Betty Loomis

Production Credits
Chief Executive Officer: Clayton Jones
Chief Operating Officer: Don W. Jones, Jr.
Executive V.P., Publisher: Tom Manning
V.P., Managing Editor: Judith H. Hauck
V.P., Design and Production: Anne Spencer
V.P., Manufacturing and Inventory Control: Therese Bräuer
Director, Interactive Technology: W. Scott Smith
Production Editor: Jennifer Angel
Associate Editor: Victoria Jones
Design and Composition: Nicolazzo Productions
Copy Editor: Jane Hoover
Cover Design: Anne Spencer (art by Bert Dodson)
Printing and Binding: Courier Kendallville
Cover Printing: Courier Companies

Originally published, in different form, by Times Books, a division of Random House, Inc.

Library of Congress Cataloging-in-Publication Data
Hoagland, Mahlon B.
Exploring the way life works: the science of biology/Mahlon Hoagland, Bert Dodson, Judith Hauck.
p. cm.
Includes bibilographical references (p.).
ISBN 0-7637-1688-X (softcover: alk. paper)
1. Life (biology) I. Dodson, Bert. II. Hauck, Judith. III. Title.

QH501 .H57 2001
570--dc21 00-067790

Printed in the United States

05 04 03 02 01 1 2 3 4 5 6 7 8 9 10

Contents

The unity that underlies diversity. Science as an organized, accessible manifestation of human curiosity.

A panorama of sixteen of life's key features.

How living creatures transform energy, and how energy flows through organisms and through communities.

Chapter 3 Energy—Light To Life 87

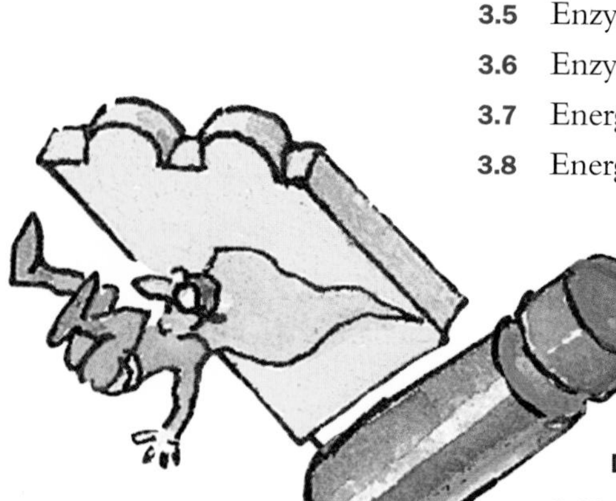

How the four-letter language of DNA spells out instructions for building millions of life forms.

Chapter 4 Information—The Storehouse of Know-How 139

How DNA's instructions are transcribed into molecules of RNA and then translated into proteins.

Chapter 5 Machinery—Building Smart Parts 183

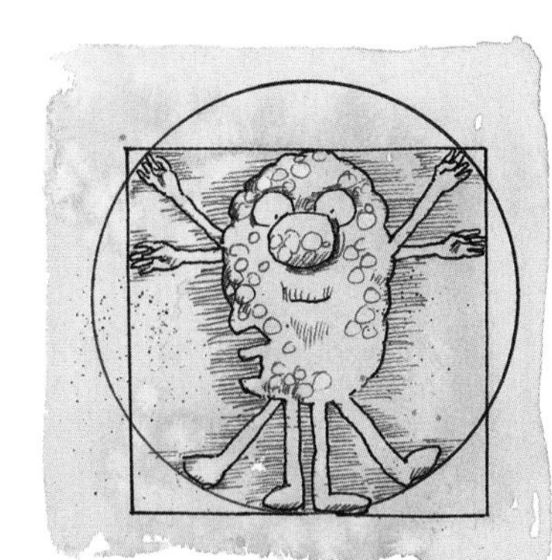

Homeostasis: how cells, organisms, and communities regulate their internal environments.

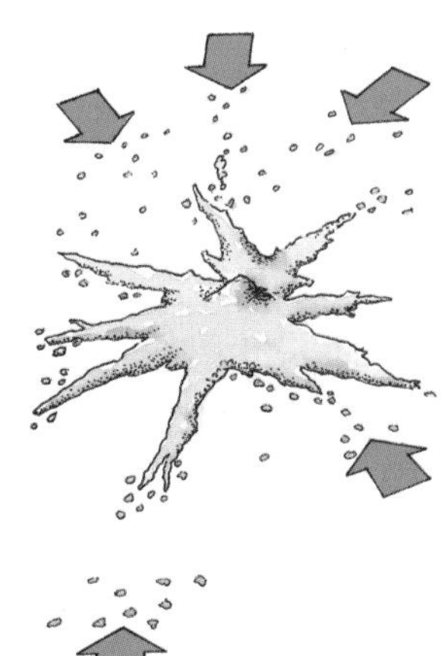

Chapter 6 Feedback—Signaling, Sensing, Reacting 209

How a cell diversifies into communities of specialized cells, organs, system feedback loops, and how individuals interact and organize.

Chapter 7 Community—E Pluribus Unum 243

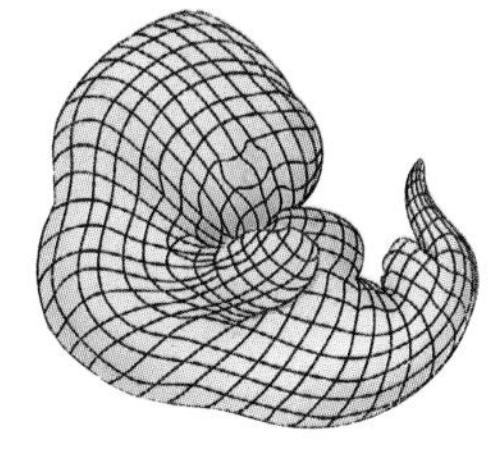

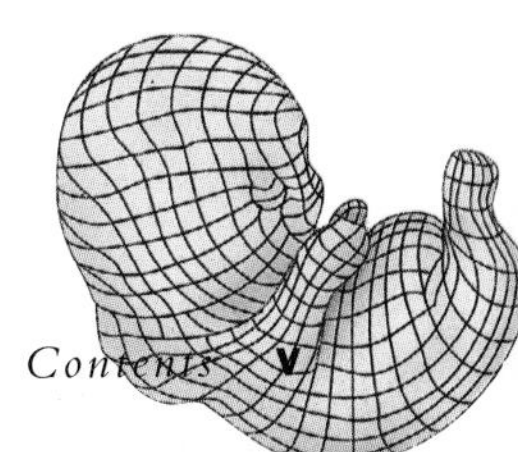

How the river of DNA information flows across generations, and how the information of DNA is sifted and sorted by cells and selected by environments.

Preface

Authors' Notes

When two of us — biologist and artist — first met in 1988, we discovered that we shared a fascination with the unity of life — how, deep down, all living creatures, from bacteria to humans, use the same materials and ways of doing things.

We began exploring ways we might share our wonder with others and came to believe that we could achieve our purpose through an intimate merging of science and art. In the process, we hoped to persuade our readers that a deeper understanding of nature would enhance their appreciation of its beauty — and thereby enrich their lives.

Scientist as teacher and artist as student explained, questioned, searched and argued. One day, Bert emerged with a two-page spread of pictures and Mahlon got a new vision of what he thought he knew; artist became teacher, scientist became student. Our confidence grew. We sifted, sorted, and pieced together our interpretation of the way life works, and turned it into a book — *The Way Life Works,* published in 1995 by Times Books. To our delight, people bought it. More especially, teachers bought it, and started using it in their classes. One teacher, Judy Hauck, Mahlon's daughter, brought many other innovative teachers into the picture and a new book began to emerge — a textbook, *Exploring the Way Life Works — The Science of Biology.*

The scientist wants to leave the reader with a feeling of awe and pride in the achievements and future of scientific exploration, in the human potential for ever-deeper understanding. The artist, on the other hand, sees the possibility that an appreciation for our oneness with the living world can guide our individual actions as we shape our collective future. The teacher wants to reach the curious and interested and inspire them to further exploration, and reach the bored and listless to communicate the accessibility and excitement of science so they'll never be bored and listless again. We hope our readers will be moved by all three missions.

To the Student

This book is about the way ALL life works.

If you want to know what all living things have in common with each other and with you, and want to understand how living things function, diversify, and evolve over time, this is the book for you.

If you want to get an idea of how people got to know what they know so far about life's processes and interrelationships, and if you want to get a sense of what the science of biology is and where it's heading, you'll be sure to enjoy the Doing Science and the Tools of Science papers, the Questions, and the References and Great Reading.

We think you'll find that this is fun and fascinating stuff, so above all, ENJOY. You are a part of the ongoing process of science, too.

To the Instructor

This book is organized around one central idea — the amazing unity that underlies biological diversity. Its chapters emphasize broad, unifying themes, and illustrate these themes with dramatic and intriguing specific cases.

The concepts are concisely articulated in the text, and explained with equal conciseness and clarity by the accompanying graphics. This balance of verbal and graphic explanation answers an often-stated need for scientifically accurate and responsible teaching that addresses students' different learning styles. The book distills a lot of important science into a very small space.

This book is permeated with the process of science. The Doing Science papers abstract current and historically interesting research papers and the Tools of Science discussions describe classic and current research tools and technologies. Throughout, there is an understanding that science is not owned by scientists — that many of the questions science asks are formalized versions of the questions most curious people ask. It focuses, too, on the context in which science happens, and shows that science arises from careful observation of nature, a province accessible to all of us, as demonstrated by the work of the artist, novelist, and poet observers included in each chapter.

While this is NOT a textbook that emphasizes vocabulary, hierarchies, or the exhaustive detailing of biological processes, it does encompass the important conceptual and factual material found in most short introductory biology textbooks. And the associated web site and instructor's tools expand on the book's content.

Because this book is unlike traditional biology textbooks in its focus and emphasis, we've provided some tools to help you make a painless transition to using it.

Instructor's ToolKit CD-ROM

- A set of illustrated PowerPoint™ lecture outline slides arranged by chapter. These slides cover all the material in the book, with additions such as more photomicrographs, more illustrations, more detail. There are also a slides that may be useful to you that cover cell structure and function in greater detail than is to be found in the text itself.
- An Image Bank that contains all of the book's illustrations. These full color digital images can be used to create your own slide set, to print to transparency acetates, or to import into tests or homework handouts.
- A TestBank with questions and answers based on the book's content.

Instructor's Manual

- The text of all the PowerPoint™ slides. Material that expands on the book's contents is indicated in boldface type.
- Chapter objectives, key terms, and hints for alternative ways of sequencing chapter material.
- Class discussion questions with suggested answers.
- Web exercises and links that help you preview the exercises and links found on the book's website.

www.jbpub.com/connections — Exploring The Way Life Works Web Site.

- Links to carefully chosen external web sites, arranged by chapter and page number. Links are accompanied by questions or exercises that focus on both book and web site material.
- An area where teachers can exchange their ideas for innovative ways of using the book and web site, or for making the transition from a more traditional book to this one.
- A communal archive of additional material to be used as class handouts. Here we encourage the book's users to contribute short (400 words or less), interesting applications of the book's concepts, or abstracts of relevant research papers.

Acknowledgments

Elaine A. Alexander
Washington University

Cydney C. Brooks
AdipoGenix, Inc.

John B. Beaver
Western Illinois University

Lesley Blair
Oregon State University

Elizabeth Godrick
Boston University

Jeffrey Goodman
Appalachian State University

Amy Handshoe
Hazard Community College

David Harbster
Paradise Valley Community College

Beth Kelley

Mark Lavery
Oregon State University

Ann S. Lumsden
Florida State University

Patricia A. Mancini
Bridgewater State College

Kevin Padian
University of California, Berkeley

Javier Penalosa
Buffalo State College

Jeffre Witherly
National Human Genome Institute

Everyone involved with making this book seemed to be inspired by its potential for giving students a better understanding of biology as a fascinating and enjoyable science. Susan Shephard has been lovingly involved in research and writing from the beginning. Dick Morel contributed text, art, and entomological enthusiasm. Mark Lavery, Lesley Blair, Amy Handshoe, Kevin Padian, John Beaver, and Jane Hoover read every word, suggested improvements and helped catch errors or potential miscommunications. We didn't take advantage of every one of their suggestions, so any errors you find are ours, not theirs.

Jennifer Angel did far more than production work. She contributed research papers, tracked down hard-to-find photos, and made the book better than the one turned over to her. Mary Hill stepped in late in the game to shape us up, thank goodness. Anne Spencer designed the front matter and cover, and Lianne Ames and Philip Regan finalized pages. Thank you. And thanks to Jason Miranda for the wasp-spider story.

CHAPTER 1

SETTING THE STAGE

IMAGINE YOU ARE WALKING ALONG A DESERTED BEACH AND YOU COME UPON THE CARCASS OF a whale. Time and tide and carrion birds have taken much of the flesh. Your first reaction might be a compassionate recognition of kinship. You might be curious about what happened—what was this whale's story?

As you examine the skeleton, a pattern strikes you. In each of the whale's front fins, the bones are arranged in three sections, with one bone in the section closest to the body, two parallel bones in the middle section, and five radiating branches of smaller bones in a more complex outer section. In fact, the bones of a whale's fin look very much like those of a human arm and hand. The proportions differ, but the pattern is remarkably similar.

How is it that a whale has arms like yours? And why does a whale have finger bones when it doesn't have fingers? Does this mean we're related to whales? Could it be that this limb pattern has been around longer than either whales . . . or humans?

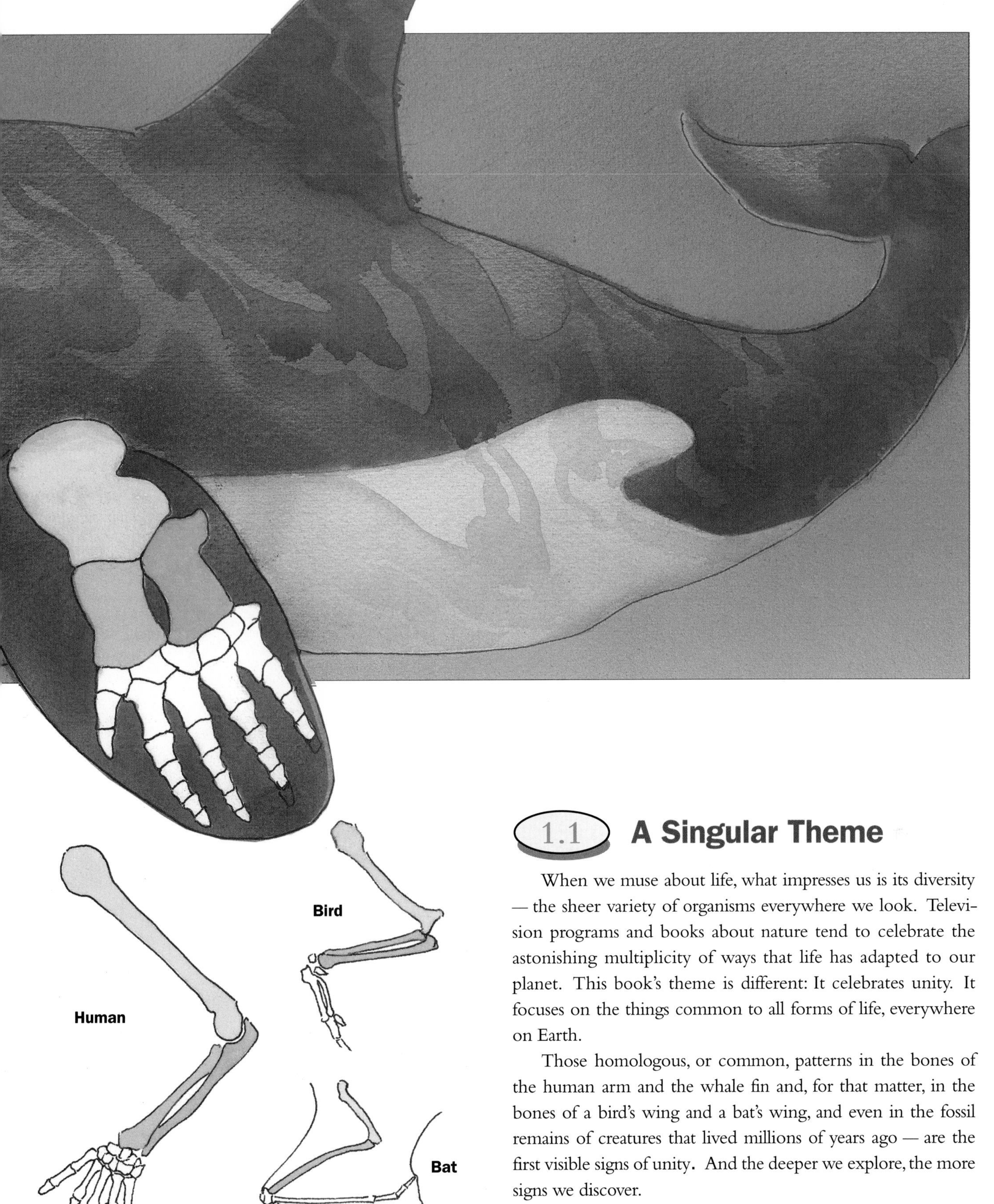

1.1 A Singular Theme

When we muse about life, what impresses us is its diversity — the sheer variety of organisms everywhere we look. Television programs and books about nature tend to celebrate the astonishing multiplicity of ways that life has adapted to our planet. This book's theme is different: It celebrates unity. It focuses on the things common to all forms of life, everywhere on Earth.

Those homologous, or common, patterns in the bones of the human arm and the whale fin and, for that matter, in the bones of a bird's wing and a bat's wing, and even in the fossil remains of creatures that lived millions of years ago — are the first visible signs of unity. And the deeper we explore, the more signs we discover.

Every living being is either a cell or is made of cells: tiny, animate entities that gather fuel and building materials, produce usable energy, and grow and duplicate.

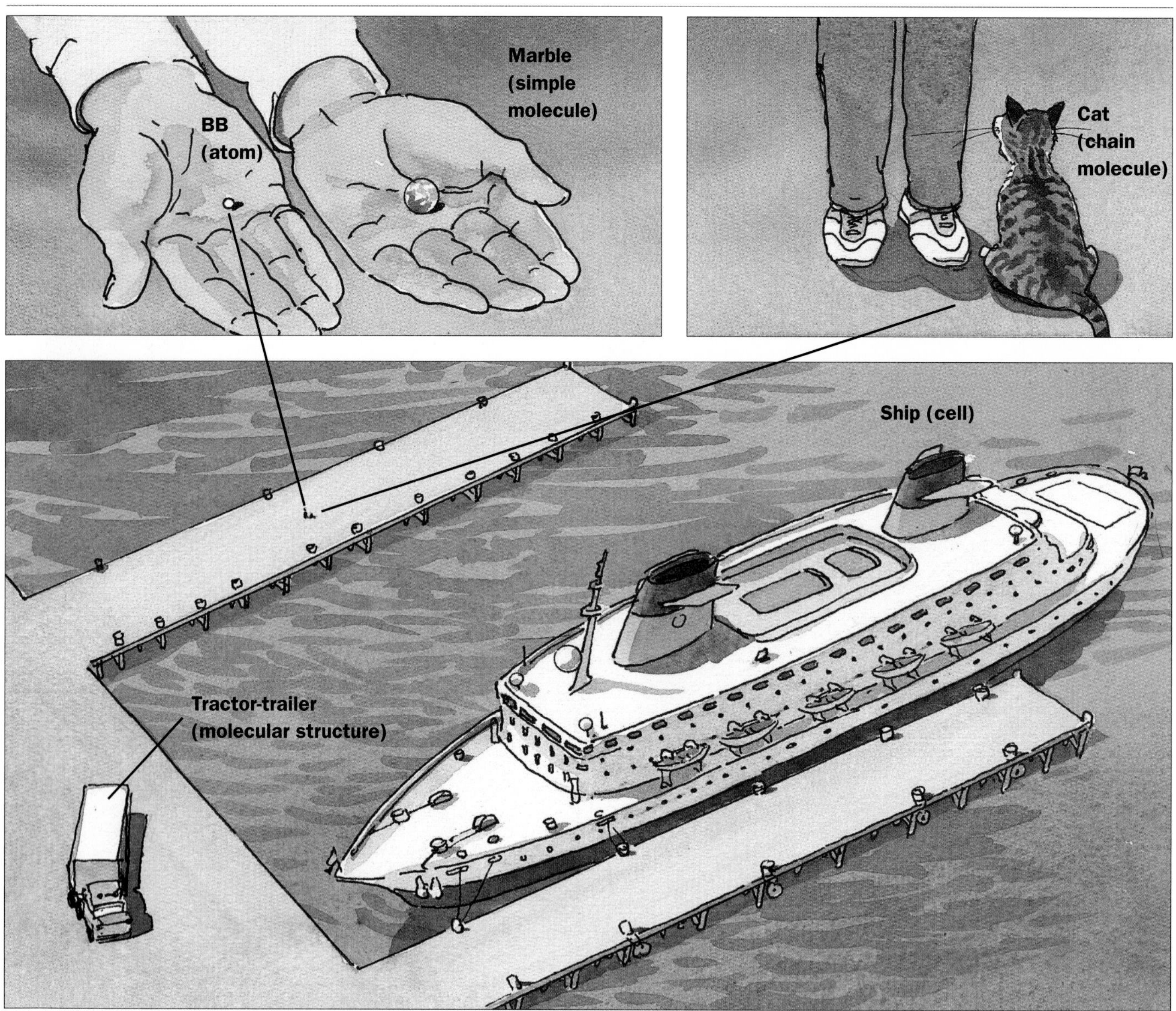

1.2 Thinking Small

Much of this book takes place inside the cell. If you are unfamiliar with this microscopic landscape, understanding just how small and how numerous molecules are requires a considerable stretch of the imagination.

The great Scottish mathematician and physicist Lord Kelvin said: "Suppose that you could mark the molecules in a glass of water; then pour the contents of the glass into the ocean and stir the latter thoroughly so as to distribute the marked molecules uniformly throughout the seven seas; if then you took a glass of water anywhere out of the ocean, you would find in it about a hundred of your marked molecules."

Size and speed are related. Generally, the smaller an object is, the faster it can move. Water molecules, and all the other thousand or so kinds of molecules you have within you, swim about at stupendous speeds, flashing past each other and bumping into each other every millionth of a millionth of a second.

Life depends upon these frequent and vigorous collisions. It becomes a little easier to grasp the speed of the life-sustaining chemical transformations that constantly occur inside your cells (at the rate of thousands of events per second) when you realize that the participants move and collide millions of times faster.

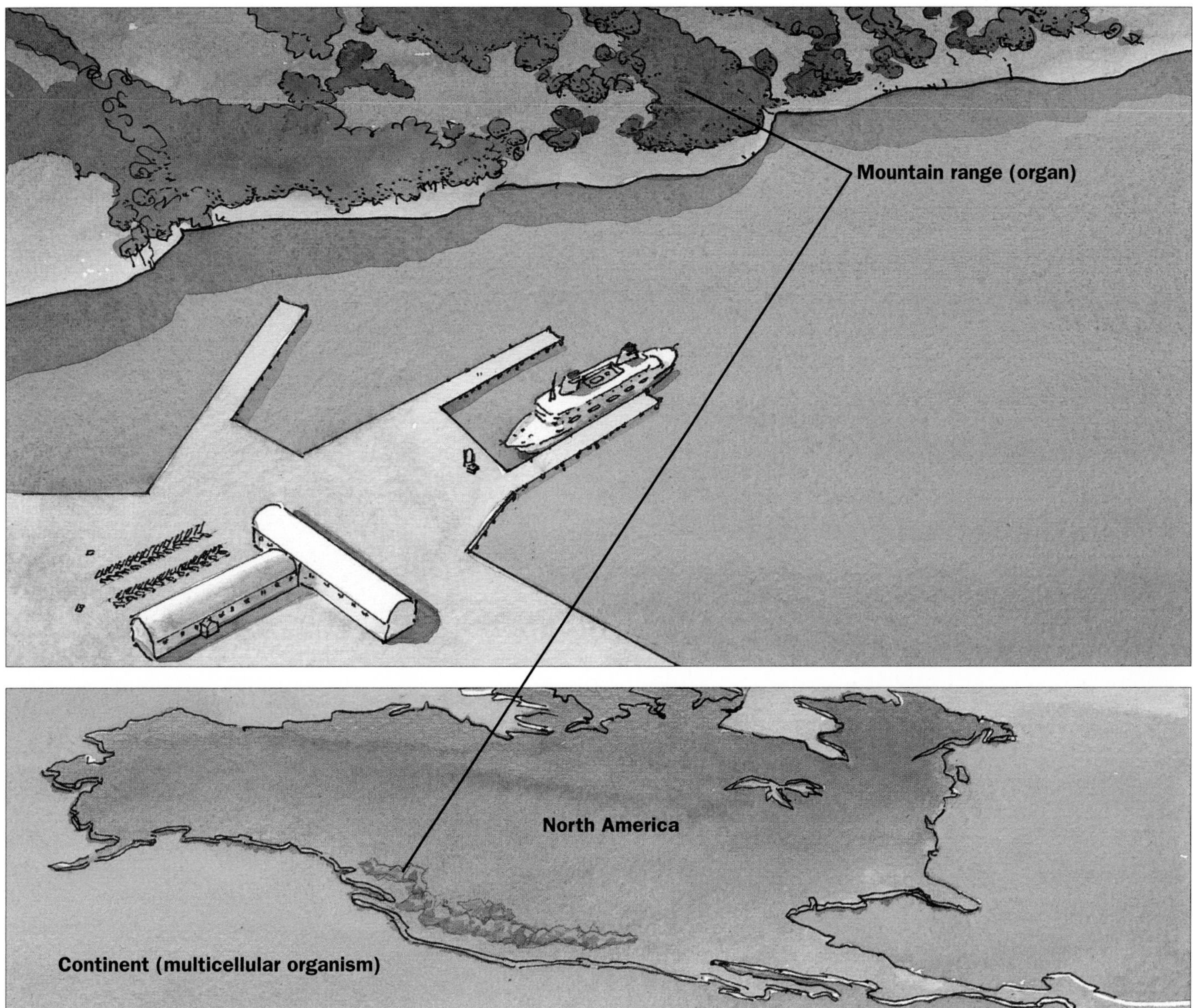

When we think about parts of the body, we tend to think of organs: lungs, heart, brain, etc. The next step down in size brings us to the cells of which those parts are made. That drop in size is immense. Human cells are about ten times smaller than the point of a pin, and your body is composed of 5 trillion of them. Within each cell are multitudes of atoms, molecules, and structures made of molecules — the principal characters in our story.

As we introduce them, the picture above might help you to grasp their relative sizes.

Imagine you are standing on a pier. In one hand, you hold a BB — its size will represent an atom. In the other hand, you hold a marble — analogous to a simple molecule. Next to you is a cat — a chain molecule. Parked nearby is a tractor-trailer truck — a molecular structure. Tied up at the pier is an ocean liner — a cell. The pier is on the coast of North America — the whole continent being analogous in size to a human being.

On the following four pages we present a visual guide for distinguishing small things. Notice that four separate scales are necessary for spanning the range of size from atom to cell (a 200,000-fold jump in size).

From Atoms to Cells — Comparative Sizes

Scale 1. Atoms and Molecules

Magnified 50 Million Times

Atoms are the elemental units of which everything in the universe, living and non-living, is made. Atomic diameters range from one to a few hundred millionths of an inch.

Molecules are atoms bonded together. Much of life depends on three tiny molecules that have 2 to 3 atoms apiece: carbon dioxide (CO_2), the ultimate source of life's carbon atoms; oxygen (O_2), the gas crucial to energy generation in most life forms; and water (H_2O), the sea inside our cells in which life's machinery is bathed, and which aids chemical events inside our cells.

Carbon dioxide

Water

Oxygen (gas)

Roughly one thousand different kinds of slightly larger molecules made of 10 to 35 atoms are also found inside cells. These small molecules are either food (fuel) or building materials, or molecules that have been or will be food or building materials. We call all these simple molecules. The important ones in this book are sugars, nucleotides, and amino acids.

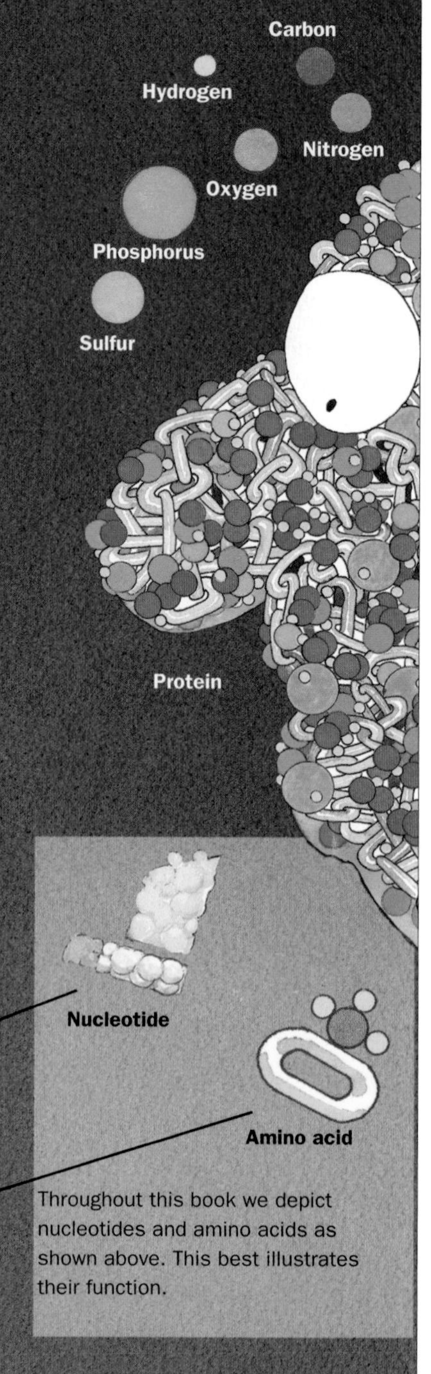

Throughout this book we depict nucleotides and amino acids as shown above. This best illustrates their function.

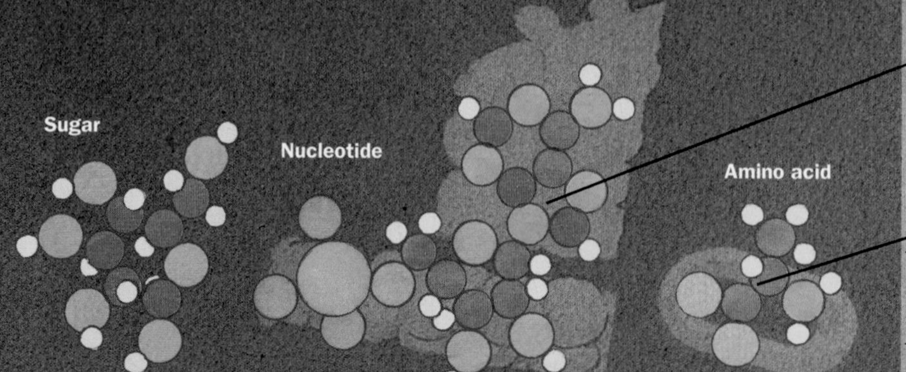

Scale 2. Chain Molecules

Magnified 10 Million Times

The vital working parts inside cells are chain molecules — very long strings of many simple molecules linked to one another. The most numerous of the chain molecules are proteins, which consist of 300 to 400 or more amino acids strung end to end. Each protein molecule — there are thousands of different kinds — has a special job to do in the cell. Cells also contain many varieties of ribonucleic acid (RNA), which can have tens of thousands of linked nucleotides, and deoxyribonucleic acid (DNA), which can have millions of nucleotides.

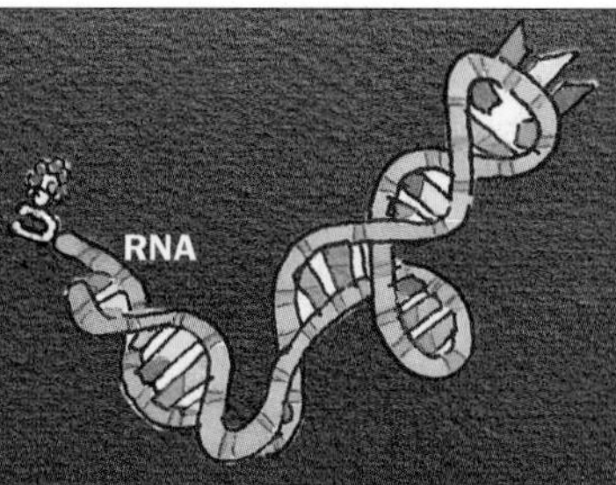

RNA

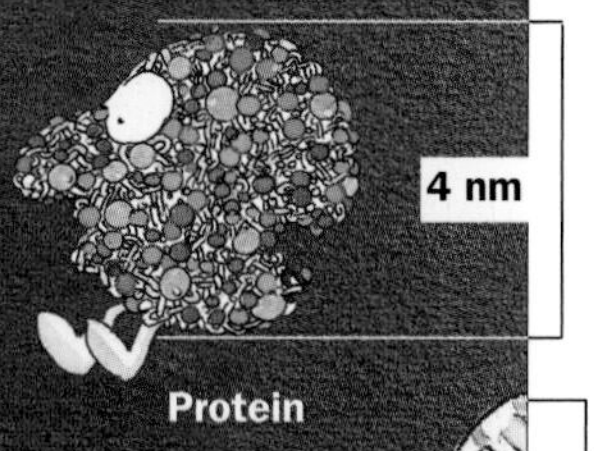

4 nm

Protein

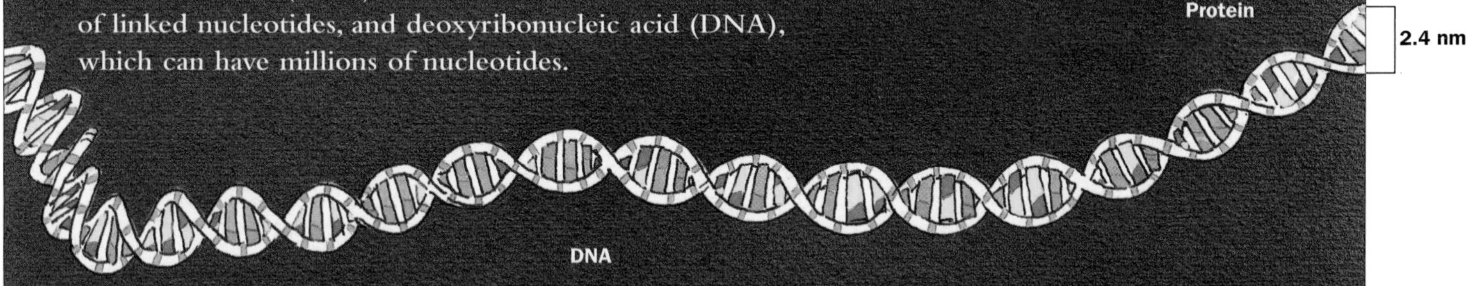

2.4 nm

DNA

Scale 3. Molecular Structures

Magnified 1 Million Times

Chain molecules can fit together forming complex architectural arrangements — called molecular structures — inside cells. These complex molecules are the cell's infrastructure, the equivalent of its roads, tunnels, power plants, factories, and libraries. Shown here are a ribosome, the cell's protein-making factory, and a bit of a mitochondrion, the cell's energy generator.

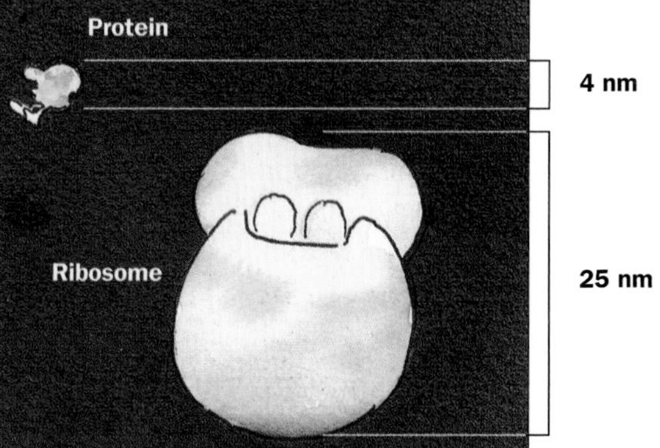

Protein

4 nm

Ribosome

25 nm

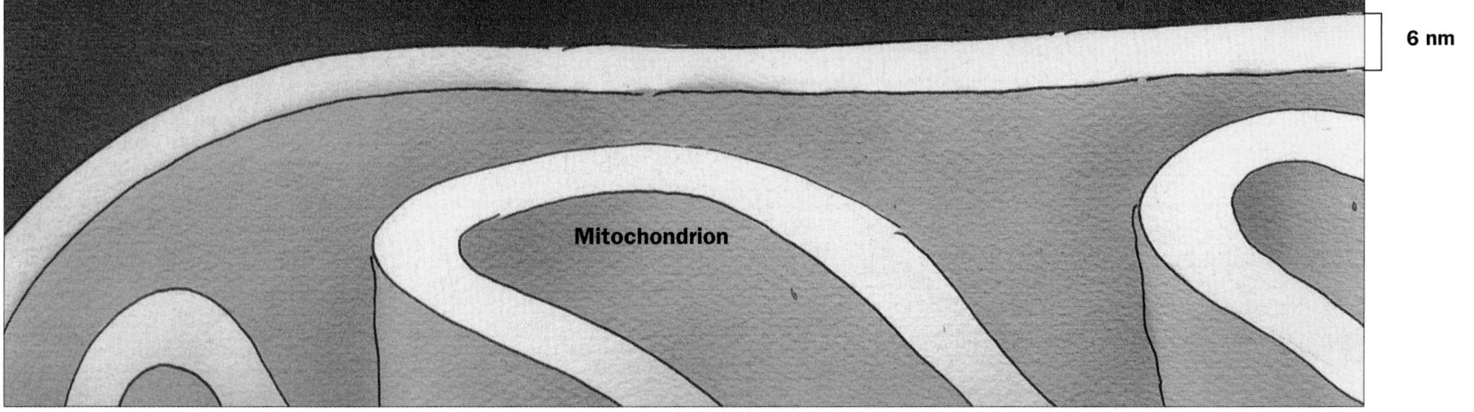

6 nm

Mitochondrion

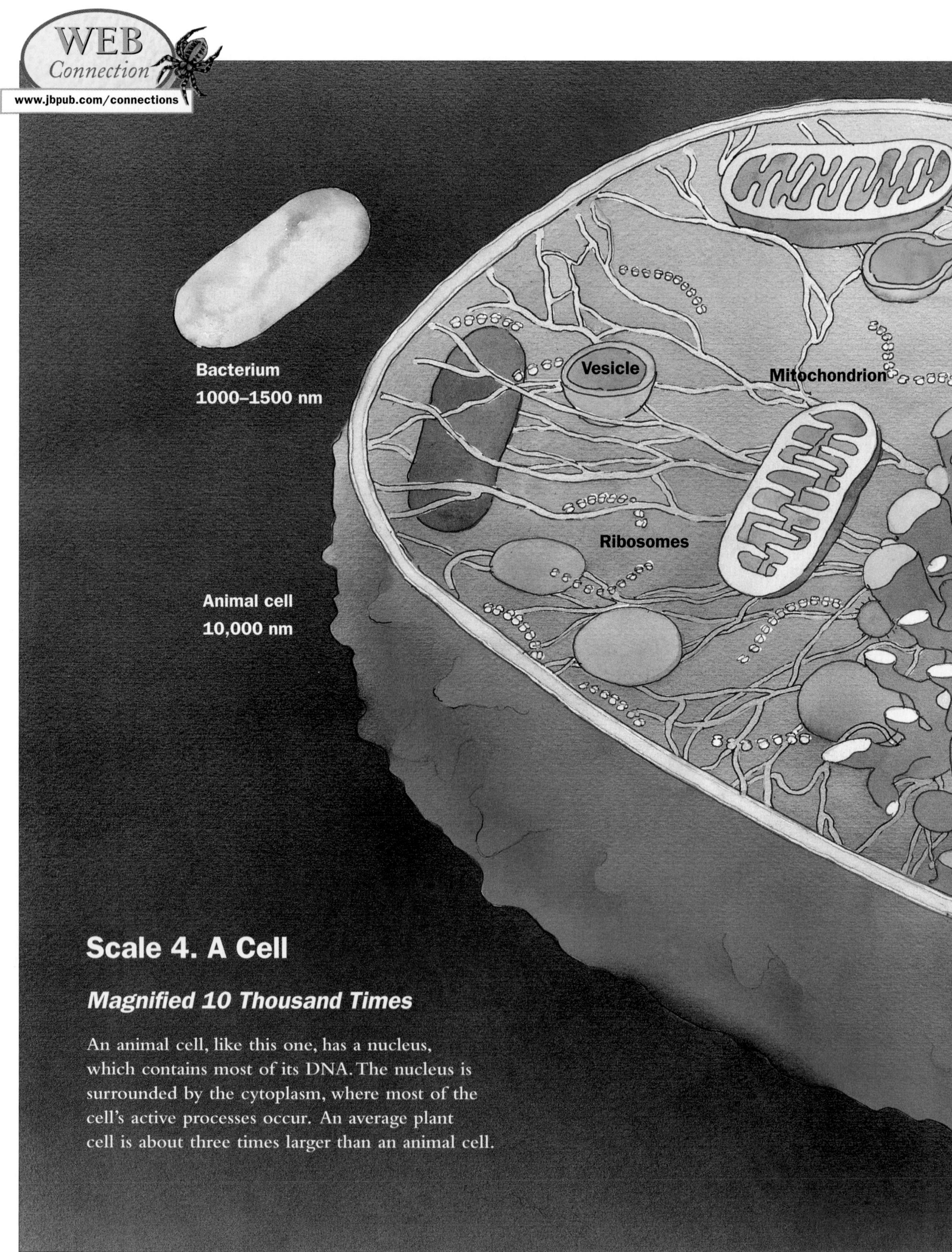

Scale 4. A Cell

Magnified 10 Thousand Times

An animal cell, like this one, has a nucleus, which contains most of its DNA. The nucleus is surrounded by the cytoplasm, where most of the cell's active processes occur. An average plant cell is about three times larger than an animal cell.

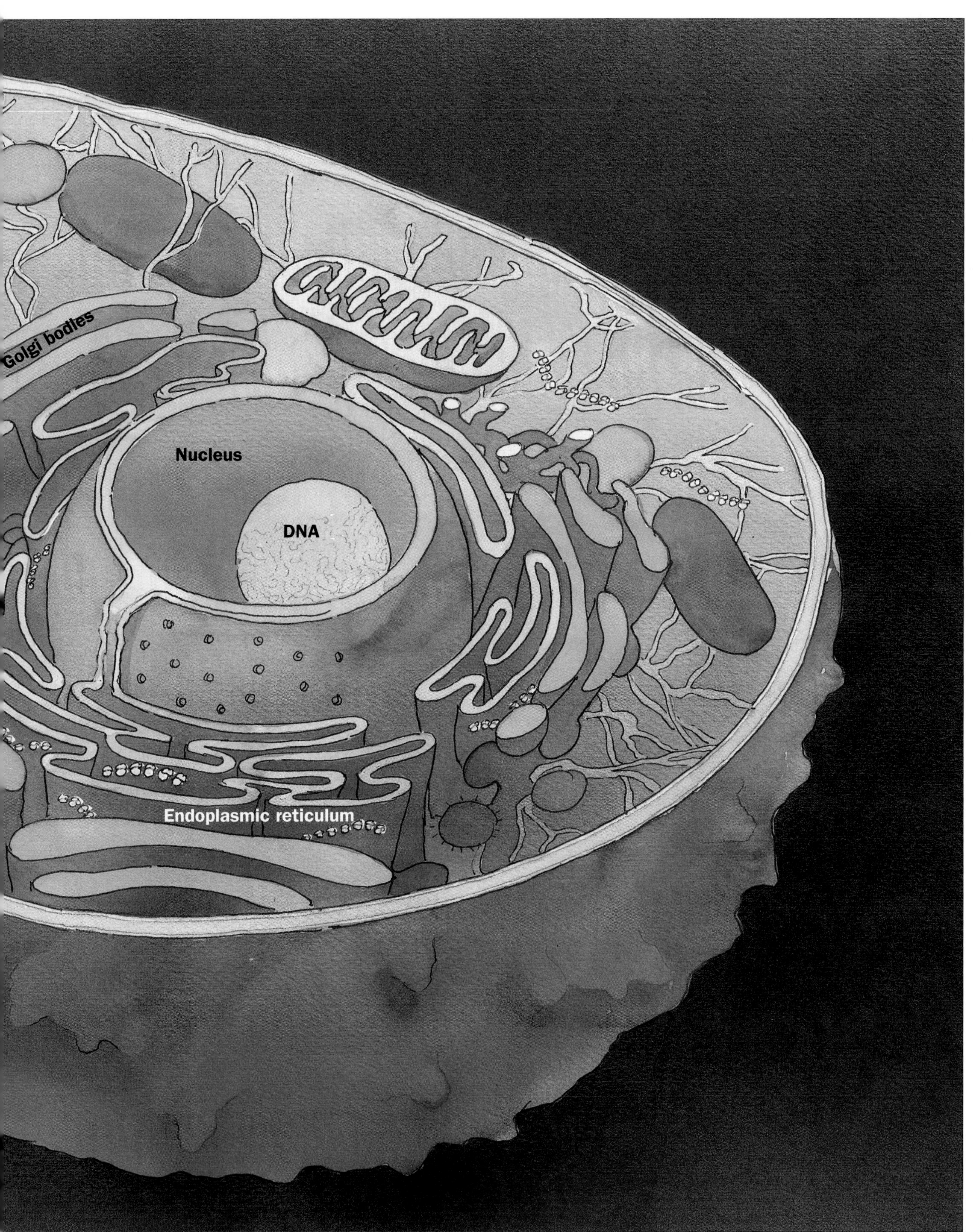
Golgi bodies
Nucleus
DNA
Endoplasmic reticulum

Robert Hooke — discovering a new world

Hooke was so enchanted by the new world of tiny things that he studied and observed anything and everything he could. Among his *Descriptions of Minute Bodies Made by Magnifying Glasses With Observations and Inquiries Thereupon,* recorded in 1665, were these snowflake structures at four different magnifications.

Using Microscopy to Explore the Cell and Beyond

Focus on the Minuscule

All the pictures you see on pages 6 and 7 of atoms, molecules, chain molecules, molecular structures — including a bit of a mitochondrion — depict things that until this century were invisible and thus unknown to us. It was not until the mid 1600s that the first evidence of the existence of things smaller than the unaided eyes could see was gathered.

In the 1660s, Robert Hooke was the first to report using a magnifying lens to systematically study the microscopic world (see above). It was he who first applied the term "cell" to each of the densely packed chambers he saw in a thin slice of cork. As he carefully drew each one, he was reminded of monks' cubicles — otherwise known as cells. This was the first documented microscopic observation of what we now know to be the basic structural unit of living things.

During the next three centuries, many improvements were made to the basic microscope (called the light microscope), and people were able to observe a previously unexplored universe of microstructures and microorganisms. Cells and many details of their surfaces and interiors, bacteria and their moving parts, and millions of other tiny living things are vividly observable under a light microscope. But the light microscope has its limits. To get a closer view, a new technology had to be created. To communicate the sizes of the minuscule things observed, a new system of measurement had to be devised.

All the units we use for measuring the sizes of these minuscule structures are subdivisions of a meter — the standard unit of length used in the metric system (or International System of Units). A centimeter is 1 hundredth of a meter (10^{-2}), a millimeter is 1 thousandth (10^{-3}), a micrometer is 1 millionth (10^{-6}), and a nanometer — the unit we use on pages 7 and 8 — is 1 billionth of a meter (10^{-9}).

The scales on a butterfly's wing magnified 70 times.

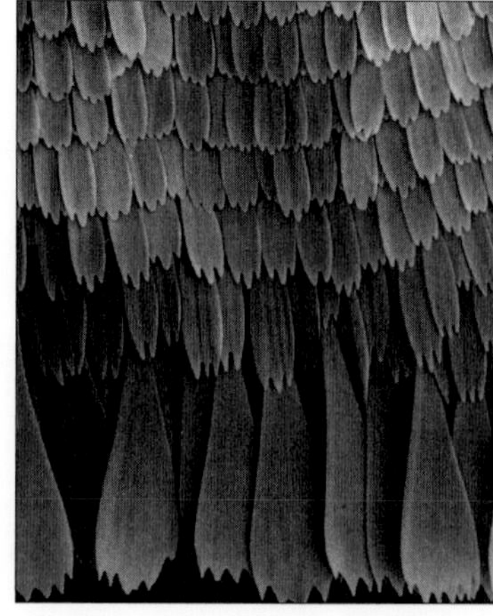

I wish [that man], before entering into larger studies of nature . . . [would] look on himself and get to know the proportions between nature and man . . . let him behold the tiniest things he knows of. Let a mite show him in the smallness of his body parts incomparably smaller legs with joints, veins in the legs, blood in the veins, humours in the blood, vapors in the drops, which dividing to the smallest things, he wears out his imaginative powers. . . .

Blaise Pascal, *Pensées,* 1669

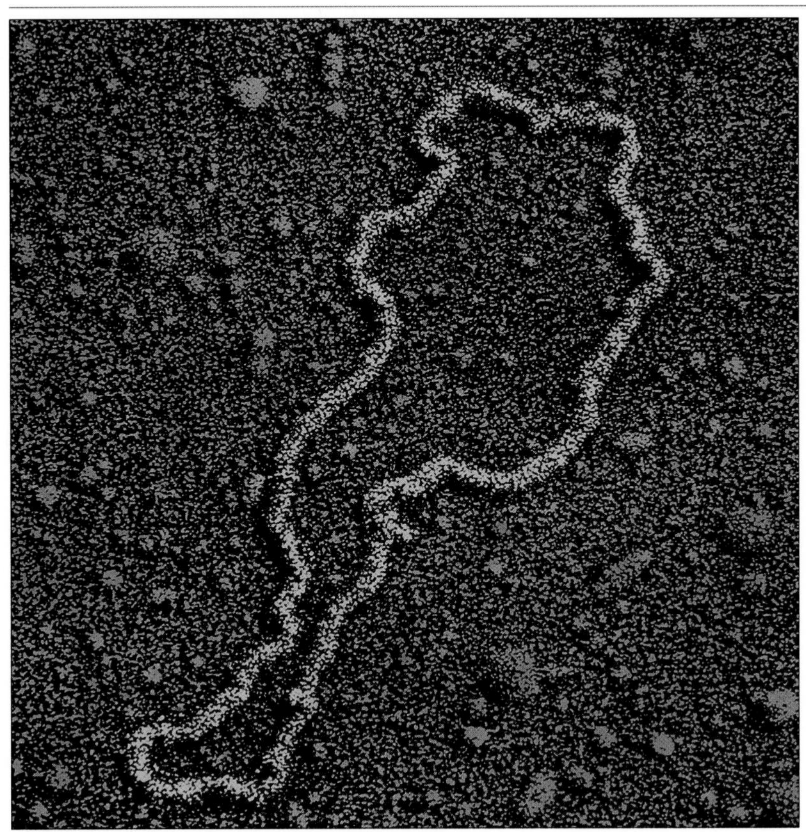

Magnified to the max

This is a strand of bacterial DNA magnified using current technology — a scanning electron microscope. Magnifying a 12-inch length of string this much would yield an image twelve miles long. We can see that DNA is a long molecule, but we get no information about the details of its structure.

The light microscope, which works by magnifying and focusing the image formed when light passes through an object, cannot distinguish objects smaller or closer together than the shortest wavelength of visible light. (Visible light is just that — the light we can see, and its shortest wavelength is small indeed — about 200 nanometers or nm)*. A protein molecule, like the one shown at different scales in all the drawings on pages 6 and 7, is only about 4 nm, and so is not visible through a light microscope). The transmission electron microscope and the scanning electron microscope use a beam of electrons controlled by electromagnetic fields. With such a microscope it is possible to see details of cell surfaces and interiors invisible with a light microscope and to discern at least the rough shapes of large molecular structures such as ribosomes.

Comparing Sizes

The scanning electron micrograph at right shows part of the inside of a cell, including a piece of a mitochondrion (marked M) and a portion of a Golgi body (marked G). The scale line is 1000 nanometers, or 1 micrometer. The mitochondrion is about the size of the common bacterial cells (*E. coli*) that live in our intestines. Ribosomes (25 nm) and individual proteins (25 nm) are far smaller, almost invisibly so, even to the most powerful microscopes. We can only visualize the details of these complex, interlaced chains of molecules through the mathematical power of computer models, using data from X-ray diffraction (see page 13).

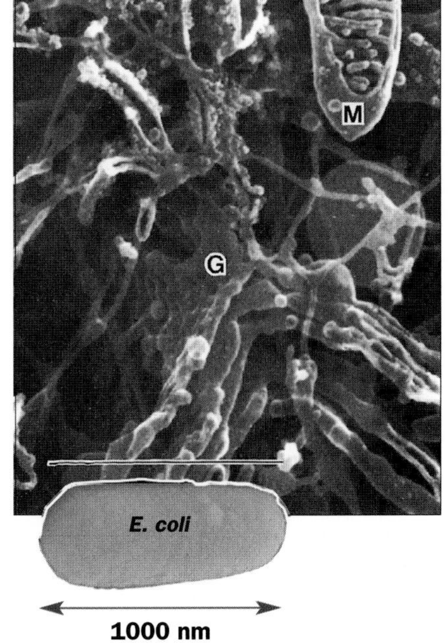

*See page 106 for more about light.

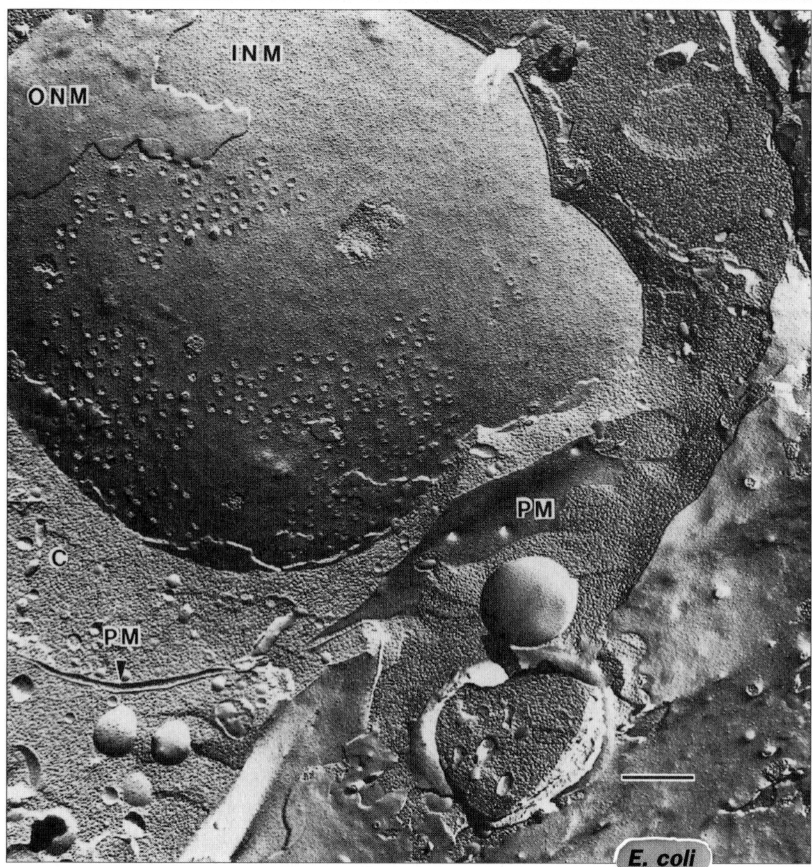

The cell's nucleus

The rounded shape in the upper left of this micrograph is the nucleus of a cell; it is covered by outer and inner nuclear membranes (ONM and INM) and surrounded by cytoplasm, in which float various organelles. PM (plasma membrane) marks the double membrane that envelopes all of the cell's contents.

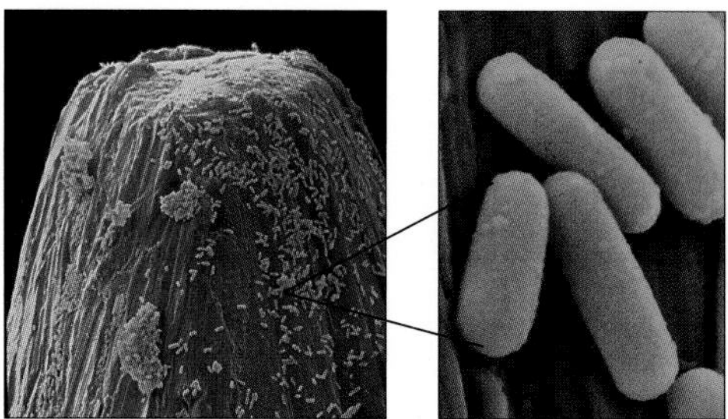

It's a small world

A magnified pin point with a population of *E. coli* bacteria.

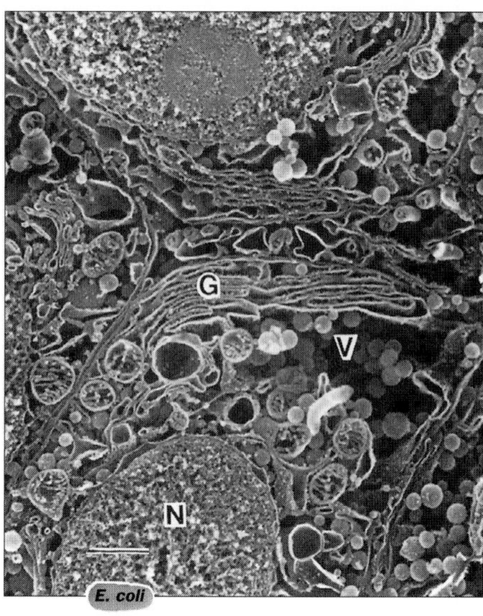

A 3-D interior

This view allows us to see many of the cell's molecular structures, or organelles.

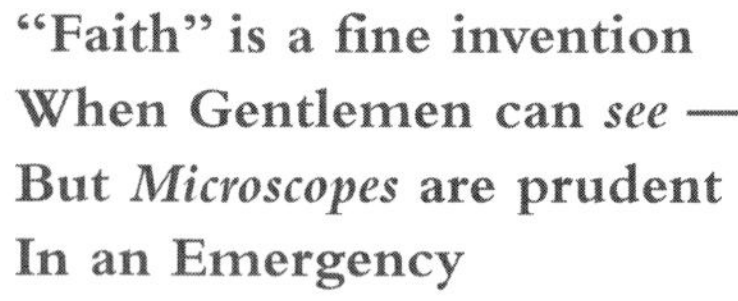

"Faith" is a fine invention
When Gentlemen can *see* —
But *Microscopes* are prudent
In an Emergency

Emily Dickinson (1830–1886)

How many bacteria can live on the point of a pin? Quite a few. The rod-shaped bacteria in the picture (top right), each a single cell, are about the size of the mitochondrion you saw on the preceding page, anywhere from 1000 to 1500 nanometers long. A typical animal cell is much larger; its nucleus (the library) — a portion of which you see above left in a cell broken open by freezing and then fracturing it — is roughly ten times larger in diameter than the bacterium.

Another view of the cell (above right) on a similar scale of magnification, but using a different magnification technique, shows part of the nucleus (N), and more of the cytoplasm's contents, such as Golgi bodies (protein-packaging factories, G) and vesicles (V).

The smallest living cells are about 1 micrometer, or 1 millionth of a meter, in size. Among the largest cells is *Acetabularia,* an algal cell that can reach 10 centimeters in length (see page 189). Some animal nerve cells are incredibly long, reaching from the spinal cord to the foot, although they are so thin that you would still need a microscope to see them.

Using X-ray Diffraction

Not only do microscopes allow us to "picture the invisible," they allow us to see them in their natural "landscape," in relation to the other structures around them. If we want to go deeper and study the structural details of the individual protein molecules that do the cell's work and are critical parts of cellular structures, we must resort to a different approach. First, we must "isolate" the protein molecules — free them of all of the associated materials that surround them. The protein molecules are crystallized, so that they stack regularly in a three-dimensional lattice. Then, we apply X-ray crystallography, which allows us to "see" the molecules at a wavelength of a few tenths of a nanometer (close to the diameter of a hydrogen atom!). This technique contributed to Watson and Crick's discovery of the double helix structure of DNA and Perutz and Kendrew's description of the structure of the blood protein hemoglobin — Nobel-Prize-winning achievements.

X-rays, like visible light (and the microwaves in your oven, for that matter), are a form of electromagnetic radiation (see page 106), but their wavelength is much smaller: 0.1 nm. If a beam of X-rays is focused on a crystallized protein, most of the rays pass through, but some are deflected, or scattered, when they hit the atoms. The deflected X-rays will produce a diffraction pattern — a pattern of exposure spots on a photographic film placed behind the protein sample. Regularly repeating atoms in the crystal structure deflect the X-rays at certain angles, creating, on the film, spots whose density and spacing correspond to the density and spacing of the atoms.

Collaboration among many scientists has combined information from X-ray diffraction, electron microscopy, and other technologies. This information, entered into enormous data banks, is the basis for accurate computer models of various, amazingly complicated protein molecules. Now we truly can visualize the invisible.

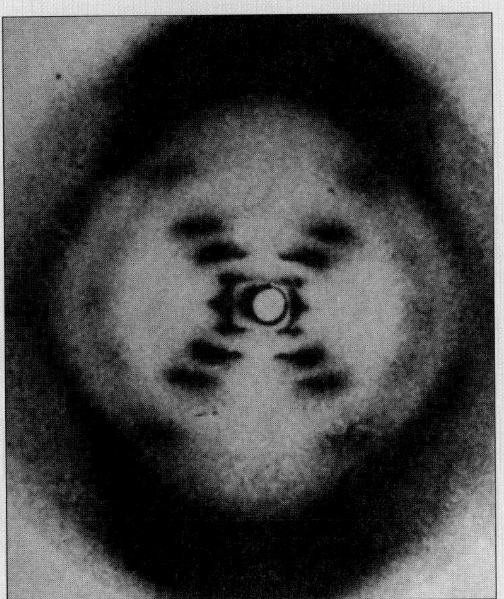

Light diffracted by DNA

In this diffraction pattern of DNA captured by Rosalind Franklin and deciphered by James Watson and Francis Crick (see page 157), the distance between spots forming the X indicates the distance between turns of the DNA's helix. The X is a reliable indicator of a helical (corkscrew-like) molecular shape.

1.4 Parts and Wholes

It's useful to think of life's organization in levels, from the simple to the complex: atoms, simple molecules, chain molecules, molecular structures, cells — and onward and upward to organs, organisms, populations, communities, and ecosystems. A higher level includes everything in the levels below it, as shown by the Russian nesting dolls above.

Scientists find that knowing a lot about a lower level produces useful explanations of what's happening at the next higher level. To understand how your car works, you must know something about cylinders and spark plugs and fuel injection and how they interact.

This way of getting to understand the whole by learning about its parts, called reductionism, has produced in the last several decades an explosion of knowledge about what genes are and how they work, and how living processes are energized, informed, operated, and controlled. They are the "what" and "how" questions we take up in this book.

When we ask why things are the way they are we need to see things from the outside, and in relationship to others and to the surroundings. For example, why do birds have different beaks? To discover the answer, we need to study not just the birds themselves but the food they eat and where they live. "Why" questions address patterns of connection in both space and time. They relate particularly to evolution — a subject touched on throughout Chapter 2, *Patterns,* and explained in depth in Chapter 8, *Evolution.*

Biochemists and molecular biologists tend to see themselves as reductionists, while naturalists and ecologists tend to take a holistic view. But, in fact, every scientist must shift his or her gaze regularly from the parts to the whole — from the trees to the forest — and back again.

We recommend that you try to be similarly fluid so that you can move back and forth with us as we shift from the tiny micro world to the larger macro world and back again.

1.5 The Way Science Works

"Why" questions have been posed ever since humans have had conscious brains, and over the course of the last 300 years or so, observant, curious people scattered everywhere on Earth have, in common, devised a verifiable, self-correcting system for answering these questions. We call that systematic asking of questions and search for answers science.

Observation of and wonder at the workings of nature are what initiate "why" questions. These activities are not the sole province of scientists. In fact, they begin in childhood and are more or less developed in all of us. Throughout this book you will find observations of nature by novelists, poets, amateur scientists, and painters, done in their own ways. Science joins art as another branch on the tree of observation and wonder.

Science is, in essence, organized curiosity. It is initiated by careful observation and nurtured by wonder, creativity, and skepticism. While the overall pattern of a scientific inquiry is common to scientific work, each individual explorer lays out his or her own itinerary into the unknown: observes an interesting event or phenomenon, identifies a particular aspect of it that can be stated as a problem, produces a hypothesis (an imagined scenario) that explains the event, and when possible, tests the hypothesis by experiment. In science *"one endlessly play[s] at setting up a fragment of the universe which the experiment . . . rudely correct[s]"* (François Jacob). The process is a wholly natural and active extension of the one we all use intuitively, from birth, to build our picture of reality.

The hypotheses scientists come up with generally lend themselves to predictive statements: If I do this, that should happen. If a prediction is borne out by experiment or observation — if the predicted event happens — this outcome builds confidence in the hypothesis but cannot prove it right (since better information or new experimental techniques may come along later and indicate that it's wrong). Thus, good hypotheses are often those that suggest ways in which they can be proved false. If a hypothesis offers no way to prove itself false, it is not useful scientifically. The conclusions scientists arrive at after extensive observation or after experimentally testing many hypotheses are statements that have a greater or lesser probability of reflecting reality; they are never certainties. They gain strength as ongoing tests and accumulating evidence continue either to verify them or to prove reasonable alternative hypotheses to be false. And the better the conclusions fit with those of other experimental approaches to the same problem, the surer we are they're right. The possibility that some of our most cherished truths may someday turn out to be false can never be ruled out. *"In the growing cathedral of science, many crumbling stones at the growing points are replaced, and the more important their position, the sooner the defect is disclosed"* (B. D. Davis).

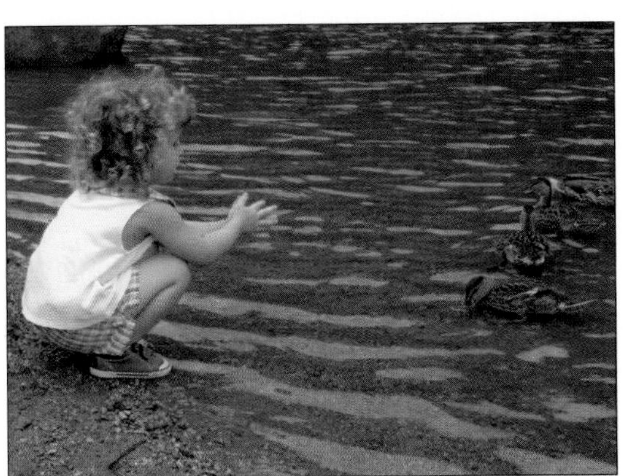

Hypotheses that are disproved by experiments have value, of course. They are signposts telling others where not to go. In science, an idea becomes substance only if it fits into a dynamic accumulating body of knowledge, a progression of understanding. Each new piece of work must fit into the bigger picture — the published work of other scientists. It is inspected, tested, tentatively accepted, modified, perhaps discarded. In the march of scientific discovery the artisans of experimentation blend into history like the builders of the great cathedrals. Scientists would have to have more than their fair share of egotism to avoid acknowledging their own expendability. This reality, as well as teaching us how little we know and how difficult what we do know was to come by, makes science a profoundly humbling experience.

WEB Connection
www.jbpub.com/connections

TOOLS *of Science*

The Ultracentrifuge

Developing tools (like the microscope) that help us to visualize, measure and define the invisible world around us is one very active and imaginative aspect of the scientific process. Biology adapts the tools of other sciences and of industry to its own purposes.

Dr. T. Svedberg, at the University of Uppsala, Sweden, was trying to find a way to make the protein molecules he was studying settle out of the solution they were floating in. Learning of work done in England in the 1880s in fluid dynamics that used a very fast-spinning centrifuge (from the Latin for "flee from the center") to increase the gravitational pull on objects in solution, he seized on this idea to develop an entirely new way of separating and comparing the sizes of cell components and biological molecules — the ultracentrifuge.

Over the succeeding 75 years, the ultracentrifuge has proved to be one of the most useful tools of cell biology and biochemistry. It is basically a rotor that spins tubes containing materials from broken cells at speeds of up to 80,000 rpm — exerting a force of as much as 500,000 times that of gravity!

Cells are first broken open by any of several means: by grinding in a glass tube and rotating pestle, by treatment with ultrasound, or by osmotic shock. These procedures are carried out in the cold in solutions resembling those in the interior of cells. The result is a sort of rich organic soup (referred to as a homogenate) in which most of the cells' organelles remain intact and functional.

The most common use of the ultracentrifuge is to prepare large quantities of cell components for use in biochemical laboratory experiments. The homogenate is centrifuged first at low speed (typically at 1000 times gravity for 10 minutes). This brings whole unbroken cells, clumps of membranes, and nuclei to the bottom of the tube in the form of a "pellet." The fluid above this pellet, the supernatant, is poured into a second tube and centrifuged at 20,000 times gravity for 20 minutes, producing a pellet of mitochondria. Next, the supernatant above the mitochondria, centrifuged at 80,000 times gravity for an hour, yields a pellet of microsomes (ribosomes attached to membranes). Finally, very high-speed centrifugation (150,000 times gravity for 3 hours) yields free ribosomes and large molecules.

Velocity sedimentation provides a finer degree of separation. Here, large molecules like the chain molecules on page 7— proteins, DNA, and RNA — are centrifuged through increasing concentrations of sugar (sucrose), where they separate into bands according to their size. The rate of sedimentation (the S value, named after Svedberg) is a standard way of describing the size of large molecules.

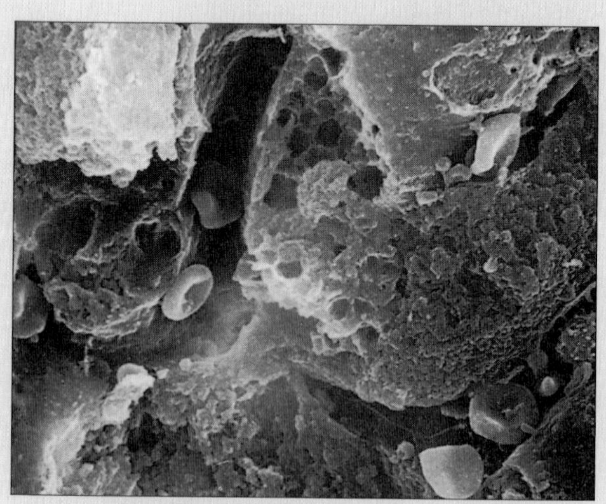

This tiny tissue section of liver cells, blood cells, and connective tissues would be ground up into "organic soup" and centrifuged to separate out the various cellular components.

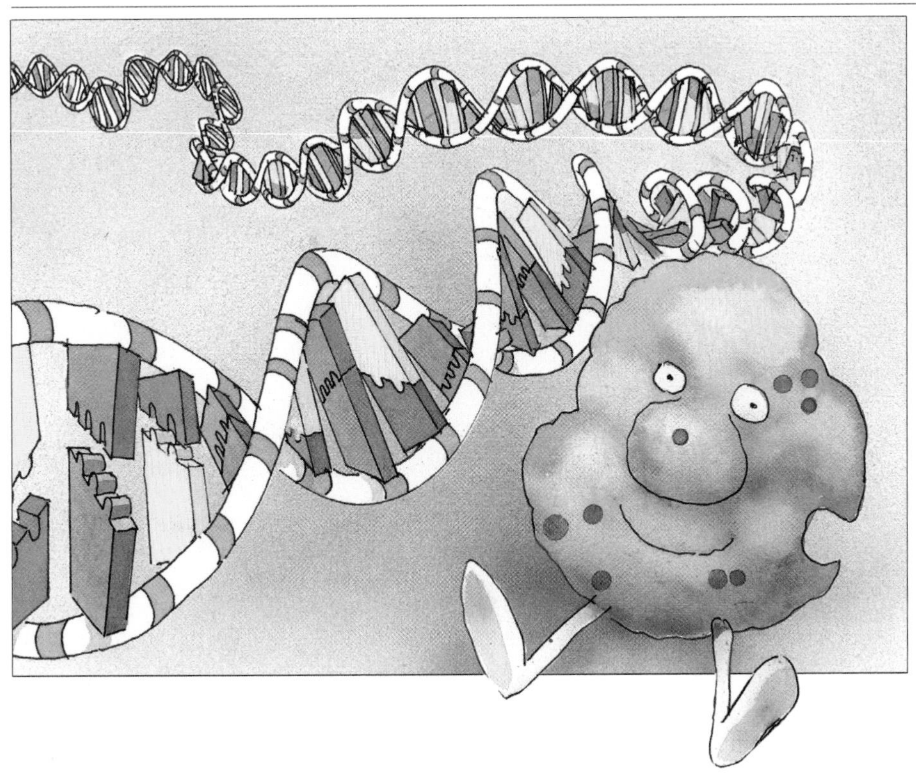

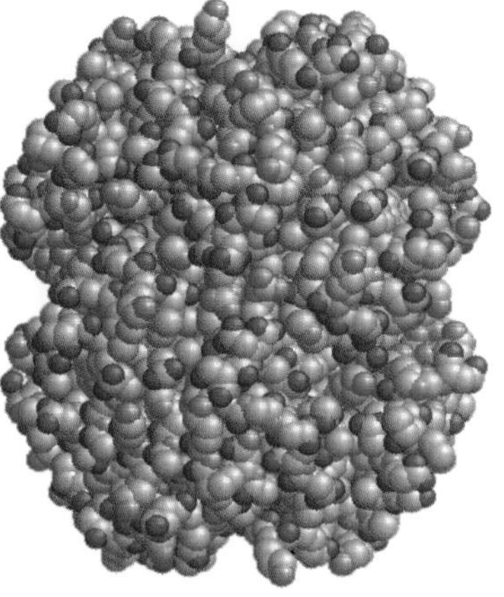

A computer-created model

A model of the oxygen-carrying protein hemoglobin, a chain molecule contained in all of our red blood cells. Each tiny sphere represents an atom of carbon, nitrogen, hydrogen, or iron. The atoms are bonded together in amino acid molecules, which in turn are strung together into chain molecules, which fold and twist into very specific three-dimensional molecular structures — in this case called globular proteins — of which four then join to form a multiple-chain molecule.

1.6 The Way Life Works — The Basic Idea

In exploring life's unity, we set out to connect the world of molecules with the world you can see around you.

Our central characters in this story are two chain molecules: One carries the information, the other does the work. To put things simply, you might say that life is played out in the interaction between these two players — DNA and protein, whose relationship can be seen as that between information and machinery.

Picturing the Invisible

Objects the size of atoms, simple molecules, and even DNA and proteins, are truly invisible because, even with the aid of the highest-magnifying microscope, our eyes can't see them. Although as you have learned scientists do have other powerful ways of finding out what very small things "look" like (see the computer model above), nobody really sees details of molecular structures exactly. Thus we have taken liberties in picturing our principal molecular characters in ways that convey clearly what they do.

We depict DNA as a kind of extended Tinker-Toy structure that readily assembles and pulls apart. Proteins — the working molecules of life — are pictured as somewhat human-like little characters. This distinguishes them — things that act — from other molecules, things that are acted upon. We don't imply that proteins are like people in any other ways — except, perhaps, for a certain obsessive tendency to do the same things over and over again. The protein's affable but blank expression should convey the idea.

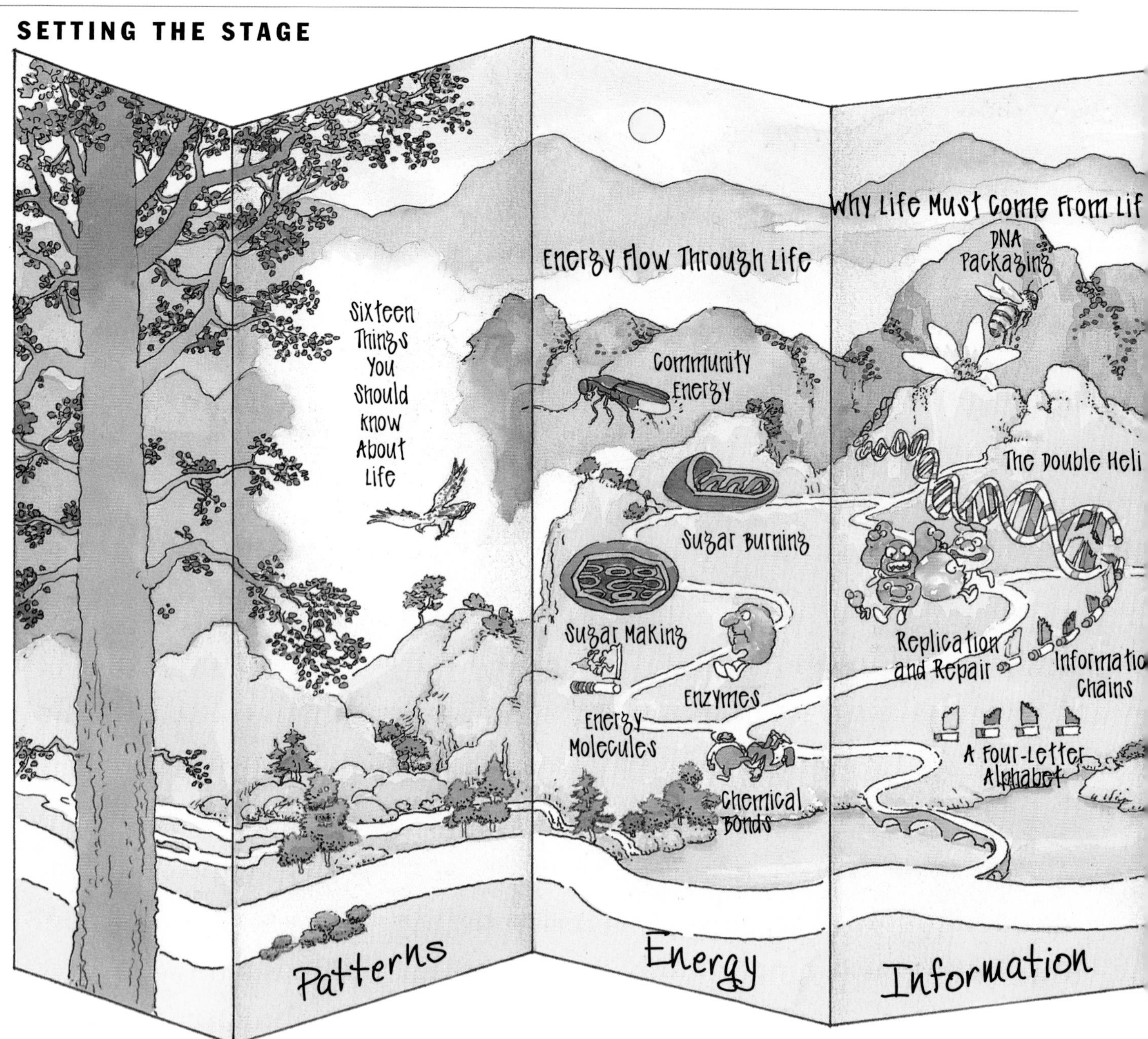

Your Itinerary — A Map of This Book

The next chapter — *Patterns* — offers a panorama of some of life's key features designed to focus your thinking and whet your appetite. Many of the questions it raises will be answered by the time you've finished the book.

Life is sustained by the conversion of sunlight into energy. The story of this flow is the subject of Chapter 3, *Energy*.

You may find it useful, in starting out, to think of *Information*, the subject of Chapter 4, as a primer on the "know-how" for life, written out in the chemical language of DNA, and stashed inside each cell of a living organism. Information, in turn, codes for life's *Machinery*, discussed in Chapter 5 — protein molecules that do all life's jobs, including constructing themselves.

Energy, information, and machinery would lead nowhere without some means by which cells can regulate rates of chemical reactions, minimize waste, promote efficiency, and ensure that multiple interlocking processes work

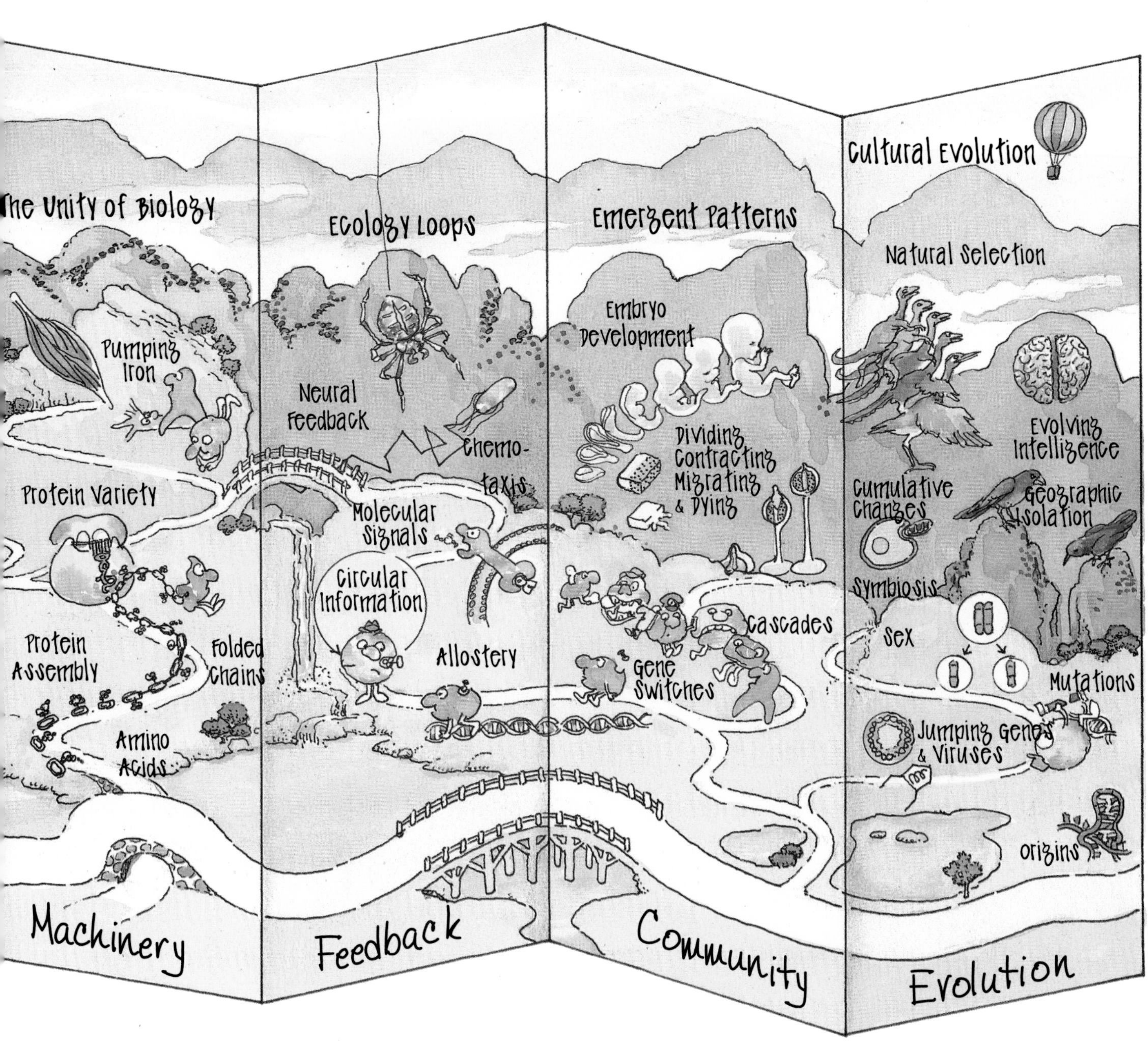

harmoniously together to promote the welfare of the whole. This is the role of life's system for coordination and control — which we call *Feedback*, the focus of Chapter 6.

All the above has to do with what individual cells need to live and function. Chapter 7, *Community*, examines principles governing how cells interact with each other in multicellular organisms and, particularly, how a single cell — a fertilized egg — becomes a multicellular individual.

Having examined the "what" and the "how" of life, we consider the "why": Why are living things the way they are? As information passes from generation to generation over vast stretches of time, it changes and, inevitably, modifies life's machinery. That machinery, the means by which every organism makes contact with the surrounding world, determines the organism's fate and, consequently, the fate of the information within it. Chapter 8 addresses the theme that knits all of biology together into a comprehensible whole — *Evolution*.

Some of the Things You Learned About in Chapter 1

Questions About the Ideas in Chapter 1

1. The theme of this book is expressed on the first page — uncovering the unity in diversity. In light of what you have read in this chapter, what does this phrase mean to you?

2. Come up with an example of unity in diversity from your own surroundings.

3. As you observe the photographs and illustrations in this chapter, what "why" questions can you come up with? Alternatively, look around, outdoors or in, and come up with some "why" questions.

4. Why bother to connect the "world of molecules" with the world you can see around you?

5. What are homologous patterns? What does the existence of homologous patterns tell us about possible relationships among apparently dissimilar living creatures?

6. If you were the person standing on the dock in the picture on page 4, would you be closer in size to a simple molecule, a chain molecule, a cell, or a molecular structure?

7. Three tiny molecules (2 to 3 atoms apiece) found in abundance on Earth, are essential to life. What are they?

8. Three larger molecules (10 to 35 atoms apiece) are the building blocks for life's energy, information, and machinery systems. What are they?

9. Which of the following best captures the meaning of the nest of dolls on page 14?
(a) Everything fits into compartments.
(b) A higher level includes everything in the levels below it.
(c) The bigger the object, the more complicated its parts.

10. Look carefully at the skeletal forelimbs of the frog, lizard, cat, and whale. See if you can identify and label the homologous bones in each. Describe some of the useful differences among these structures that make them well suited for the environment and needs of that organism. Can you hypothesize as to why the thickness of the bones of the cat and the whale differ so much? What might be one explanation for the structural differences between the cat's forelimb and the lizard's?

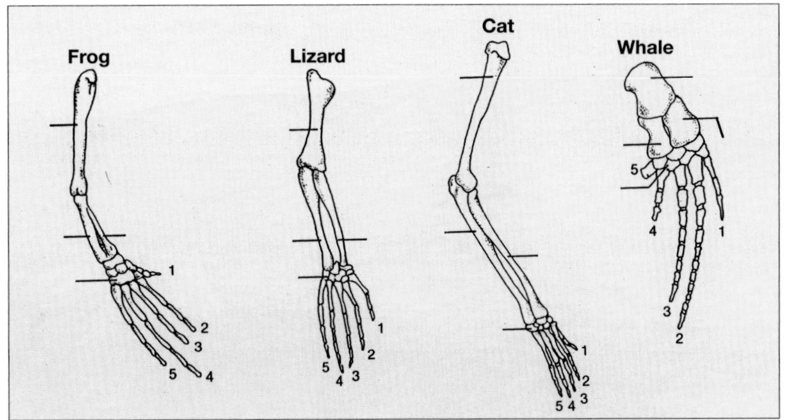

11. Unity and diversity frequently go hand-in-hand even with ordinary household objects. Choose two completely different types of chairs with which you are familiar, and describe how they are the same, and yet different. Find the one single *unifying element* for all chairs, then show how multiple variations can be made using different types of chair components (wood, stone, metal, plastic, bamboo) arranged in different ways.

12. To understand cellular fractionation, answer the following questions based on the size, weight, and density of a substance: Gold sinks to bottom of a prospector's pan, while dirt particles remain suspended in water — Why? Why does the nucleus separate out and sediment on the bottom of the tube in a low-speed centrifuge run, while membrane vesicles (smaller, sealed pieces of membrane) require much faster speeds (higher gravitational forces) to form a pellet on the bottom?

References and Great Reading

Alberts, B., D. Bray, J. Lewis, M. Raff, K. Roberts, J.D. Watson. 1994. *Molecular Biology of the Cell,* 3e. New York & London: Garland Publishing, Inc.

Bozzola, John L., and Lonnie D. Russell. 1999. *Electron Microscopy,* 2e. Sudbury, MA: Jones and Bartlett Publishers.

Bronowski, Jacob. 1973. *The Ascent of Man.* Boston: Little, Brown and Company.

Davis, B. D. 2000. The Scientist's World. *Microbiology and Molecular Biology Reviews,* 1-12.

Robin, Harry. 1992. *The Scientific Image, From Cave to Computer.* New York: Harry N. Abrams.

Strickberger, Monroe. 1999. *Evolution,* 3e. Sudbury, MA: Jones and Bartlett Publishers.

Svedberg, T., and R. Fähraeus. 1926. A new direct method for the determination of the molecular weight of the proteins. *J. Am. Chem. Soc.* 48: 430-438.

For more questions and links to web resources, go to

CHAPTER 2

PATTERNS

An Overview of the Basic Concepts of Biology

TO SEE LIFE AS A WHOLE — TO OBSERVE WHAT ALL LIFE HAS IN COMMON — REQUIRES A SHIFT in the way we normally look at things. We must look beyond the individual insect or tree or flower and seek a more panoramic perspective. We need to think as much about process as we do about structure. From this expanded viewpoint, we can see life in terms of patterns and rules. Using these rules, life builds, organizes, recycles, and re-creates itself.

Here we describe sixteen of life's patterns. Most apply to the smallest organisms and their molecular parts as well as to the most complex of us. We make no claim that our list is definitive. We simply invite the reader to think about life from the standpoint of not just what makes each living thing unique and different, but also what it is that unites us all.

The Sixteen Patterns:

1. Life Builds from the Bottom Up
2. Life Assembles Itself into Chains
3. Life Needs an Inside and an Outside
4. Life Uses a Few Themes to Generate Many Variations
5. Life Organizes with Information
6. Life Encourages Variety by Recombining Information
7. Life Creates with Mistakes
8. Life Occurs in Water
9. Life Runs on Sugar
10. Life Works in Cycles
11. Life Recycles Everything It Uses
12. Life Maintains Itself by Turnover
13. Life Tends to Optimize Rather Than Maximize
14. Life Is Opportunistic
15. Life Competes Within a Cooperative Framework
16. Life Is Interconnected and Interdependent

2.1 Life Builds from the Bottom Up

The Influence of Small Things

Early debate about evolution centered around the then horrifying notion that humans and apes had a common ancestor. Charles Darwin's idea (at left) had far more radical implications: Every individual is a colony of smaller individuals (cells), which are in turn made up of smaller nonliving bits. Further, these smaller bits were the first to develop in our evolutionary history. Occasionally these were usefully incorporated into cells, which, over great gulfs of time, assembled into multicellular organisms. Our ancestors were microscopic, wriggling, squirming creatures similar to what we now call bacteria, whose own ancestors were bits of self-replicating molecules.

> **Each living creature must be looked at as a microcosm — a little universe formed of a host of self-propagating organisms, inconceivably minute and as numerous as the stars in the heaven.**
>
> Charles Darwin, 1856

Before a single plant or animal appeared on the planet, bacteria invented all of life's essential chemical systems. They transformed the Earth's atmosphere, developed a way to get energy from the Sun, evolved the first bioelectrical systems, originated sex and locomotion, worked out the genetic machinery, and merged and organized into new and higher collectives. These are ancestors to be proud of!

Given the complexity of the tasks above, we can see why the first multicellular organisms did not appear until the most recent one-eighth of life's duration on Earth (see page 297). So we exist as "corporate elaborations" — composite communities of cells built out of the accomplishments of our one-celled forebears.

Cooperating communities of cells

Small communities of cells — like the taste buds on our tongues — work together as an army of specialists. They create a unique structure, with nerve connections to our brain, that allows us to taste the world around us. (The picture at right represents an enlargement of the human tongue.)

Small things are made of yet smaller things.

The bumps on the surface of our tongues, called papillae, contain our taste buds. These, in turn, are formed of clusters of about fifty cells.

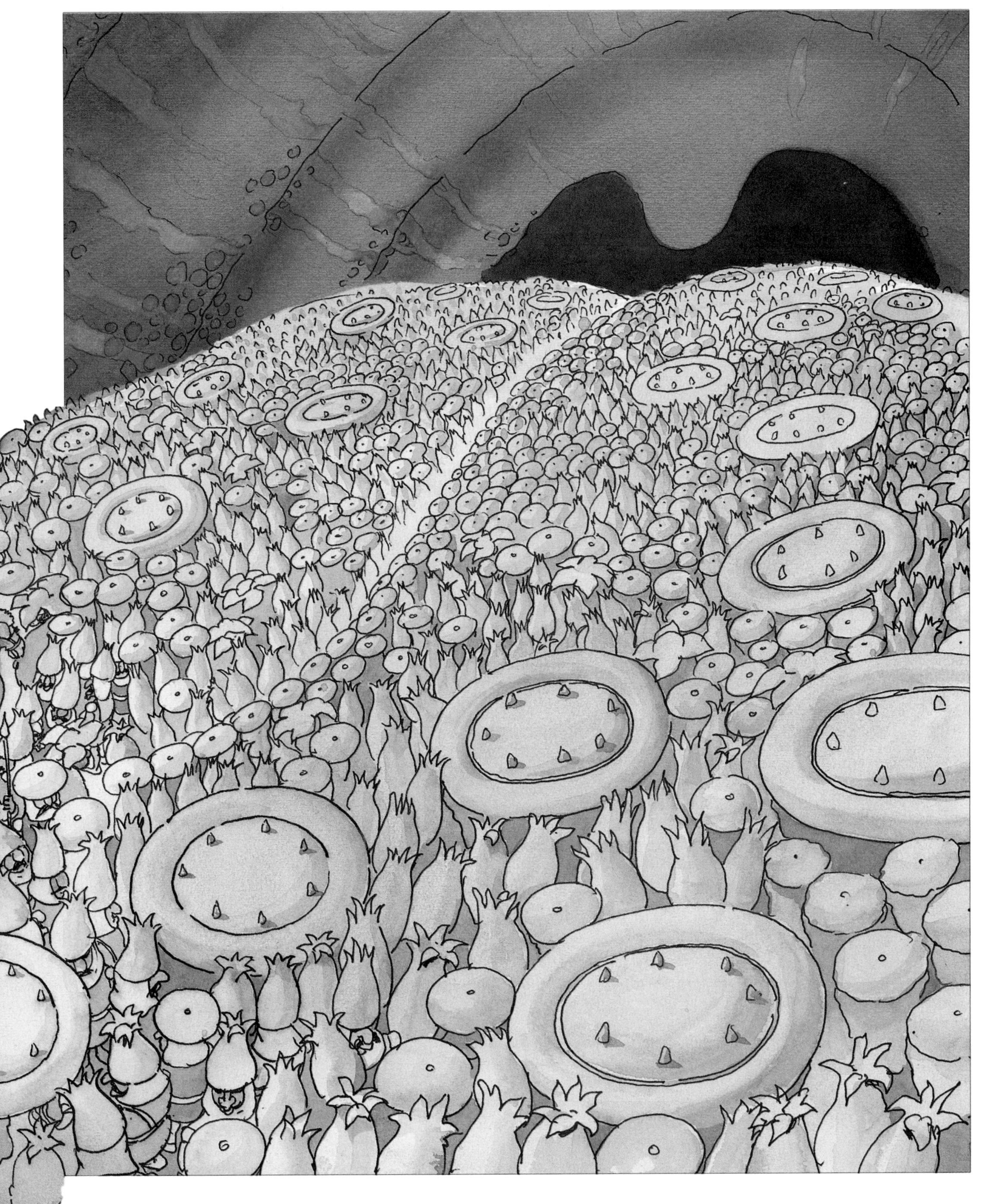

From Bottom Up to Top Down

The building of life from the bottom up (i.e., from single-celled creatures into multicelled creatures) suggests a one-way evolutionary direction — from the simple to the more complex. This, however, is only part of the story. As multicellular creatures evolved, they created new environments for the already existing simpler creatures. For example, the unicellular bacteria residing in the guts of all animals live in a mutually evolving dance with their larger hosts. They provide important benefits, including making the host's gut inhospitable to disease-causing organisms and producing necessary substances such as vitamins. In some animals, bacteria also secrete powerful digestive enzymes that break down food to prepare it for the host's own digestive resources. Though the ancestors of these helpful microbes clearly existed before animals, it is highly probable that their hosts contributed to the direction of their later evolution.

This is also the case with parasites, organisms that harm their hosts. They follow a simple rule: "Why make a product yourself when you can easily get it from someone else?" Bacterial and viral parasites, in particular, must have evolved after their hosts. It is likely that many parasites have actually become simpler than their ancestors were.

Finally, consider organisms whose evolution humans have genetically engineered: bacteria that eat oil or attack crop pests. Beyond demonstrating our growing ability to manipulate nature, such creatures exemplify the ongoing worldwide coevolution of micro and macro environments.

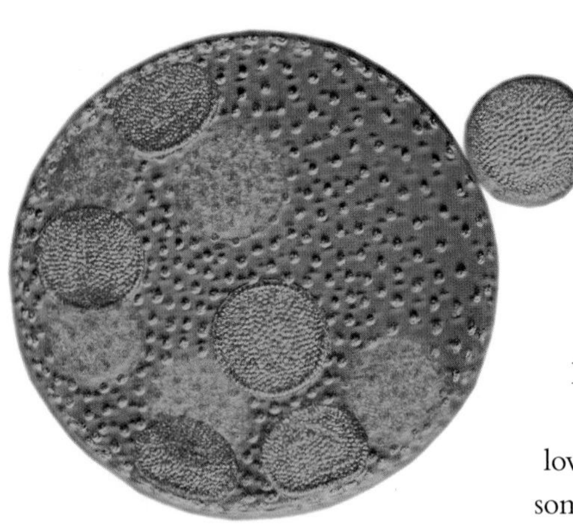

Volvox

A spherical colony of many single cells rolls through the water of a pond. Inside the sphere grow daughter colonies, which from time to time break out to spin away on their own.

The colony of cells called *Volvox* shown above left can be made up of as many as 50,000 individual cells, each with two whip-like propellers (flagella). The cells are held together in a gelatinous sphere, not actually connected to each other as are the cells of your tongue, let's say, or those of any multicellular organism. Still, the flagella move in a coordinated way to roll the colony through the water, and the colony does seem to have a forward and backward orientation, as well as an inside where new colonies form.

Question.

Why do you suppose such colonies might be more likely to survive and reproduce than might free-living single cells over time? What might be the advantage of a colony's remaining just a colony, rather than evolving into a multicellular volvox organism?

Answer...

In a fairly calm water environment (where the spheres are likely to remain whole), a colony has an advantage over single cells in that it has daughter colonies that develop in a relatively protected place. Remaining a colony might also be an advantage, because a bite from a predator would just downsize the colony, not kill it.

Virus Attack

The ranks of balloon-shaped objects lined up along the outer edge of this bacterial cell are viruses — bacteriophages ("bacteria-eaters"). About 30 minutes before this picture was taken, they injected all of their DNA into the bacterium. The bacterium's molecular machinery leapt into action and translated the information in those DNA molecules into new viruses, which you can see bursting through the cell membrane at the lower right. Viruses are so simple that they cannot exist without using another organism's molecular machinery to reproduce.

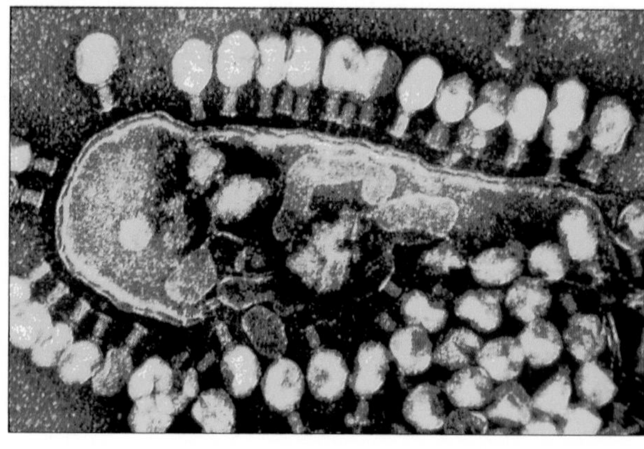

A Corporate Elaboration

A sponge is perhaps the simplest "corporate elaboration." Its cells function as a cooperating group of individuals. Unlike other multi-celled creatures, sponges have no true tissues (groups of differentiated cells that work in concert, as in taste buds or muscles), let alone tissues organized into organs. Sponges have eight to ten different types of cells (some of which you can see in the illustration at right), cooperate to maintain constant water flow through the pores, to trap food, to create fibers and mineral structures that maintain the sponge's shape, and to transport nutrients and wastes.

Probably because of their relative simplicity, sponges regenerate easily: chop a living sponge into pieces and each piece will become a new sponge. Even pressing a live sponge through a fine sieve won't kill it. Deposited on a culture medium, the tiny fragments will begin to migrate and clump together in mounds that eventually become miniature new sponges.

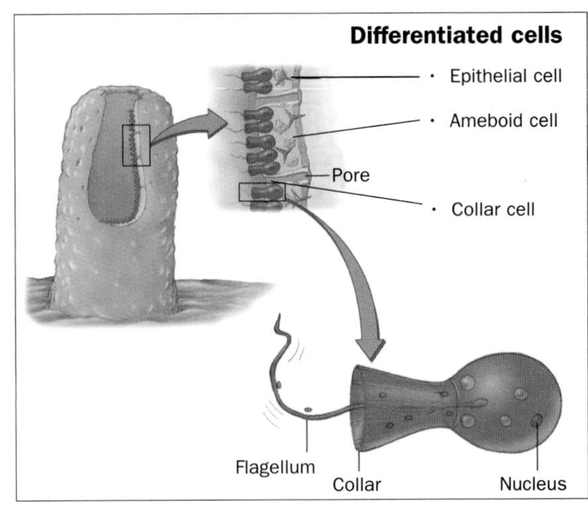

Sponge cell specialists

A colony of a very ancient type of tropical sponge

Notice the tubular, porous structure, ideal for enhancing water flow. Specialized cells with long appendages called flagella create currents that pull nutrient-filled water in from outside and up through pores in the tube. Each individual cell exchanges nutrients and waste with the outside world. The human intestine and trachea (both tubular organs) are also lined with cells that perform a similar kind of absorbing and transporting function.

Autumn, Kellar et al. 2000.
Adhesive force of a single gecko foot-hair.
Nature 405: 681-685.

Imagine a hair one-tenth the diameter of a hair on your head. A gecko lizard has a half-million such hairs on the bottom of each of its four feet. On each of these hairs are from 100 to 1000 still smaller spatula-shaped structures.

Kellar Autumn and collaborators have discovered that these very small structures give the gecko a "leg up" in climbing vertical walls. They have measured intermolecular forces sealing the gecko's foot spatulas to the wall surface. In other words, each spatula's tiny size allows it to press so close to the wall, within a distance comparable to an atom's diameter, that the molecules of each spatula attract to the molecules of the wall. The intermolecular force generated, multiplied by the millions of spatula structures on each foot, is more than enough to keep the gecko on the wall.

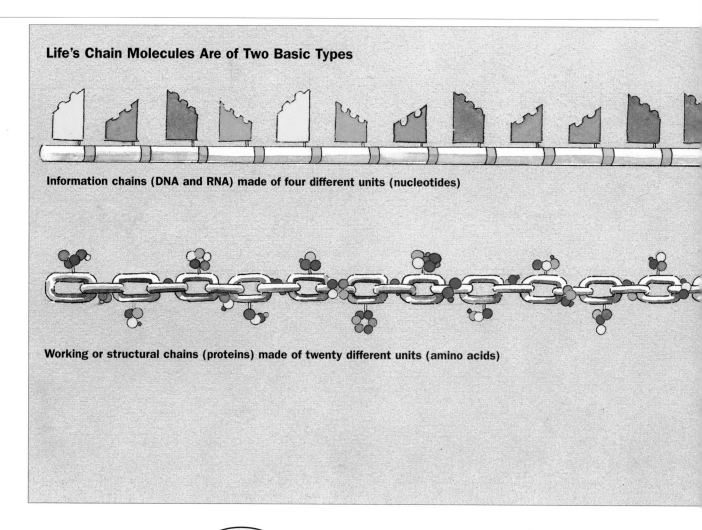

Life's Chain Molecules Are of Two Basic Types

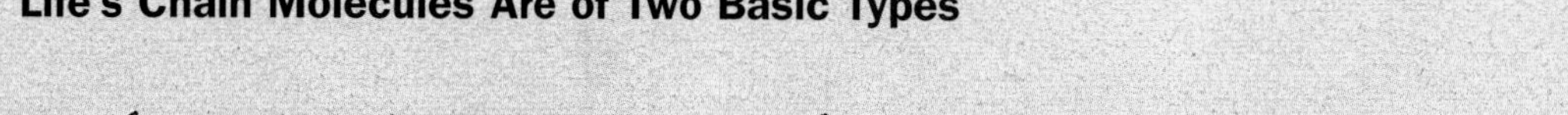

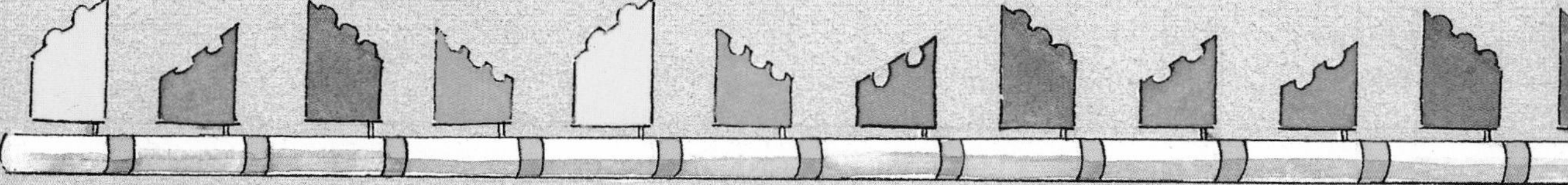

Information chains (DNA and RNA) made of four different units (nucleotides)

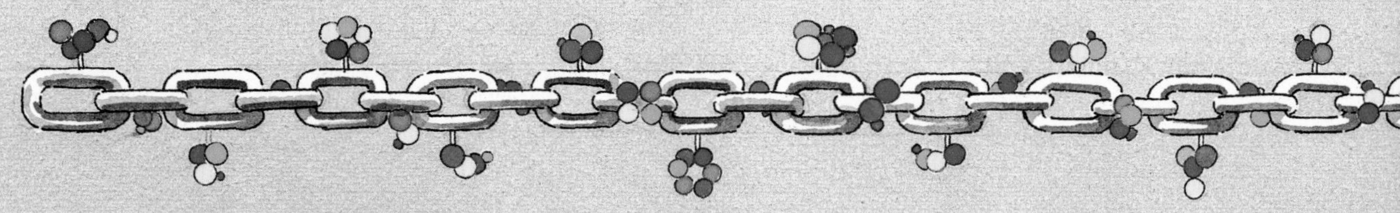

Working or structural chains (proteins) made of twenty different units (amino acids)

2.2 Life Assembles Itself into Chains

When Difference Becomes Information

At the molecular level, life has adopted the chain as its organizing principle. Chains are made of simple units connected together in long, flexible strands. In an ordinary chain, the links are all the same. In contrast, life's chains are molecules containing different links. In this respect, the links are the alphabet of life. Letters, in appropriate order, form meaningful words, sentences, paragraphs. Similarly, the sequence of individual links in a chain molecule conveys information.

Chain molecules fall into two main classes: information chains, which store and transmit information, and working chains, which carry out the business of living. Specific lengths of the information chain, called genes, carry the information that becomes specific working chains, called proteins. The two kinds of chains work together in a cooperative loop: information chains provide the genetic prescription or recipe that is translated into working chains; these in turn make it possible to copy the information chains so they may be passed on to the next generation. All of this is explained in much greater detail in Chapters 4 and 5, *Information* and *Machinery.*

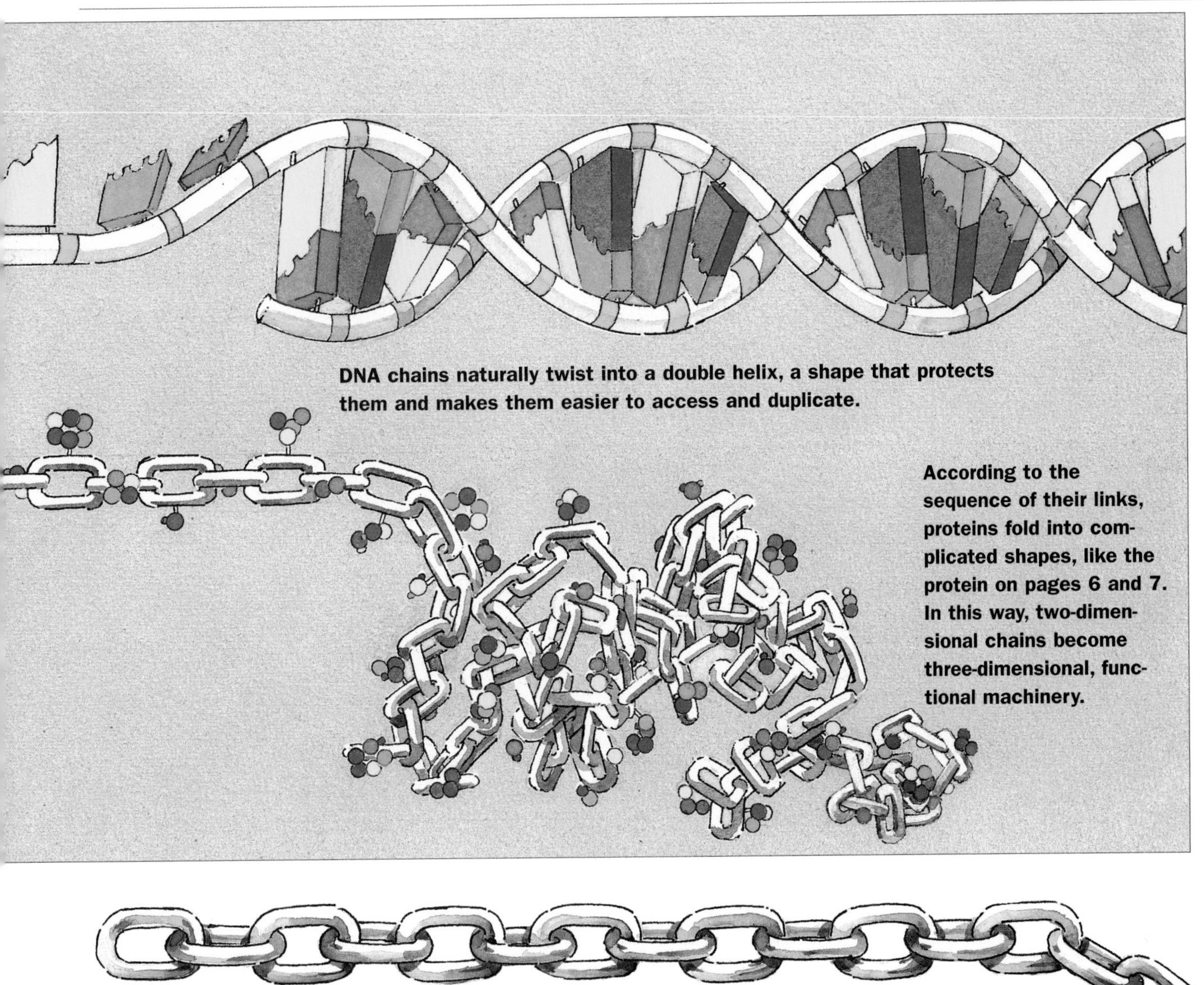

DNA chains naturally twist into a double helix, a shape that protects them and makes them easier to access and duplicate.

According to the sequence of their links, proteins fold into complicated shapes, like the protein on pages 6 and 7. In this way, two-dimensional chains become three-dimensional, functional machinery.

A chain of uniform links is simply a chain —

but a chain of different links can carry information:

Morse code is a chain of two different units (dots and dashes),

10011000111101000110001011110001100100101

computer language is also a chain of two units (ones and zeros), and

Now is the winter of our disco

an English sentence is a chain of twenty-six units (letters).

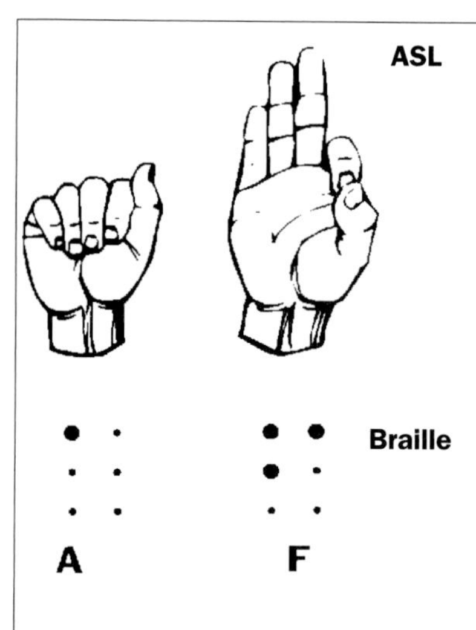

Two Simple Examples of How Information Requires Difference

Your hand can become an eloquent communication device if you use it to create the letters of American Sign Language (ASL). For the different hand shapes of ASL to qualify as information, they must be recognized by someone else—they must be "read." As Helen Keller "read" the changing shapes of her tutor's hand, those shapes become information about the world. And, to be useful over time, information must be stored — placed in memory. For any system to have a memory, it must map differences in the world into coded sequences and keep these secure for later reading. Braille does this: it codes letters into distinct patterns of raised dots impressed on paper or other media.

Folding into Shapes That Work – More About Three-Dimensional Machinery

Information chains (DNA) are translated into working chains (proteins). The shapes that protein chains fold into depend on the multiple interactions among the amino acid molecules that make up the chain. Chemical groups of the amino acids attract each other, making the chain stick together at various points (the black dots in the illustration at right depict the ways that hydrogen bonds can connect one amino acid to another at a distant place on the chain). Once the sequence of amino acids in a chain is dictated by DNA, the shape of a protein inevitably follows. That protein then takes up one of many thousands of different functions in a living organism. It may be the kind of protein that provokes chemical reactions — an enzyme or catalyst.

You can see this kind of invisibly small enzyme at work when you slice into an apple or an avocado — the pristine slice rapidly turns an unappetizing brown as its enzymes cause oxygen in the air to react with the disorganized content of damaged cells to create a molecule that reflects color (a pigment molecule). It turns out that this kind of enzyme, called a polyphenol oxidase, is very ancient and can be found in everything from amoebas to mushrooms to people (where it is responsible for, among other things, suntans and hair and eye color).

Reading Information

Living creatures have incredible capabilities for extracting information from — that is, reading differences in — the world. Monarch butterflies apparently navigate the 1500 miles from Canada to a small area in Mexico by reading differences in the Earth's magnetic field.

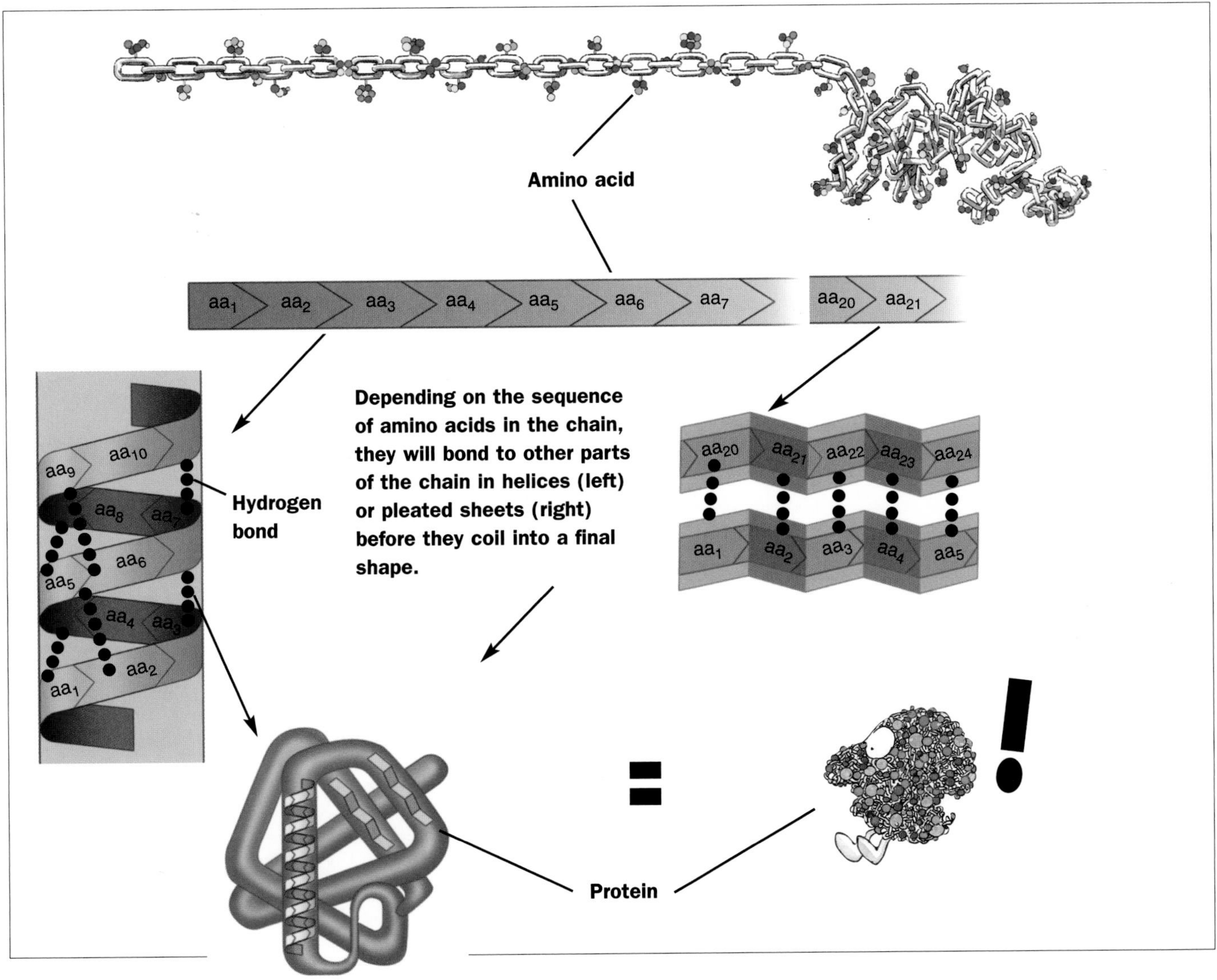

Bats maneuver in darkness using echolocation: responding to differences in the echoes of the high-frequency sounds they emit. Trees "know" when to withdraw the nutrients from their leaves at the approach of winter, in part by reacting to differences in day length.

The ability of widely various creatures to read environmental information has a common source. Embedded in living cells are specialized chain molecules (proteins), which are activated and altered by tiny differences in their surroundings. These complexly coiled proteins act as information receptors and processors, picking up distinctive information from the environmental stream and reporting it to other working proteins for appropriate action. The ultimate information — i.e., the information for making all these information-gathering proteins — is found in DNA.

Life Needs an Inside and an Outside

Heads Out — Tails In

When danger threatens, musk oxen gather in a circle — heads and horns to the outside, tails to the inside — sheltering their vulnerable calves in the center. This circle of protection is a memorable analogy for one of life's most fundamental organizing principles — a difference between inside and outside. Life's chemicals must be kept close together — concentrated — so that they can meet frequently and react readily. To function, the inner environment must maintain a stable level of saltiness, acidity, temperature, etc., different from the outside. These differences are maintained by some form of protective barrier, e.g., a baby's skin, a clam's shell, or a cell's membrane.

A cell membrane

Cellular membranes are formed by combining two layers of regimented phospholipid molecules. On the outer row, the water-liking heads face outward toward the watery surroundings.

On the inner row, the water-liking heads face toward the inside of the cell. The two rows effectively isolate the inner environment. Protein pumps, like the one shown at top, move molecules in and out.

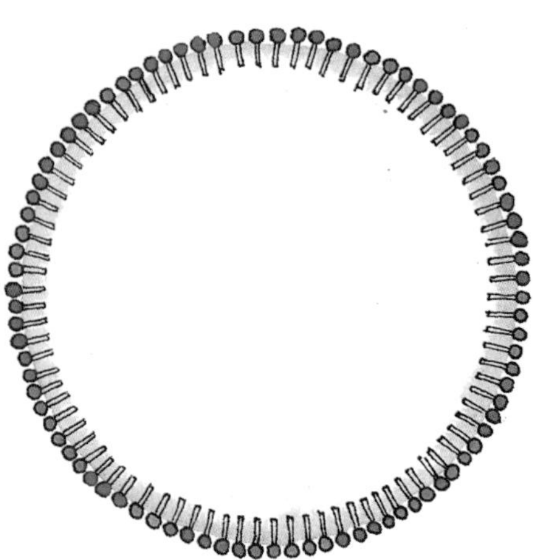

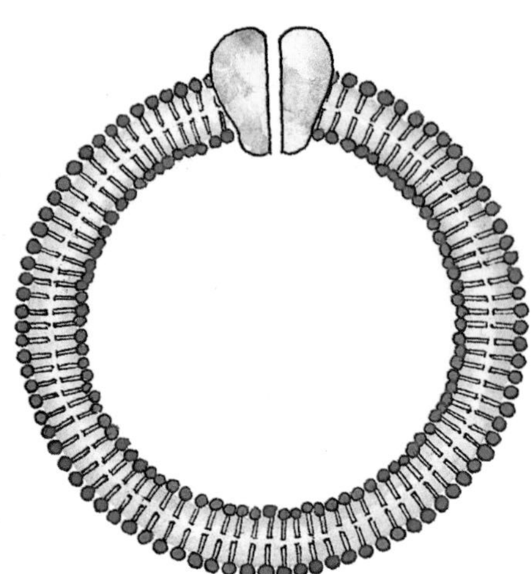

The membranes surrounding each of our cells behave something like the threatened musk oxen. The constituent fat molecules have a water-liking head and a fat-liking tail. Heads face outside toward the watery environment beyond the cell; tails face inward (above left). Since the inside of a cell also has a watery environment, a second row of fat molecules aligns itself tail-to-tail with the outer layer, heads facing inward (above right). With this protective structure creating an inside and an outside, plus several pumps embedded in the membrane to move materials in and waste out, life can do its work.

Larger "membranes"

Bark safeguards the living part of the trunk of a tree (usually the outermost ring) from insects, disease, and harsh weather.

The atmosphere helps regulate the Earth's temperature as it protects life from the Sun's harmful ultraviolet rays and insulates us from the cold vacuum of space.

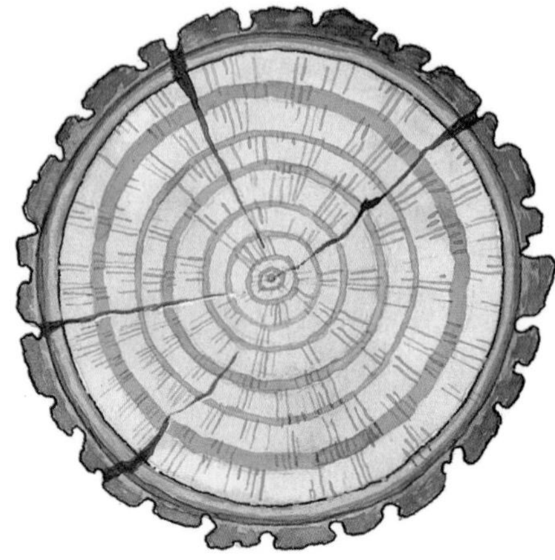

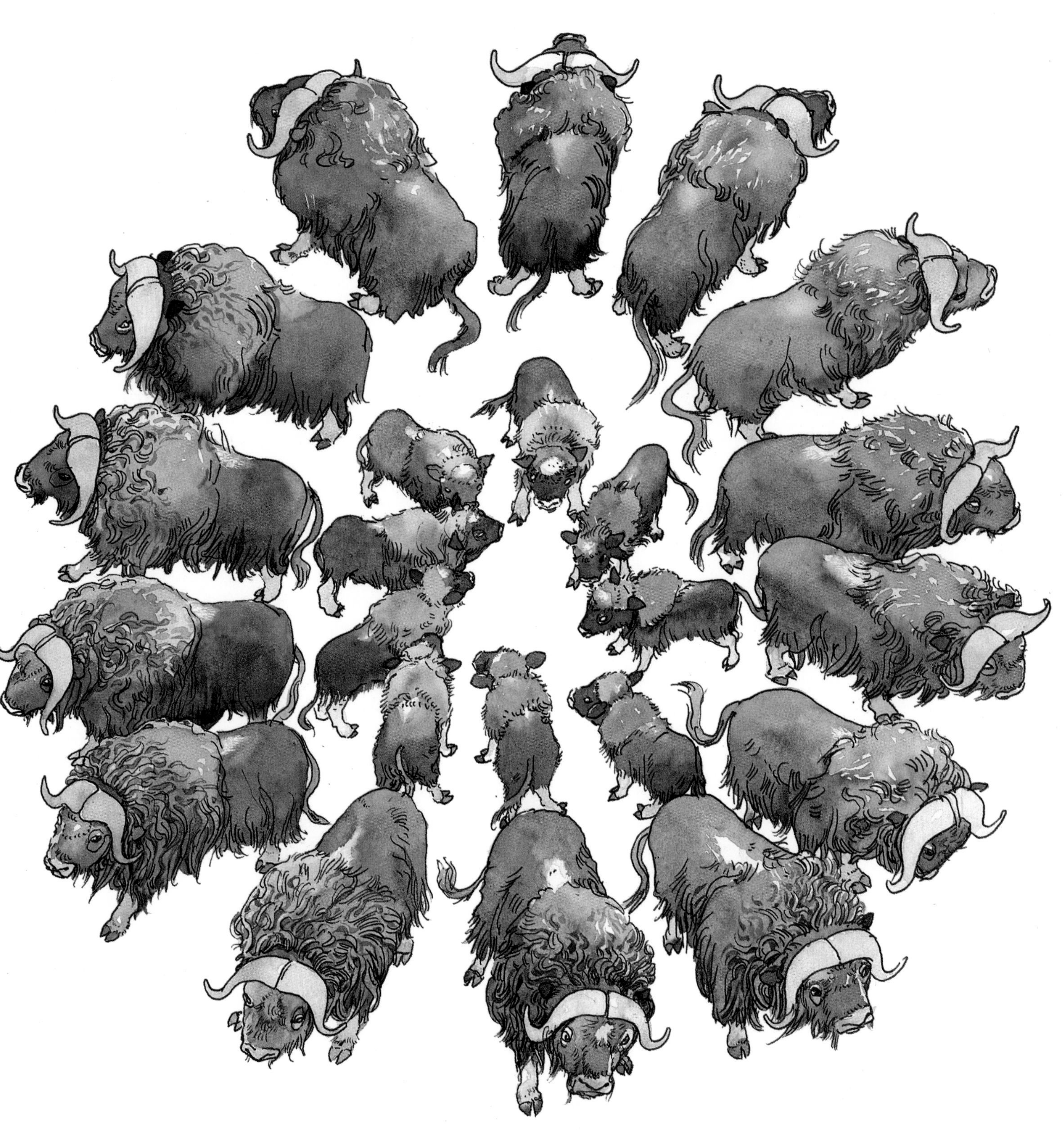

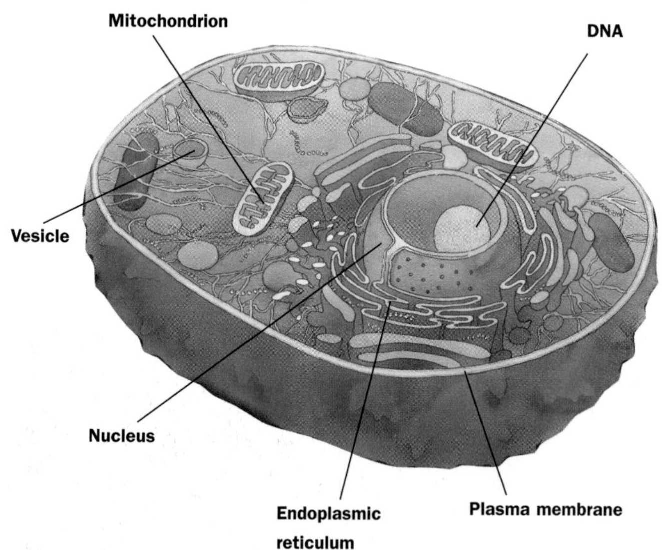

Our Debt to Fat

In cells other than bacteria, the cell's outer surrounding membrane, the plasma membrane, has its counterparts inside the cell. The nucleus is surrounded similarly by a membrane enclosing the DNA, and the machinery for its reading and duplication (Chapter 5). Mitochondria, the bacteria-sized bodies inside cells are membrane-enveloped. And inside them, membranes and associated mobile molecules act as electron conductors in the process of converting sugar to useable chemical energy. Chloroplasts of plants, too, use membrane electron conduction as they convert the energy of sunlight into sugar (Chapter 3).

Throughout cells there are elaborate networks of membranous tubes and channels and conveyor belts which are closely connected with the cell's elaborate machinery for making protein (Chapter 5). This endoplasmic reticulum serves to convey newly made proteins to their proper locations.

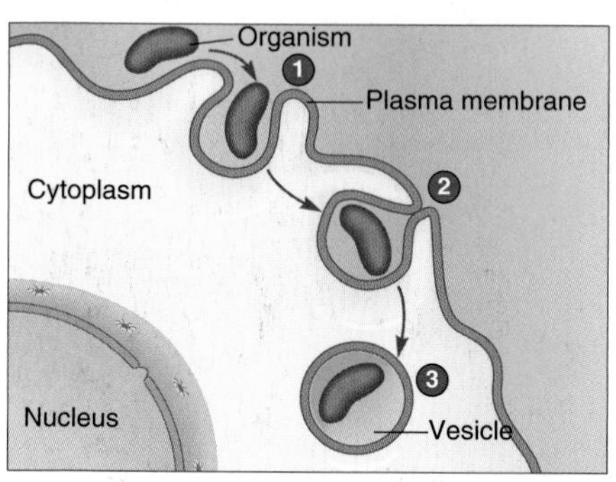

Membranes even act like cell mouths. A segment of membrane on the surface of the cell can engulf chunks of material it wants to "eat," enclose them in membrane, and move the package into the cell. Inside the package, the contents are broken down; the breakdown products are released and then used to build new cell material.

Of course, membranes are barriers too: they keep things that aren't wanted out of cells. They also bring things that are needed into cells, and they keep them there until they're used. And they conduct waste material out of cells. These functions are mediated mostly by proteins which sit in the membrane and act as channels or gates.

Many more additional functions of membranes are mediated by proteins lodged snugly into the membrane. Special proteins govern the connections between cells, insuring their coordinated interaction. Proteins that penetrate the full width of the membrane receive signals arriving at the outside of the cell, and by changing their shape, thereby convey a response to the inside of the cell (see pages 218–219).

So these fatty molecules called phospholipids, banding together to shun and preserve ubiquitous water, create for us an inner world in which both water-sustained and water-avoiding chemical events can exist side-by-side to make life possible.

At right are the shapes taken by layers of certain dried protein molecules when they are heated slightly and mixed with water. Notice the double layers in (a) and the complex interior structures in (b). As it turns out, these membranes let certain substances through and repel others.

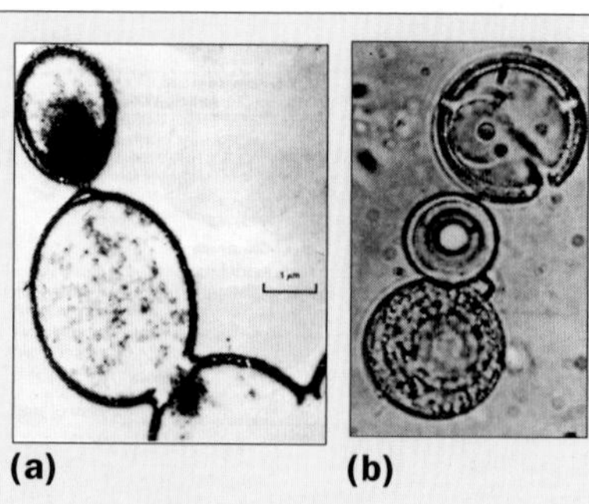

(a) (b)

Question.

Chemists studying the properties of phospholipid molecules and of fairly simple protein molecules found that when they were mixed with water, millions of them joined spontaneously to form small bubble-like spheres. When the mixture was shaken, the spheres broke up into even smaller spheres, always self-sealing. To scientists curious about the mystery of how cellular life might have started on our planet, this was an exciting discovery. Why?

Answer...

The discovery provided a possible explanation of how the first cells might have arisen. A fundamental problem in understanding the origin of life is explaining how the vital chemicals might have been confined in a small enough space to promote continuous proximity of reactants. If natural substances like proteins or phospholipids spontaneously create self-sealing spaces, they might, in a primitive soup, have trapped these chemicals inside, allowing them to react and form molecules like nucleotides and proteins.

Earth as Organism

WEB Connection
www.jbpub.com/connections

Thinking about life on a much grander scale than that of molecules and cells, some imaginative scientists — James Lovelock and Lynn Margulis were early proponents — have suggested that the Earth itself can be viewed as a life form. In this view, called the Gaia ("Mother Goddess") hypothesis, the atmosphere, oceans, soils, and living organisms comprise a biosphere — a global self-regulating system that works to maintain its own internal balance (homeostasis) in much the same way a cell or an organism does. Although this hypothesis is hotly contested in the scientific community, viewing the Earth as a life form provides a useful model for thinking about living systems and their need for protective and containing membranes.

Terrestrial vegetation acts as a protective membrane for the land and its living contents. Vegetation absorbs atmospheric CO_2 and gives off a great deal of the water vapor, partially responsible for cloud formation and subsequent rainfall. When the membrane of trees and other plant life is removed from a region, water vapor is no longer given off and the surrounding land may become a desert. In coastal areas assured of plentiful rainfall, deforestation (removal of the vegetative membrane) leads to a result just as harmful; once the protective layer of plants and roots is disrupted, erosion is magnified with every rainstorm, and the nutrient layer of topsoil soon is washed away. Eventually little is able to grow or live there.

The Earth's membrane

This NASA satellite photograph shows the relatively thin outer envelope (the Earth's atmosphere) that surrounds and protects the planet much like the membrane that surrounds and protects a cell.

Earth itself is surrounded by a membrane that is both fragile and tough — the atmosphere. It admits light, vital to the existence of life on this planet, and emits excess infrared radiation (heat) produced by the activities of living things. The atmosphere protects us from the deadly cold of space, from meteorites, and from the Sun's harmful ultraviolet rays; it also moves and cleanses the air we breathe, and replenishes our fresh water supplies.

When the atmospheric membrane is perturbed (major volcanic eruptions, for instance, can launch particulates into the upper atmosphere), serious climatic changes can occur. After the violently explosive eruption of Krakatau in the South Pacific in 1883, volcanic dust in the stratosphere caused cooler temperatures and spectacular sunsets worldwide. The cooling effect was so great that 1884 was known as the "Year Without a Summer" in much of the Northern Hemisphere.

Atmospheric perturbation

A recent volcanic eruption seen from the Space Shuttle.

WEB Connection
www.jbpub.com/connections

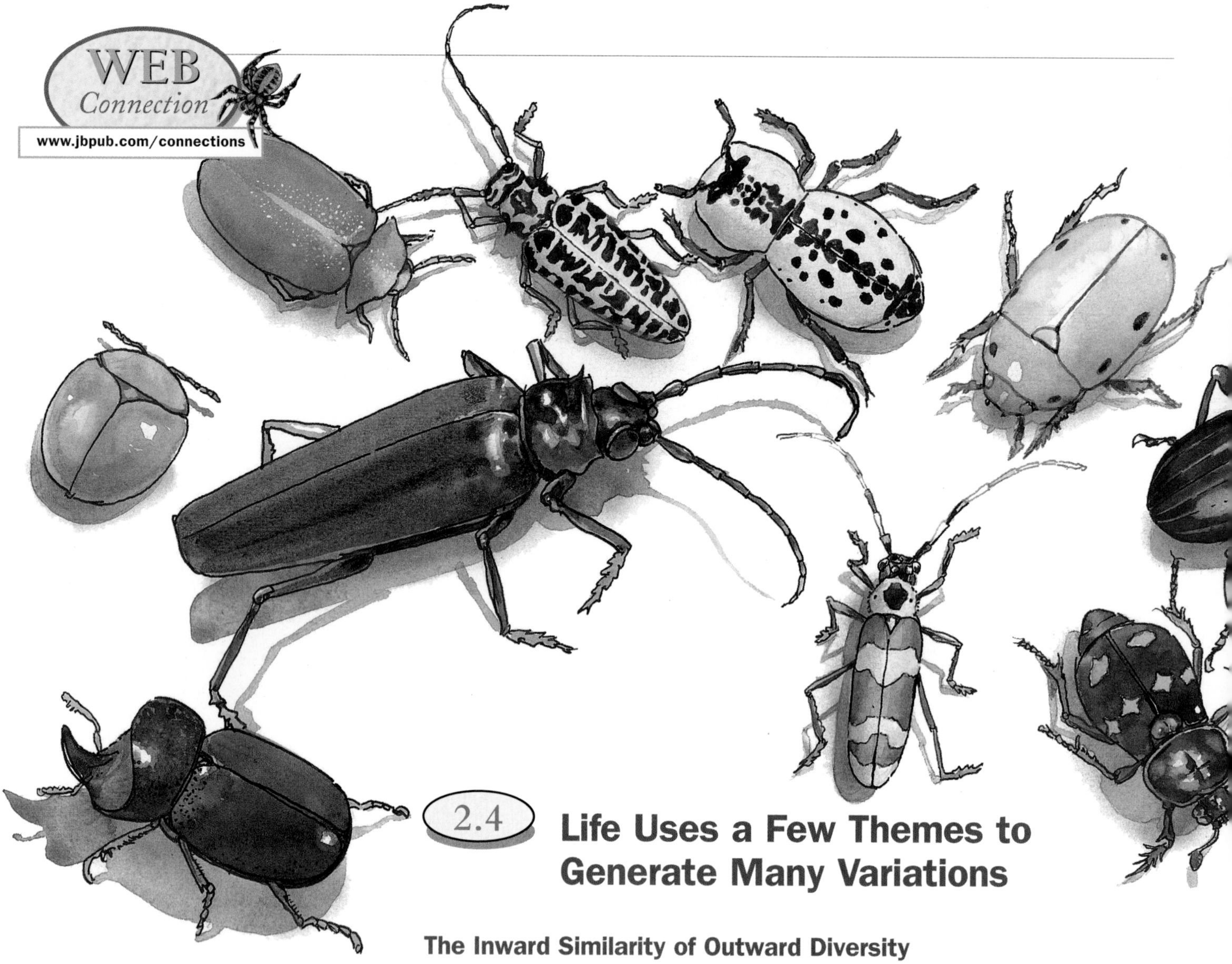

2.4 Life Uses a Few Themes to Generate Many Variations

The Inward Similarity of Outward Diversity

Life hangs on to what works. At the same time, it explores and tinkers. This restless combination leads to a vast array of unique living creatures based on a considerably smaller number of underlying patterns and rules. For example, when cells divide and grow, they do so in a mere handful of ways. New cells can form concentric rings, as they do in tree trunks and animal teeth. They can form spirals, as in snails' shells and rams' horns; radials, as in flowers and starfish; or branches, as in bushes, lungs, and blood vessels. Organisms may display several combinations of these growth patterns, and the scale can vary; but for all life's diversity, few other growth patterns exist.

Life, in striving for the most economical use of space, borrows mathematical rules. For instance, count the branches coming off a stem for a given number of full turns around the stem, and with surprising consistency the numbers of turns and branches relate to each other as in the series 1 1 2 3 5 8 13 21. . . — the so called Fibonacci series — in which each successive number is the sum of the two preceding it. For example, in a pine cone, there are thirteen scales for every seven turns. Similar patterns occur in the spirals of florets in sunflowers and daisies, the sections of the chambered nautilus, even the branchings of the bronchial tubes in our lungs. Such similarities in pattern give us some insight into how simple rules, used in different contexts, can produce great variety. From few notes, nature creates many symphonies.

Variations on a theme

The beetle, with some 300,000 separate species (the world's most numerous order), displays every imaginable color, decorative motif, and proportional distribution of body parts — yet the pattern of relationships that makes beetles is constant.

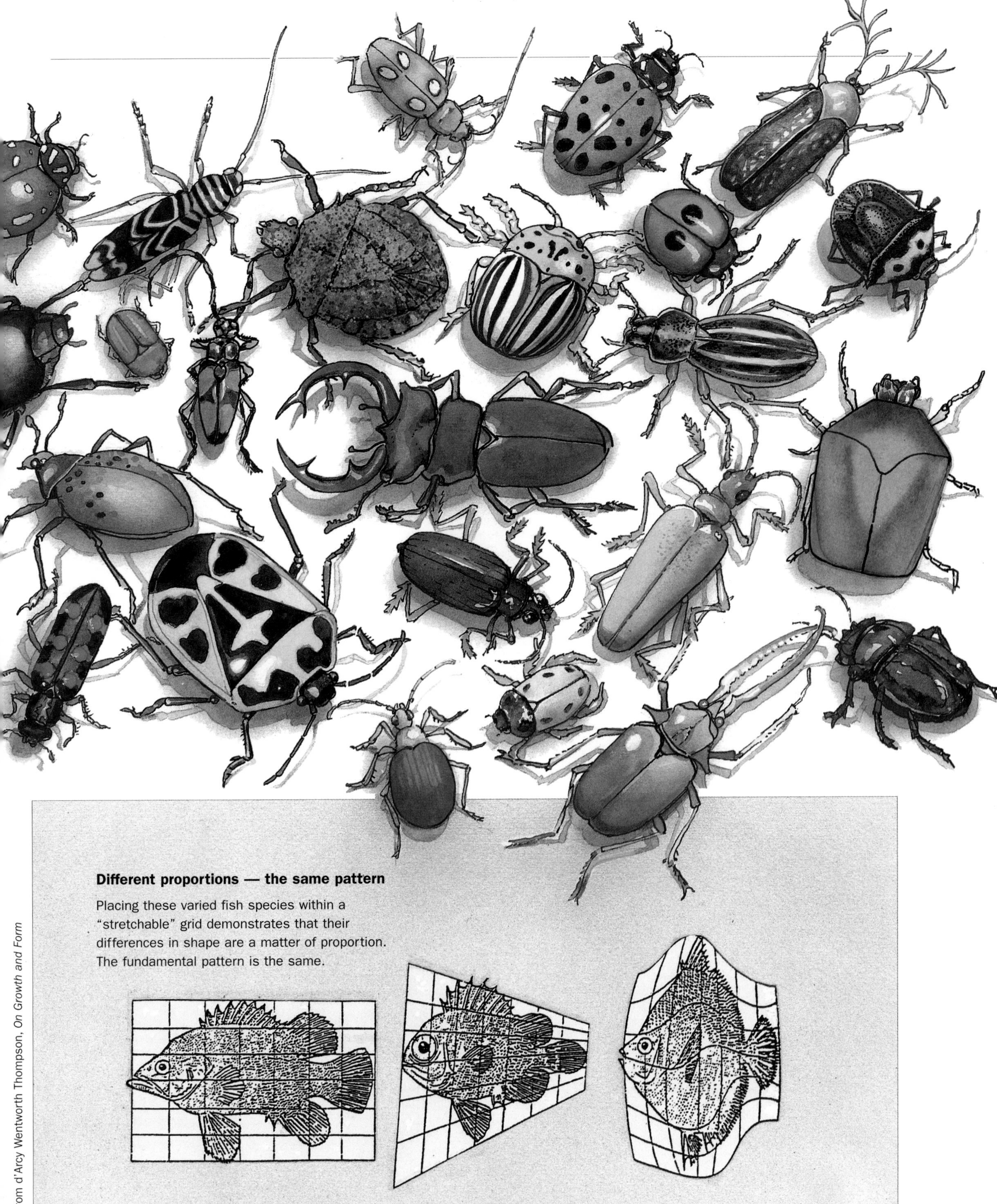

Different proportions — the same pattern

Placing these varied fish species within a "stretchable" grid demonstrates that their differences in shape are a matter of proportion. The fundamental pattern is the same.

From d'Arcy Wentworth Thompson, *On Growth and Form*

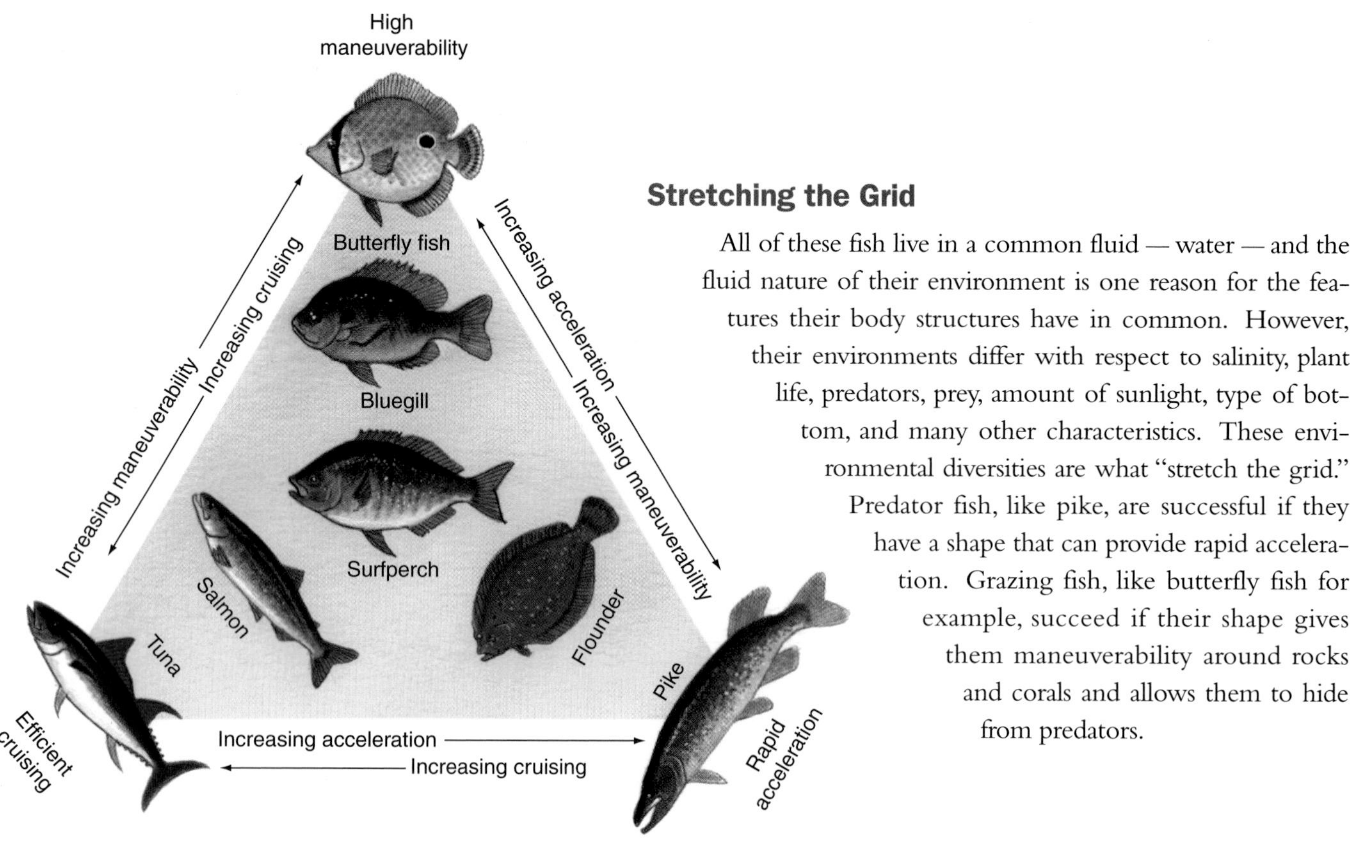

Stretching the Grid

All of these fish live in a common fluid — water — and the fluid nature of their environment is one reason for the features their body structures have in common. However, their environments differ with respect to salinity, plant life, predators, prey, amount of sunlight, type of bottom, and many other characteristics. These environmental diversities are what "stretch the grid." Predator fish, like pike, are successful if they have a shape that can provide rapid acceleration. Grazing fish, like butterfly fish for example, succeed if their shape gives them maneuverability around rocks and corals and allows them to hide from predators.

How a Slime Mold Makes Its Living

We have already seen that anything living has to have an inside protected from the outside. It's also true that the shapes living things take over time are indirectly molded by specific survival needs and by the forces of the world outside. Most living things that are the wrong color or shape or size or have the wrong kind of teeth or breathing apparatus for their environment don't survive long enough to reproduce. Ones that are better adapted to their environment do reproduce and succeed.

A branching growth pattern

A plasmodial slime mold, *Physarum polycephalum.*

The successful growth pattern in the yellow slime mold shown at left is an adaptation that allows it to absorb food from and exchange gases with the outside. It presents a very large surface to the world, and possesses a branching pattern of veins to move materials throughout its volume. The mold lives in a moist, dark, forest underlayer, so it doesn't need a thick shell or skin to protect it from its environment. In fact, in some sense, the forest underlayer can be thought of as the mold's "skin."

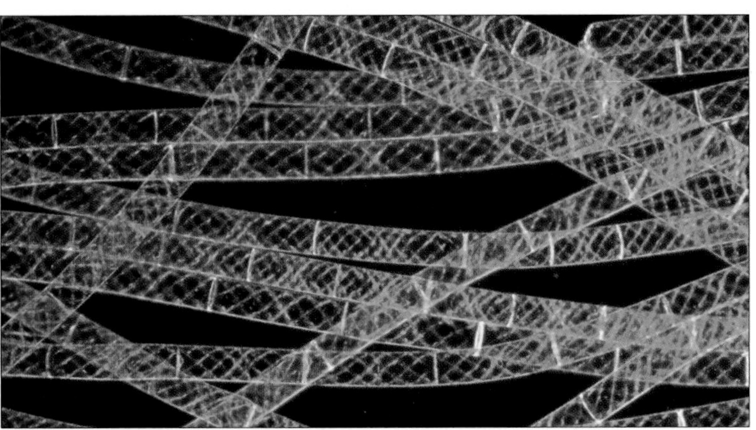

A linear growth pattern

Strands of algae—cells divide in only one plane.

Multiple growth patterns

Think of all the different patterns of cell division that created this dragonfly's shapes.

Patterns of Multiplication

When a cell divides in two, which is how it reproduces itself, the two resulting daughter cells can go their separate ways as unicellular organisms, or they can stick together and function as a multicellular organism. Dividing and adhering cells can occupy space in only four basic ways (right): (1) They can grow in one plane of space, say north-south, creating a single long chain of cells. (2) They can keep extending in that one direction, with occasional offshoots east or west. (3) They can grow consistently in two directions, making a thin sheet of connected cells. (4) Or they can grow in all three spatial planes, adding up and down to east-west and north-south, making chunks, cylinders, and spirals.

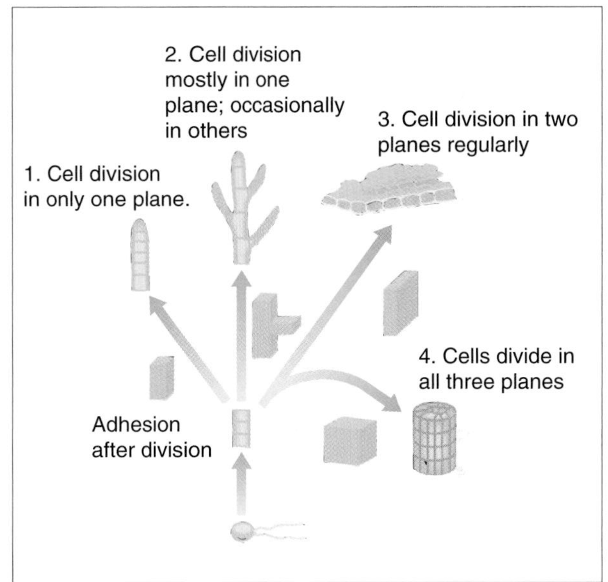
2. Cell division mostly in one plane; occasionally in others
3. Cell division in two planes regularly
1. Cell division in only one plane.
4. Cells divide in all three planes
Adhesion after division

The highly complex system of airways of a human lung (a) has a pattern very similar to that of the simpler slime mold. A very different kind of organism, the seaweed *Fucus* (b), also has a similar pattern.

Question.

Explain the similarity of the patterns you see here in terms of the way each structure supplies the needs of the living organism. How are the outsides of these structure adapted to protecting them (or not) from their environment?

Answer...

(a)

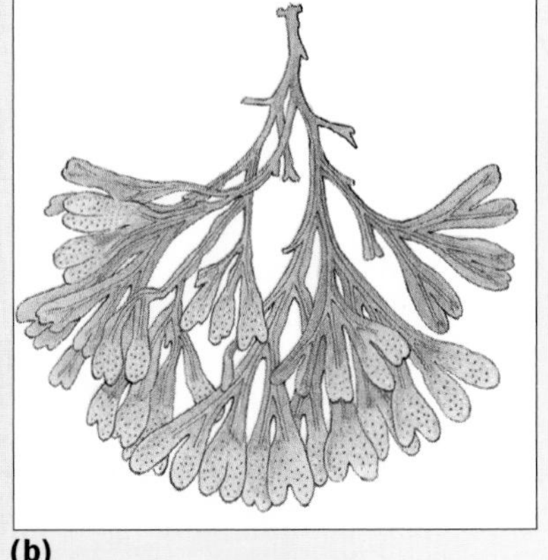

(b)

The branching pattern of airways in the lungs creates a large surface area for air to enter all parts of the lung. That same branching pattern exposes a large area of the seaweed to its fluid environment, allowing it to absorb nutrients and exchange gases. The "outside" of the lungs' airways is really their interior, and this is covered with cilia and mucus for protection. The seaweed, too, is covered with a kind of mucus, and has tough, leathery skin.

2.5 Life Organizes with Information

Making the Parts That Make the Whole

By itself, a sub-robot could never make a complex working part.

For a team of specialists, however, each completing a single step, the task becomes manageable.

The business of living requires a lot of information. An organism needs to know how to maintain a constant temperature, how to replace worn-out parts, how to defend against invaders, how to get energy out of food, and so on. It has been estimated that the information a human being needs for all of his or her functions would fill up 15 encyclopedias. It might be many times greater than that but life has developed a strategy of dealing with this large amount of information: it stores only a certain kind. The nature of this information might be best understood by the following analogy: Suppose you decided to build a complex robot requiring millions of individually handcrafted working parts. Presumably, this task would require instructions for the making of each part, plus instructions for the overall assembly, as well as operating instructions. But now imagine that you had another option: You could acquire the instructions to make several thousand tiny sub-robots, each of which knew how to fabricate one stage of one of the parts. And by working collectively, these sub-robots could assemble and operate the entire robot. In other words, an extraordinarily complex robot would result from complicated interactions among many sub-robots, each of which performs a relatively simple task.

This is the kind of information that life stores in its DNA. Sections of DNA — genes — contain no information on maintaining temperature, defending against invaders, decorating a home, choosing a mate, etc. They contain only information on how (and when) to make proteins. The rest is up to the proteins — life's subrobots.

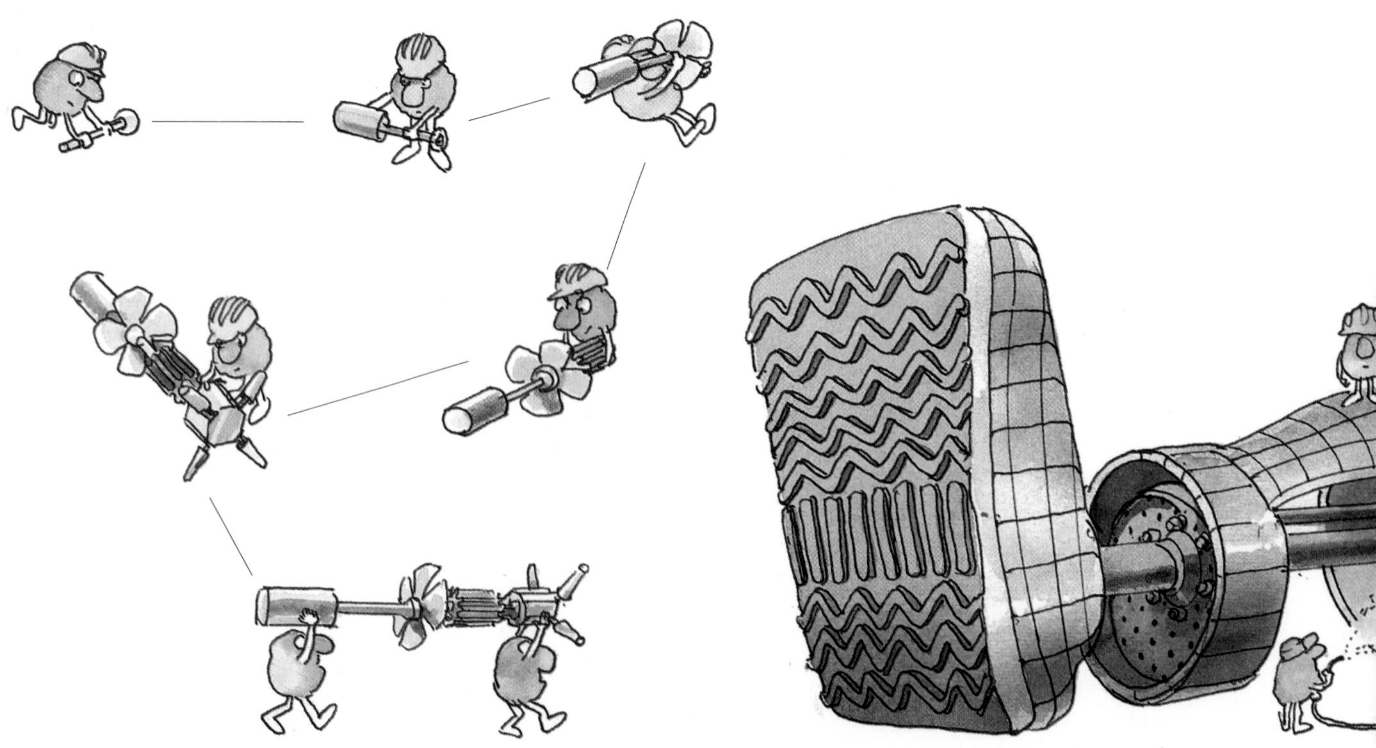

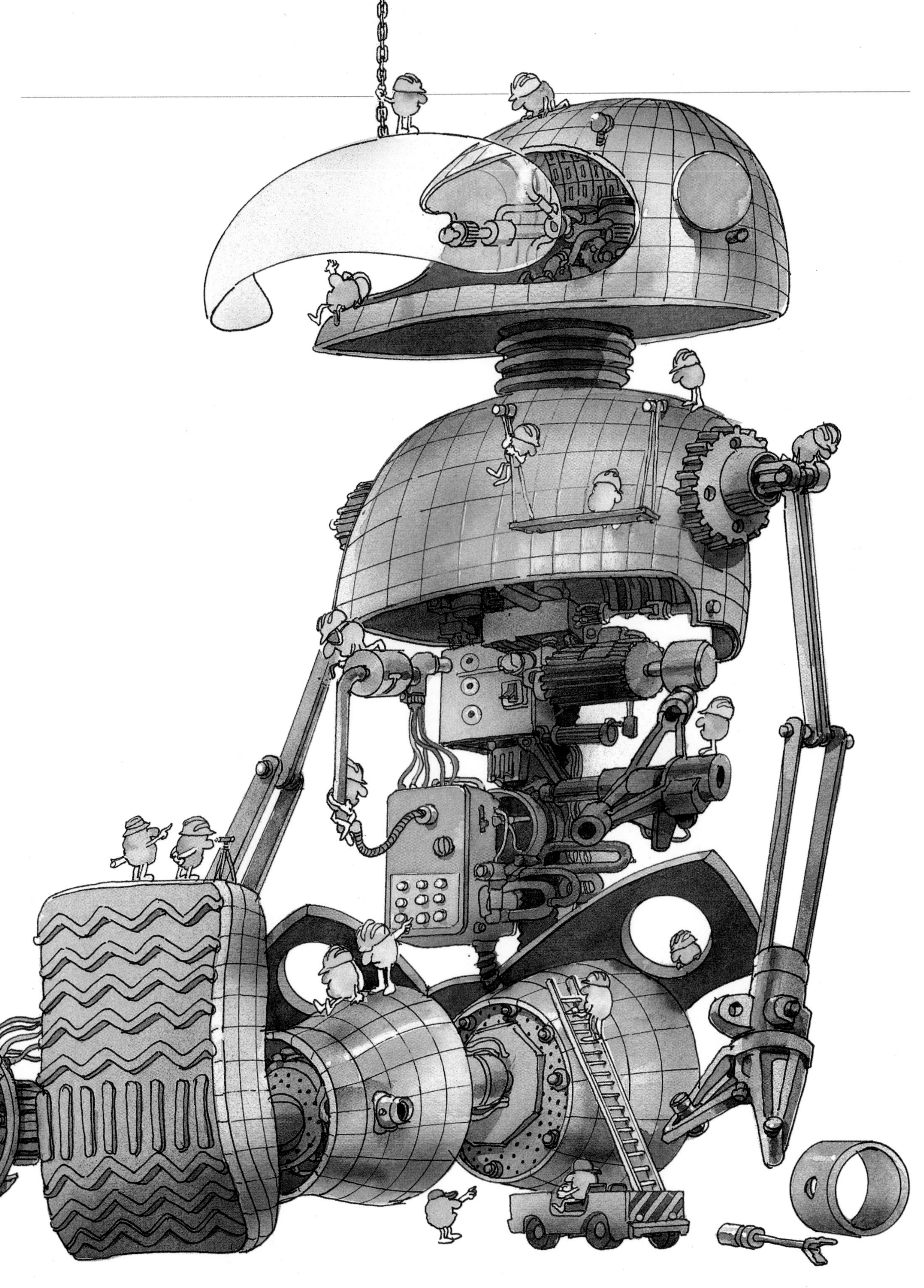

Enzyme Workers

The cell's workers — those sub-robots — are protein molecules (in this book, we often show them as the lumpy characters at right). They are able to perform relatively straightforward chemical tasks, like transforming a specific kind of molecule into a slightly altered version. They do this at incredible speeds — thousands of molecules processed per second — without being changed in the process. This kind of protein is called an enzyme — a biological catalyst, or a chemical reaction facilitator. Teams of enzymes work inside cells in a coordinated fashion to convert simple molecules, such as sugar, into the essential building blocks life uses to assemble its own substance: amino acids, nucleotides, fats, etc. Enzymes can also convert cholesterol to hormones such as estrogen and testosterone. They can, with the help of chemical energy, accomplish movement, as in the action of muscle or the transport of substances throughout cells. They can, in sum, by their many coordinated interactions, maintain a human life!

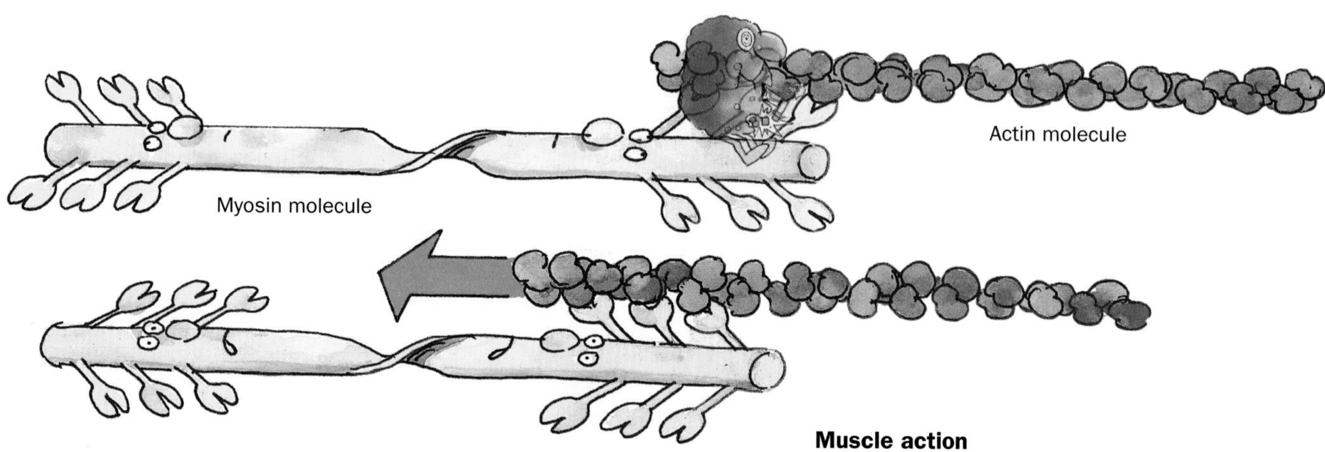

Muscle action

An enzyme attaches to the myosin molecule, causing the myosin to change shape. The shape change attaches myosin to the actin molecule and pulls it alongside the myosin, causing a shortening of the actin-myosin complex.

Sweetly Splitting Sugar

A good example of a critically important life task accomplished by a team of enzyme workers is glycolysis (from the Greek for "sweet splitting"). Glycolysis is the breakdown of the six-carbon molecules of glucose to two molecules of the three-carbon pyruvate (shown in the figure top left on the facing page). This process involves ten enzymes, each accomplishing tiny steps, and the end result is the net production of two energy-rich ATP (adenosine triphosphate) molecules from one sugar molecule. ATP is the energy coinage of life, the molecule that is used by enzyme workers to drive all cellular activity.

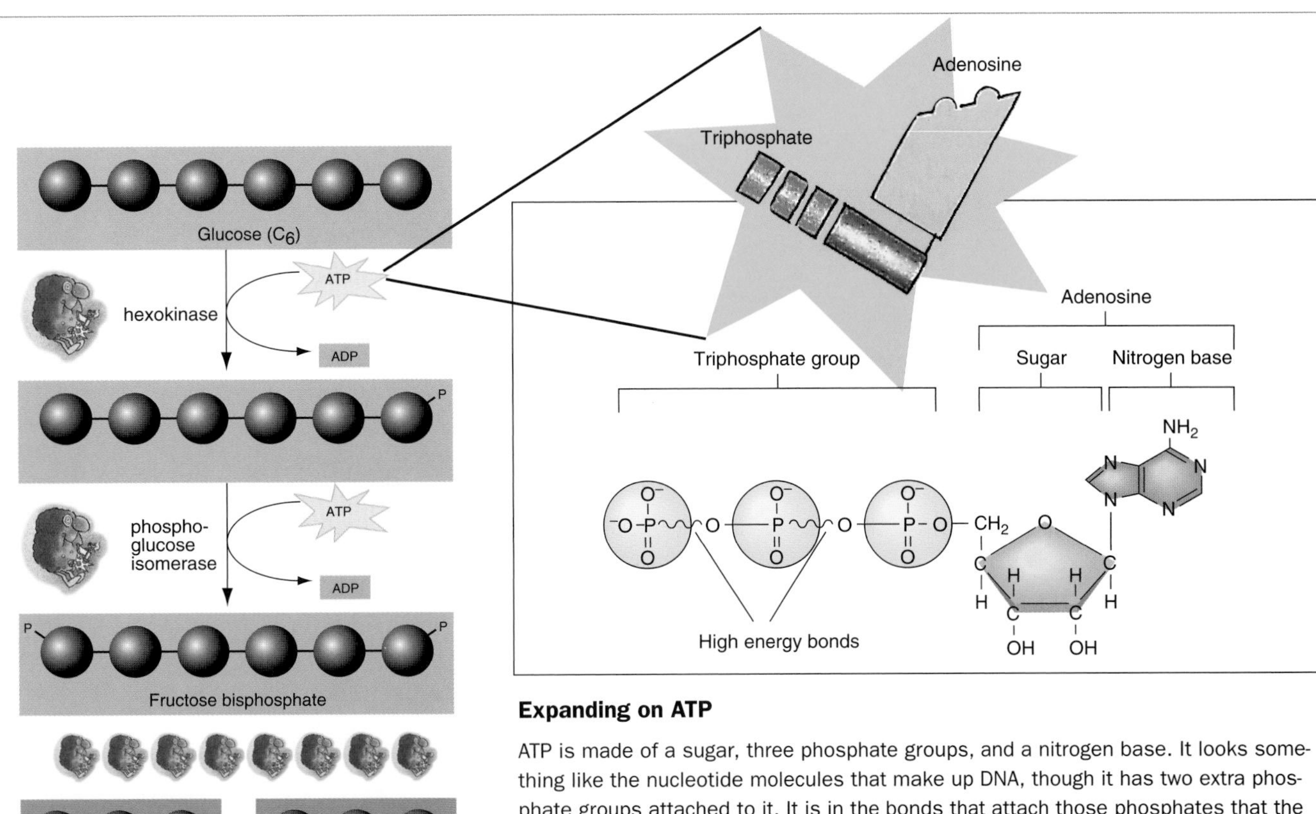

Expanding on ATP

ATP is made of a sugar, three phosphate groups, and a nitrogen base. It looks something like the nucleotide molecules that make up DNA, though it has two extra phosphate groups attached to it. It is in the bonds that attach those phosphates that the energy that drives glycolysis is stored.

Count the enzymes in glycolysis

This picture shows a simplified model of sugar—just its six carbon atoms. The enzyme hexokinase makes the first small change possible, the addition of a phosphate group (here shown as P). In the next tiny step, phosphoglucose isomerase, another enzyme, makes the next phosphate addition possible. Eight more enzymes orchestrate the remaining small chemical changes to glucose that turn it into two three-carbon molecules called pyruvate.

Glycolysis is a way of getting energy from sugar when oxygen is not available. It was used by bacteria-like organisms billions of years before there was free oxygen in the Earth's atmosphere. Organisms eventually evolved the ability to use oxygen to "burn" the waste product of glycolysis, pyruvate, to produce much larger amounts of ATP. But glycolysis proved to be so useful throughout evolution that it remains a central feature of almost all creatures alive today. You'll examine this process in much greater detail in Chapter 3, *Energy*.

Glycolysis is but one example of many very simple parts working together in a coordinated fashion to create complexity. Seemingly impossible tasks are accomplished by doing one small chemical conversion at a time.

You're a robot designer and you've automated an extremely complex machine by designing little sub-robots to do various simple tasks. The problem is that these sub-robots are doing so many simultaneous tasks that they're getting in each other's way. If you could build a new feature into some of the sub-robots, they could orchestrate their various tasks and thereby reduce the chaos.

Question.

What new feature would you add? Can you come up with a cellular analogy for this robot design problem?

Answer...

The new feature could be a switch that turns a sub-robot on and off in response to a signal. The switch would be installed in at least one member of every production line or operating team so that it would function only when necessary. Similarly, body cells might reach a point where they were producing too much pyruvic acid from glucose. The solution to this problem might be another enzyme that reacts to that overload by interfering with the action of the enzyme hexokinase until the amount of pyruvic acid in the cell decreases.

2.6 Life Encourages Variety by Recombining Information

Recombining Instructions

Nature creates new combinations by recombining information. The earliest life-forms, simple bacteria-like organisms, found a way to inject bits of information (DNA) into each other — a primitive form of sex. Over time, life acquired the ability to exchange ever-larger chunks of information, thus evolving sexual reproduction, which is a more elaborate form of information mixing.

Sexual reproduction, in simplest terms, involves lining up two long chains of information from two individuals, randomly cutting them, exchanging the pieces, and then passing a mixed chain to the next generation.

It's easy to imagine that the longer the chains, the more the possibilities for information exchange. The number of possibilities is truly staggering. Just with the ten-unit charm bracelet we use here, (each charm represents a gene), there are 1233 possible exchanges. With a chain only two units longer — 12 — there are 4086 possible mixes. And we humans have some 70,000 genes!

It's important to understand that matching pairs of genes are not always identical. For instance, your gene for the oxygen-carrying hemoglobin protein in your blood may be slightly different from mine. The difference may be trivial, or it can account for poorly functioning hemoglobin in one of us. All the differences in our genes account for why each of us is genetically unique. Between any two individuals, about one-third to one-half of all our genes are different in this way.

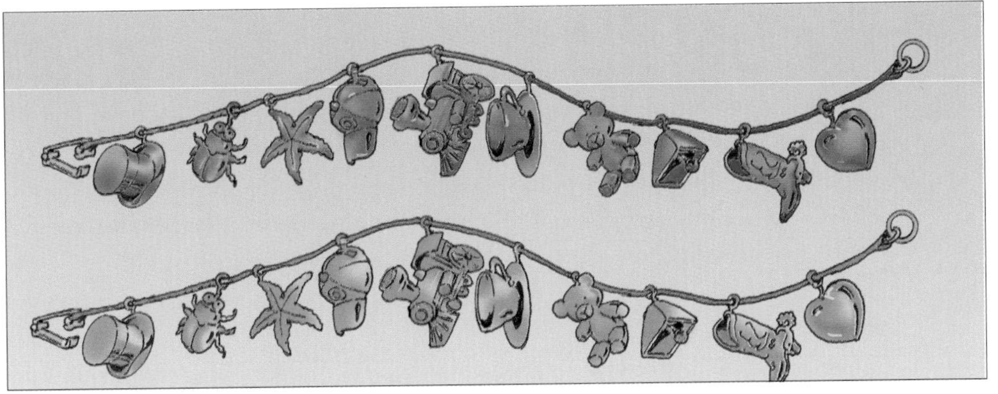

Some of the matching charms differ in small ways from one another, accounting for the genetic differences between individuals.

Our DNA consists of sequences of genes represented here by charms on a chain. One sequence of charms is from our mother, one from our father. The charms are always in the same order on both chains.

When sperm and egg cells are made, the chains are brought together, lined up, cut . . .

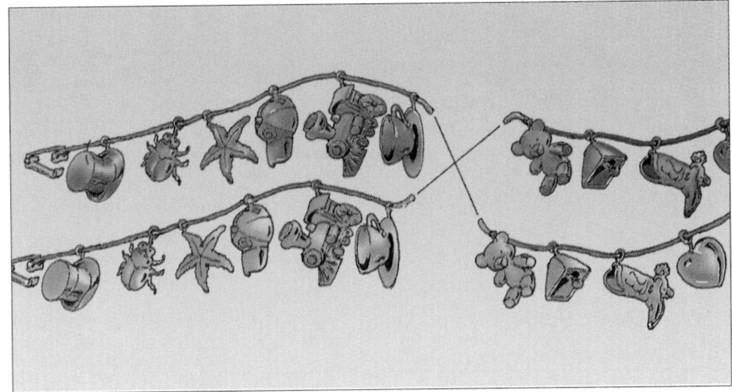

...and joined crosswise, each to the opposite strand...

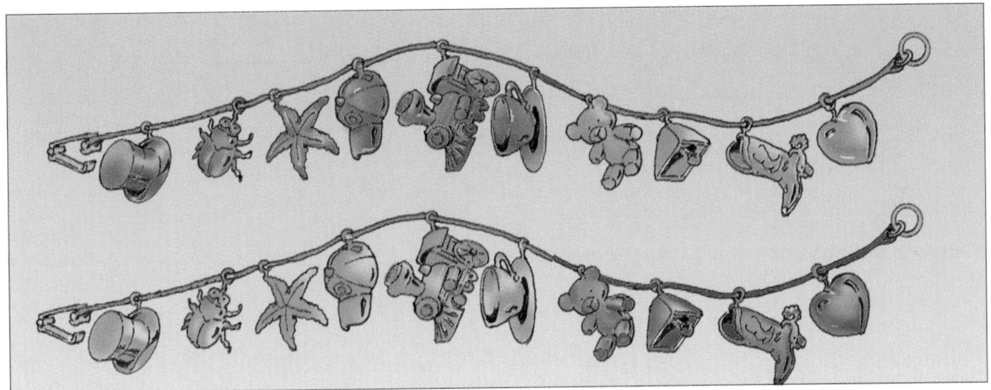

...making mixed chains. These are now ready to be separated and passed to the next generation.

Choosing chickens

By selecting only chickens with elaborate varieties of head plumage to breed with one another, breeders can quickly vary the appearance of successive generations of these showy chickens.

Having It Your Way with Genes

With generation after generation of information mixing, changes in genes accumulate and the appearance or function of succeeding generations of a living organism can change dramatically. In nature, changes in genes arise from accidental alterations — mutations — and from the exchange of genetic information when organisms reproduce. With each alteration of the information that specifies their form and function, creatures become better or worse suited to their environment. The better suited tend to survive and reproduce (to mix and pass on their genes again).

Gene reshuffling creates diversity, causing animals of the same species to come to vary greatly in appearance. Such variation within a species often occurs naturally, but it can be accelerated when genes are purposefully recombined through selective breeding. Throughout history, humans have taken advantage of the possibilities for variation allowed by gene reshuffling to control the characteristics of other animals. Dog breeders do this, for example. All domestic dogs, from Mexican Chihuahua to Great Dane, belong to the same species, though they look radically different. Cats, horses, sheep, and cows have also been manipulated genetically by humans to emphasize certain characteristics that make them more useful or decorative.

Horticulturalists have been very selective in plant breeding, as well. U.S. grain crops, with their large seeds and huge yields, would be unrecognizable to ancient farmers. The flowers in almost every garden today are mostly selectively bred strains that didn't exist a century ago, and the same is true for most of the fruits and vegetables you eat.

The decorative chickens you see above are widely varying descendants of much plainer ancestral chickens. At Plimoth Plantation in Massachusetts, agricultural scientists are selectively breeding highly specialized modern chicken species to try to produce ones with the characteristics of their seventeenth-century ancestors — a kind of reverse selective breeding.

Question.

What characteristics would the scientists try to achieve, given the environmental challenges to the original Plimoth chickens?

Answer...

The traits that would be selected for are likely to be muted colors, ability to thrive on sparse food, hardiness to cold and damp, quickness, and good vision.

Taking Gene Mixing into Our Own Hands

Since the mid-1970s we have been able to accomplish an entirely new kind of purposeful gene transfer, almost unimaginable before the discovery of the structure and function of DNA and proteins. We have learned to take specific lengths of DNA (genes from one organism) and insert them into another, in some cases turning that other organism into a "factory" for making specific proteins. The first such successful gene transfers were accomplished in 1979, when the genes that describe two human proteins, insulin and human growth hormone, were inserted into bacteria. These "genetically engineered" bacteria multiplied, producing large amounts of the two proteins. Now these proteins and others useful to humans are produced industrially in huge fermenter tanks that can hold thousands of gallons of bacterial cultures.

The same idea lies behind the genetic engineering of new food plants such as the New Leaf potato. With an inserted gene from a bacterium (*Bacillus thuringiensis*), this plant makes its own "natural" pesticide. The bacterial gene produces a protein that is toxic almost exclusively to the plant's principal predator, the potato beetle. Along with the rest of the plant's DNA, the gene for the toxin is reproduced in every cell of the potato plant as it grows, protecting it from being eaten by the beetles. Thus it is unnecessary to spray the potato plants with insecticides that might kill other, more benign insects.

Rice is a major food staple in the developing world, where hundreds of millions of people suffer from vitamin A and iron deficiencies. Lack of vitamin A leaves people susceptible to disease and progressive blindness. Lack of iron affects even more people and causes anemia, as well as playing havoc with the immune system. Recently, a new strain of rice has been engineered that includes bacterial genes for producing beta-carotene, a protein that the human body converts into vitamin A. This golden-grained rice also has genes that promote the accumulation of iron. Thus, people who eat this grain get, at one bite, their staple food and a vital vitamin and mineral supplement.

Mixed-up corn

Here you can see the progressively larger edible seeds of selectively bred corn plants. The wild form, called teosinte, is on the left of the picture.

Question.

Who is engineering organisms, and why? It is important to note that there is a lot of controversy surrounding genetically engineers organisms and crop plants. What might be some of the objections to their use?

Answer...

This one is up to you.

Wen-Jing Hu et al. 1999. Repression of lignin biosynthesis promotes cellulose accumulation and growth in transgenic trees. *Nature Biotechnology* 17: 808-812.

Dr. Hu and his colleagues report in this paper that they have genetically engineered aspen trees to make them into more useful and faster-growing paper producers. (Aspen trees' structural cellulose chains are used to make paper, and the lignin molecules that "glue" the chains together gum up the papermaking process.)

Hu's research demonstrates a way to block the gene that produces an enzyme that aspen cells use to make lignin. The genetically engineered trees end up with half as much lignin in proportion to cellulose as normal aspens have.

With a lower proportion of lignin to cellulose, aspen trees are not only more easily and cheaply made into paper — the process uses less energy and fewer chemicals — they also grow faster. The lower proportion of lignin also makes it more practical to use these trees to produce ethanol and other biofuels.

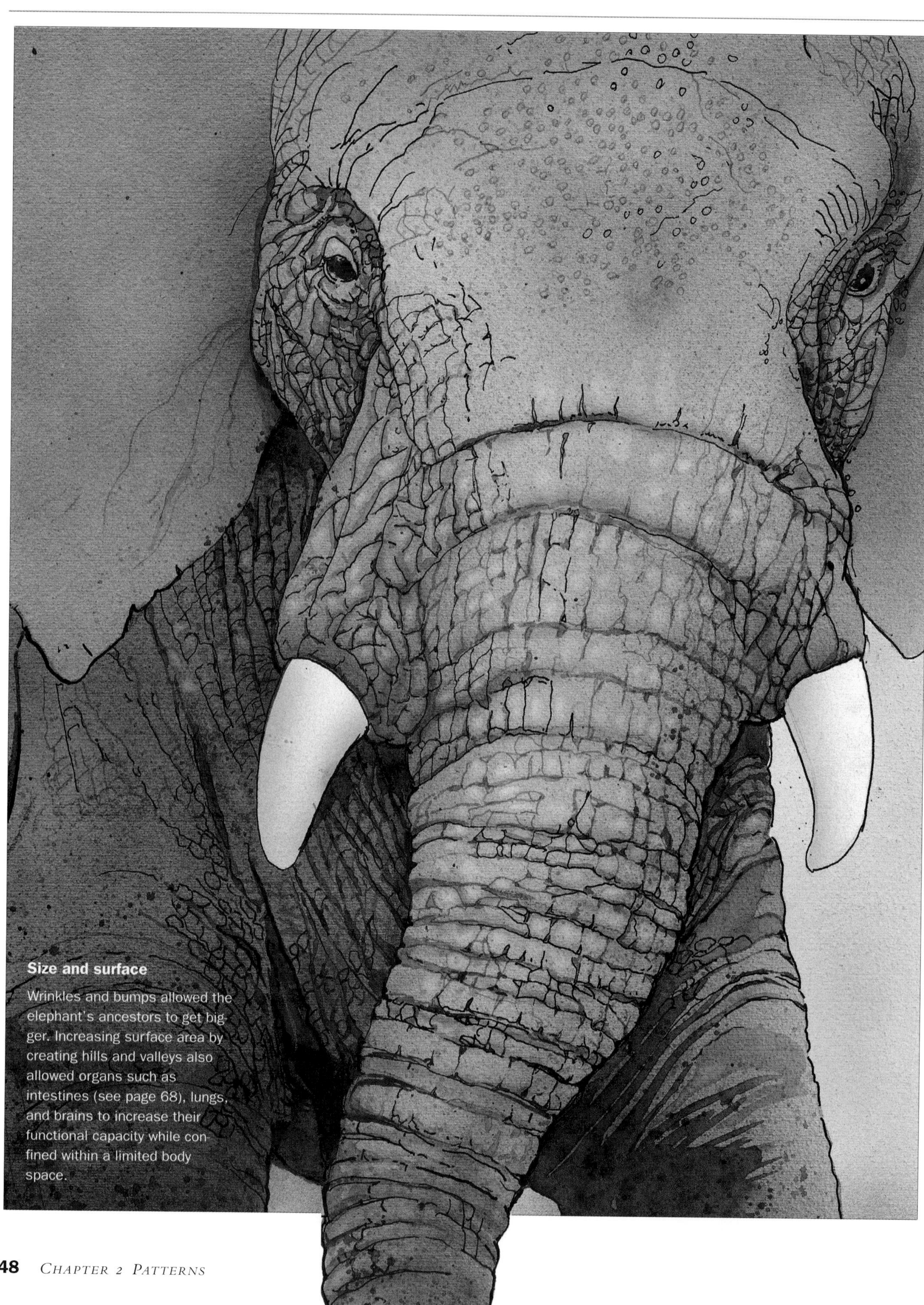

Size and surface

Wrinkles and bumps allowed the elephant's ancestors to get bigger. Increasing surface area by creating hills and valleys also allowed organs such as intestines (see page 68), lungs, and brains to increase their functional capacity while confined within a limited body space.

2.7 Life Creates with Mistakes

Accidents Ensure Novelty

When individual cells reproduce, they first make a copy of the information they carry in their DNA. Usually this copy is exact, so the information is transmitted perfectly to the next generation. But every so often, cellular mechanisms make errors in nucleotide sequences — sometimes by only a tiny bit. Miscopying even a single nucleotide in a gene, like dialing a single wrong digit in a phone number, alters the gene sequence (see page 28) and therefore changes the piece of information being transmitted. The altered information may have no effect, or may show up in the offspring as a defect. But every once in a while, it shows up as an improvement — something that makes the offspring better adapted for survival than its parents.

As an example, take the elephant. Scientists speculate that its early ancestors were small and smooth-skinned. Imagine a copying error in the distant past that jumbled the instructions for the elephant's skin cells, making them assemble into wrinkly and bumpy patterns. It happens that wrinkly skin provides more surface area than smooth skin, a fact of geometry that came in handy for the elephant. Large animals generally have a problem with overheating. A wrinkled skin exposes more surface to the air or water and thereby cools the animal more efficiently. Thus wrinkled skin helped make it possible for the elephant to grow larger and to enjoy the advantages that come with increased size.

As you come to appreciate the evolutionary role of copying errors, it is apparent that calling them "mistakes" oversimplifies. We may, in a larger context, view them as nature's way of introducing randomness, an essential feature of all creative processes.

A mistake for one organism can be an advantage for another.

Albinism, a defect in pigmentation, occasionally shows up in many kinds of plants and animals. Most albinos find themselves at a disadvantage in life, since they don't blend into their surroundings, and albino offspring in many species do not survive infancy. Snowy white polar bears, ptarmigans, arctic foxes, and snowshoe hares, however, owe their camouflaging white coloring (and their very existence) to their albino ancestors.

Nature is, above all, profligate. Don't believe them when they tell you how economical and thrifty nature is . . . Extravagance! Nature will try anything once. This is what the sign of the insect says. No form is too gruesome, no behavior too grotesque. If you're dealing with organic compounds, then let them combine. If it works, if it quickens, set it clacking in the grass; there's always room for one more; you ain't so handsome yourself. This is a spendthrift economy; though nothing is lost, all is spent.

Annie Dillard, *A Pilgrim at Tinker Creek, 1999*

Mutations — Good or Bad?

Consider this paradox. The genes of all organisms have evolved to their present state through mutations — random changes in life's information chains (DNA). Mutations have gotten us here. At the same time, a few of these mutations produce effects that are detrimental to the organism that inherits them. There are over 4000 known genetic diseases that are attributable to defects in a single gene. Huntington's disease, Down's syndrome, and sickle cell disease are examples of the harmful effects of mutated genes. Like a fine Swiss watch, perfected over centuries, an organism does not tolerate change easily. Modifications are likely to make it work less well, rather than better. Thus, it's not surprising that life has evolved mechanisms to limit mutations. Cells have effective machinery for monitoring their DNA, finding errors and correcting them. So the mutation rate — the rate at which inherited, uncorrected changes accumulate in the genes of living creatures — is kept to a minimum.

There's an important distinction to be made between the mutations that occur in the trillions of cells that make up the body and those that occur in the special cells that become sperm and egg cells. Mutations in body cells affect only the individual who sustains them. Mutations in sperm and egg cells, on the other hand, are passed on to another generation so a mutation rate substantially higher than the current one would compromise the future of the entire species.

Mutations are generally caused by: (1) mistakes made by the DNA duplicating machinery when cells divide, (2) X-rays or UV and cosmic radiation impinging on DNA, or (3) certain toxic chemicals interacting with DNA. Viewed more broadly, a mutation could be considered to be any change in DNA, such as a mistake in duplication resulting in extra genes or the acquisition of genes from a virus. It is important to realize, however, that although a small change in a gene can affect a critical function of a protein, many changes in genes affect non-crucial parts of a protein, and therefore have no noticeable effect. These neutral differences in protein can accumulate over time to produce the variations that we detect by gene sequencing (see pages 168-169).

It might seem ironic that genetic mutation, the effects of which can be disease or malformation, is also the source of evolutionary creativity. When the venerable and trustworthy Swiss watch underwent the substitution of a quartz crystal for its spring, it made a quantum leap forward in timekeeping. So it is with genes. A mutation that allowed cells to stick together opened the door for all multicellular organisms. Such a change might have arisen when the gene for a receptor molecule in a single cell's membrane mutated and instead produced a protein that recognized and bound firmly to a receptor molecule on another cell. Proteins called adherens act in just this way.

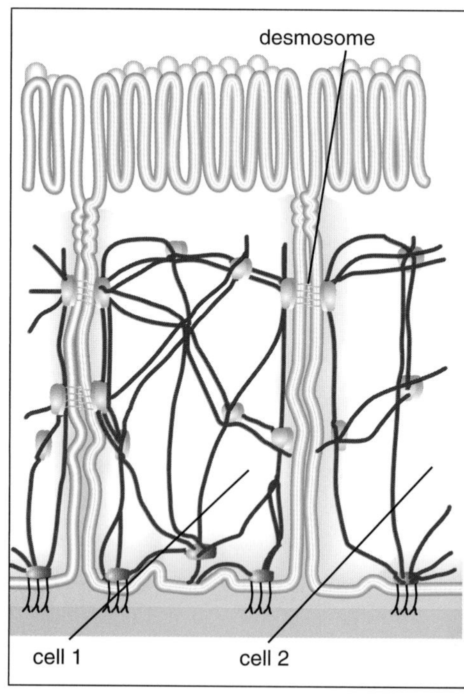

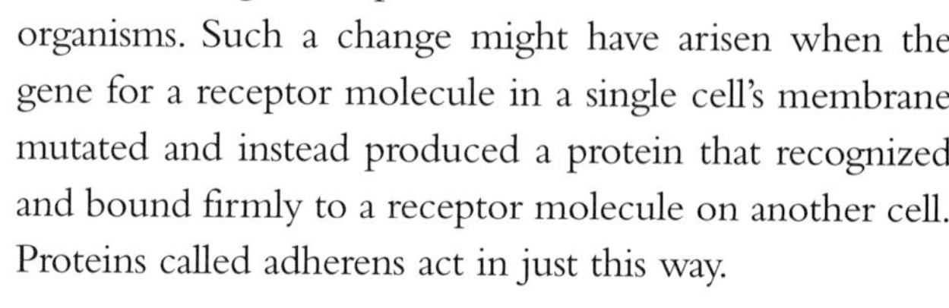

Sticky proteins

Cells stick together by adherens proteins at places on their membranes called desmosomes.

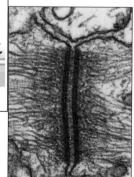

Housekeeping mutation?

The lens of the eye, as seen by a scanning electron microscope.

Mutations affecting the gene for an ordinary housekeeping protein may have resulted in the transparent tissue that forms the lens of the eye. That protein, called lactate dehydrogenase, has been involved in cellular energy production for millenia. It is structurally identical to lens proteins called crystallins, which stack like lumber inside lens cells.The crucial genetic event in the past was probably a change that allowed large quantities of the housekeeping protein to be made. Somehow, cells packed with an excess of the transparent protein provided a simple organism of the past with the ability to detect and respond to light. An advantage that made that organism survive and reproduce better than other, non-mutant types led, step by step, to the eye.

Whether a Mistake Is Good or Bad May Depend on Where You Are

Among humans, one genetic mistake shows up as sickle cell disease, a painful and debilitating hereditary condition. When depleted of oxygen (deoxygenated), normal human red blood cells retain their familiar round shape. In sickle cell disease, some of the deoxygenated red blood cells become elongated and curved in shape (like the tool called a sickle). When this happens, the sickle cells begin to clog blood vessels, and inflammation and tissue destruction occur. All of this damage is the result of a small change in the gene for the oxygen-carrying protein hemoglobin.

Interestingly, however, the sickle cell condition provides some protection against malaria, a blood parasite. Sickle cell disease is very common in equatorial Africa, where malaria is endemic. Scientists speculate that the disease, which might have been expected to remain uncommon because it decreases its victim's chances for survival and reproduction, is common precisely because it protects against malaria, a serious killer in Africa. A person who is homozygous for sickle cell disease, meaning that the defective gene was inherited from both parents, usually dies at a young age; one who is heterozygous, or inherits the gene from only one parent, suffers a much milder form of the disease and thus lives long enough to enjoy its protective effects against malaria, and to pass the gene on to descendants. This offers a good example of how a beneficial trait that evolved in one environment (tropical Africa) may prove detrimental in another environment (temperate America and Europe).

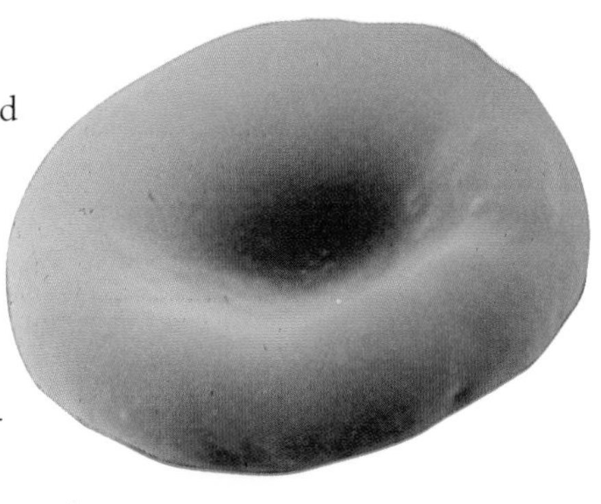

Normal and sickled red blood cells

These are scanning electron micrographs of a normal red blood cell (top) and of a cell with just one incorrect amino acid in its hemoglobin protein (see that protein on page 17).

2.8 Life Occurs in Water

The All-Purpose Molecule

Of all the molecules of life, none is so omnipresent as water. Our cells are 70 percent water. Life began in water. When our ancestors arose from the sea to become land dwellers, we brought water along with us, within our cells and bathing them. Most of the essential molecules of life dissolve and transport easily in water.

The most abundant fluid on Earth is, happily, the one most suited for encouraging living chemistry.

Water participates in all kinds of chemical reactions. Bounded by water-insoluble membranes, cells owe their shape and rigidity to water. Water provides an inexhaustible supply of the hydrogen ions needed for converting the Sun's energy into chemical energy.

What is it about water that makes it so special? The key is its polarity. Composed of a single oxygen atom sharing electrons with two hydrogen atoms like a head wearing a pair of Mickey Mouse ears — a water molecule looks quite ordinary. While the molecule's overall electric charge is neutral, the oxygen tends to pull negatively charged electrons toward it, leaving the hydrogen "ears" slightly positively charged relative to the more negative oxygen "head."

This means that an ear of one water molecule will form weak bonds with the head of another and vice versa, so that water molecules continuously stick and unstick to each other, thus forming dynamic, evanescent lattices. This self-embracing quality of water accounts for its tendency to remain liquid when most other substances with molecules its size are gases.

Most of life's important molecules are readily soluble in water: they tend to form weak bonds with water as easily as water bonds with itself. The random motion of all molecules, and their tendency to spread out evenly in a solution ensures that, once dissolved, they rapidly diffuse throughout the body's watery environment.

Luckily for us, water also has the unusual property of expanding when it freezes, so that the less dense ice floats. This provides an insulating layer that prevents further freezing of our lakes, rivers, and oceans. If water were like most natural materials, whose solid state is denser than their liquid state, ice would sink, and bodies of water in colder climates would freeze solid, making life untenable.

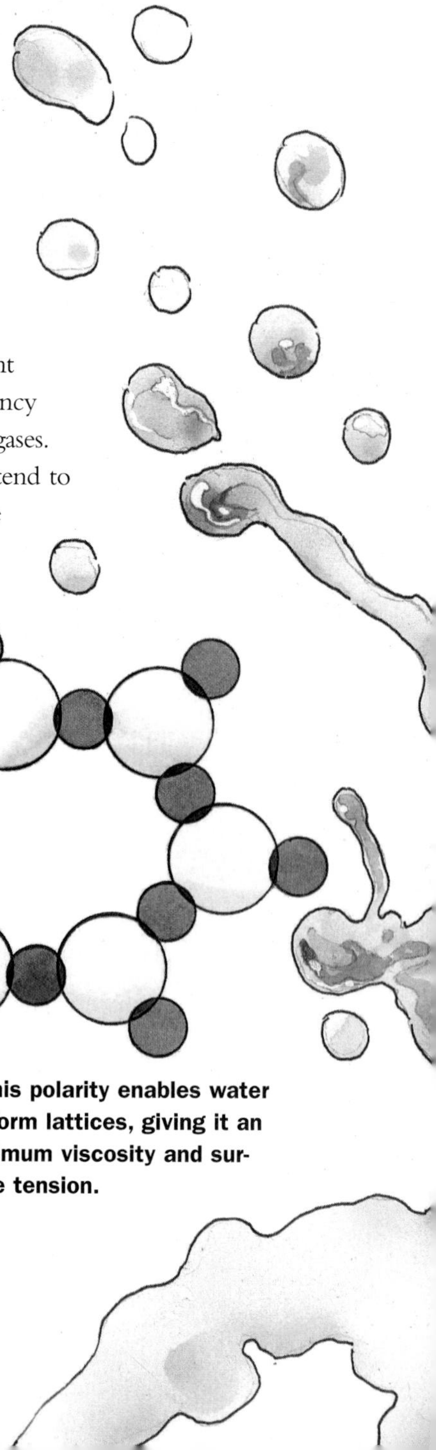

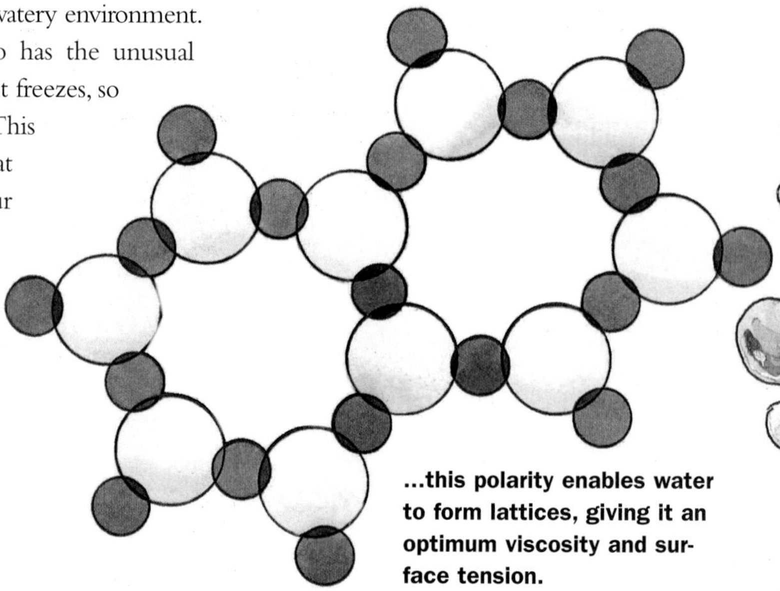

...this polarity enables water to form lattices, giving it an optimum viscosity and surface tension.

+ +

\-

Water's specialness is due to its molecular structure. The two hydrogens (Mickey's ears) have a positive charge, the oxygen, a negative charge...

The Just-rightness of Water

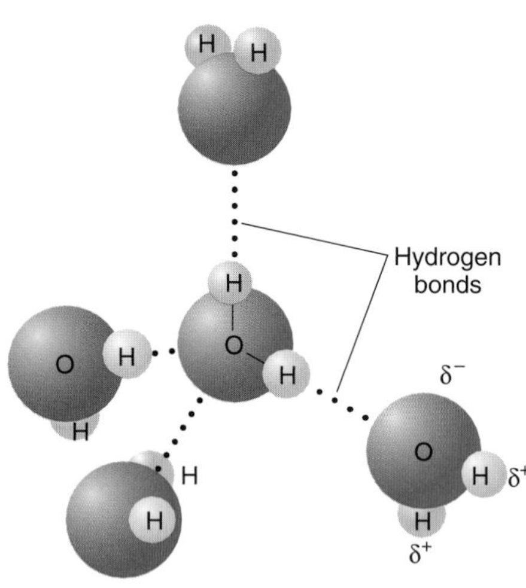

Surface tension
The forces between the water molecules (shown above) are strong enough to support the weight of this water strider on the surface of a pond.

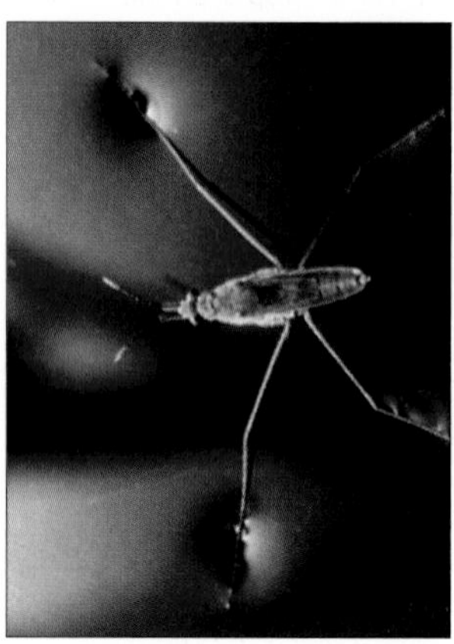

Water is by far the most plentiful chemical constituent of living creatures; it is the medium in which life came into being on our planet; and it is a pervasive part of the environment in which we now find ourselves. That molecules consisting of nothing more than an oxygen atom bonded to two hydrogen atoms could be so essential to sustaining life seems hard to believe, but water has some surprising properties that make it optimally suited to its role. 1) Its tendency, already mentioned, to become lighter (less dense) as it approaches freezing, and even lighter when it crystallizes as ice. (2) Its relatively high heat conductivity as a liquid but poorer conductivity as ice or snow. (3) Its thermal capacity — the fact that it takes a lot of energy to change its temperatures. (4) Its high surface tension — cohesiveness or stick-togetherness (illustrated at left). (5) Its capacity to remain liquid over a wide range of temperatures. (6) Its relatively low viscosity — low resistance to flow and consequent ease with which substances diffuse through it. And finally (7) its activity as an almost universal solvent.

This unusual collection of properties has a number of consequences for our environment: Water is preserved on the Earth's surface (it doesn't fly off into space), and even in the coldest climates remains liquid under an insulating layer of ice and snow. Consider the impact water has on rocks. Water has, for billions of years, crept into crevices of rocks (high surface tension), cracked them (when water froze and expanded), ground them up (in glaciers), dissolved their minerals, and carried these minerals to the sea in streams and rivers where they became essential constituents of living things.

Water contributes importantly to the general environment because of its resistance to temperature change; water vaporizes as the temperature rises (absorbing heat energy) and condenses as the temperature falls (releasing heat energy). If water's density weren't close to that of the creatures floating in it, they would sink to the dark cold of the bottom or rise to the surface where they would be more readily exposed to the damaging effects of ultraviolet light.

Within and among living cells, water's optimum viscosity protects delicate structures from the sheering forces of shape-change and motion — it acts as a kind of lubricant. Evaporation of water from the leaves of plants and trees constantly pulls water upward — water's surface tension makes this possible. In multicellular creatures, where a circulatory system is needed to move materials to all cells, and to conduct away waste and heat, water's low viscosity ensures that tiny-diameter capillaries will conduct it and its dissolved chemicals to and from the remotest parts of the body. It is especially interesting that the viscosity of watery fluids containing cells (such as blood) drops as the pressure forcing it through a vessel rises, making the distribution of materials even easier.

It is hard to imagine a more perfect environment for life. This is why scientists searching for clues to the existence of life elsewhere (as on the Moon or Mars) keep a sharp eye out for evidence of past and present collections of liquid water.

Water molecules' tendency to stick together means that water disrupted by dissolved salts on one side of a membrane will attract water molecules from the other side of the membrane to dilute the salt solution — a tendency called osmotic pressure.

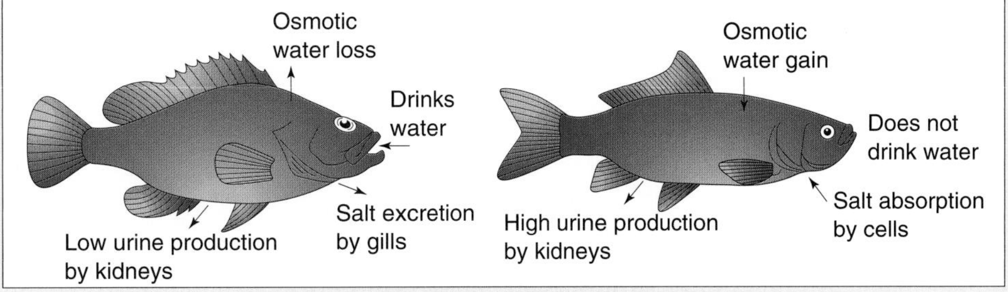

Question.

One of these fish lives in the ocean; the other lives in fresh water. The labels describe each fish's water loss or gain and how it regulates the movement of water across its cell membranes. Which one of the fish must live in the ocean? What significance does the amount of urine production have?

Answer...

The fish on the right does not drink water, but still gains water through its cell membranes (osmotic water gain). Its exterior environment must be less salty than the interior environment of its cells; it is a freshwater fish. High urine production is a way of concentrating salts in body cells to keep the internal environment from becoming as dilute as the exterior environment.

Water Organizes and Orients Other Molecules

Almost immeasurably tiny and opposing electric charges accumulate on the single oxygen and two hydrogen atoms of a water molecule. The attractions and repulsions set up by these electric charges create an environment for life. Water molecules cling to one another's oppositely charged ends, and so influence one another's orientation in space. They also attract or repel and therefore orient other kinds of charged molecules. This interplay between electrical attraction and repulsion sets the scene for the development of molecular containers (cells) that maintain an inner and an outer environment.

The molecules (called phospholipids, see page 32), that make up most of a cell membrane have one charged end. The instant such molecules are in the presence of water, they orient their charged (phospho-) end toward the hydrogens of the water molecules. The fatty (lipid) tails of the molecules are not charged at all, and so tend to stay away from any water molecules. This hydrophilic ("water-liking") and hydrophobic ("water-shunning") molecular behavior provides a clue to how bilayered spheres might first have formed.

How salt dissolves in water

Table salt crystals are composed of oppositely charged sodium and chloride atoms (called ions). In water, the negative chloride ion attracts the positively charged end of water molecules, and the positive sodium ion attracts the negatively charged end. These attractive forces are strong enough to pull the salt crystals apart. Notice how the dissolved salt components can now diffuse quickly through the water.

Bilayered sphere formation

When a layer of phospholipid molecules like this one on the water's surface is agitated by wind, spheres with water droplets inside can form, and when the sphere drops back to the surface, the fat-liking ends of the molecules on its outside join with those extending from the water surface to create bilayered spheres.

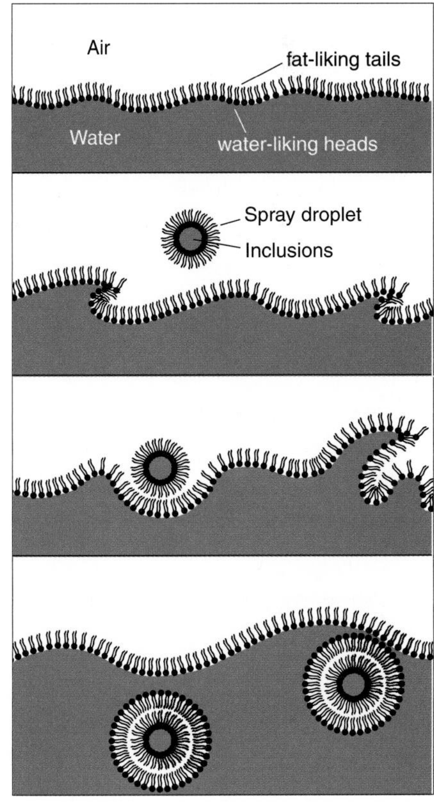

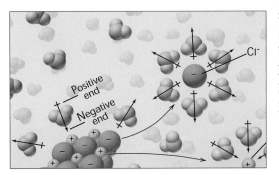

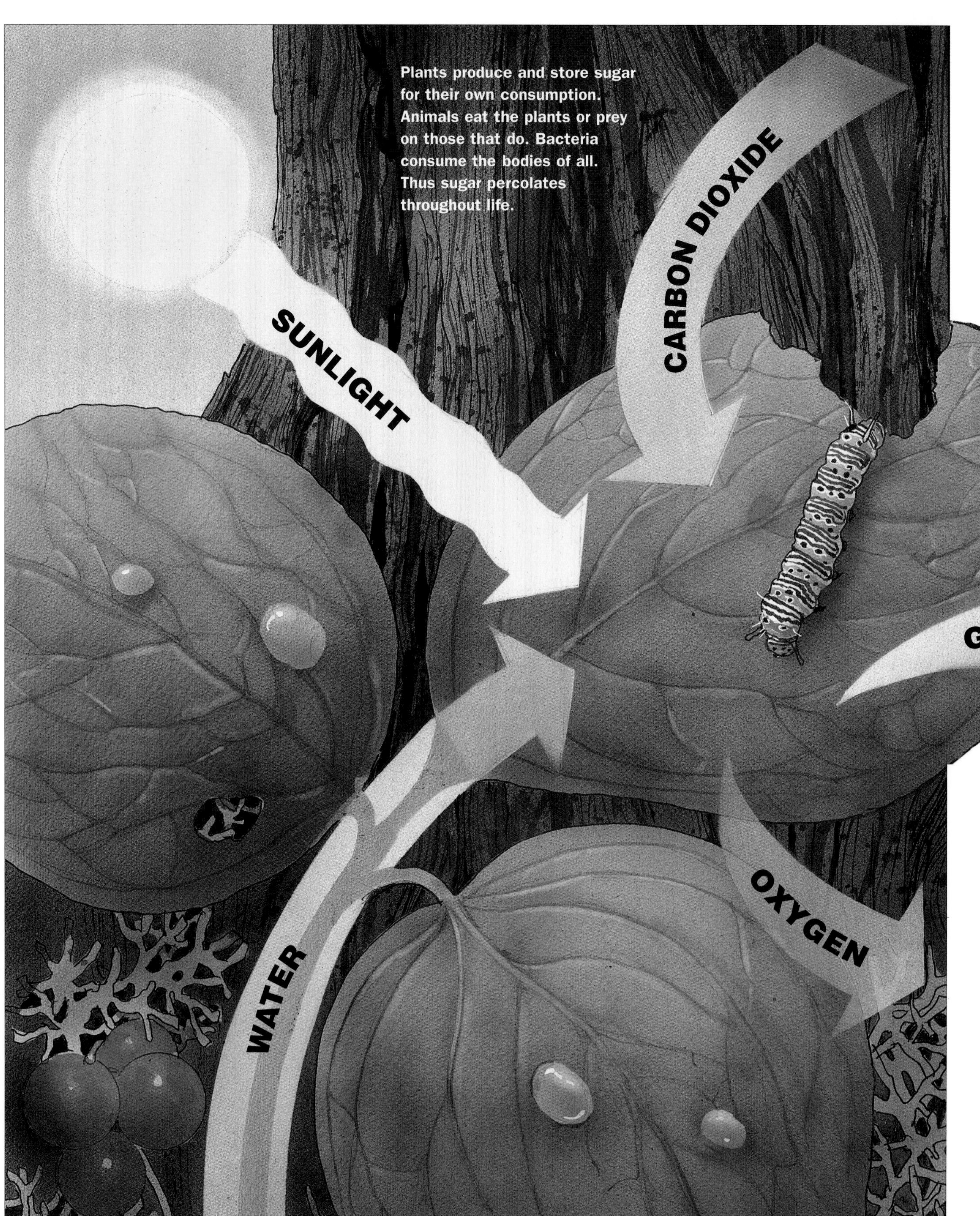
Plants produce and store sugar for their own consumption. Animals eat the plants or prey on those that do. Bacteria consume the bodies of all. Thus sugar percolates throughout life.
SUNLIGHT
CARBON DIOXIDE
GL
OXYGEN
WATER

2.9 Life Runs on Sugar

A Molecule to Burn

Sugars are simple, energy-packed chains of three to seven carbon atoms festooned with hydrogens and oxygens. Life's central sugar is glucose. It is the fuel that drives the engine of life and the basic material from which much of life is constructed. Each year, plants, marine algae, and certain kinds of bacteria convert 100 billion tons of atmospheric carbon dioxide (CO_2) and hydrogens extracted from water (H_2O) into sugar — using energy from sunlight in a process called photosynthesis. The by-product of this massive conversion is oxygen.

Plants, algae, bacteria and animals all "burn" sugar. That is, inside their cells they transform the energy in sugar's chemical bonds into an especially potent form of chemical energy — ATP. In this living combustion process, called respiration, sugar's carbons and oxygens are discarded as CO_2 and its hydrogens are linked to oxygen from the air and discarded as H_2O. Thus the very substance of life materializes from air and finds its way back to air. The constantly generated ATP powers all life's work, such as moving, breathing, and laughing. Sugar also serves as the starting material for the assembly of the simple molecules — amino acids and nucleotides — from which large molecules are assembled.

Several hundred million years ago, the rate of photosynthesis was greater than respiration, and enormous quantities of the remains of trees, plants, animals, and bacteria were buried deep in the earth, subjected to intense heat and pressure, and transformed into coal, petroleum, and natural gas. Much of this material was initially chains of sugar molecules — cellulose and other related chain molecules. So sugar reemerges as the basic ingredient of the fuels that drive the engines of civilization.

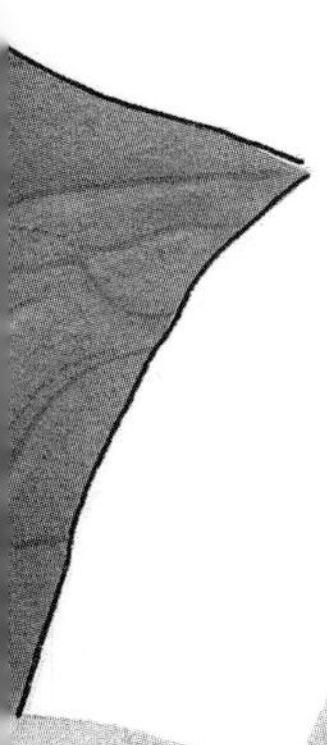

Glucose, life's key sugar molecule, is broken down — metabolized — by living cells, and its parts used to make life's essential molecules.

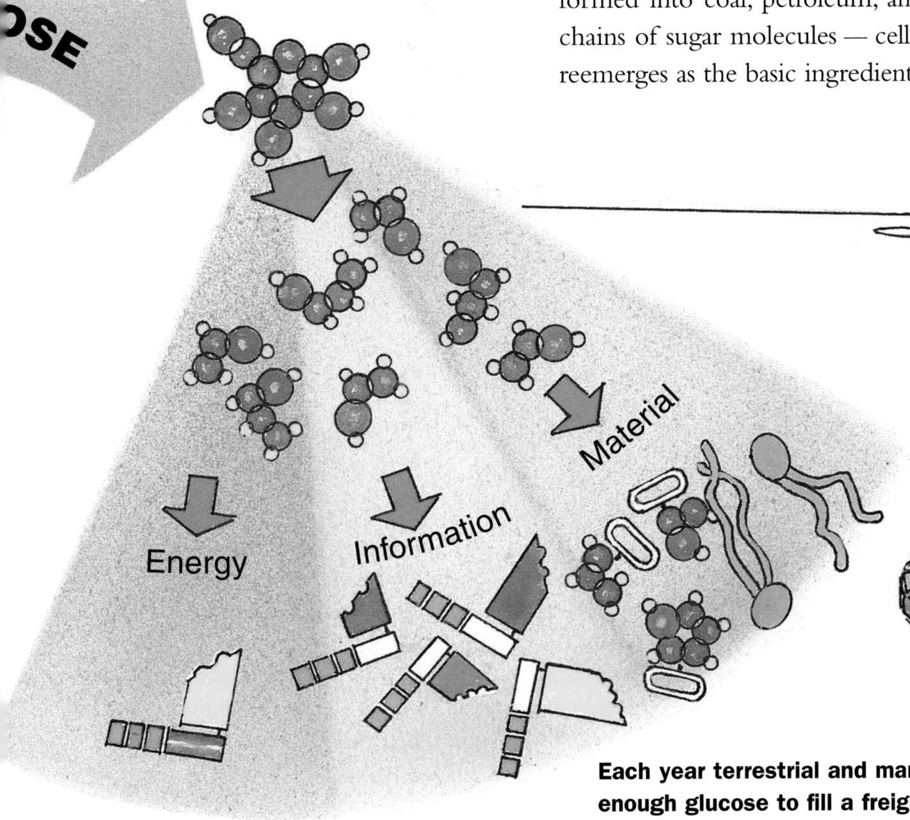

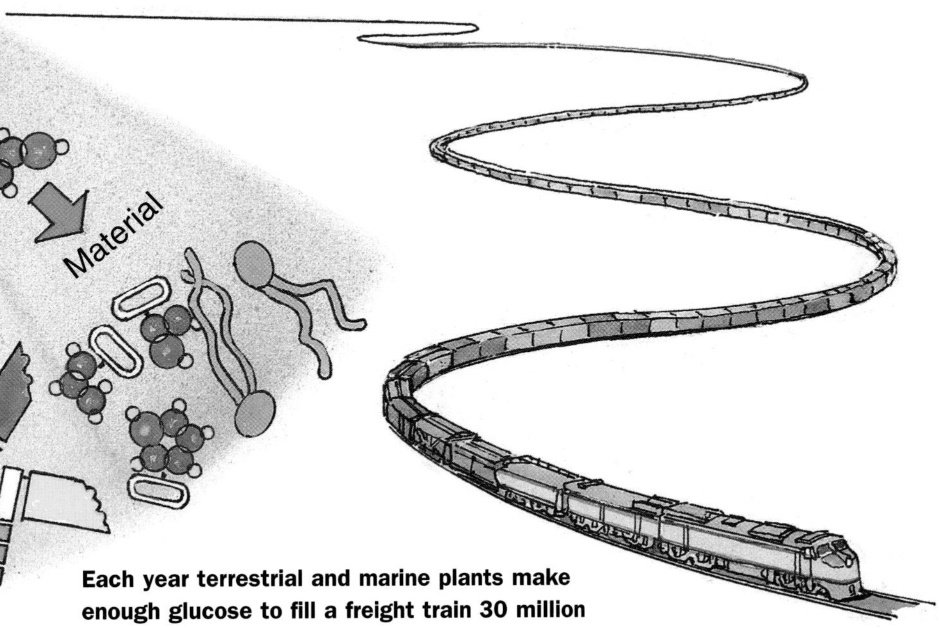

Each year terrestrial and marine plants make enough glucose to fill a freight train 30 million miles long.

2.10 Life Works in Cycles

A steam engine

The engine's main wheel is turned by steam. A belt from the wheel causes the governor — a spinning ball system — to rotate. The faster the wheel turns, the faster the governor's shaft turns, the farther outward fly the balls. This lifts the disk, raising the lever, and closes the steam input line, slowing the engine.

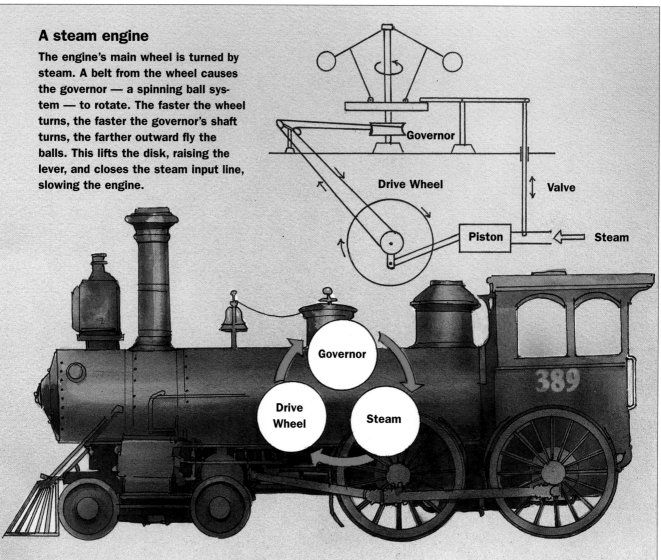

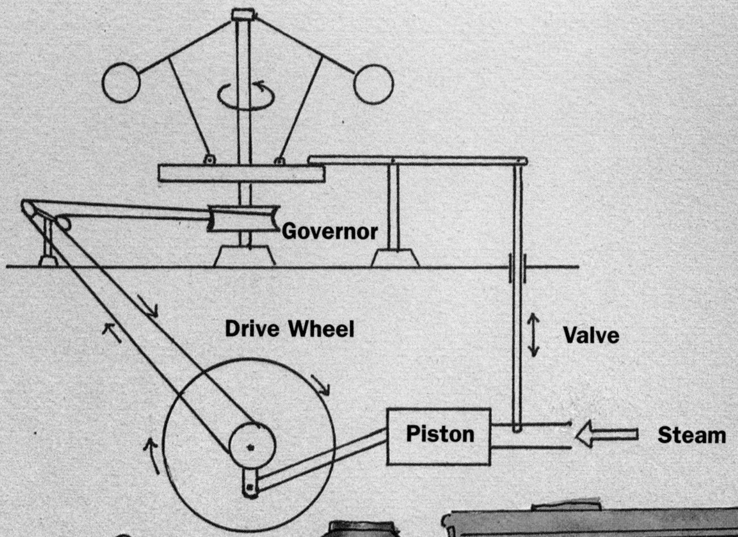

Governor

Drive Wheel

Valve

Piston

Steam

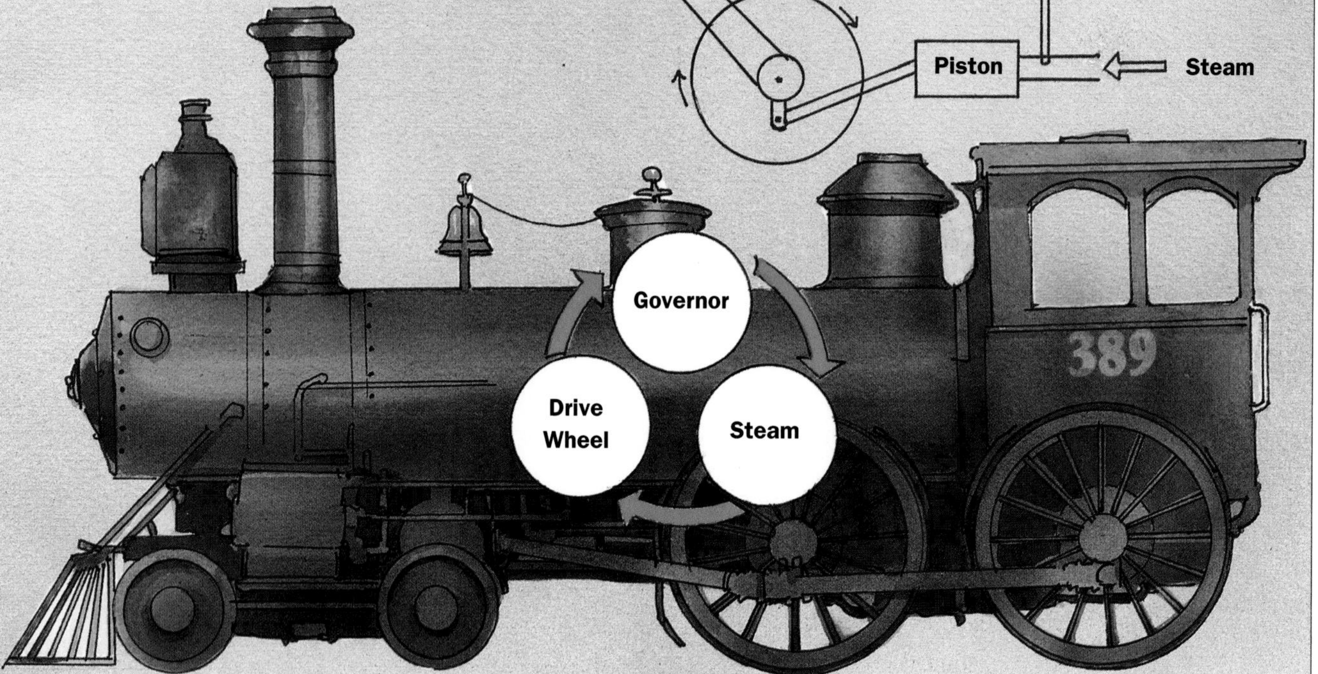

Governor

Drive Wheel

Steam

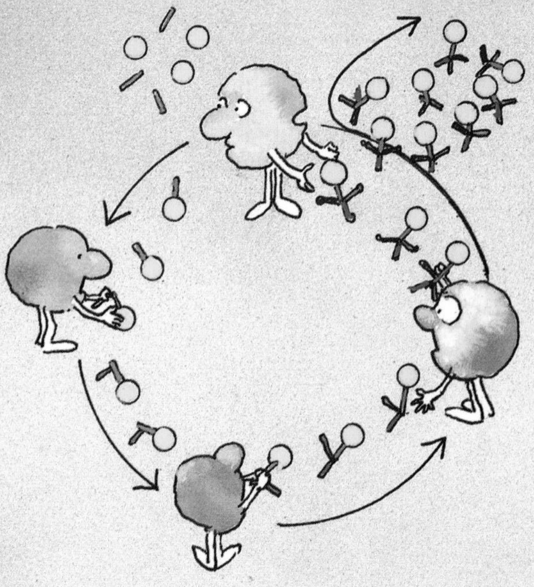

389

Circular Control

In the simplified steam engine above, a fire heats water, making steam, which activates a piston, which turns the engine's drive wheel, which spins the governor, which controls the steam supply. Such a three-component loop passes information from part to part so that the engine is able to self-correct by way of the governor.

A similar self-correcting system comes into play when a protein makes a chemical product. Each protein performs a simple task (e.g., adds a part) in assembly-line fashion. The circular arrangement allows the initial protein to keep track of the overall output. As products either pile up or become scarce, it adjusts the speed of the overall operation. Chapter 6, *Feedback,* discusses how it does this.

A Circular Flow of Information

Life loves loops. Most biological processes, even those with very complicated pathways, wind up back where they started. The circulation of blood, the nervous system's sensing and responding, menstruation, migration, mating, energy production and consumption, the cycle of birth and death — all loop back for a new start.

Loops tame uncontrolled events. One-way processes, given sufficient energy and materials, tend to "run away," to go faster and faster unless they are inhibited or restrained. The steam engine with a governor illustrates the principle: As steam pressure rises, the engine goes faster. The governor, consisting of two rotating arms that lift higher as its shaft spins faster, progressively reduces the steam input; the engine slows; the governor slows; the steam input increases; the engine speeds up. Thus information courses around the circuit to produce action in the opposite direction. The system self-corrects; the parts self-adjust. If such self-generated restraints and inducements occur in small steps, the overall system appears to maintain itself in a steady state.

Every biological circuit, whether a sequence of proteins in the act of consuming a sugar molecule or a complex ecosystem exchanging material and energy, exhibits self-correcting tendencies like those of the steam engine.

Information flows around the circuit and feeds back to the starting point, making necessary adjustments along the way. It's easier to understand how molecular systems assemble into complicated, apparently purposeful organisms when we look at events in terms of multilayered loops of control and creation — and substitute the term "self-correcting" for "purposeful."

A large number of valuable scientific hypotheses were inspired by the Industrial Revolution (including the governor on steam engines).

T. H. Huxley (1825–1895), a British anatomist, describes an early reductionist approach to understanding biological processes in just such terms. (Huxley was the first and one of the strongest supporters of Darwin's theory of evolution by natural selection and was often called "Darwin's bulldog."

After men acquired a rough and general knowledge of the animals about them, the next thing which engaged their interest was the discovery in these animals of arrangements by which results, of a kind similar to those which their own ingenuity effects through mechanical contrivances, are brought about. They observed that animals perform various actions; and, when they looked into the disposition and the powers of the parts by which these actions are performed, they found that these parts presented the characters of an apparatus, or piece of mechanism, the action of which could be deduced from the properties and connections of its constituents, just as the striking of a clock can be deduced from the properties and connections of its weights and wheels.

T. H. Huxley, *The Crayfish,* 1880

Self-correcting maneuvers

As an owl tracks a fleeing mouse, she quickly translates the mouse's zigzags into movements of her wings and tail. The owl gets her dinner by maintaining a feedback loop between her eyes, brain, wing and tail muscles, and the mouse's movements.

A continuing feedback loop keeps the micro environment of this pond clean and nurturing for all kinds of organisms **(a).** Aquatic plants photosynthesize and provide food and oxygen for aquatic animals, which produce carbon dioxide as waste, and eventually die, providing further nutrients for microbes and plant life. When too many nutrients are pumped into such an environment from fertilizer used on crops or lawns, an uncontrolled event occurs; the loop is interrupted. In response to the nutrients, huge numbers of plants grow **(b),** far more than the animal life can process. The plants die and fall to the bottom. In decomposing, the plants use up all of the water's dissolved oxygen.

(a)

(b)

Question.

What would be the overall effect of too little oxygen in the water? How could the interrupted loop be reestablished?

Answer...

With too little oxygen, fish and other aquatic animals suffocate. Their decomposition uses up even more oxygen. Limiting nutrients such as phosphorus and nitrogen that enhance plant growth and/or aerating the water to introduce more oxygen would help reestablish the pond's balanced ecology.

Fever as a Feedback Loop

Fever is the body's biological response to viral and bacterial infection. Proteins called pyrogens, produced by the invading organisms and by the body's own white blood cell defense system, cause an area in the brain called the hypothalamus to "reset" the body's temperature higher. This reset signal constricts the blood vessels and causes shivers (which produce heat internally). The body's core temperature rises to the new set point, and the higher temperature either kills the invaders outright or stimulates the body's immune system to dispatch them.

When the invaders have been destroyed, pyrogen levels drop. The body responds by dilating the blood vessels and sweating profusely. Evaporation of water from the skin surface cools the body, bringing its temperature back to normal. This is a classic feedback loop — a self-correction par excellence.

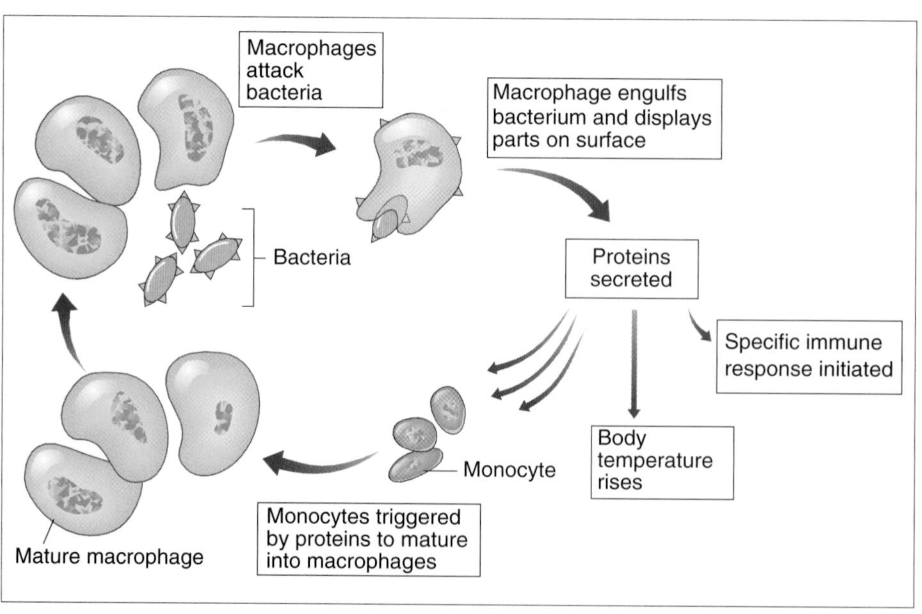

Big eaters

Here is another part of the feedback loop that suppresses infection. The green invading bacteria are engulfed by large white blood cells called macrophages ("big eaters"). The bacteria and the macrophages produce pyrogens and other proteins that cause immature white blood cells to mature into macrophages that can engulf more bacteria. This cycle continues as long as there are bacteria to trigger it.

Interfering with Self-Adjusting Systems

The elaborate anatomical loop you see below — the human gas transport and exchange mechanism (the respiratory and circulatory systems) — has many feedback functions: It regulates the blood's oxygen and carbon dioxide levels, helps maintain blood's acid-base balance, and helps protect us from airborne toxins and disease.

Biological circuits like this one exist in a delicate balance with their extracellular environments. If something coming from outside interferes with the functioning of one part of the loop, the system will call for greater input from other parts of the loop — heavier breathing, faster heart rate, and so on.

The gas transport loop works effectively if the body's gaseous environment remains within certain limits — for example, if the lungs take in normal atmospheric gases (mainly oxygen, nitrogen, and carbon dioxide). Add an uncontrolled event to the mix — carbon monoxide, sulfur oxides, nitrogen oxides, or ozone (produced by cars, power plants, factories, home heating and air conditioning systems, and cigarettes), and blood chemistry goes awry, along with the respiratory system's cellular functioning.

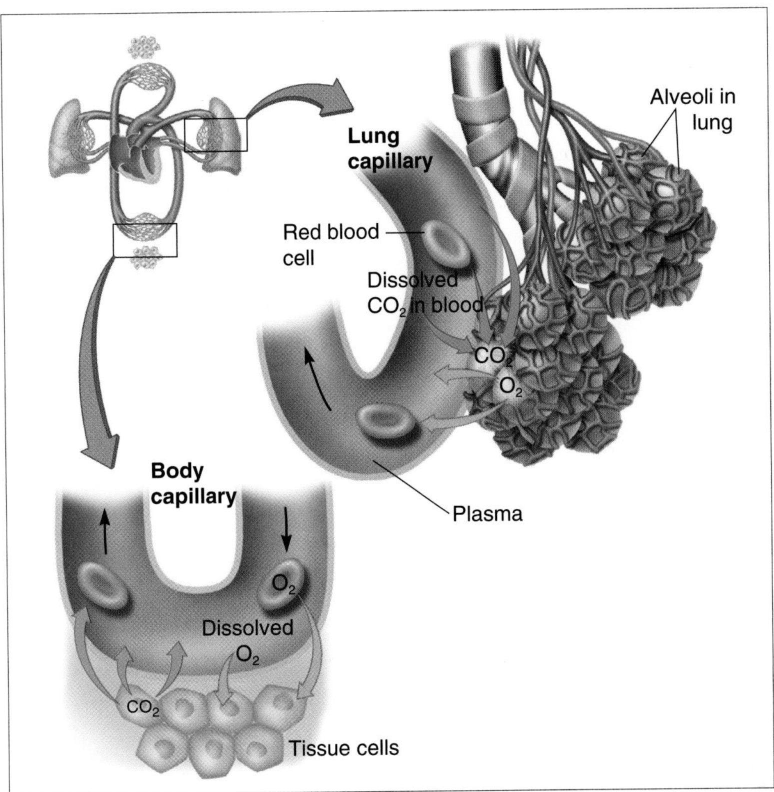

Breathing as a feedback loop

This is how we exchange gases with the environment: Oxygen, absorbed by alveoli in the lungs, enters the bloodstream and travels to body tissues. CO_2, the product of respiration, diffuses into the blood from tissue cells, travels to the alveoli, and is exhaled.

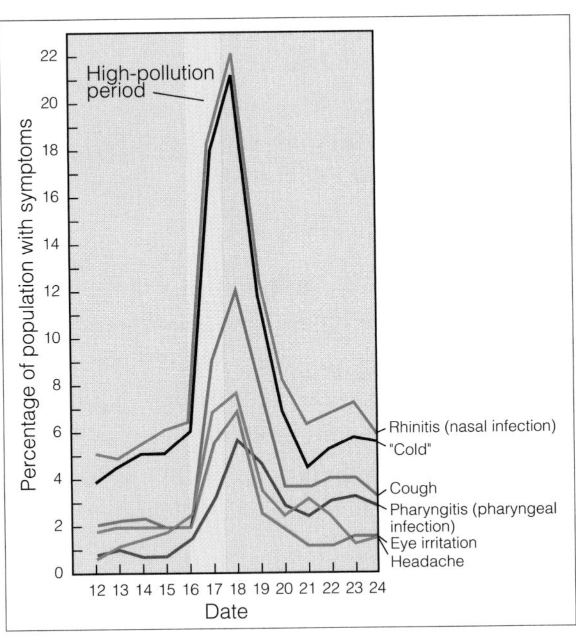

The pollution connection

This graph is a dramatic display of the relationship between incidents of respiratory illnesses and an episode of high air pollution in New York City in 1962. In a two-day period — days 16 and 17 — sulfur dioxide levels rose from 0.2 ppm to 0.8-0.9 ppm, and immediately the numbers of nasal infections, colds, coughs, pharyngeal infections, eye irritation, and headaches all rose dramatically. [Redrawn from J.R. McCarroll et al., Health and the Urban Environment: Health Profiles versus Environmental Pollutants in *American Journal of Public Health* 56 (1966): 266-275.]

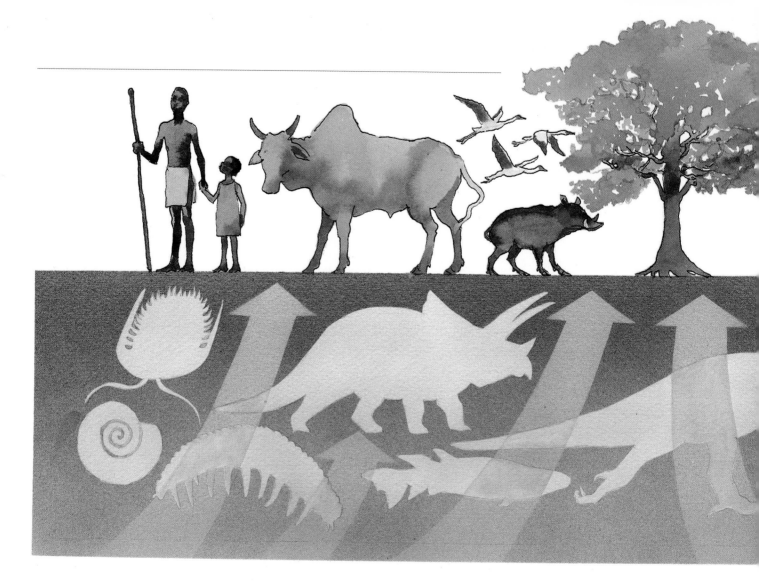

 Life Recycles Everything It Uses

For every molecule that the living world makes or uses...

...there exists an enzyme somewhere to break it down.

A Circular Flow of Materials

We humans are unique among animals; we leave behind us a trail of accumulating, unusable products. Everywhere else in the living world, intake and output are balanced, and one organism's waste is another's food or building materials. Waste from a cow circulates from bacteria to soil, to earthworms, to grass, and back to the cow. Crabs need calcium, which they normally get from the ocean, to build their shells. Land crabs, lacking an ocean source, extract calcium from their own shells before discarding them during molting. Hermit crabs save energy by moving into shells cast off by other species, trading up when the shell gets too small.

At the molecular level, key atoms pass from molecule to molecule in a succession of small steps. The end product of one process becomes the starting point of another, the whole train of events bending around into a circle. One creature's "exhale" becomes another's "inhale." Oxygen, dumped by plants as a by-product of photosynthesis, becomes an essential key to combustion in animals' respiration. The carbon dioxide waste that animals exhale is taken up by plants for sugar-making. From the standpoint of the whole ecosystem, these interchanges occur so smoothly that the distinction between production and consumption, and between waste and nutrient, disappears.

Each generation of living things depends on the chemicals released by the generations that have preceded it.

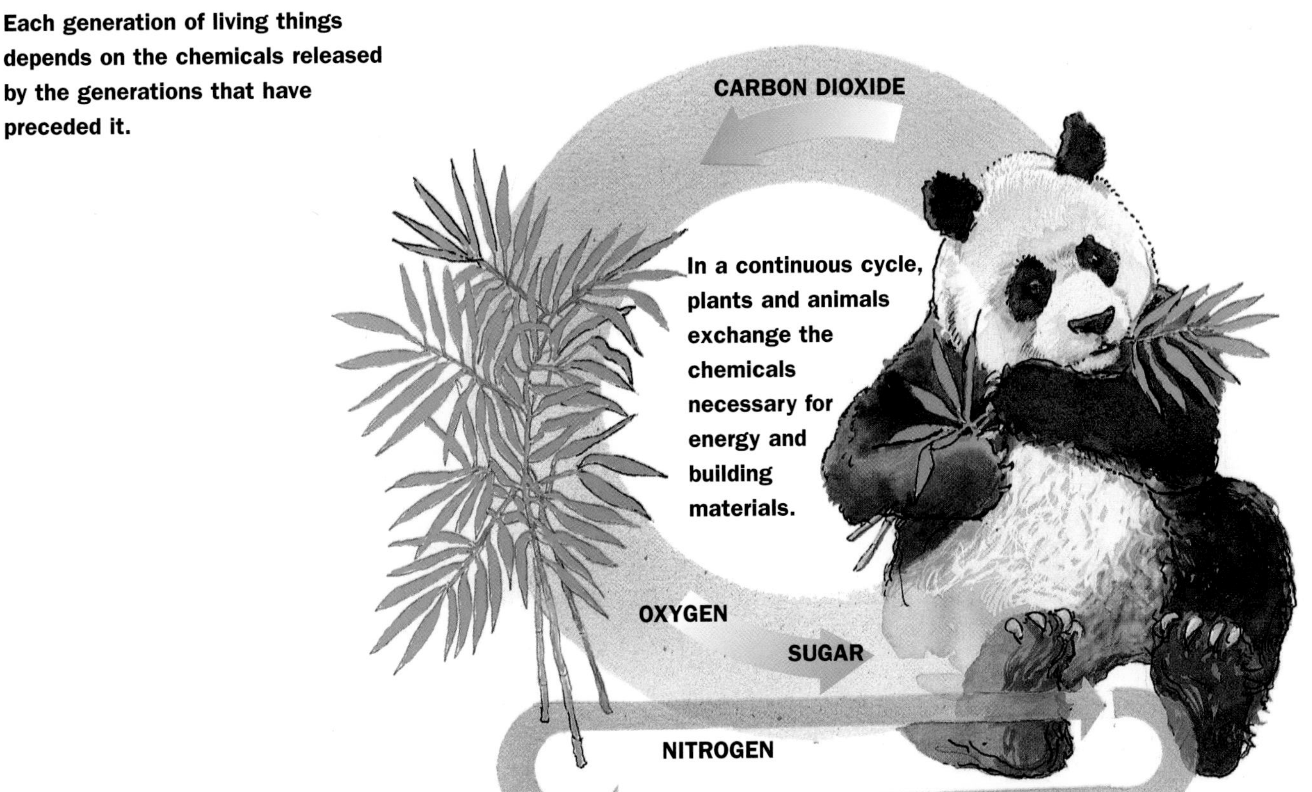

In a continuous cycle, plants and animals exchange the chemicals necessary for energy and building materials.

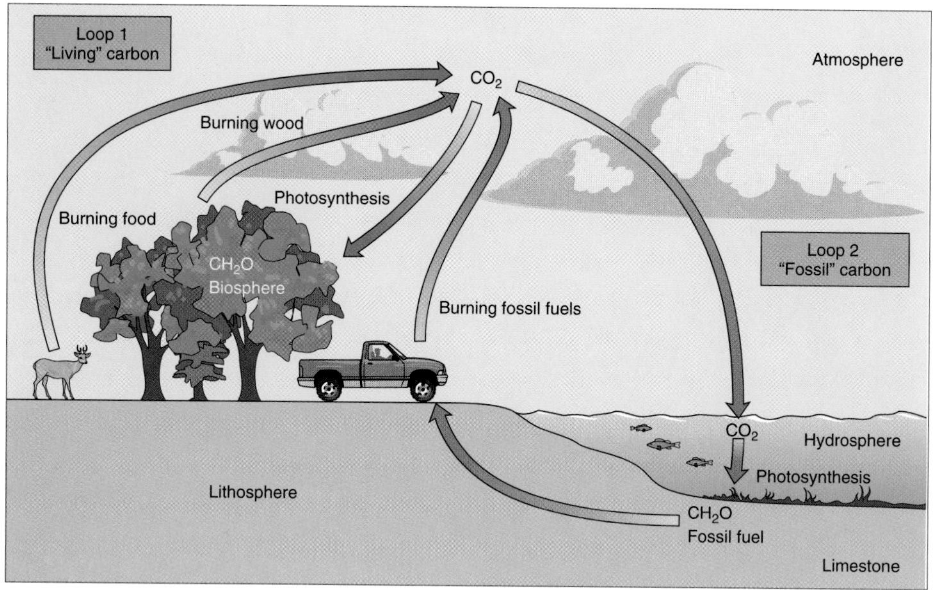

Carbon exchange

Carbon atoms are the basic structural units of the molecules that compose the cellular structures of all living organisms. The purple arrows show how, by photosynthesis and respiration, carbon circulates through organisms and their environments. The carbon of long-dead organisms may be stored for millions of years in limestone (the remains of shellfish) or in oil and coal deposits (the remains of other living things). Eventually, even that carbon is used as fertilizer, or burned and released to the atmosphere as carbon dioxide, once more becoming part of life's carbon loop.

Ultimate Recycling

When things die, they don't go to waste. Death is a natural process for recycling life's raw materials. Life lets nothing go to waste. The carcass of a bison, left behind first by wolves and then by coyotes and crows, continues to be digested and broken down by insects, bacteria, and fungi. When each of these organisms dies, it decomposes as well. Every living thing has other (often many) living entities that feed on it, and one organism's waste products are another organism's source of nutrients or shelter.

Cremation is simply combustion (burning), which turns the complex molecules that make up a living organism into simpler ones such as water and carbon dioxide gas. Thus the body's carbon, hydrogen, oxygen, nitrogen, calcium, and phosphorus atoms return to the atmosphere and the soil. Burial in an embalmed state, in a lined coffin, slows down the decomposition process for hundreds of years, and mummification can slow it for thousands. Eventually, though, every organism returns to the earth or atmosphere as a dispersal of atoms and molecules.

Every atom in your body is unimaginably old, dating back to the origin of the universe. However, during their last few billion years spent here on Earth, your atoms have cycled through a great many mineral and organic forms, over and over again. At one time or another, they may have been part of the atmosphere or the Amazon River, diatoms or dinosaurs, rocks or rabbits, trees or trilobites. This is ultimate recycling.

Wood recyclers

Fungi and bacteria are hard at work decomposing the intricately complex molecular structures of these fallen trees.

Ferreting Out the Final Facts

Along with bacteria, fungi and insects are among the natural world's most persistent and enthusiastic recyclers. For at least a century, our growing understanding of their diverse and predictable habits of decomposing once-living organisms has played an important role in establishing the time and (in some cases) place of death of a human or other organism. Forensic entomologists study the succession of insect types that find, lay eggs on, and eat dead organisms. By observing the kind of insect feeding on a dead animal, and the larval or pupal stage its offspring have reached (many insects go through four or five different developmental stages), they can tell how long the organism has been dead. For instance, one of the first kinds of insects that detects and arrives to colonize a dead vertebrate animal is the blowfly. The fly will lay eggs on a dead body within two days. The egg then hatches to form, in succession, three larval stages and two pupal stages before the adult emerges. Each of these stages takes a predictable length of time to develop, so the time since the flies' arrival can be determined, and from this the time of death. The insect species found on a body can also tell the trained observer where the body has been kept, whether it has been moved from its original position, and whether or not it has been frozen.

In recycling a body, the insects also ingest any drugs or poisons the once-living organism may have used or been given. This recycled body chemistry thus can become useful evidence of drug use or poisoning, even when the body itself is too far decomposed for testing. Even DNA can be retrieved from the digestive tracts of blood-sucking insects, and used to identify a specific person's presence at a specific location during a specific time period.

"... I think your [victims] were killed during the day, or at least the bodies were exposed during daylight hours for a while before burial. I found larviposition by *Sarcophaga bullata*.... Indeed, I also suspect that the bodies were exposed outside, at least for a short period. The Sarcophagidae aren't quite as willing to enter buildings as some other groups."

Kathy Reichs, *Déjà Dead*, 1999

National Briefs

Oregon

Forest fungus called the largest organism

CORVALLIS — U.S. Forest Service researchers report that a fungus that has been weaving its way through the roots of trees for an estimated 2400 years has become the largest living organism ever found. The fruiting bodies of *Armillaria ostoyae* are linked by an underground network extending a length of 3.5 miles and burrowing as deep as 10 feet underground.

WEB
Connection
www.jbpub.com/connections

2.12 Life Maintains Itself by Turnover

Put It Together — Take It Apart

Protein mechanics

Keeping a living system in a state of high organization necessitates the continuous building and destroying of its parts.

Consider the following dilemma. To exist, life requires organization. Organization requires energy. Life's complex molecules have lots of energy in the bonds that hold them together, but these bonds don't hold together indefinitely. They tend to fall apart — dissipate. Now, a system that is unstable when it's organized has a problem. How can it avoid inevitable breakdown? Living systems have answered this question with an ingenious strategy. Day in and day out, round the clock, organisms routinely take apart their own perfectly good working molecules and then reassemble them. Each day about 7 percent of your own molecules are "turned over." That means virtually 100 percent have "turned over" in about two weeks. In this way, no molecule lingers in your system long enough to "unintentionally" dissipate.

Turnover also provides flexibility. A change in the environment often calls for a switch in proteins. New proteins can be made from disassembled old ones.

In turnover we can sense life's continuous "flow-through" of energy. A high-information/high-energy state must be dynamically maintained by the ceaseless building and destroying, ordering and disordering, of life's parts.

Cell turnover

Whole cells also turn over; i.e., they have a short or long life, die, and get replaced by new ones.

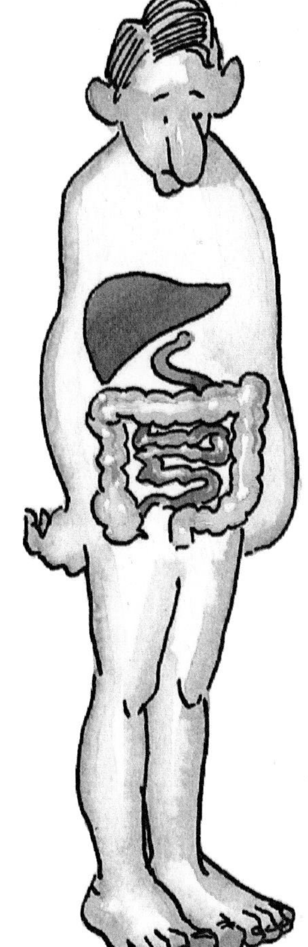

Cells that turn over in days or weeks — liver, intestine, skin

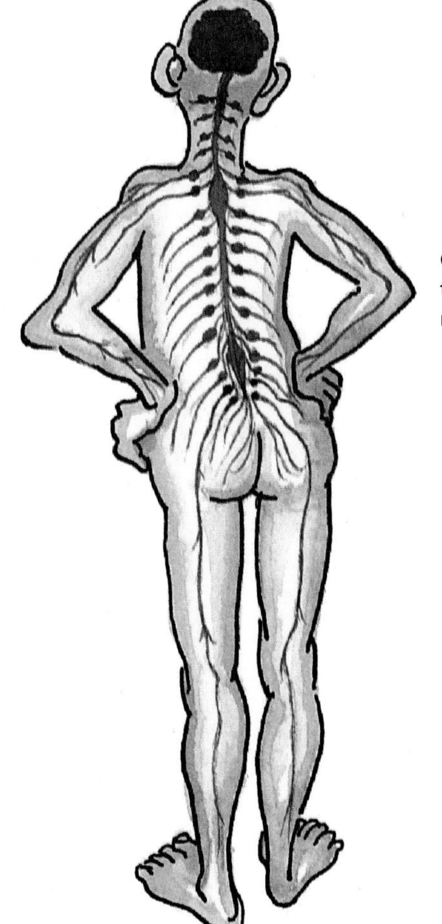

Cells that rarely turn over — neurons

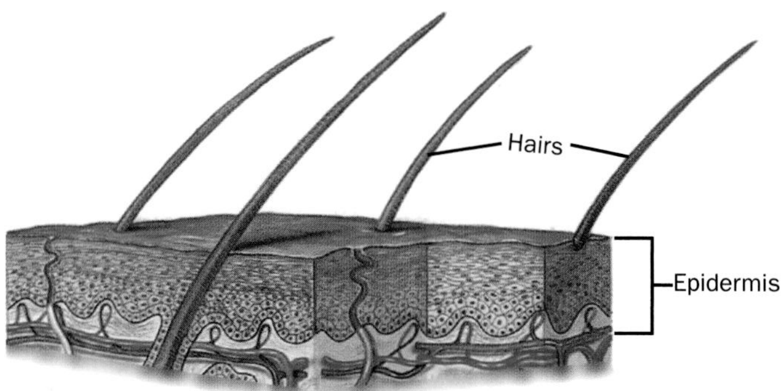

Skin renewal

The bottom layer of these skin cells is fed by tiny blood vessels looping up from larger vessels, providing a constant supply of raw materials.

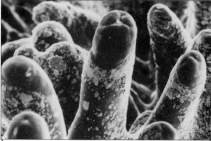

Parts Renewal

Living cells' ongoing turnover requires a constant supply of raw materials and the energy to assemble them. The energy, originally supplied by the Sun and trapped in plants' molecular bonds by photosynthesis, enters your body in the form of food. That food provides most of the raw materials for synthesizing new proteins and other components of cells. Other raw materials come from the breakdown of these same complex molecules.

In some parts of the body, existing cells reproduce themselves by division. Your skin (or dermis) constantly replenishes itself, for instance, by shedding its surface layer (the epidermal cells). The cells at the base of the dermis (the basal cells) divide; as new cells are formed at the bottom and old ones are shed at the top, the cell layers advance toward the surface and their eventual death. In fact, the surface of your entire body is covered with dead cells, forming a protective layer for the living cells beneath. What are they protecting you from? Ultraviolet radiation, chemicals, dehydration, hydration, and abrasion. Living animal cells would die rather quickly in the presence of any of these.

You have another dermal layer lining your gut (digestive tract) from mouth to anus. Your gut is effectively a tube that runs right through your body. Food goes in one end and is processed, and waste is expelled from the other end. The processing involves acids and enzymes that break down food substances into smaller molecules, and the dermal layer lining the gut protects your other cells from these digestive acids and enzymes. Each section of the gut is lined with specialized cells that variously protect, secrete chemicals, or absorb nutrients after food has been digested. The basal cells of the gut's dermal layer also divide continuously and the surface cells are scraped and sloughed off with the rest of the waste.

Villi of the intestinal mucosa

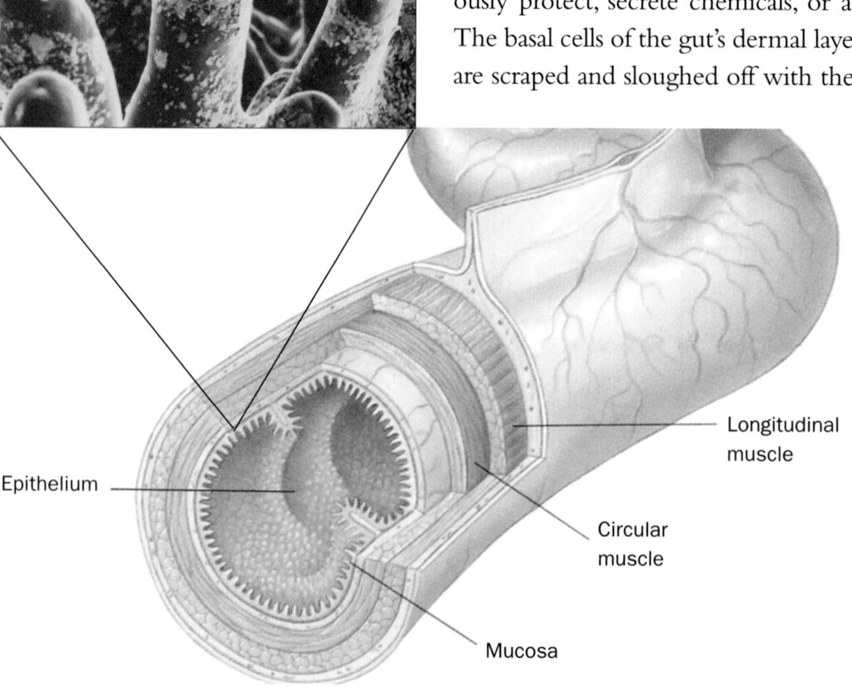

In animals, many cells turn over (i.e., are born and die) but not most nerve cells (including those in the brain associated with long-term memory).

Question.

Speculate on why nerve cells rarely turn over.

Answer...

One plausible theory is that nerve cells' function — including memory — depend on patterns of connections with other nerve cells. These connections develop over time. Cell death and replacement could break existing connections, thereby disrupting memory and other functions.

Blood cells, both red and white, originate in the marrow — the soft fatty tissue in the hollows of large bones. In the marrow, stem cells divide continuously, producing a constant supply of blood cells. A single drop of human blood typically contains 5 to 6 million red blood cells, and 5 to 10 thousand white blood cells. Extra blood cells are stored in the spleen, and old ones are broken down in the liver — a constant turnover.

Provided with enough nourishment, your body can produce a sufficient supply of skin, gut, and blood cells. Usually the new cell is an exact copy of the old one. As you know, though, mistakes can happen when any of your cells divide. For example, too much sunlight can damage the DNA of skin cells, resulting in abnormal cells. They lack the protein that tells them when to *stop* reproducing. When these abnormal cells divide, they often do so too fast, piling up into tumors.

A single abnormal cell...

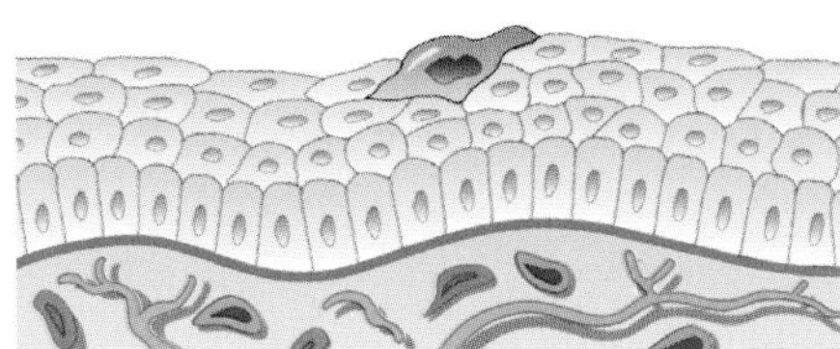

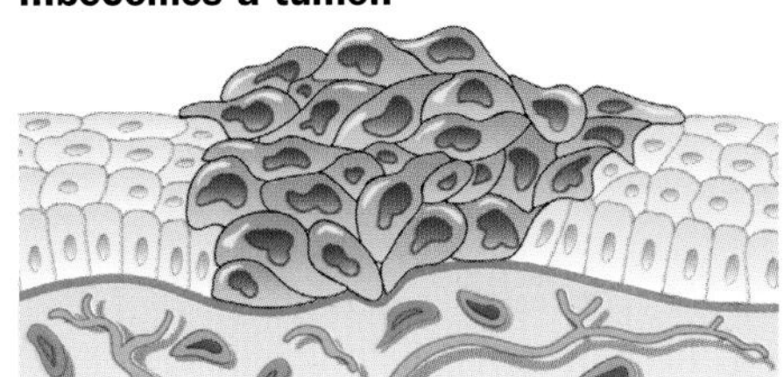
...becomes a tumor.

The value of science remains unsung; ours it not yet a scientific age. Perhaps one reason for this is that you have to know how to read the music. A scientific article, for instance, might say something like this: "The radioactive phosphorus content of the cerebrum of the rat decreases to one-half in a period of two weeks." Now, what does that mean? It means that the phosphorus in the brain of a rat (and also in mine and yours) is not the same phosphorus that was there two weeks ago. All the atoms that are in the brain are being replaced, and the ones that were there before have gone away. So what is this mind? What are these atoms with consciousness? Last week's potatoes! Which now can remember what was going on in your mind a year ago — a mind that we long ago replaced. When we discover how long it takes for the atoms of the brain to be replaced by other atoms, we come to realize that the thing I call my individuality is only a pattern or dance. These atoms come into my brain, dance a dance, and then go out, always new atoms but always performing the same dance, remembering what the dance was yesterday.

Richard Feynman, *The Value of Science,* 2000

When a human liver is damaged in an accident and part of it has to be removed surgically, the remaining portion grows rapidly, producing a full-sized liver in a week or so. Experiments show that, normally, most of the liver's substance is being regularly broken down and rebuilt (i.e., its big molecules are turning over inside its cells). Also, in the early stages of regeneration, the liver begins to increase in mass, without any increase in the rate of production of new cells.

Question.

How can cellular mass increase without a pickup in the rate of production?

Answer...

If there are increasing numbers of liver cells and the rate of production has not increased, it must be that fewer liver cells are being broken down.

2.13 Life Tends to Optimize Rather Than Maximize

When Less Is Better

To optimize means to achieve just the right amount — a value in the middle range between too much and too little. Too much or too little sugar in the blood will kill. Everyone needs calcium and iron, but too much is toxic. The rule of optimization generally holds true for minerals, vitamins, and other nutrients the body requires, as well as for behaviors such as exercise and sleep.

At the molecular level, life operates elaborate signaling and management systems to maintain optimum levels. Certain proteins have the ability to regulate precisely concentrations of essential chemicals, shutting down production when optimum quantities have been reached, starting up again when concentrations fall below critical levels.

At the level of the organism, optimizing is an intricate dance involving many interacting parts and values. Deer antlers require an optimum mix of strength, shock absorption, weight, and growing ability (since they must be regrown every year). A change in any one of these variables might adversely affect the others. Something that might make the antlers stronger, like a higher mineral content, might also make them heavier or unable to grow quickly enough. Thus, maximizing any single value (i.e., pushing it to the extreme) tends to reduce flexibility in the overall system, so that it may not be able to adapt to adverse environmental change.

Maximizing can be seen as a form of addiction, in that more leads to more. Occasionally, over generations, an organism may drift from optimizing to maximizing, from adaptation to addiction. The peacock's tail has been cited as an example of the maximizing of one variable trait. If female peacocks choose males who display the most luxuriant tail feathers, the next generation of peacocks will have a greater representation of "big tail" genes. If this process continues unabated, each generation will have a larger average tail size until the tails reach the upper limit of physical practicality. A tail can only grow so large in relation to body size before it impedes a bird's ability to get around. Likewise, a redwood tree can only grow so tall without toppling over; a walrus's tusks can grow only so long without overstraining the animal's neck muscles.

Every once in a while, a sudden change in the environment can catch a species that has drifted too far into maximization and push it into extinction. More often, as the costs of maximization rise, the species self-corrects. Larger-tailed peacocks may be unable to run as fast or hide as well. Because these peacocks are more vulnerable to predators, the survival advantage shifts back toward their smaller-tailed rivals. Thus, life persistently tends toward optimal balance, illustrating one of nature's cardinal rules: "Too much of a good thing is not necessarily a good thing."

There is, however, one value that life can be said to maximize. Every organism has as its most elemental goal the transfer of its genetic information to the next generation. In this sense, all optimizing of function aims at this ultimate maximization — the survival of DNA.

ELKS, WHELKS, AND THEIR ILK

The monarchs of the Irish bogs
Succumbed to neither men nor dogs
But (most ecologists agree)
To calcium deficiency.
They scoured the base-deficient peat
For antlers and old shells to eat
Around the Celtic countryside
And finding all too few, they died.
Then mourn the passing of the elks
But not the wisdom of the whelks
That roam the shore — their native heath —
With silver-indurated teeth.
And bore to death their mollusc friends,
Who come to sad, unsuccored ends.
Without the need for extra lime
The whelks survive to modern time.
Thus ungulate and gastropod,
And all that live by sea or sod,
Are doomed to be or not to be
By biogeochemistry.

Ralph A. Lewin, *The Biology of Algae and Diverse Other Verses*, 1987

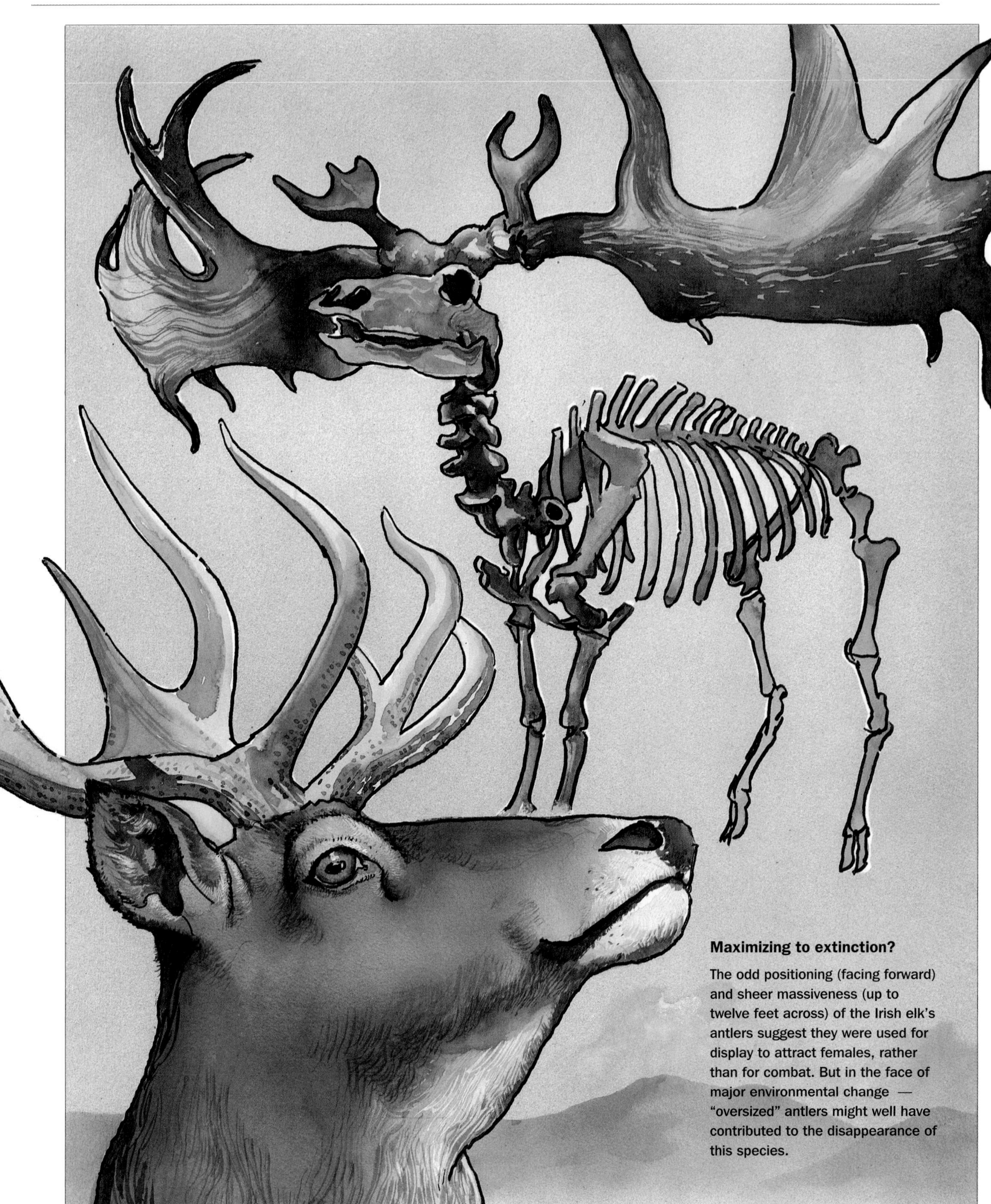

Maximizing to extinction?

The odd positioning (facing forward) and sheer massiveness (up to twelve feet across) of the Irish elk's antlers suggest they were used for display to attract females, rather than for combat. But in the face of major environmental change — "oversized" antlers might well have contributed to the disappearance of this species.

With all hair removed, the bodies of a gibbon and a human are remarkably similar. Humans have adapted successfully to far more environments than have gibbons.

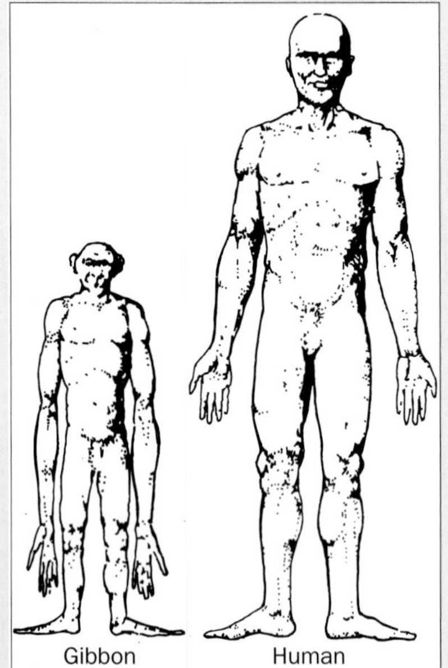

Question.

How is the gibbon maximized compared to the human? How might the human's optimal body structure explain its success in adapting?

Answer...

The gibbon's arms, hands and legs are optimized only for living in and moving through treetops — outside a forest these traits would be maximized, and a gibbon would be largely helpless and easy prey for fast ground-based predators. Also, the use of the hands for locomotion interferes with their usefulness as a tool for manipulation of objects in the environment. The human, on the other hand, while not specialized for a single environment, can climb trees if necessary, run across open land, climb mountains, and use tools to modify the environment.

Being Adaptable Pays Off

In times of crisis, the most specialized (maximized) organisms tend to become extinct; the most adaptable (least specialized) survive. Specialization always has a price: loss of adaptability. In stable times, maximization sometimes works; in changing times, optimization rules.

Over the course of life's evolution on Earth, there have been periods of major global climate change and mass extinctions. Geologists have devised a time scale based on the sedimentary rock and fossil records. Two of the most famous documented mass extinctions came at the end of the Permian period, approximately 250 million years ago, and at the end of the Cretaceous period, 65 million years ago. It is estimated that more than 90 percent of marine animals and a large percentage of land animals became extinct at the end of the Permian. The Cretaceous extinction, most famous for the demise of the dinosaurs, saw the number of species decline by perhaps 50 percent.

What survived? The least specialized, most adaptable organisms. In other words, the optimized plants and animals capable of living in altered environments, adapting to changes in climate, air or water composition, and diets. Following a mass extinction, adaptive radiation (the evolution of many different species from a few ancestors) occurs on a large scale as life rushes to fill the vacated niches of vanished species.

Some of the most spectacular fossils are those of maximized species, such as the largest dinosaurs, the saber-toothed tiger, the woolly mammoth, and the Irish elk pictured on the previous page. But maximization is not confined to large animals. Certain plant species (many orchids, for example) depend entirely on a single species of insect for pollination. The two, plant and insect, are said to have coevolved. The disappearance of the one is likely to lead to the extinction of the other.

Certain aphids demonstrate a sort of optimization. One species (pea aphids) produces individuals of two colors, red and green. Both colors exist together, feeding on the same plants (peas and other legumes). The pea aphids have two main predators, ladybugs and parasitic wasps. The ladybugs primarily eat the red aphids, presumably because they're easier to see. The wasps more often lay eggs in the green aphids, perhaps because eggs laid in the red individuals get eaten before maturity. The dual coloration appears to be an optimization strategy, permitting more individuals to survive in the presence of either predator.

An exclusive orchid

Only certain insects can find their way into this orchid's distinctive blossom, and the orchid's pollen is deposited on very specific parts of such an insect. When the insect visits the next orchid, pollen is rubbed directly onto the plant's reproductive part, the stigma.

Optimizing from the Bottom Up

Just as life has to organize from the bottom up, so must it optimize its parts from individual cells on up through cellular communities to whole organism. Individual, free-living cells and the ones that make up multicellular living things are almost all microscopically small, as you saw in the first chapter, and most are similar in size. There is an important reason for the size limit on most cells: in order to function they must constantly take in useful materials from their environment and dispose of waste materials to the environment. The only way they do this is through the cell membrane, and the materials have to get where they're going pretty quickly to do their jobs. Since most molecules move through a cell simply by being bumped around by other molecules, they don't move very fast.

If you look at the picture (above right) of the difference in surface to volume ratio between a larger and a smaller object, it becomes pretty clear that the smaller the cell, the larger the proportion of area it has to move molecules in and out, and the shorter distance they have to travel once inside. A diameter of one to ten thousand nm seems to be an optimal size for an animal cell, and most of them fall within this range. Larger cells need to be very long and thin, or have very convoluted surfaces (the same adaptation as the elephant's skin), or devise more elaborate mechanisms for moving molecules. You'll read more about this in Chapter 5, *Machinery.*

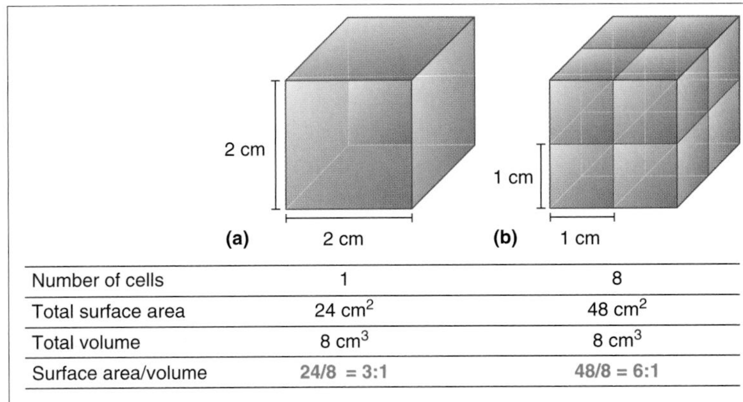

	(a)	(b)
Number of cells	1	8
Total surface area	24 cm^2	48 cm^2
Total volume	8 cm^3	8 cm^3
Surface area/volume	24/8 = 3:1	48/8 = 6:1

Antler growth and extinction of Irish elk.
Ron A. Moen, John Pastor, and Yosef Cohen.
Evolutionary Ecology Research, 1999, 1: 235-239.

Adult male Irish elk grew antlers that averaged 40 kg (88 lbs) in weight, the largest antlers of any of the deer species, alive or dead. Fossil remains of the elk tell us that they all died out over a relatively short period of time. They were all gone about 100 years after a major and continued temperature drop forced a change in their environment, and diet — from mineral-rich willow and spruce to less mineral-dense tundra plants.

In this paper, Ron Moen and his colleagues hypothesize that the amount of minerals required to grow antlers yearly was so great that when the elks' food supply changed to a less mineral-dense forage, they were unable to get enough minerals from their food to support building both skeleton and antlers. The investigators devised a computer model that compared the known nutritional and mineral needs of the modern moose to those of a simulated Irish elk. The simulated Irish elk depleted the mineral reserves in its skeleton to support antler growth during the summer but was able to replenish those reserves in the fall, when antlers were shed after the mating season. Thus, the bigger its antlers, the more likely the elk was to suffer from temporary osteoporosis. With climate and vegetation change, the Irish elk were less and less able to replenish their skeletal minerals, leading to either permanent osteoporosis or reduced antler growth, either of which would interfere with the elk's reproductive success, and with the long-term survival of the herd. The authors conclude: "Sexual selection pressures for larger antlers and larger body size were opposed by selection pressures for smaller antlers and smaller body size imposed by environmental change. We suggest that the inability to balance these opposing selection pressures in the face of rapid environmental change contributed to extinction of the Irish elk 10,600 years [ago]."

2.14 Life Is Opportunistic

Making the Most of What Is

A rotting tree on the forest floor may look like life at a dead end. In actuality, it marks the beginning of an explosive new stage — more varied and bustling than when the tree was alive. Early on, mosses and lichens establish themselves on the decaying surface. Carpenter ants, beetles, and termites initiate a succession of invasions by tunneling through the rotting wood. Fungi, roots, and microbes follow these paths. They in turn become food for grazing insects. Spiders feed on the grazers. Roots of seedling trees and shrubs take hold in the emerging humus as moles and shrews burrow through the soft wood to feed on the newly grown mushrooms and truffles.

The "living dead" tree illustrates not only life's tenacity, but also life's universal tendency to "make do" with whatever is available in its surroundings. Because of this tendency, life flourishes even in the world's harshest places. In Africa's Namib Desert, surface temperatures soar to 150 °F, and rain may not fall for three or four years at a stretch. Few plants can survive, yet just under the barren sand live a host of insects, spiders, and reptiles — even several types of mammal. The smallest creatures get moisture from wisps of fog and nutrients from tiny bits of plant and animal detritus blowing across the sands. The larger creatures live on the smaller.

Self-burial

To avoid winter's harsh dry winds, the mescaxl cactus withdraws completely into the ground.

In the arctic ice, 100-year-old lichens grow in temperatures of −11 °F. Some antarctic fish have a natural antifreeze running through their blood vessels, enabling them to thrive where others would perish. Tubeworms live in darkness 8000 feet underwater, depending on minerals streaming from hot water vents on the ocean floor. The world's champion adapters, fast breeding generations of bacteria, can adapt to virtually any environment — from near-boiling sulfur springs to the acid guts of termites. And so on.

Together, over time, the genetic code and the protein structure of all living things permit a marvelous flexibility. Hence, life forms are opportunists. Generations of opportunists don't wait around for the right conditions. They adapt to what is, and they make use of whatever they find around them.

Growing toward darkness

In order to find a tree to climb, the monstera vine must first grow toward darkness. Once it reaches a trunk, it switches strategies and grows toward light.

Adapted to fire

Resin in the seedbearing cone of the lodgepole pine prevents the scales from opening. Fire not only melts the resin and releases the seed, it also leaves a fertile bed of ashes in which the seedling can take root.

An invitation to sex

With the right odor, pattern, and degree of hairiness, the bee orchid entices the male bee into an attempt at copulation. The bee leaves, covered with pollen, to be enticed by another bee orchid.

Living stones

Lithops are plants that look like stones, which helps them avoid being eaten by foraging animals.

Hollow leaves

Moisture condenses on the inside of the pitcher plant's leaves and is then carried directly to the roots, which need to be kept wet because they are exposed to the air.

Like rotting meat

With an evil smell, the *Rafflesia* plant encourages pollination by flies.

2.15 Life Competes Within a Cooperative Framework

Strategies for "Fitting In"

1. Every creature acts in its own interests.
2. The living world works through cooperation.

These two statements may appear to be contradictory; they are not. Creatures are self-interested but not self-destructive. Selfish behavior, pushed to the extreme, usually has unpleasant costs. A dominant animal engaging in too-frequent combat may sustain injuries. A parasite may kill its host and have nowhere to go. These self-defeating strategies generally get weeded out by evolution, so that in the long run most organisms tend to adopt some form of "getting along."

Being eaten may not feel much like getting along, but in fact when predators take only the smallest, weakest, or most unhealthy of their prey species, they leave the fittest members to survive and reproduce. We can recognize this as being competitive at the individual level, cooperative at the group level. (Although we don't suggest that creatures generally think in terms of the group.)

Noncompetitors

Although these different species of wading birds feed side by side, they might as well be on separate planets. Each eats a different diet with its unique bill. The fact that each species occupies its own special niche may be taken as evidence for nature's tendency to "get along."

Plants and animals evolved from predator/prey truces among bacteria. The ancestors of chloroplasts and mitochondria (the sugar-making and sugar-burning components of plant and animal cells, respectively) originally acted as small predators, invading larger bacteria. They exploited but did not destroy their host. Such "restrained predation" is a recurring theme in evolution, and in it we see the beginnings of cooperation. In time, the host developed a tolerance for the invaders, and each began to share the other's metabolized products. Eventually they became full-fledged symbionts — i.e., essential to each other's survival. This progressive cooperation set the stage for all higher life forms. The lesson, as biologist Lewis Thomas has stated, is not "Nice guys finish last," but rather "Nice guys last longer."

From predation to cooperation

A parasitic mitochondrion invades a larger bacterium.

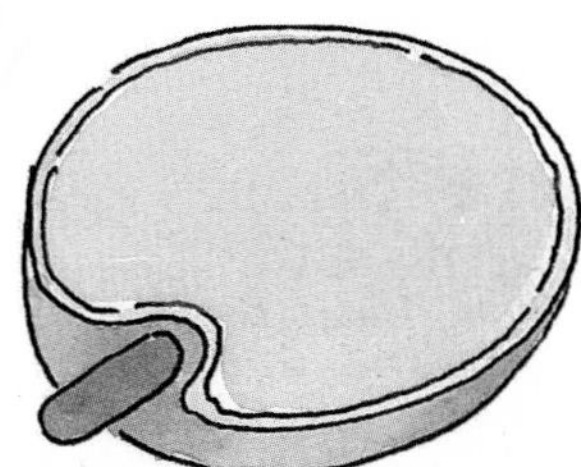

Many generations later, invader and host begin to share metabolized products.

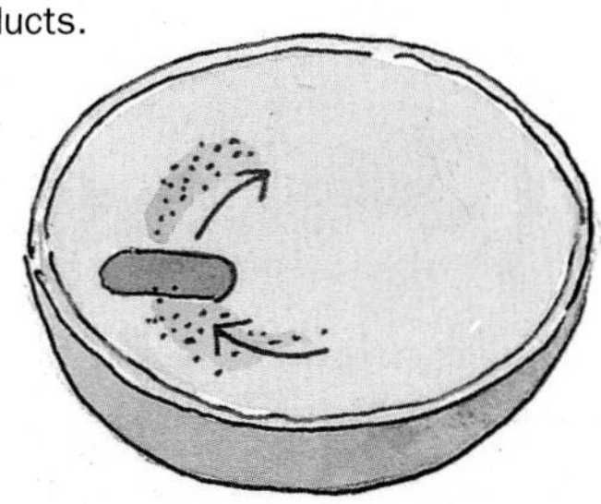

After many more generations, they've come to need each other.

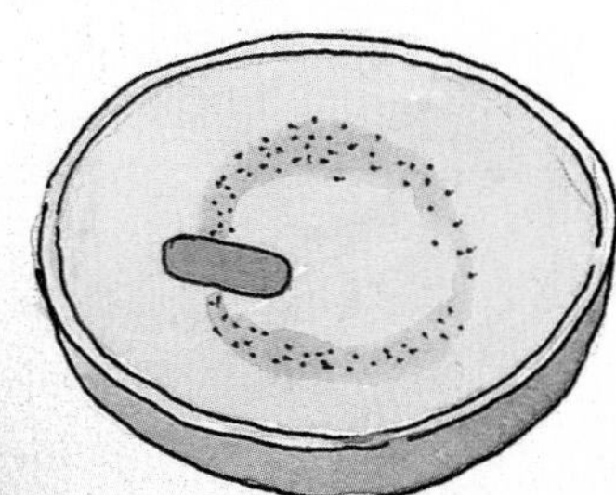

Ritualized aggression

Animals compete to establish dominance. Such fights rarely result in serious injury and frequently involve only "displays." This can be seen as cooperative behavior.

Question.

Why did the opposing sides during trench warfare in World War I refrain from shelling each other's meal wagons?

Answer...

The restraint derived from mutual self-interest. If one side left the other's meal wagons alone, the other side would tend to reciprocate and so both would get to eat. This is a good example of how cooperation can evolve — even among mortal enemies.

Shrimps' Selfish Cooperation

In the June 6, 1997 issue of *Nature,* J. Emmett Duffy reported his observations of several different colonies of Caribbean snapping shrimp. He found that all of the shrimp in a single colony (they live inside sponges) are closely related — descendants of a single mother, or "queen," and a single father. The shrimp are fierce defenders of their colonies and will chase off or kill any intruders from other colonies. Duffy set up some small experimental colonies in his lab and discovered that the shrimp welcomed former members of their own colony, even when space and food were in short supply. This welcoming behavior might seem to be foolish at first, but it is a beautiful example of combined self-interest and cooperation. Purely individualistic self-interest might dictate that any intruder would use up precious resources and should be killed or chased off. The longer-term interest of the colony, though, would consider any shrimp with the same DNA just as valuable for carrying on the life of the colony. Thus, the welcoming behavior is in the interest of the colony's survival.

Wormy Opportunists

In the absolute darkness of the ocean bottom, no photosynthesis can occur. Even so, dense communities of gigantic tube worms, mussels, and clams cluster around towering hot water vents on the ocean floor. These communities are something of a mystery. There are absolutely no plants here. What do these creatures use for food? Why are they found only near the hot water vents?

It seems that the answer lies in the relationship between bacteria and the molecules dissolved in the hot sea water. As sea water seeps into the ocean floor in places where hot magma from the Earth's interior is close to the surface, it dissolves subterranean minerals such as iron, calcium, sulfur and copper. These minerals then rise with the heated water through the vents in the ocean floor. They provide food and energy (in the form of the bonds in molecules of hydrogen sulfide gas) for enormous communities of bacteria that live on the inner surfaces of the vents. Just as the chloroplasts in plants use solar energy to turn carbon dioxide and hydrogen into sugar, these bacteria use the chemical energy in hydrogen sulfide molecules to make sugar. The tube worms, mussels, and clams filter these bacteria out of the sea water, ingest them, and use their sugars as nourishment. There would be no opportunistic animals in this dark biosphere without the opportunistic bacteria.

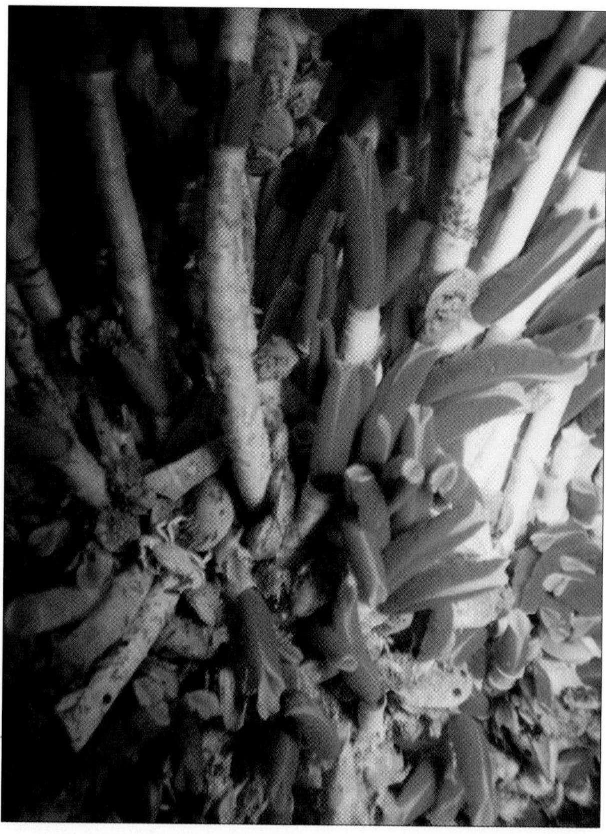

A colony of tube worms

Making the Most of Toxic Waste

Bacteria, among the most versatile and opportunistic of living things, can populate what may seem to be impossible environments and can turn many kinds of chemical bonds into energy for themselves. Using a lot of creative imagination, we human beings are learning to cooperate with bacteria to address many of our environmental cleanup problems.

The Savannah River nuclear materials site in Georgia has been contaminated for 50 years with the toxic solvents tri- and tetrachloroethylene (or TCE). TCE is so toxic that five gallons of it can contaminate up to a billion gallons of groundwater, and TCE molecules are very stable, unlikely to break down into harmless component molecules without outside help. Certain bacterial proteins, though, can do the degrading job, specifically those of bacteria that use the bonds in methane (natural gas) molecules to supply their energy needs. These methanotrophs ("methane-eaters") were already living in the contaminated Savannah River site soil and were very slowly breaking down the TCE molecules when their preferred methane wasn't sufficiently abundant.

To the creative minds of the cleanup crew at Savannah River came the idea of taking advantage of the methanotrophs' food preference. They figured that more methanogens would mean faster TCE degradation, so they ran pipes through the soil that bubbled out a constant supply of bacterial nourishment — methane and oxygen. The bacterial population burgeoned, and then, when the supply of methane was lowered, this huge population turned to the resident TCE molecules, breaking them down into harmless component molecules. The cleanup was a success, and it took months instead of years.

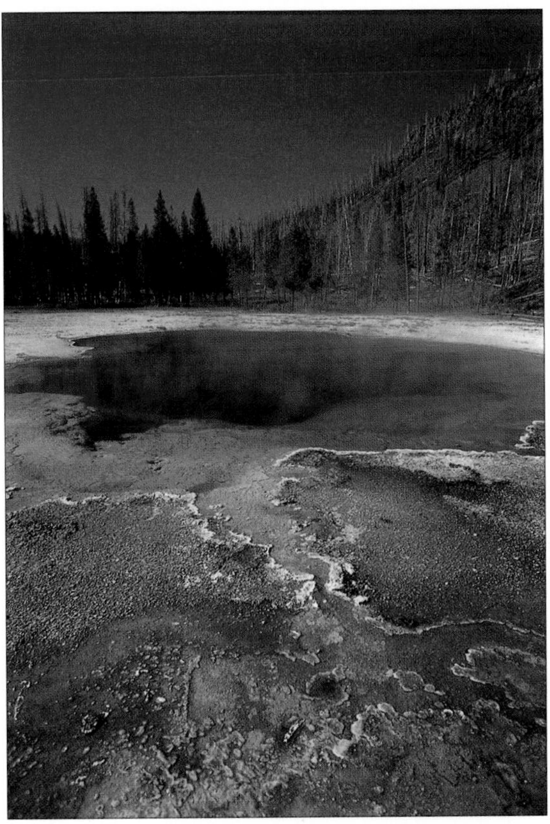

An extremophile swimming pool

Organisms that live in conditions at the extremes of temperature, acidity, alkalinity, or salinity are called extremophiles. This 75 ºC alkaline spring in Yellowstone National Park is a congenial home for the mat of bacteria in the foreground.

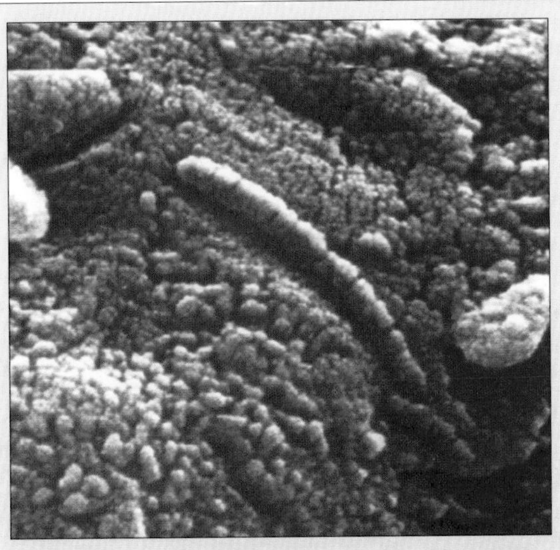

Any competition here?

Electron microscopic views of Mars meteorite ALH84001 found in Antarctica in 1984 showed these 20 nm long tubular shapes, which suggested the possibility of microorganismic life on Mars. (See page 7 for comparative sizes.)

Question.

What might be one strong argument against the possibility of these shapes being microorganisms?

Answer...

Looking at the sizes of the molecules on page 7, it seems clear that 20 nm is too small a space to include all of the basic chemical machinery of life.

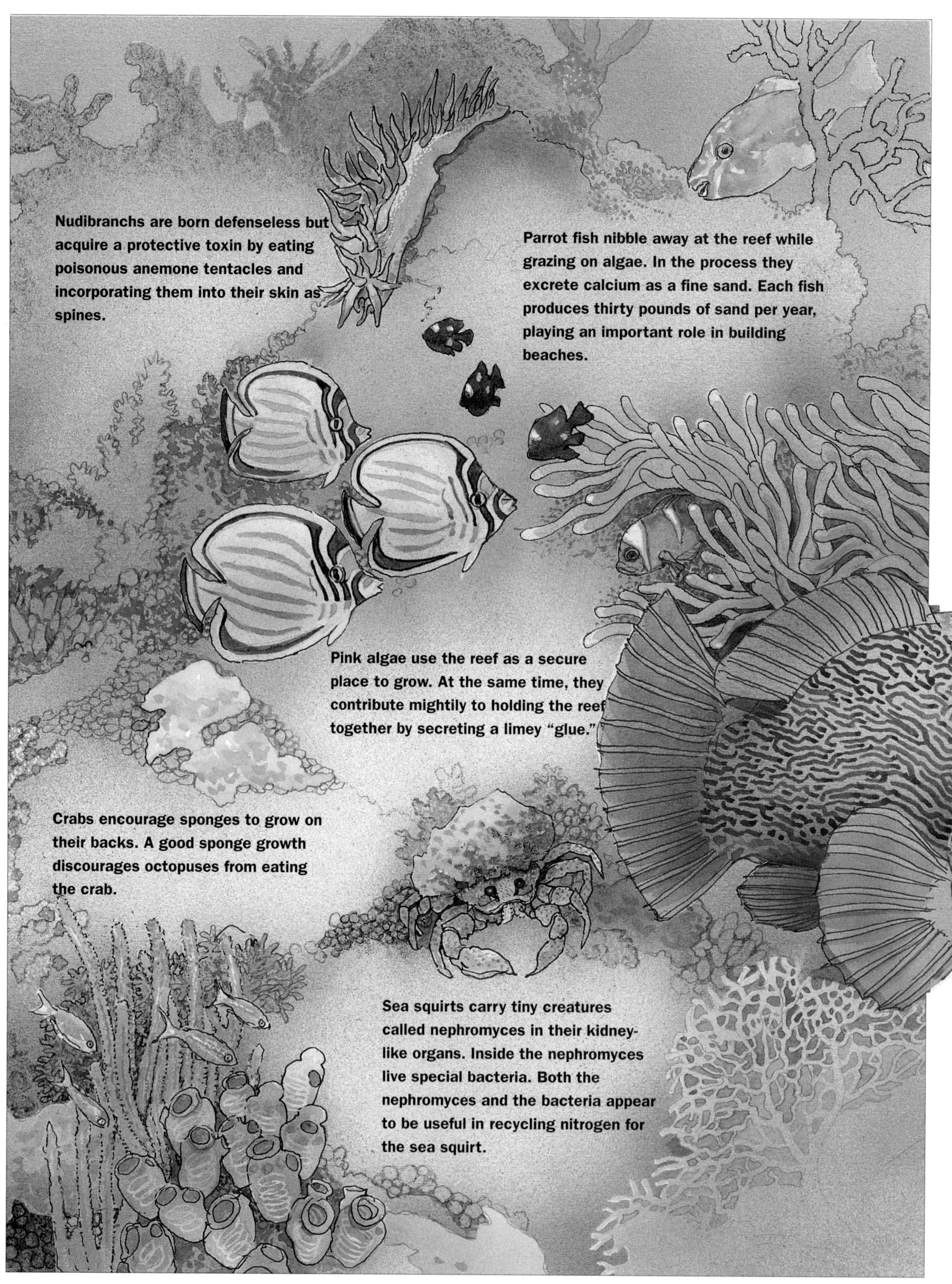
Nudibranchs are born defenseless but acquire a protective toxin by eating poisonous anemone tentacles and incorporating them into their skin as spines.
Parrot fish nibble away at the reef while grazing on algae. In the process they excrete calcium as a fine sand. Each fish produces thirty pounds of sand per year, playing an important role in building beaches.
Pink algae use the reef as a secure place to grow. At the same time, they contribute mightily to holding the reef together by secreting a limey "glue."
Crabs encourage sponges to grow on their backs. A good sponge growth discourages octopuses from eating the crab.
Sea squirts carry tiny creatures called nephromyces in their kidney-like organs. Inside the nephromyces live special bacteria. Both the nephromyces and the bacteria appear to be useful in recycling nitrogen for the sea squirt.

2.16 Life Is Interconnected and Interdependent

Reef-building coral polyps harbor tiny algae within their cells. The algae promote the coral's growth and receive carbon dioxide and nutrients in exchange.

A Network of Interactions

The stony coral, a pea-sized animal that resembles a miniature flower, might easily go unnoticed were it not for the tiny limestone cup it secretes for its home site. As the multiplying coral add on their cups, they form vast apartment complexes — the largest life-made structures on earth. Pink algae, taking hold in the crannies, "mortar in" the loose and broken sections with a limey secretion of their own. Turtle grass, sea fans, sponges, and mollusks attach themselves to the reef surface. Moray eels take up residence in the dark crevices. Starfish arrive to feed on the coral, and triton conches feed on the starfish. Hundreds of species of fish — some grazers, some predators — move in, along with crabs, octopus, shrimp, and sea urchins. Competitive and cooperative relationships emerge.

Damselfish flit with complete immunity among the poisonous tentacles of the large sea anemones. Crabs place sponges on their backs where they grow and act as a protection from octopuses. Cleaner fish and shrimp remove parasites from predator fish and eels, even entering their gills and mouths with complete safety. Algae live comfortably inside the coral's cells, and large sponges offer housing to thousands of minute creatures.

Look at the coral reef as a multilevel, integrated system. Ultimately, everything in the reef connects with everything else. The survival of the reef shark is closely tied to the survival of the coral polyp, even though the two may have no direct contact and no particular awareness of each other. What survives and evolves are patterns of organization — the organism plus its strategies for making a living and for fitting in. Any successful change of strategy by one organism will create a ripple of adjustments in the reef community. Called coevolution, this is the kind of creative force at work everywhere life has taken hold.

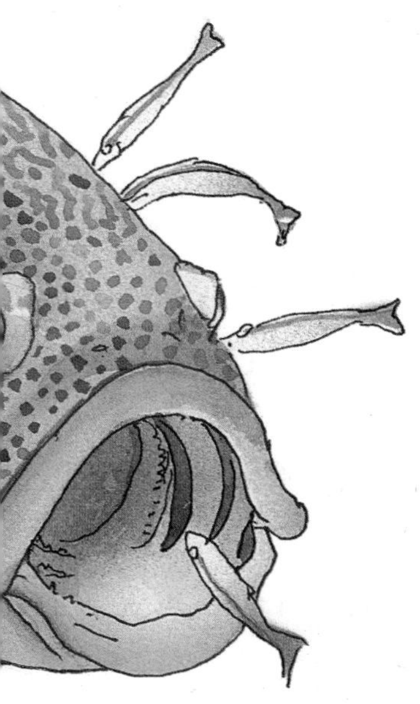

Cleaner fish live safely in the mouths and gills of larger fish, removing parasites.

Surface differences

The surface differences of Earth and its moon, as seen by the Galileo spacecraft, show the vast difference that bacterial life has made to the surface of our planet.

Small Creatures, Big Effects

New evidence points to the astonishing role that life, particularly microscopic life, has played in establishing and maintaining the Earth's atmosphere, temperature, and climate. Life has acted both as a stabilizing force, dampening the effects of solar fluctuations and volcanic activity, and as a creative force, setting the stage for new and more complex organisms.

For example, estimates of the Sun's energy output in life's earliest phase suggest that the Earth would have frozen over but for the small amounts of ammonia released by primitive bacteria. These were simple creatures that thrived in a low-oxygen environment. Later, photosynthetic bacteria and other organisms secreted enough oxygen into the atmosphere to create an environment in which aerobic (oxygen-breathing) creatures, including the first animals, evolved.

Since the Earth's beginning, the Sun has been getting hotter. At the same time, volcanoes have been adding carbon dioxide to the atmosphere, contributing to the atmosphere's ability to absorb energy from the Sun (the well-known greenhouse effect). Although the Earth's temperature would be expected to rise steadily, this has not happened very fast — until the widespread burning of fossil fuels in this century. Instead, average temperature has remained relatively constant, primarily because, over eons, microbes have steadily and dramatically reduced the amount of carbon dioxide in the air to its current level of 0.03 percent. Tiny plants, as they grow and die, "nibble away" rocks and trap carbon dioxide in the soil. There, it dissolves in water and washes eventually into the sea where it is used by marine life to build shells. Ocean algae also trap large quantities of carbon dioxide as they photosynthesize, which finds its way into marine shells (as calcium carbonate) and ends up on the ocean floor, ultimately becoming limestone and chalk.

This photo of Mars' surface was taken in 1997 during the Pathfinder landing. This mission confirmed that Mars had once been covered by large amounts of liquid water, but so far there is no undisputed evidence that there has been life on the planet.

And what about water? Planets such as Mars and Venus have lost all of their surface liquid water, probably because certain elements such as iron bonded with the water's oxygen to form oxides. The leftover hydrogen was too light to be held by the planet's gravity and was lost to space. On Earth, microbes prevented this loss of hydrogen by taking volcanic hydrogen sulfide, extracting the hydrogen for energy, and excreting the sulfur as pellets. Also, photosynthetic organisms kept producing water as they make ATP — enough for the Earth to retain its oceans, and thus, provide an environment for the further evolution of life.

A recent and plausible theory proposes that microorganisms are responsible for cloud formation over the oceans. Cloud formation is aided by the presence of tiny particles. Marine algae emit vast quantities of sulfide particles, which serve as "seeds" for cloud condensation. The creation of clouds over oceans covering two-thirds of the Earth's surface significantly affects the global climate.

So microscopic forms of life have worldwide physical and chemical effects — they create much of their own environment.

And it is strange thing that most of the feeling we call religious, most of the mystical outcrying which is one of the most prized and used and desired reactions of our species, is really the understanding and the attempt to say that man is related to the whole thing, related inextricably to all reality, known and knowable. This is a simple thing to say, but a profound feeling of it made a Jesus, a St. Augustine, a Roger Bacon, a Charles Darwin, an Einstein. Each of them in his own tempo and with his own voice discovered and reaffirmed with astonishment the knowledge that all things are one thing and that one thing is all things — a plankton, a shimmering phosphorescence on the sea and the spinning planets and an expanding universe, all bound together by the elastic string of time.

— John Steinbeck, *Log from the Sea of Cortez,* 1958

All mammals' skeletons are said to be fundamentally alike (homologous). These pictures show a horse's hind leg and a human leg. The two different legs appear to bend in the opposite direction.

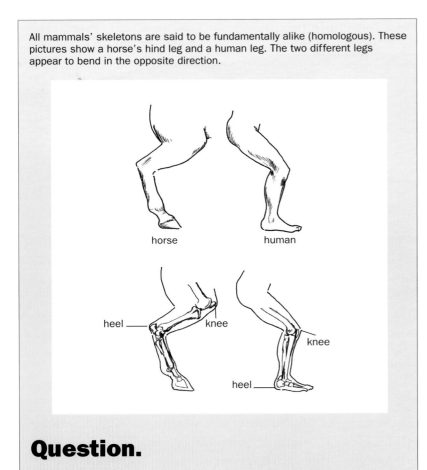

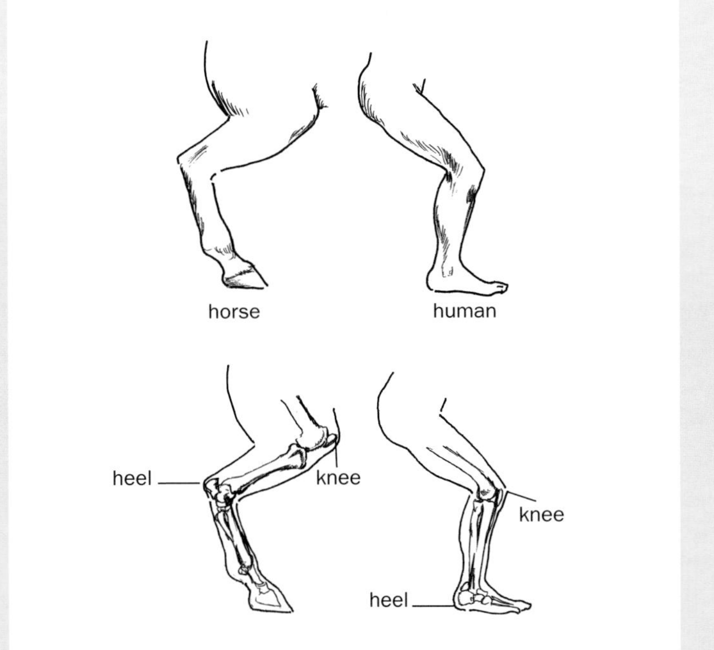

Question.

Does this suggest a lack of homology, or is there some other explanation?

Answer...

Appearances are deceiving. A horse's hind leg bends forward at its knee, just as the human leg does. However, the proportions of the various leg bones differ — the horse actually stands on its toe.

Some of the Things You Learned About in Chapter 2

Questions About the Ideas in Chapter 2

1. What is the fundamental feature of information?
2. What is the smallest number of units required to make a code?
3. Given that cells both contain and are surrounded by a watery environment, what special features must membranes have to enclose their contents?
4. Name at least three features that all the beetles on pages 34-35 have in common.
5. How does the expression "similarity precedes difference" apply to the beetles on pages 34-35?
6. If you were to take all of the beetles on page 34 out of the widely diverse environments in which they actually live and place each, with a mate, on a rosebush with red blooms in the middle of a garden full of sharp-eyed songbirds, which of the beetles do you think would live to produce offspring? Which would be among the first to be eaten by the birds? Describe what the population of beetles might look like three years from now.
7. How are proteins like tiny robots?
8. Genes do not contain information for seeing, breathing, thinking, defending the body against invaders, and so on. What information do genes contain, then, that allows us to do all these things?
9. In what sense is error essential to the creation of new forms of life?
10. The molecules of CO_2 (carbon dioxide) and CH_4 (methane) are about the size of H_2O (water) molecules, but unlike water, carbon dioxide and methane are gases at room temperature. What accounts for the ability of water to remain liquid at room temperature?
11. All biological systems are "self-correcting." What is meant by this?
12. Which of the following exhibit self-corrective behavior: An autopilot, a stopwatch, a thermostat, car's cruise control mechanism, an automatic teller machine?

13. Give some examples of "cycles" of life.

14. What is meant by "turnover"? Give examples of two levels of turnover in living systems.

15. Why do biological systems turn over?

16. In nature, too much of a good thing is not necessarily a good thing. Why?

17. Weeds growing up through cracks in the sidewalk, mold covering bread left out too long, and birds nesting in the eaves of a barn are all examples of what tendency of life?

18. How might mitochondria have evolved from bacteria?

19. Give an example of coevolution.

20. Animals regulate their temperature partially by the amount of surface area they have to dissipate internally generated heat. What features of the elephant on page 48 are likely to be useful in temperature regulation? What features of the polar bear serve the same purpose? Explain how we humans regulate our surface area and thus, help control our temperature.

References and Great Reading

Alters, Sandra. 2000. *Biology, Understanding Life* 3e. Sudbury: Jones and Bartlett Publishers.

Autumn, K. et al. 2000. Adhesive force of a single gecko foot-hair. *Nature* 405: 681–685.

Darnell, J., H. Lodish, and D. Baltimore. *Molecular Biology of the Cell.* New York. Scientific American Books, Inc. 1986.

Evans, C., 1996. *The Casebook of Forensic Detection, How Science Solved 100 of the World's Most Baffling Crimes.* New York: John Wiley & Sons, Inc.

Feyman, Richard. 2000. *The Pleasure of Finding Things Out.* Cambridge: Helix Perseus Books.

Huxley, T. H., 1880. *The Crayfish.* An Introduction to the Study of Zoology. New York: D. Appleton and Company.

Lewin, Ralph A. 1987. *The Biology of Algae and Diverse Other Verses.* California: Boxwood Press.

Moen, Ron A., et.al. 1999. Antler growth and extinction of Irish elk. *Evolutionary Ecology Research* 1: 235–249.

Needham, C., M. Hoagland, McPherson, and B. Dodson. 2000. *Intimate Strangers; Unseen Life on Earth.* Washington: ASM Press.

Reichs, Kathy. 1999. *Déjà Dead.* New York: Pocket Books.

McGee, Harold. 1990. *The Curious Cook, More Kitchen Science and Lore.* San Francisco: North Point Press.

Wen-Jing Hu et al. 1999. Repression of lignin biosynthesis promotes cellulose accumulation and growth in transgenic trees. *Nature Biotechnology* 17: 808-812.

for more questions and links to web resources, go to
WEB
Connection
www.jbpub.com/connections

CHAPTER 3

ENERGY
Light to Life

Plants, animals, fungi, protists, and microbes make up a vast cellular carpet spread over the globe. This carpet requires a constant supply of energy from sunlight to maintain itself; it ultimately releases that energy as heat.

EVERY DAY, COUNTLESS TINY PACKETS OF LIGHT CALLED PHOTONS radiate from the Sun, travel 93 million miles through space, and strike the Earth. There, the energy of that light is turned into the energy of heat, which is the stirring of the molecules in air, water, sand, and stone. Life evolved in this sunlight-to-heat energy stream, diverting part of its flow into structures that can move, grow, and duplicate themselves. Life accomplished this by finding a way to use the energy of sunlight to make energy-rich molecules, which, in turn, could be used to bond simple molecules into more complex, long-chain molecules.

This last was a monumental step. Chain molecules are stable sequences in which the particular order of units have information value. All of life's essential "ideas" were captured in these information chains — made possible by a constant light to heat energy stream. Think of plant and animal life on Earth, then, as an ordered collection of molecules joined by bonds made of captured energy.

This chapter is divided into two equal halves. In the first half, we review some basic chemistry, introduce the key molecular players, and present an overview of life's energy flow. The second half of the chapter (beginning on page 110) shows, in step-by-step sequences, how life captures, stores, and consumes energy.

Sometimes forceful collisions can bond atoms together into molecules...

...theoretically, successive collisions could form a chain of molecules.

3.1 Making Bonds

A Chaos of Collisions

In New York City's Grand Central Station, busy travelers dash about in seemingly random fashion on their various missions. Collisions inevitably happen. Imagine that some of these commuters collide so forcefully that they stick together permanently! Now imagine these commuters as atoms, which also bump into one another constantly. When they meet with the right fit and sufficient force, they form a chemical bond — and a molecule is born. Such chemical reactions underlie everything that's happening around us and inside us.

1.

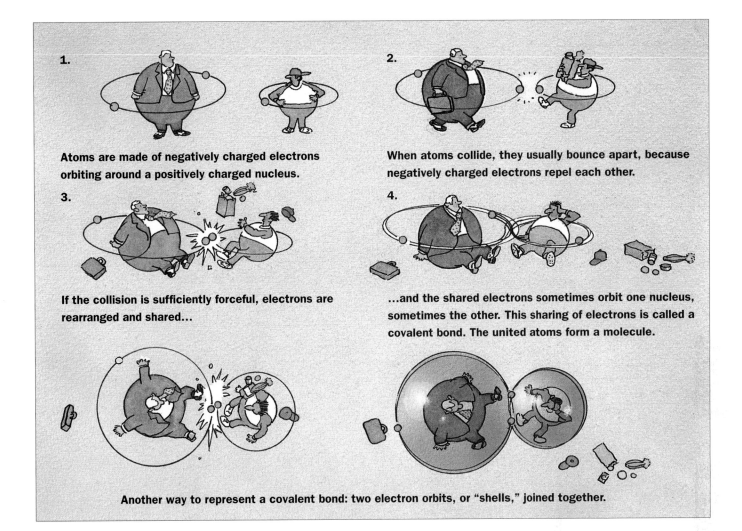

Atoms are made of negatively charged electrons orbiting around a positively charged nucleus.

2.

When atoms collide, they usually bounce apart, because negatively charged electrons repel each other.

3.

If the collision is sufficiently forceful, electrons are rearranged and shared...

4.

...and the shared electrons sometimes orbit one nucleus, sometimes the other. This sharing of electrons is called a covalent bond. The united atoms form a molecule.

Another way to represent a covalent bond: two electron orbits, or "shells," joined together.

How Atoms Stick Together

Let's back up a step and take a closer look at the atom. It consists of a positively charged nucleus — containing positively charged protons and uncharged neutrons — which is orbited by energetic, fast-moving, negatively charged electrons. When atoms collide, like the Grand Central commuters, their orbiting electrons push them apart, because like charges repel. However, as atoms careen through space, they possess what is called kinetic energy — the energy of motion. If the kinetic energy of two colliding atoms is great enough, it overcomes the repulsion of their electrons and a chemical reaction occurs, causing a rearrangement of electrons and uniting the atoms. Some of the atoms' electrons become *shared* by the two of them, producing what is called a covalent bond. These are strong bonds. They hold life's key atoms — carbon, hydrogen, oxygen, nitrogen, phosphorus, etc. — together in simple molecules, and they join those simple molecules together in chains.

Bonds are also a reservoir of the energy that went into making them. That energy, like fuel, can be put to work in cells to accomplish life's feats of moving, growing, and reproducing.

For every atom, the number of positively charged protons in the nucleus equals the number of negatively charged electrons orbiting the nucleus; therefore, overall the atom is neutral. Each kind of atom has a different number of protons in its nucleus — and, consequently, a matching number of electrons in orbits — which accounts for atoms' different diameters and masses. Oxygen, for example, contains eight protons; carbon, six; hydrogen, one. There are over 100 known kinds of atoms in the universe. Only about 20 are abundant in living organisms.

3.2 Molecular Changes

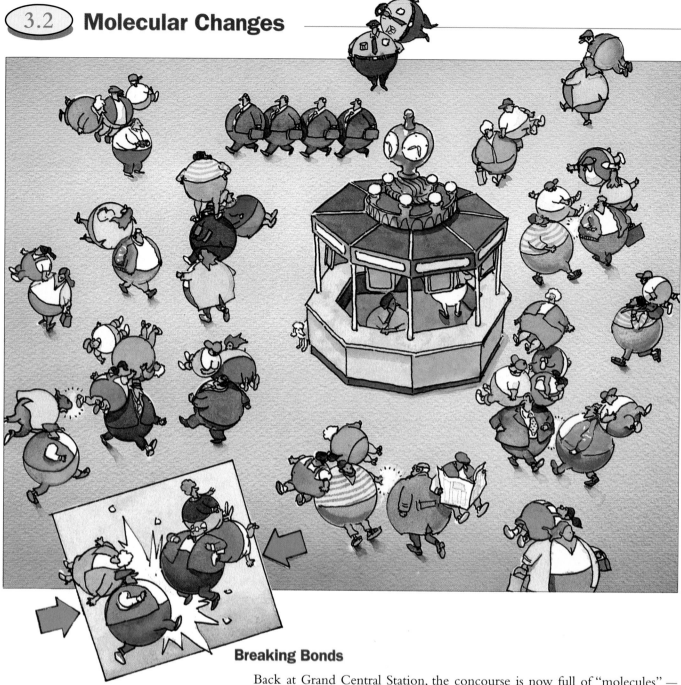

Sometimes forceful collisions can break bonds apart.

Breaking Bonds

Back at Grand Central Station, the concourse is now full of "molecules" — human "atoms" stuck together as a result of earlier collisions. In their ongoing mad dash for their trains, these molecules will frequently bump into each other without effect. But, now and then, a collision with more than the usual force and at just the right angle will cause the bonds between atoms to break. When a bond breaks, the shared electrons fall back into the original orbits around the separated atoms, releasing the energy in the bond as heat.

Cells need to be able to break bonds to rearrange molecules in all sorts of ways and to dispose of molecules no longer needed.

The energy is released as heat.

Energy Flows from One Bond into Another

A high-energy molecule is about to collide with the molecule approaching from the right.

If the collision occurs at the right place and at just the right angle, the key bond in the high-energy molecule will break...

...transferring energy to the new molecule and discarding the displaced atoms of the high-energy molecule.

Life is possible because of the great variety of molecular combinations. Using mostly carbon, hydrogen, oxygen, nitrogen, phosphorus, and sulfur, life fashions all its simple molecules and a near-infinite variety of large chain molecules.

Transferring Energy

Specific "key" bonds in certain kinds of molecules can produce an unusual amount of energy. When these high-energy bonds are broken, a substantial part of the energy in them can be *transferred* to other molecules, instead of all being lost as heat. That energy is captured and preserved in a new bond between part of the high-energy molecule and the new molecule to which the energy has been transferred. All the important activities of cells, such as constructing and moving, are carried out by large molecules of protein.

Proteins, the worker molecules of life, manage energy through this kind of transfer. Every time a bird flaps a wing, a maple tree sprouts a branch, or a clam opens its shell, bond energy is being transferred. Everything that happens in living cells is the result of various combinations of bond-breaking, bond-making, and energy transfer.

3.3 Life and the Laws of Energy

Running Uphill in a Downhill Universe

Incredibly, all the chemical processes of life and, indeed, all the energy and matter in the universe obey two simple laws: the *laws of thermodynamics.* The first law says that energy can be gained or lost in chemical processes — shifted from one form to another — but it can't be created or destroyed. Income and expenditure of energy have to balance. The second law says that energy inevitably disperses, dissipates, scatters — that is, it is transformed from more usable forms such as photons and bonds to a less usable form, namely heat. The tendency of energy to disperse, and of ordered structures to become disordered, is called entropy, and physicists say that the entropy of the universe is increasing.

I gotta get organized!

It seems strange that in a universe where matter and energy disperse — "run downhill" — life congregates and organizes — "runs uphill." This contradiction, symbolized at the right, is more apparent than real, as we explain on page 94.

And this brings up a puzzle. If the universe is dispersing its energy, if things are generally running down, how is it that life seems to be going the other way? Paradoxically, while energy has been spreading out, life appears to have gotten increasingly more ordered and complex over time. How can life build uphill with energy that runs only downhill?

In considering this question, we begin with the basic truth that life never contravenes, outwits, or otherwise gets around the fundamental laws of nature. It simply finds ways of using those laws to its own advantage.

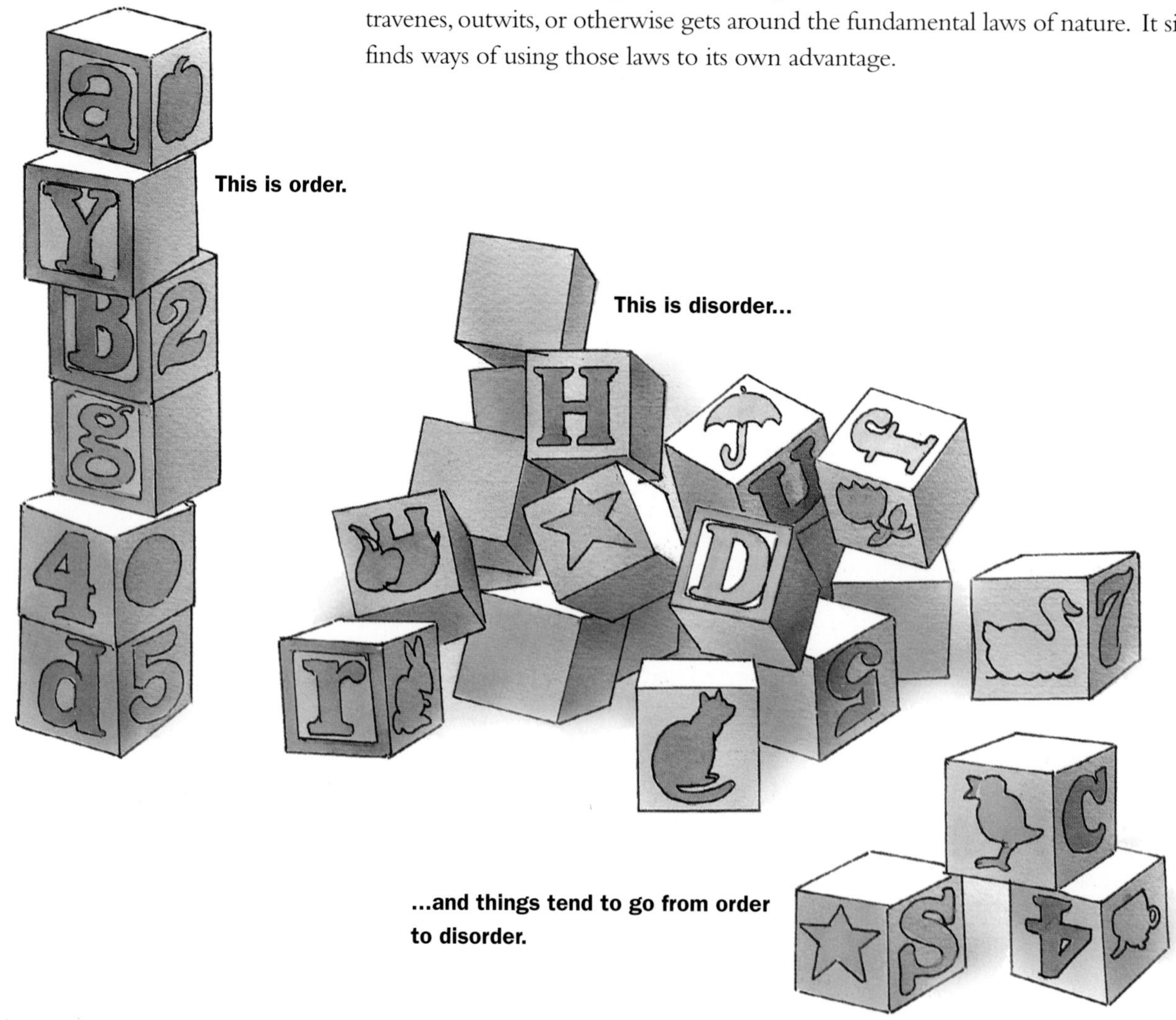

This is order.
This is disorder...
...and things tend to go from order to disorder.

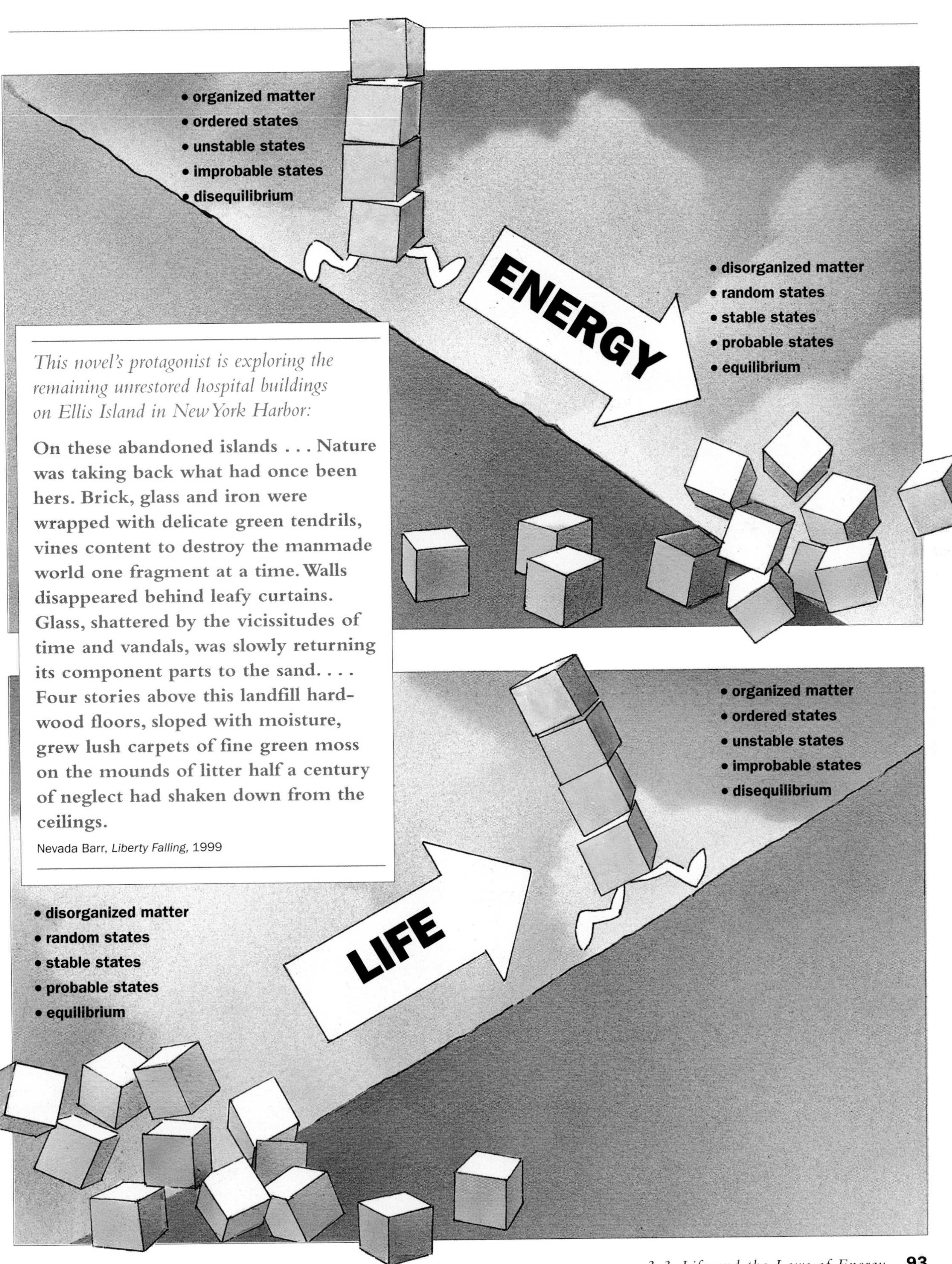

This novel's protagonist is exploring the remaining unrestored hospital buildings on Ellis Island in New York Harbor:

On these abandoned islands . . . Nature was taking back what had once been hers. Brick, glass and iron were wrapped with delicate green tendrils, vines content to destroy the manmade world one fragment at a time. Walls disappeared behind leafy curtains. Glass, shattered by the vicissitudes of time and vandals, was slowly returning its component parts to the sand. . . . Four stories above this landfill hardwood floors, sloped with moisture, grew lush carpets of fine green moss on the mounds of litter half a century of neglect had shaken down from the ceilings.

Nevada Barr, *Liberty Falling,* 1999

The Good News About the Second Law of Thermodynamics

Some of the energy that goes into the making of the bond is scattered as heat.

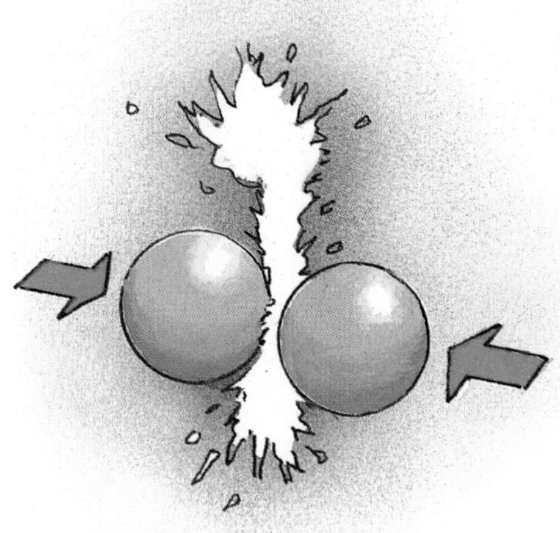

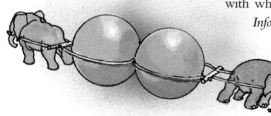

Consider Earth's lucky circumstances. Our planet orbits just near enough to the Sun to take advantage of its unlimited, steady output of energy without becoming too hot for stable chemical bonds to form. A constant flow of energy, described by the second law of thermodynamics, keeps our planet in a comfortable yet energized state in which bond-making, bond-breaking, and energy transfers occur readily. Action and change occur as energy moves toward a more scattered state. The Sun's light and heat flow over the Earth and then on into the quiet and cold of outer space, where the temperature is near -273 °C — what scientists call absolute zero. At absolute zero, nothing can happen: nothing moves, nothing has direction, time itself stops.

Let's take a close-up look at bond-making to see how this dynamic state of affairs works. Each time a bond connecting the simple atoms of life is created, some of the energy put into the bond is used to make it and some is dispersed into the surroundings as heat. In other words, more energy goes into making a bond than actually ends up *in* the bond; the excess is spread out into the surroundings. This seemingly wasteful dispersal of energy as heat, which is described by the second law of thermodynamics, has a beneficial effect. Think of it this way: If some of the bond-making energy didn't disperse but stayed nearby, it could readily flow right back and *unmake* the bond. The heat dispersal is necessary to ensure that what gets put together stays together — that, at least for a time, the building process is one way. The construction of bonds between atoms makes possible the creation of information (DNA). Information, in turn, brings order in its wake. Thus, as energy flows downhill, information accumulates, resulting in an uphill snowballing of complexity.

This ensures that the bond will be stable enough for the construction work of life. Breaking the bond would require at least as much energy as it took to make it.

Thus, the second law of thermodynamics does not threaten life, but instead guarantees: (1) a steady stream of usable energy dispersed by the Sun, (2) stable molecules with which to build, and (3) the assembly of information chains (see Chapter 4, *Information*). Running uphill is a highly creative energy — and information — driven process that depends upon a dogged and persistent rebuilding at the molecular level (like the castle-building crabs at the right).

Question.

Why do cats frequently like to sit on the hood of a car that has just been driven for a while? How does this result of a car's fuel use parallel the energy transfer and dispersal during cellular respiration (burning of sugars)?

Answer...

When a car burns its fuel, it is breaking down the organized bonds of the molecules that make up gasoline, using some of the released energy (about 20%) to push the pistons of the engine up and down, and dispersing much of it (about 80%) into the surroundings as random molecular movement (i.e., heat) that warms the engine block and the hood of the car but doesn't move the pistons. Cats take advantage of this heat source.

A Sand Castle Analogy

A sand castle is a vivid analogy for the effects of entropy. Inevitably, powerful natural forces — waves — will reduce the castle to the random disorder of the sand grains from which it arose.

In the inanimate world, what gets dispersed stays dispersed.

Life can neither circumvent nor otherwise escape the second law of thermodynamics, but it can, for a time, resist the tendency to disperse. Suppose, as a fanciful example, that after each wave, a colony of crabs rushes in and makes repairs so feverishly that the castle is completely restored before the next wave.

Of course, crabs don't actually behave this way, but in living systems, proteins perform the job of rebuilding. Their activities require a steady input of energy supplied by the Sun and then converted to high-energy bonds. In the animate world, what gets dispersed generally gets rebuilt.

Energy Flow and Equilibrium

Life is a big bag of chemical reactions.

Imagine that you've shrunk to the size of a cell and can watch a chemical reaction take place. A cell is about to put together a bigger molecule out of some smaller molecules. We call the molecules (or atoms) present at the beginning of a chemical reaction *reactants* and the resulting atoms and molecules *products.* When we talk about chemical reactions, we're usually talking about millions of atoms in a confined space constantly rushing around and colliding with each other. The more atoms there are — the more people crowding Grand Central Station — the more collisions there'll be and, therefore, the more likely that chemical rearrangements will happen.

A chemical reaction starts with lots of reactants and no products. Within seconds, reactants get converted into products. As the products pile up, the reactions begin to slow down. Finally, when the energy stored in reactants and products is equalized, no further products accumulate. The atoms have not stopped reacting with each other, however. Collisions continue to convert reactants to products, but now an equal number of collisions convert products back into reactants. When the energy flows as readily backward as forward, no further overall change takes place. This state of affairs is called equilibrium. (The flea-bitten dogs at the right illustrate the principle.) Life generally abhors equilibrium, because that's when cells become inactive and die. By ceaselessly adding reactants and removing products, living cells maintain themselves in far-from-equilibrium conditions.

Initially, cream molecules and coffee molecules are separate (as shown in the cutaway section).

Random movement and collisions begin to disperse the cream in a process called diffusion.

Question.

What might be the possible outcome if bonds break and re-form every time one atom or molecule collides with another? How can enzyme proteins help regulate reactivity?

Answer...

Stability is an important property of all molecules, especially those on which life depends. If everything that could react did react, all living organisms and much of the non-living world would burn up instantly or at least exist in a very unstable environment where life would be impossible. Luckily, molecules are far enough away from one another in most cases and have insufficient energy to react spontaneously except in rare situations. For two molecules to react, a certain threshold amount of energy is needed to break existing bonds (and therefore allow new ones to form). Enzymes help insure both stability and reactivity by overcoming this energy threshold in a controlled way, thus promoting the union of cellular molecules of the right type, in the right place, and at the right time. This helps keep the cellular interior organized despite many random collisions of these same molecules.

In time, the cream molecules will disperse throughout the coffee.

No need to stir in the cream

The second law of thermodynamics is illustrated by the tendency of cream to disperse in coffee. Once the cream molecules thoroughly disperse, they stay that way. The chance that they'll all float back to the surface is virtually zero. Even though they continue to move and bump into other molecules, they remain more or less evenly dispersed.

How a Dog Shares Its Fleas

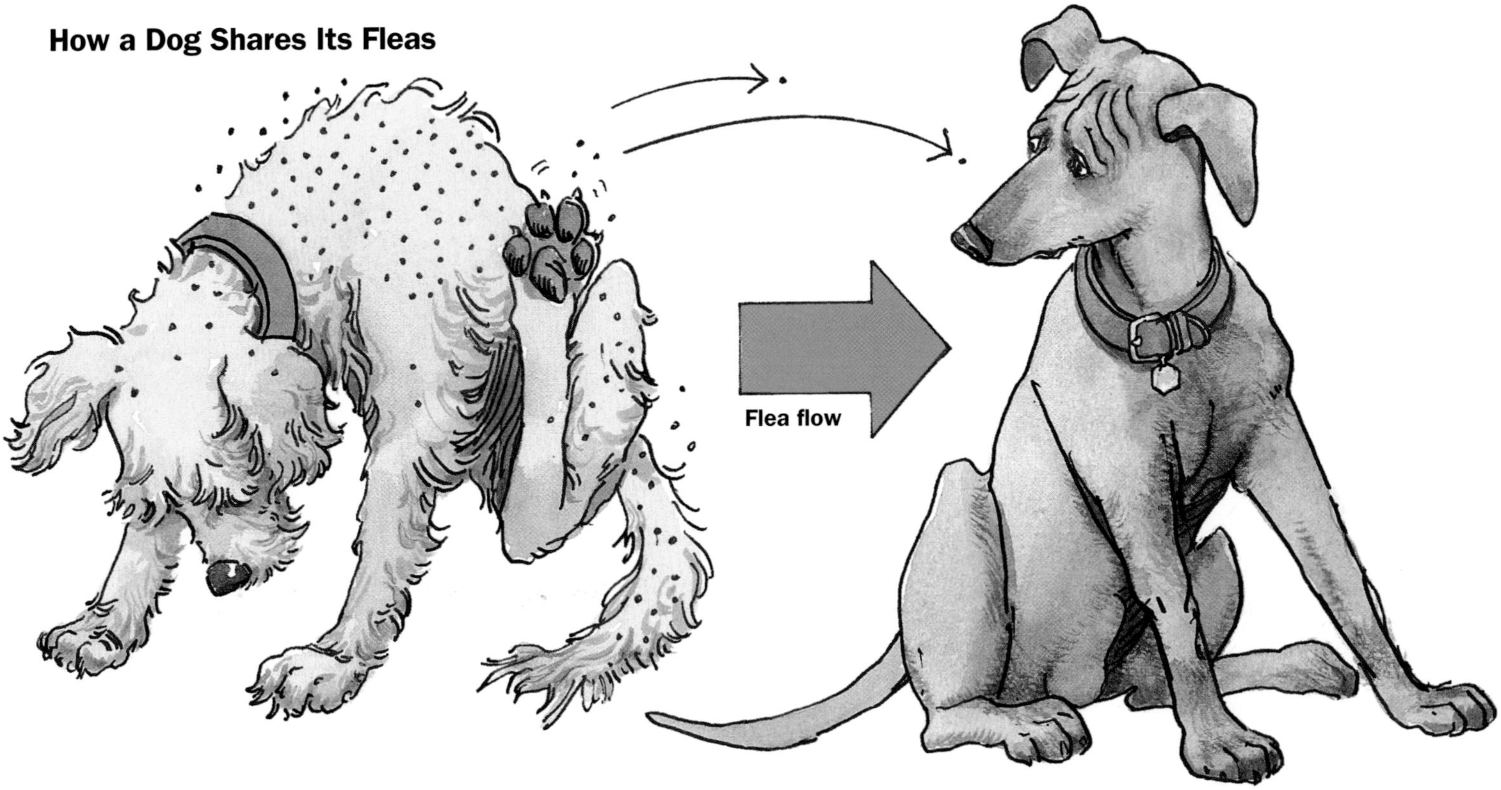

Flea flow

Assume the fleas will, with equal readiness, jump from one dog to another. If all of the fleas are initially on the left-hand dog, the overall flow of fleas will go from left to right.

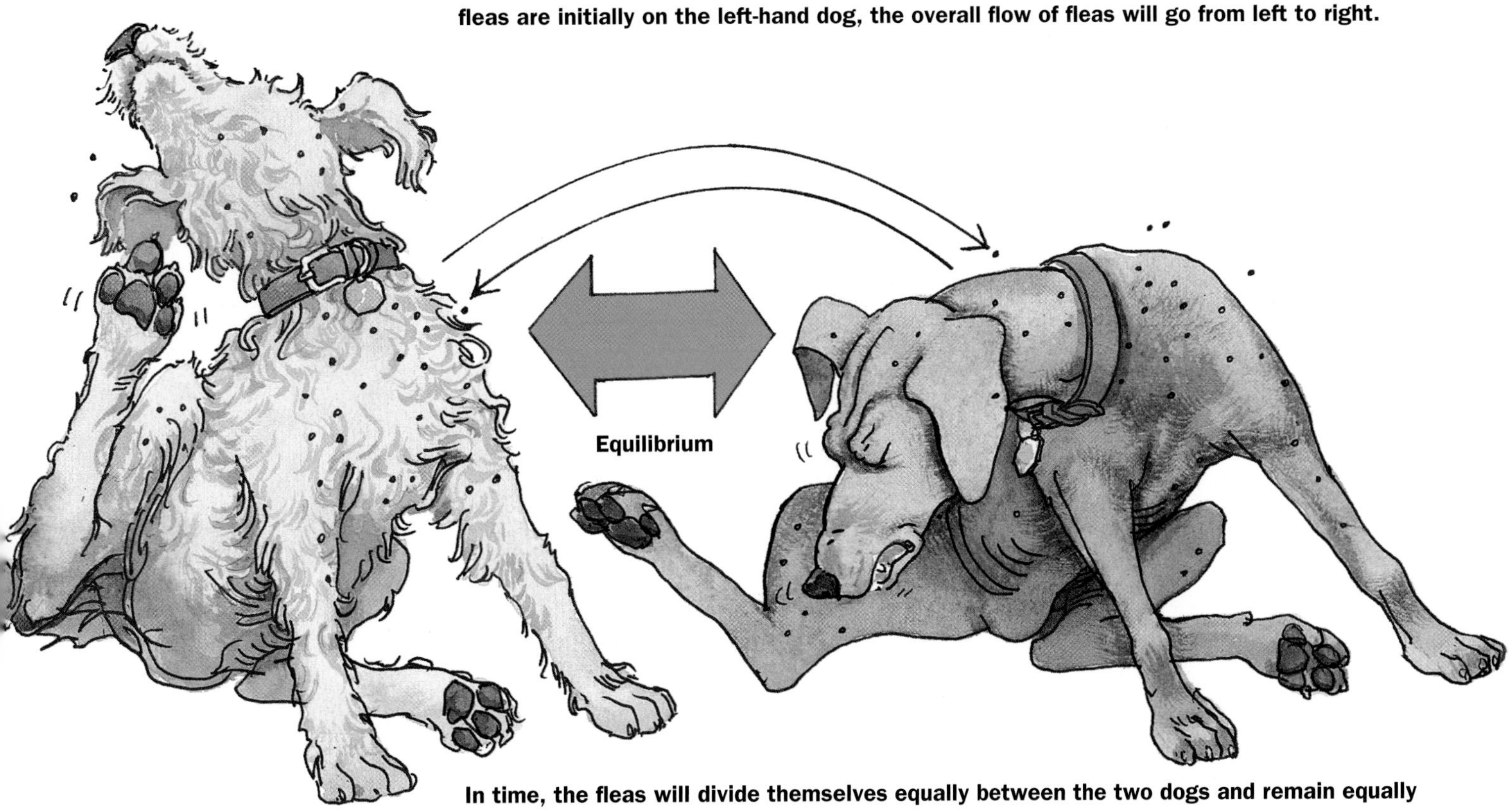

Equilibrium

In time, the fleas will divide themselves equally between the two dogs and remain equally divided even though individual fleas will continue to jump back and forth at the same rate as before. This is equilibrium. To keep the fleas flowingfrom left to right, we would have to put more fleas on the left dog or take fleas off the right dog.

Chemical Energy Makes Electrical Energy

As you read this page, you're generating heat. The act of reading consumes chemical energy and releases heat. Light and shadow — the letters on this page — register on your retina, and nerve cells conduct this information to your brain. Your brain, in response, sends messages to the muscles that control the movements of your eyeballs. Other messages go forth on nerves to other parts of your brain where conscious sense is made of what you see on the page. All of these activities dip into the universal stream of energy, extract what useful energy they need, and throw back their debt in dissipated energy.

A nerve cell and its long axon, or fiber, along which messages are conducted, is electrically charged, i.e., the inside of the axon's membrane is negatively charged relative to the fluid outside it. This is because sodium ions (Na^+) are concentrated outside, and a variety of negatively charged ions are on the inside. This state of electrical readiness — called polarization — is maintained by protein pumps embedded in the membrane which keep forcing sodium out of the cell. This constant pumping action requires energy supplied, of course, by ATP. And the by-product of this effort is heat.

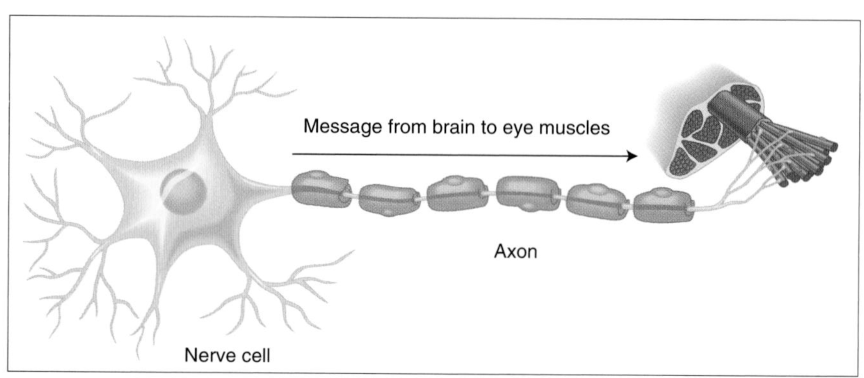

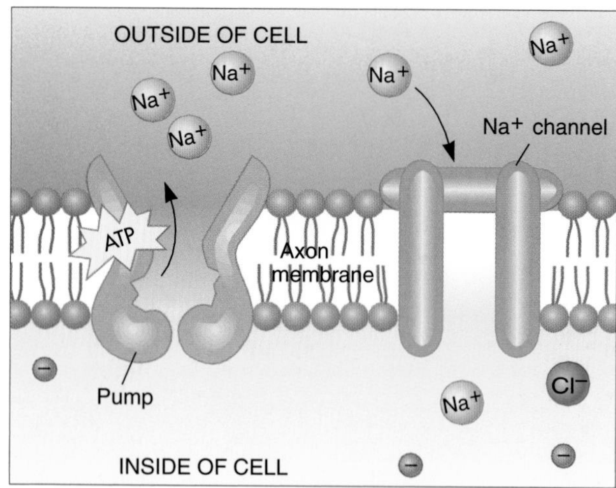

Polarization

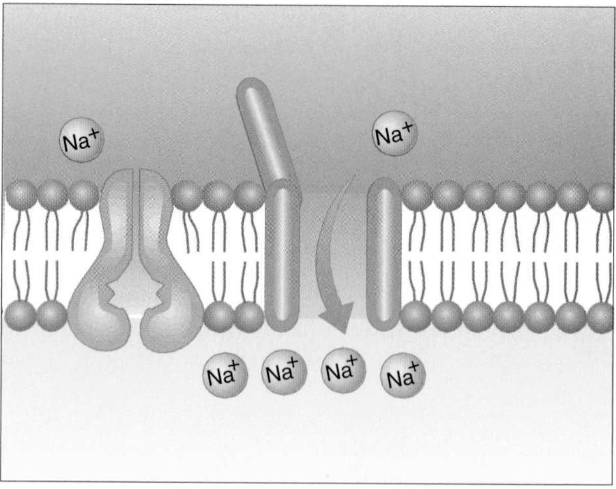

Depolarization

Sending the message

The nerve cell's ending in the eye is sensitive to light. Light abruptly changes the cell's permeability, causing sodium ions to flow inward through special channels. The membrane's charge drops, and this depolarization is propagated along the fiber up to the brain as a wave of voltage drop — the nerve impulse. Immediately after the passage of the impulse, the pumps in the membrane, energized by ATP, restore the nerve's original charge, insuring its readiness to fire again. This cycle of stimulation, depolarization, and recovery takes place in a few thousandths of a second.

Getting Rid of Heat

The dissipated energy is carried from the cells via your bloodstream to your lungs and skin, where it escapes, from your body to the atmosphere. The churning of the atmosphere eventually radiates this heat to outer space, where it is lost to us forever. The universe's total entropy has increased.

Distributing the heat 

The circulatory system, a nerve- and muscle-powered network of arteries and veins, distributes nutrients and heat throughout the body. As long as there is an input of chemical energy (food) to the body's cells, they produce heat, which is carried to the lungs and skin and, from there, radiates into the surroundings.

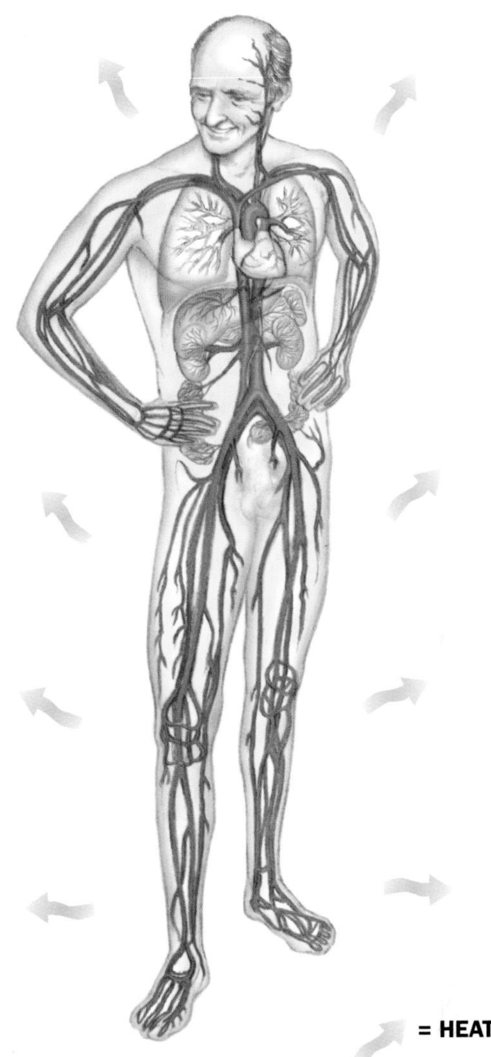

Heat regulation ▼

A specific structure in the brain's community of cells (the *hypothalamus*) works in a cycle with the muscles of the circulatory system to regulate the amount of heat radiated, keeping body temperature very near 98.6°F (37°C). In cold surroundings, when body temperature starts to decrease, these brain cells send messages signaling the small muscles that surround tiny blood vessels (arterioles) in the skin to tighten up. Blood flow to the skin is reduced, and the body retains more heat. When conditions lead to an increase in body temperature, these same brain cells signal the muscles to relax, sending more blood to the surface. More heat diffuses out of the body.

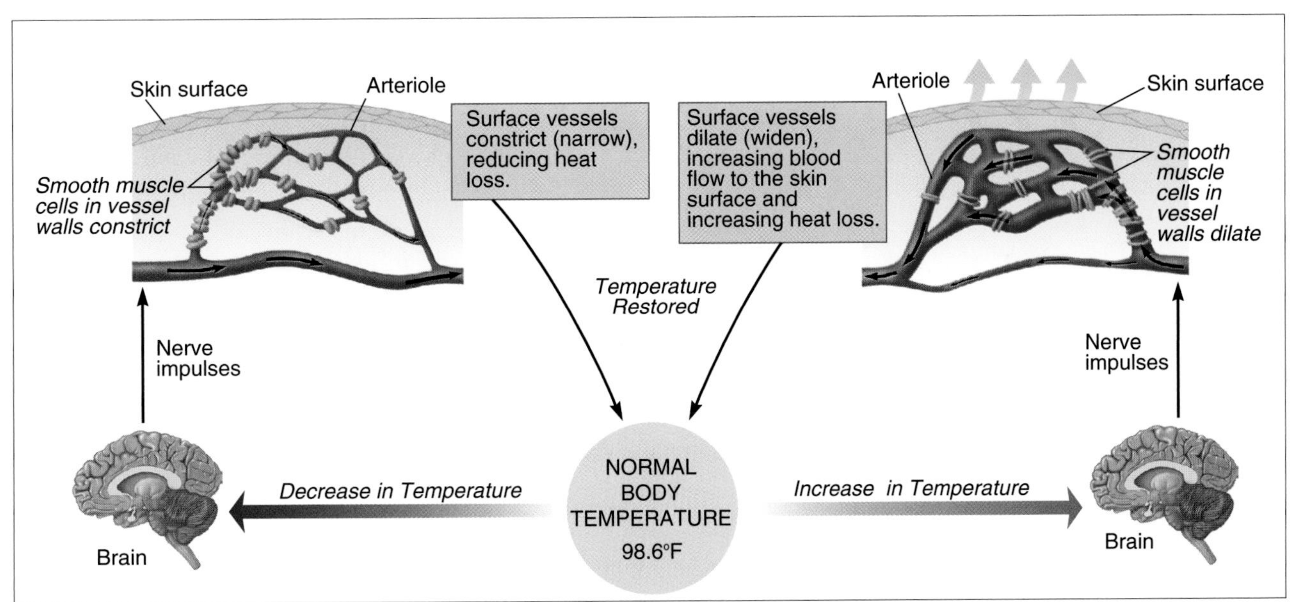

3.4 ATP — The Energy Molecule

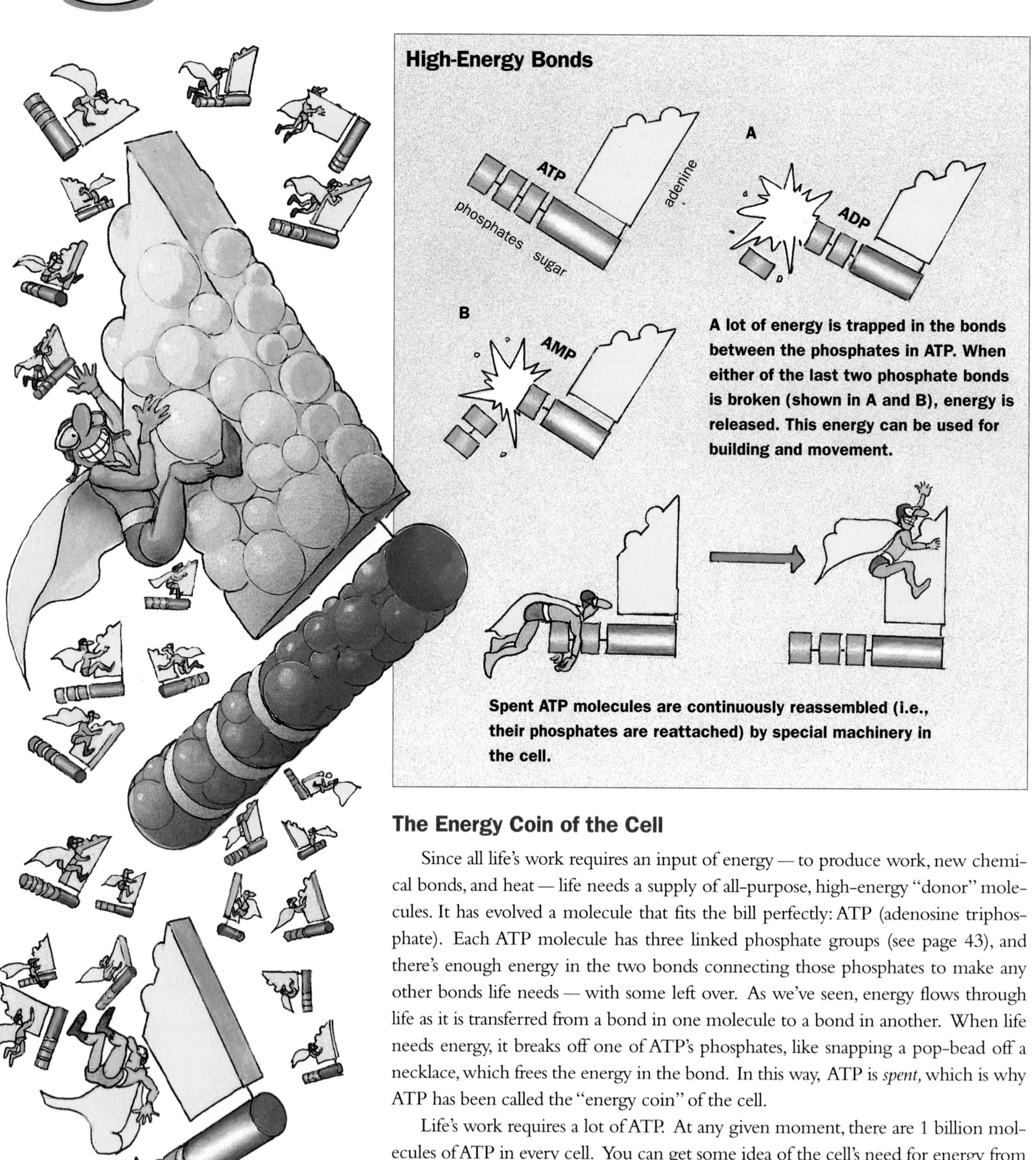

A lot of energy is trapped in the bonds between the phosphates in ATP. When either of the last two phosphate bonds is broken (shown in A and B), energy is released. This energy can be used for building and movement.

Spent ATP molecules are continuously reassembled (i.e., their phosphates are reattached) by special machinery in the cell.

The Energy Coin of the Cell

Since all life's work requires an input of energy — to produce work, new chemical bonds, and heat — life needs a supply of all-purpose, high-energy "donor" molecules. It has evolved a molecule that fits the bill perfectly: ATP (adenosine triphosphate). Each ATP molecule has three linked phosphate groups (see page 43), and there's enough energy in the two bonds connecting those phosphates to make any other bonds life needs — with some left over. As we've seen, energy flows through life as it is transferred from a bond in one molecule to a bond in another. When life needs energy, it breaks off one of ATP's phosphates, like snapping a pop-bead off a necklace, which frees the energy in the bond. In this way, ATP is *spent,* which is why ATP has been called the "energy coin" of the cell.

Life's work requires a lot of ATP. At any given moment, there are 1 billion molecules of ATP in every cell. You can get some idea of the cell's need for energy from the fact that the phosphate bonds at the business ends of a billion ATP molecules get used and replaced every two to three minutes. This means that you recycle two to three *pounds* of ATP every day!

A Versatile Player: Some of ATP's Jobs

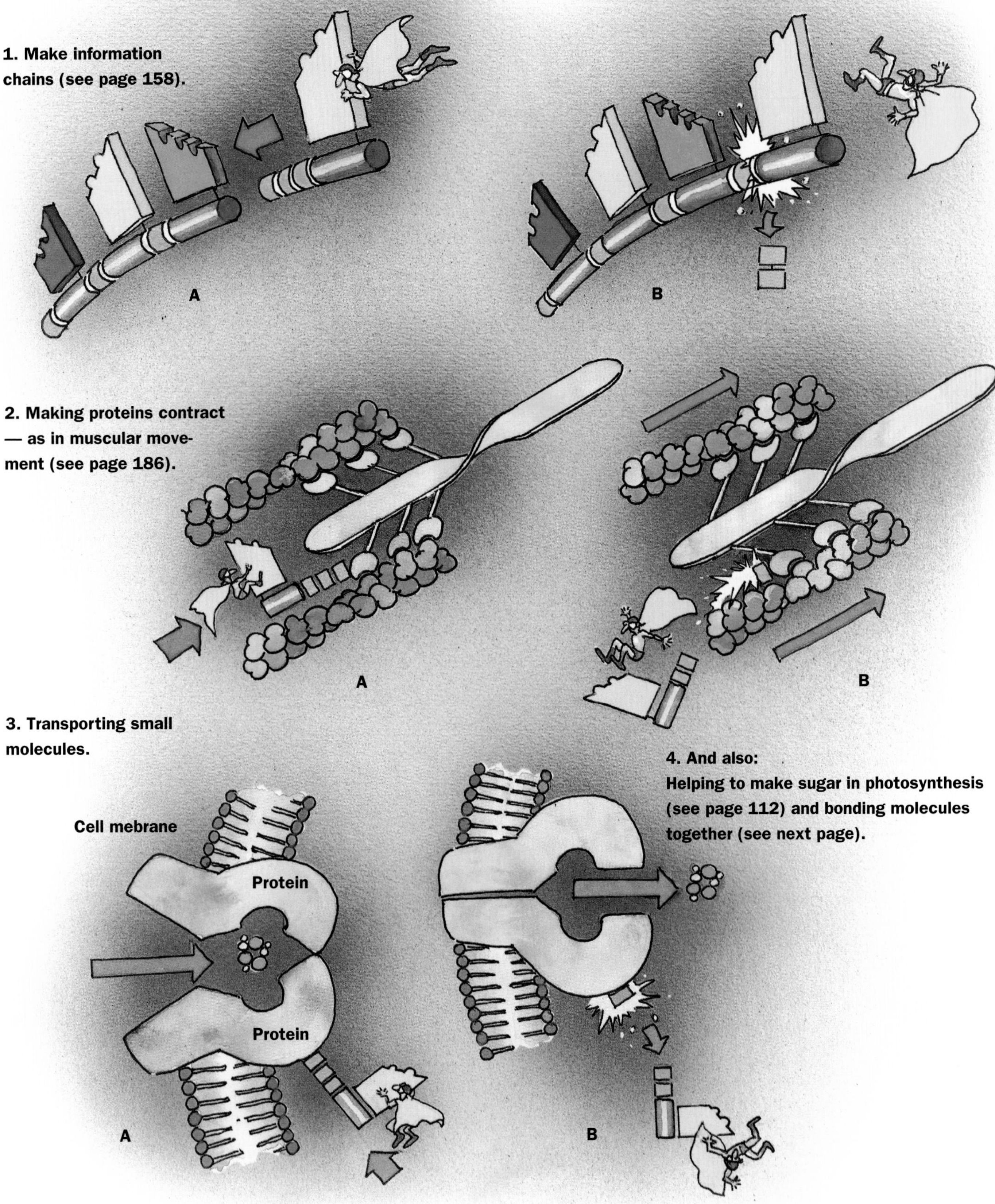

1. Make information chains (see page 158).
A
B
2. Making proteins contract — as in muscular movement (see page 186).
A
B
3. Transporting small molecules.
Cell mebrane
Protein
Protein
A
B
4. And also: Helping to make sugar in photosynthesis (see page 112) and bonding molecules together (see next page).

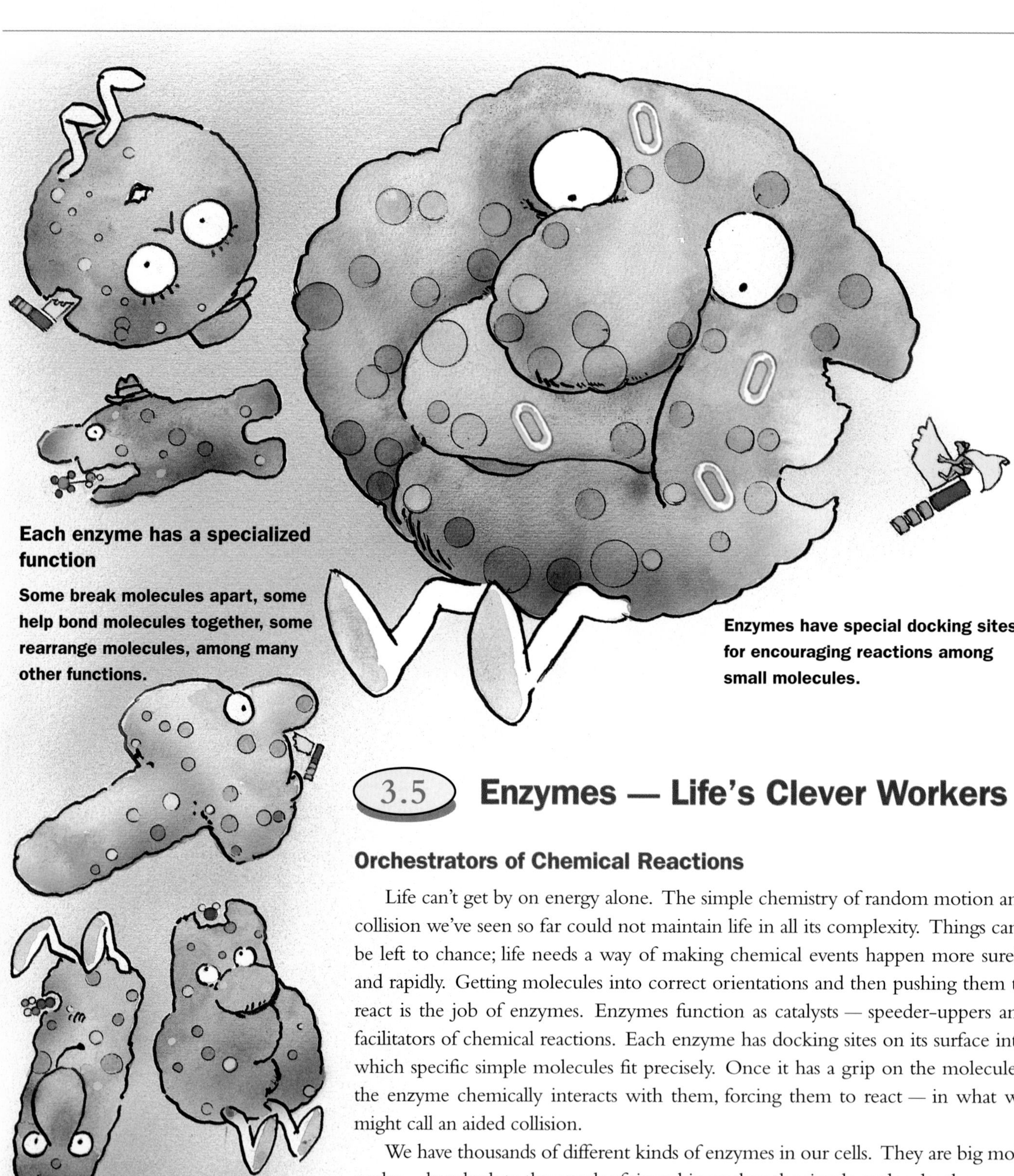

Each enzyme has a specialized function

Some break molecules apart, some help bond molecules together, some rearrange molecules, among many other functions.

Enzymes have special docking sites for encouraging reactions among small molecules.

3.5 Enzymes — Life's Clever Workers

Orchestrators of Chemical Reactions

Life can't get by on energy alone. The simple chemistry of random motion and collision we've seen so far could not maintain life in all its complexity. Things can't be left to chance; life needs a way of making chemical events happen more surely and rapidly. Getting molecules into correct orientations and then pushing them to react is the job of enzymes. Enzymes function as catalysts — speeder-uppers and facilitators of chemical reactions. Each enzyme has docking sites on its surface into which specific simple molecules fit precisely. Once it has a grip on the molecules, the enzyme chemically interacts with them, forcing them to react — in what we might call an aided collision.

We have thousands of different kinds of enzymes in our cells. They are big molecules — hundreds to thousands of times bigger than the simple molecules they work on. They are almost always protein — long chains of simpler molecules (amino acids, see page 28) that twist, bend, and fold themselves into many different shapes, most often resembling gnarled, lumpy potatoes. Their variety and versatility are awesome. They manipulate other molecules (i.e., act as catalysts), regulate production lines, "read" DNA's instructions, receive and react to chemical signals, and more.

Enzymes and ATP — A Dynamic Duo

How Life Joins "Reluctant" Molecules

An enzyme and an ATP molecule form a dynamic duo, a sort of Batman and Robin of the cell, working together to accomplish life's tasks of moving and building. Here we see how they solve the problem of joining two reluctant molecules together.

1. Molecules A and B can bump into each other indefinitely but will rarely form a bond. They are "reluctant" partners.

2. An enzyme takes molecule A into its docking site along with an ATP molecule.

3. With careful positioning, the enzyme transfers one of ATP's phosphates to the reluctant molecule.

4. The enzyme then discards the rest of the ATP, which is later rebuilt so that it can provide energy for other reactions.

5. Next, the enzyme takes molecule B into a nearby docking site.

6. Again, with careful positioning, the enzyme breaks the phosphate off molecule A, simultaneously transferring the energy to a bond between A and B.

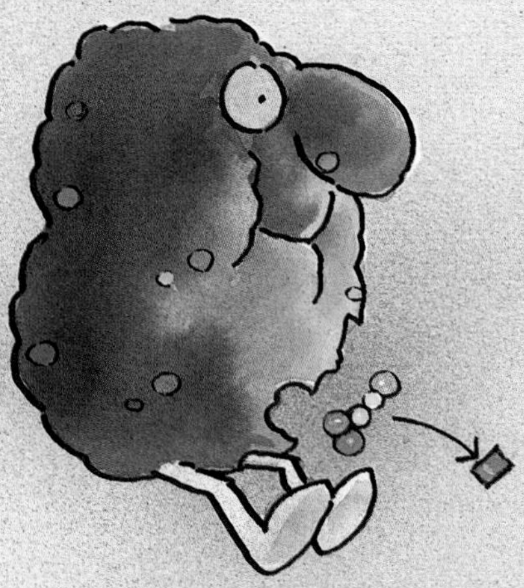

7. Now the two reluctant molecules are bonded together, and the spent phosphate is discarded.

3.7 Energy Flow Through Life — A Macro View

Producers

By far the largest portion of biomass on Earth belongs to the photosynthetic organisms: green plants, algae, phytoplankton, and photosynthetic bacteria. These are the sugar-makers. They are called autotrophs, or "self feeders."

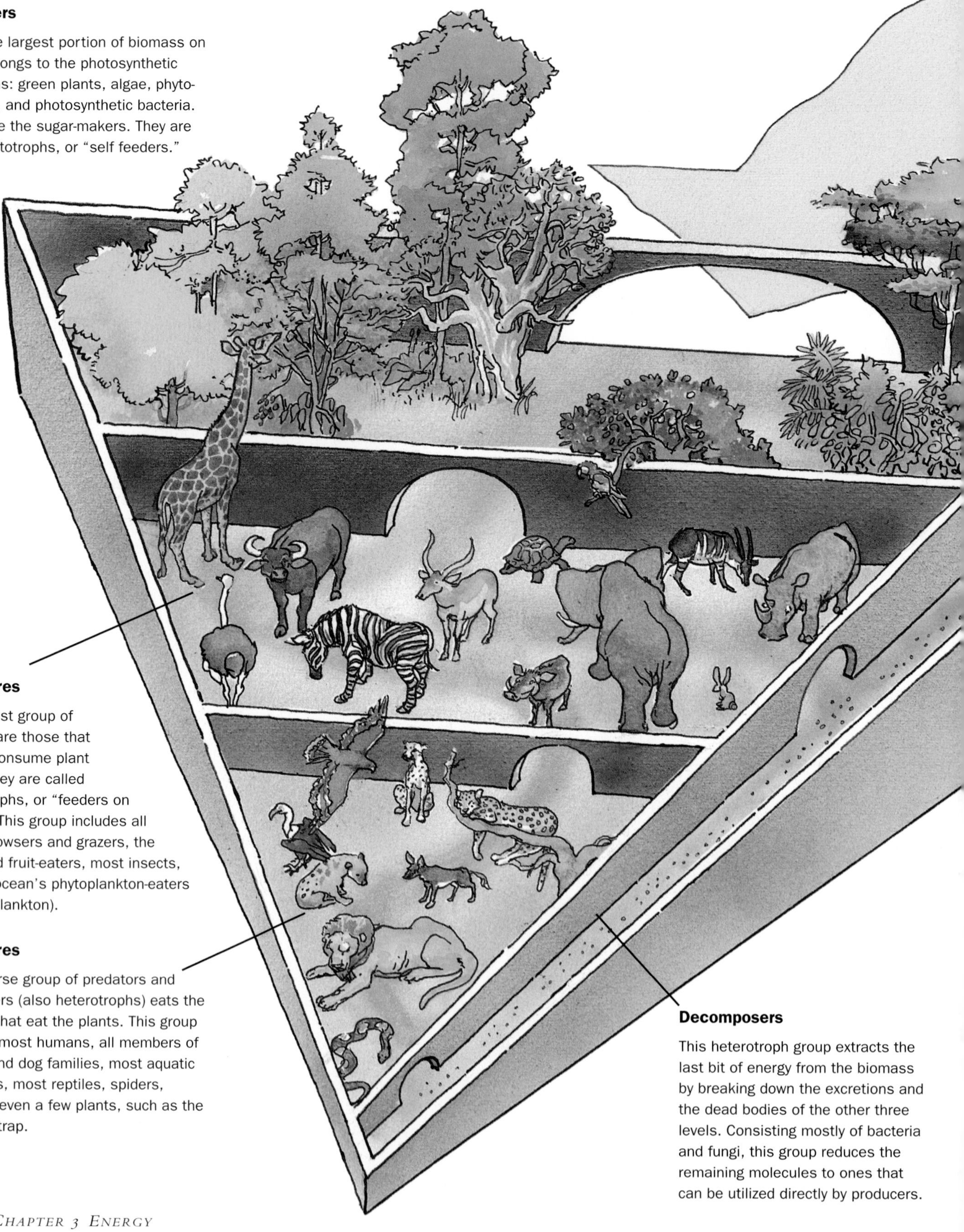

Herbivores

The largest group of animals are those that directly consume plant sugar. They are called heterotrophs, or "feeders on others". This group includes all of the browsers and grazers, the seed- and fruit-eaters, most insects, and the ocean's phytoplankton-eaters (the zooplankton).

Carnivores

This diverse group of predators and scavengers (also heterotrophs) eats the animals that eat the plants. This group includes most humans, all members of the cat and dog families, most aquatic mammals, most reptiles, spiders, starfish, even a few plants, such as the Venus flytrap.

Decomposers

This heterotroph group extracts the last bit of energy from the biomass by breaking down the excretions and the dead bodies of the other three levels. Consisting mostly of bacteria and fungi, this group reduces the remaining molecules to ones that can be utilized directly by producers.

From Plants to Herbivores to Carnivores

Energy flows through each individual and indeed through the whole carpet of life, one way and downhill. All the energy that enters the Earth's biosphere eventually goes out again, dispersed into outer space as heat. Along the way, however, energy percolates through several heterotroph levels (symbolized as the herbivore, carnivore, and decomposer compartments at the left).

At the first level, plants, photosynthetic bacteria, and algae capture the energy from sunlight and put it into the chemical bonds of sugar — life's universal food — in the process called photosynthesis. Plants make sugar for their own use, and make enough of it to support all other life as well. Herbivores get their sugar directly from plants, and carnivores get theirs from the flesh of herbivores. A fourth group, the "decomposers" (mostly bacteria and fungi) get sugar by breaking down the waste products and dead bodies of the other three groups. These organisms complete the cycle of materials by converting the substance of all other life into reusable forms, which are taken up once again by plants. All life on Earth would quickly cease if the decomposers stopped work.

Most of any living creature's energy gets consumed in metabolism — the internal process of building up and breaking down materials — or lost as heat. The amount left in its body as chemical bonds represents only a small part of the energy that has passed through it. Therefore, the food "chain" in fact looks more like an inverted pyramid, because each successive level extracts only a portion of the energy flowing through it. A square mile of grassland feeds perhaps 100 gazelles, which is about the population needed to sustain a single lion. So each gazelle's take is only about 1 percent of all the grass that is consumed, and the lion gets only about 1 percent of the gazelles (hardly a lion's share!). In a sustainable system, there can never be more lions than gazelles or more gazelles than grass.

Plants, then (and, incidentally, bacteria long before plants existed), are the vanguard of life. We animals could only come into being after plant-made sugar (and oxygen, which is a waste product of sugar-making) became plentiful on Earth. In fact, we are triply indebted to plants: (1) for our fuel, (2) for the oxygen to burn it, and (3) for saving us from being cooked by the effects of the carbon dioxide produced when we burn that fuel. Carbon dioxide, accumulating in the atmosphere from life and the industrial processes we have invented, limits heat from escaping the Earth. Plants consume that carbon dioxide in enormous quantities, thereby protecting us from overheating.

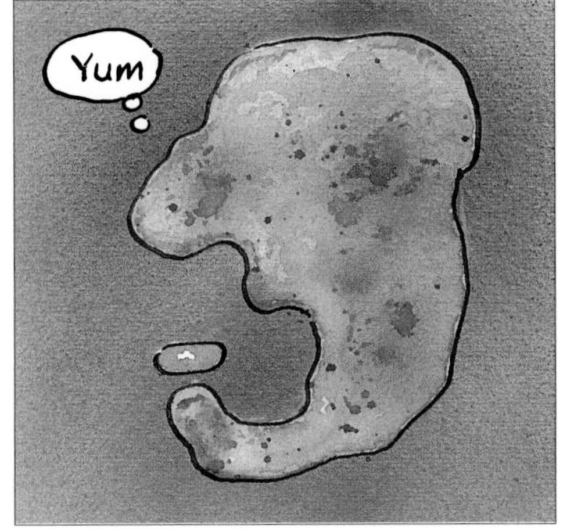

All need the Sun

At every level, organisms that cannot convert the Sun's energy are dependent on those that can. Here an amoeba eats a photosynthetic bacterium.

Visible Light Waves

Visible light, the form of energy photosynthetic plants absorb, reaches Earth's atmosphere as a whole spectrum of colors combined as white light. Molecules in the atmosphere and in Earth's non-living and living things absorb specific wavelengths (or photons) of the spectrum. What light those molecules don't absorb is reflected and becomes what we see as the color of an object or the sky. Thus, in the Annie Dillard quotation below, the goldfish's side absorbs all the light that *isn't* gold (green, blue, indigo, violet) and bats the red, orange, and yellow light Annie's way (see below).

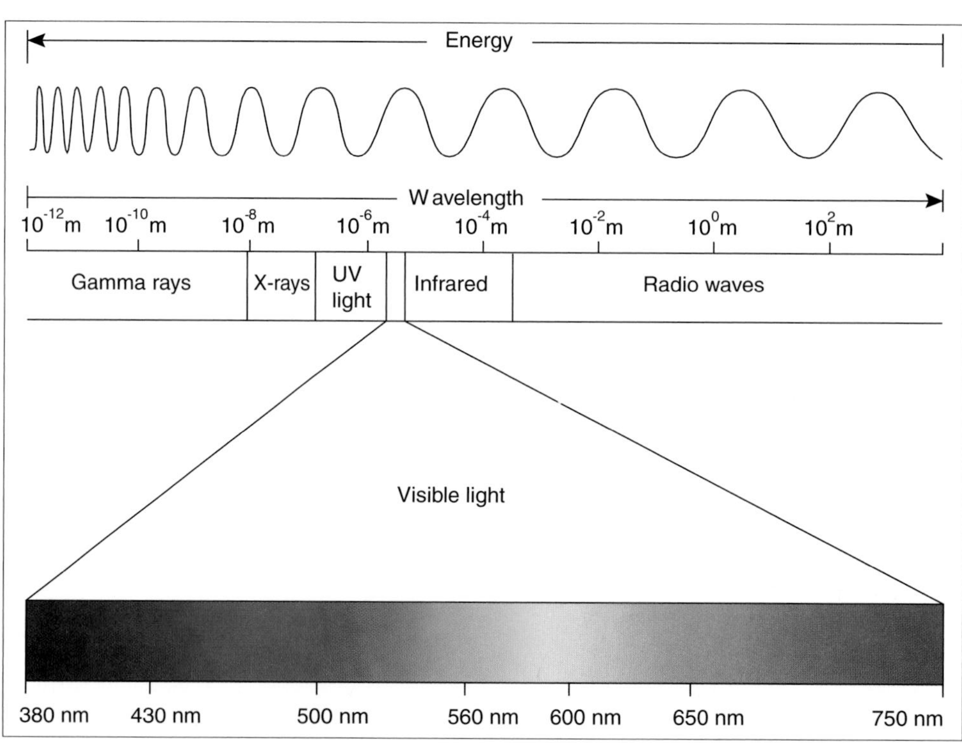

Seen and unseen energy

The angling light Annie Dillard describes in the passage quoted below is only a small part of the energy the Sun sends us. This diagram shows the entire spectrum of electromagnetic energy emitted by the Sun — you've already learned about how life uses some of these visible light wavelengths. Earth is constantly bombarded by all of this radiation, but the small portion that is visible light is all that our eyes can perceive.

An explosion of complex light fills my kitchen at the end of these lengthening June days. From an explosion on a nearby star eight minutes ago, the light zips through space, particle-wave, strikes the planet, angles on the continent, and filters through a mesh of land dust; clay bits, sod bits, tiny wind-borne insects, bacteria, shreds of wing and leg, gravel dust, grits of carbon, and dried cells of grass, bark and leaves. Reddened, the light inclines into this valley over the green western mountains; it sifts between pine needles on northern slopes, and through all the mountain black-jack oak and haw, whose leaves are unclenching, one by one, and making an intricate, toothed and lobed haze. The light crosses the valley, threads through the screen on my open kitchen window, and gilds the painted wall. A plank of brightness bends from the wall and extends over the goldfish bowl on the table where I sit. The goldfish's side catches the light and bats it my way; I've an eyeful of fish-scale and star.

Annie Dillard, *Pilgrim at Tinker Creek*, 1999

White Sunlight, *Newton* saw, is not so pure;
A Spectrum bared the Rainbow to his view.
Each Element absorbs its signature:
Go add a negative Electron to
Potassium Chloride; it turns deep blue,
As Chromium incarnadines Sapphire.
Wavelengths, absorbed, are reemitted through
Fluorescence, Phosphorescence, and the higher
Intensities that deadly *Laser Beams* require.

John Updike, *Midpoint*, 1969

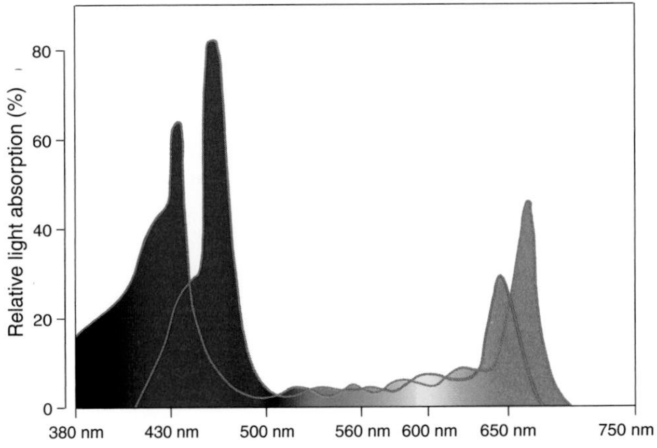

Measuring the light absorption of plant pigments

Of the wide range of energy available from the Sun, chlorophyll molecules in plants absorb primarily violet-blue and red visible light.

Frederick Edwin Church
***Aurora Borealis*, 1865**

Church's painting shows a sprectrum of light emitted not from the Sun, but from atmospheric gases. Sunlight has been absorbed by gas atoms, and that energy raised atomic electrons to higher orbits (see page 134). As the electrons fall back to their usual orbits, the atom emits the specific wavelength of energy it had absorbed. We see the overall effect as streamers of spectral colors. Church, a great traveler and enthusiastic naturalist, sketched such auroras during a trip to Labrador and Greenland.

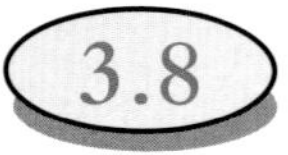

Energy Flow Through Life — A Micro View

Energy *flows through* life in a one-way stream.

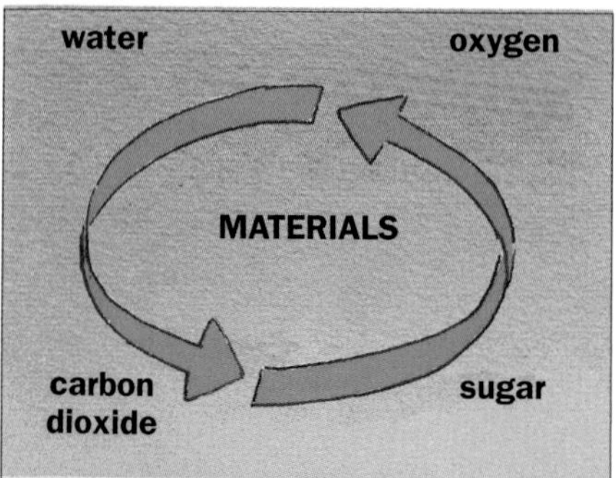

Life's molecules are cycled in a *continuous loop*.

These two processes — *flow-through of energy* and *cycling of materials* — are superimposed in living systems.

. . . He had been eight years upon a project for extracting sun-beams out of cucumbers, which were to be put into vials hermetically sealed, and let out to warm the air in raw inclement summers . . .

Jonathan Swift, *Gulliver's Travels,* 1726

From Sugar-Making to Sugar-Burning

Energy flows through the biomass of plants, animals, and other organisms in complex food chains. At the level where energy is actually captured, transferred, and put to work, a simpler pattern prevails. Incredibly, the entire living world runs on the work of just two kinds of bacteria-sized organelles within cells. Chloroplasts in plant cells make sugar using the energy of sunlight, in the process called *photosynthesis.* Mitochondria, in both plant and heterotroph cells, break down (burn) the sugar and make ATP, in the process called *respiration.* So the following general sequence represents the real flow of energy through life: Sunlight → ATP → heat (released when ATP is used).

The system is actually one step more complicated: In order to make sugar, a chloroplast must first make its own ATP. Its sugar-building enzymes need energy from ATP to work. You may wonder, then, if chloroplasts can make ATP, why do they bother to make sugar? Because sugar provides not only the energy, but *building material* as well. As we mentioned in Chapter 2, *Patterns,* cells can convert sugar (glucose) into an array of molecules with which to build, notably amino acids to make proteins and nucleotides to make RNA and DNA.

So, if we follow building material through life, rather than following energy, we see it flowing in an ever-renewing cycle: Chloroplasts take in the simplest of molecules — carbon dioxide and water — make sugar, and release oxygen (photosynthesis). Mitochondria do the opposite: They take in oxygen and sugar, merge them, and discard water and carbon dioxide (respiration). Together these cellular operations yield a simple and beautiful circularity: water and carbon dioxide in, water and carbon dioxide out, and an almost unimaginable complexity between.

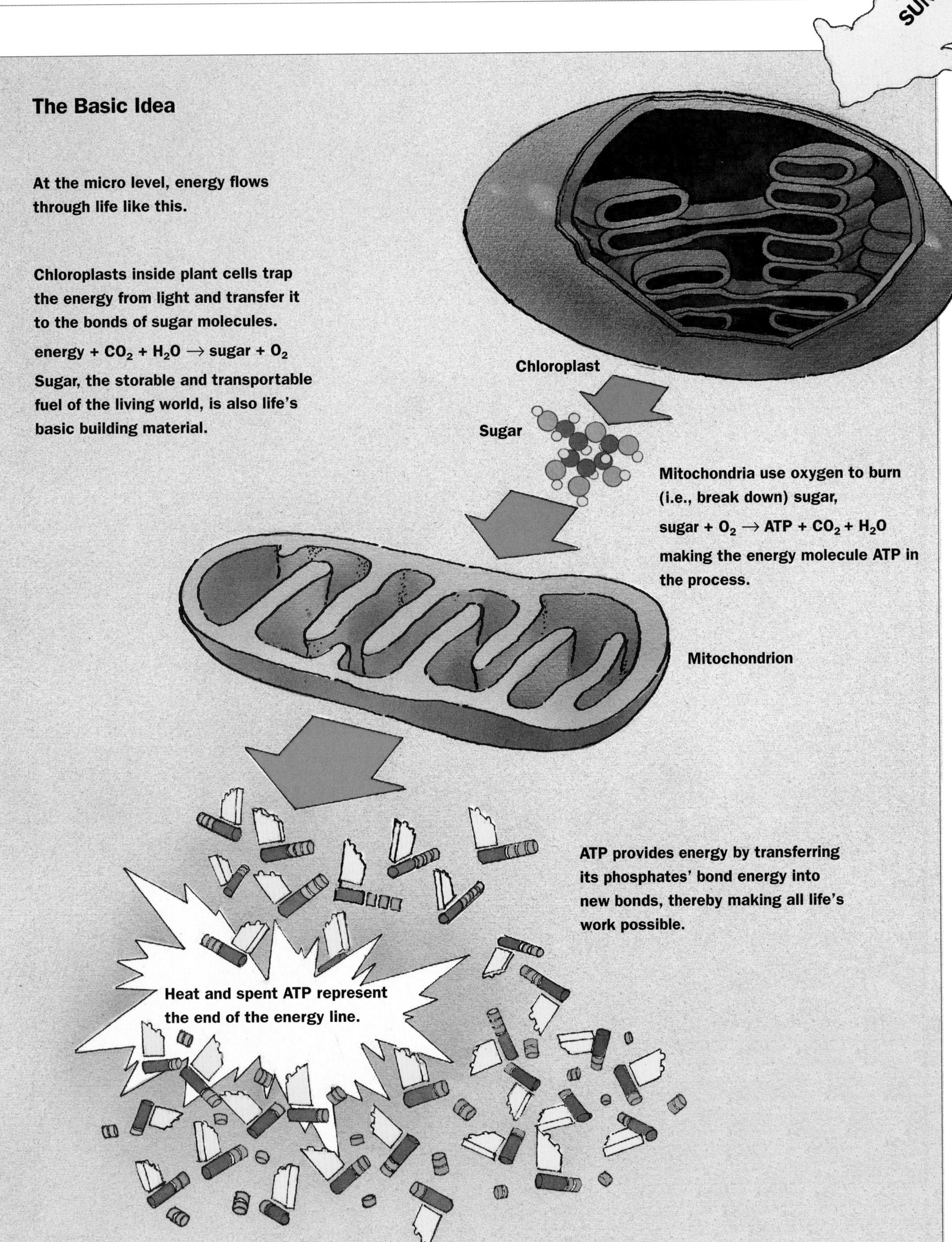
SUNLIGHT
The Basic Idea
At the micro level, energy flows through life like this.
Chloroplasts inside plant cells trap the energy from light and transfer it to the bonds of sugar molecules.
energy + CO_2 + H_2O $\rightarrow$ sugar + O_2
Sugar, the storable and transportable fuel of the living world, is also life's basic building material.
Chloroplast
Sugar
Mitochondria use oxygen to burn (i.e., break down) sugar,
sugar + O_2 $\rightarrow$ ATP + CO_2 + H_2O
making the energy molecule ATP in the process.
Mitochondrion
ATP provides energy by transferring its phosphates' bond energy into new bonds, thereby making all life's work possible.
Heat and spent ATP represent the end of the energy line.

An overexcited electron flies off one of the dancers . . .

3.9 The Chloroplast Ballroom

...and onto a bystander...

...who, in turn, is energized.

The Electron Bounce

Glittering strobe lights animate The Chloroplast Ballroom. As the lights spin and the band breaks into "Sugar Jump," the dancers go wild. Suddenly a bystander inspired by a dancer on the floor gets dancing feet. This in turn excites a second bystander to dance and before long a chain reaction takes place, each new dancer energizing the next bystander.

These jitterbugging molecules are demonstrating the initial steps in converting light energy into chemical energy — ATP. Energy from light can excite certain electrons in molecules, boosting them into higher-energy orbits. Such energized electrons will actually jump from one molecule to another to another, setting up an electron flow. This leads to the next step — the "Ion Shuffle."

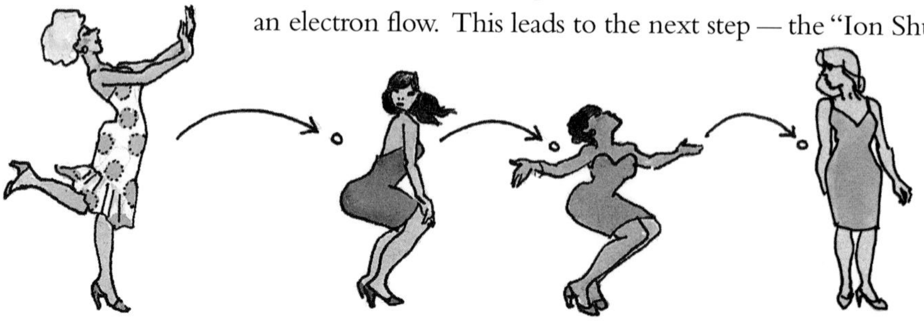

In this way, the energy jumps from dancer to dancer.

The Ion Shuffle

Energized female dancers (negatively charged because they have picked up flowing electrons) whirl by males (positively charged hydrogen ions — protons — that have lost their electrons) and dance them by a burly bouncer — who grabs each man and throws him into a nearby lounge room. The more the men crowd together, the more desperately they want out. They can only exit through a revolving door, cranking up a machine that assembles ATP.

Hydrogen Ions

Hydrogen, , the smallest of all atoms, can readily give up its single electron, , leaving behind its positively charged nucleus — a hydrogen ion, .

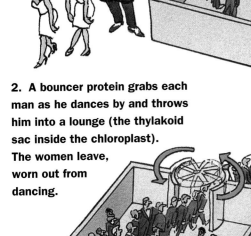

1. The female dancers (carrier molecules) team up with male partners (hydrogen ions) — because opposites naturally attract.

3. The more men in the lounge, the more they want out of the crowded room. (Remember the second law of thermodynamics, page 92.)

2. A bouncer protein grabs each man as he dances by and throws him into a lounge (the thylakoid sac inside the chloroplast). The women leave, worn out from dancing.

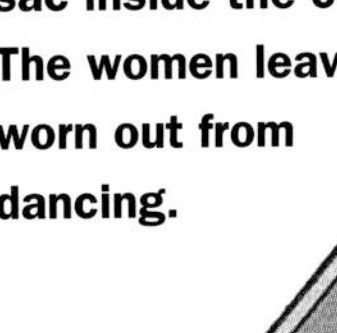

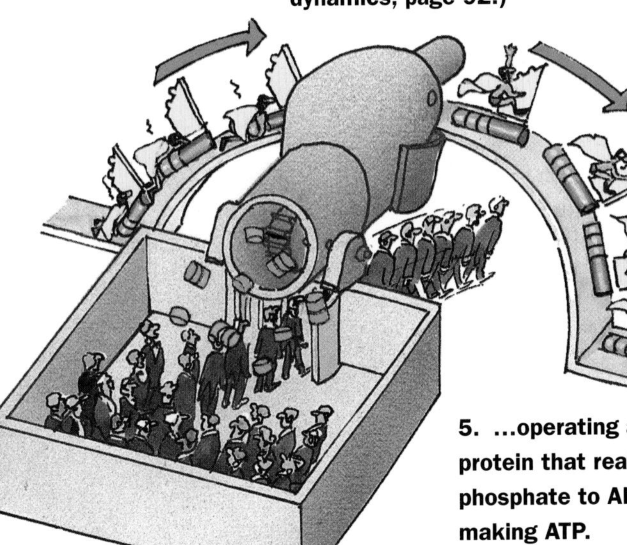

4. The only way out is through a revolving door enzyme, which spins as they escape...

5. ...operating a machine protein that reattaches a phosphate to ADP molecules, making ATP.

A sugar machine

It is estimated that a full-grown, healthy maple tree has about 500 square feet of leaves weighing about 500 pounds. This represents a total chloroplast surface area of about 140 square miles. A single maple can make two tons of sugar on one good sunny day!

A cross-section of a leaf

Shows the chlorophyll-containing cells sandwiched between protective layers of surface cells.

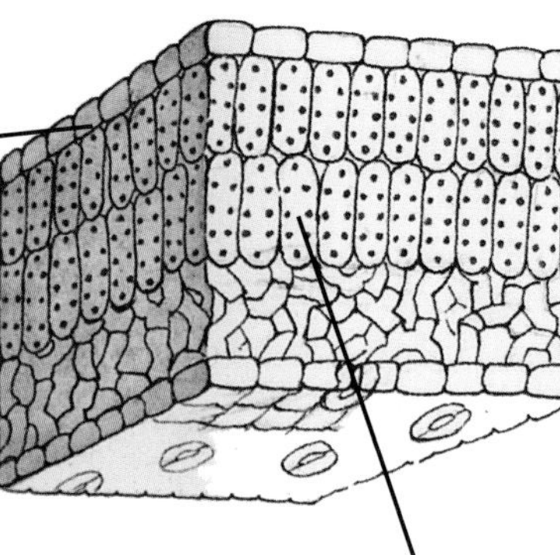

Benjamin Franklin in a letter to John Lining in Charleston, South Carolina:

Sir,

. . . I have been rather inclined to think that the fluid fire, as well as the fluid air, is attracted by plants in their growth, and becomes consolidated with the other materials of which they are formed, and makes a great part of their substance. That when they come to be digested, and to suffer in the vessels a kind of fermentation, part of the fire, as well as part of the air, recovers its fluid, active state again and diffuses itself in the body digesting and separating it. . . .

New York, April 14, 1757

A single leaf cell

Has about 50 chloroplasts — the factories that do the work of producing sugar.

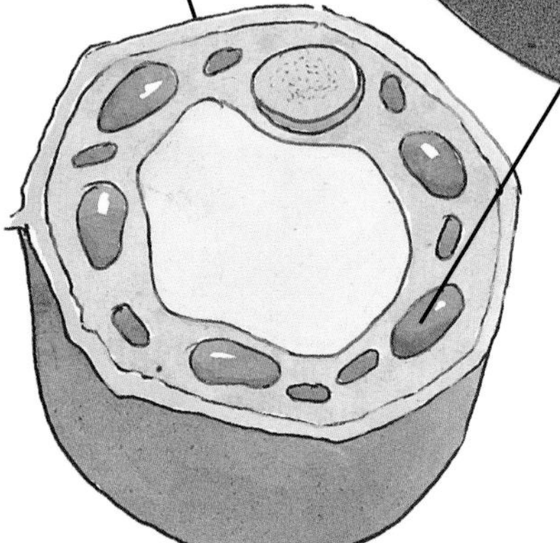

3.10 Photosynthesis — Using Sunlight to Make Sugar

It's Not Easy to Make Sugar

Here's a short-hand version of how green plants make sugar. Energy-filled packets of sunlight (photons) hit chlorophyll molecules in leaves, kicking electrons in those molecules into higher-energy orbits (1). These energetic electrons bounce along a series of chlorophyll molecules (our female jitterbuggers) and onto small carrier molecules (other female dancers) (2). The electrons lost from chlorophyll are replaced by electrons from water readying the chlorophyll for more action (3). The carriers pick up hydrogen ions (our male partners) and escort them (4) to a protein (our bouncer), which ejects them into a thylakoid sac (the lounge) within the chloroplast (5). The ions, crammed together, force their way out of the sac through a channel in an enzyme (our lounge's revolving door), a process that empowers the enzyme to make ATP (6). The electrons, after getting another energy boost from light (7), finally unite with more hydrogen atoms on a special molecule, NADP, forming highly reactive "hot" hydrogens (8). Finally a team of enzymes, using the ATP for energy, grabs carbon dioxide molecules from the surrounding air, combines them with the "hot" hydrogens, and links them together to produce sugar (9).

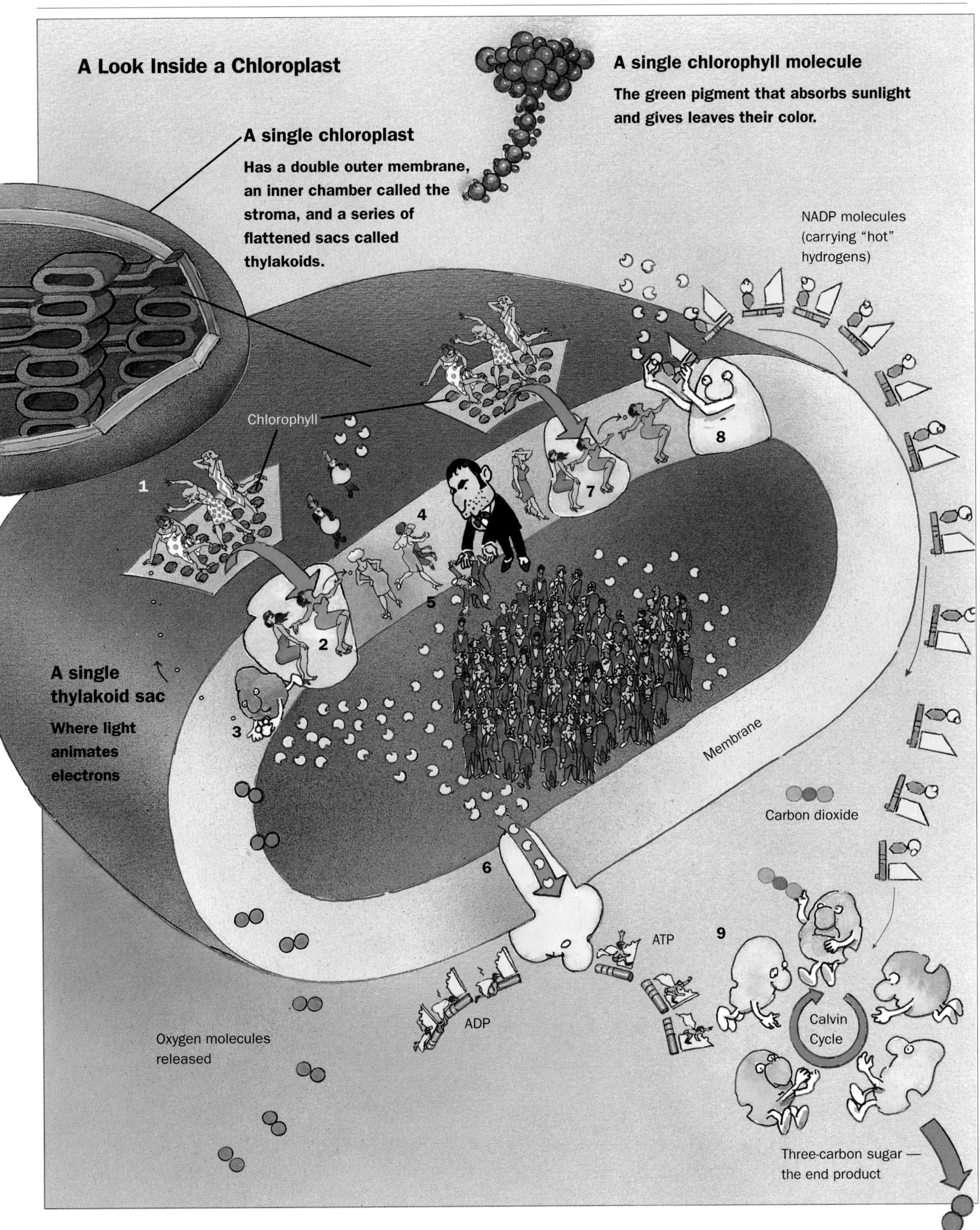
A Look Inside a Chloroplast
A single chlorophyll molecule
The green pigment that absorbs sunlight and gives leaves their color.
A single chloroplast
Has a double outer membrane, an inner chamber called the stroma, and a series of flattened sacs called thylakoids.
NADP molecules (carrying "hot" hydrogens)
Chlorophyll
1
2
3
4
5
6
7
8
9
A single thylakoid sac
Where light animates electrons
Membrane
Carbon dioxide
ATP
ADP
Calvin Cycle
Oxygen molecules released
Three-carbon sugar — the end product

Step by Step

These pages depict the steps in photosynthesis: first, in a simplified sequence (shown below): then, in a more detailed version (beginning on the following page). We show photosynthesis as a series of steps, but you need to imagine it as a continuous, rapid flow.

The Basic Idea

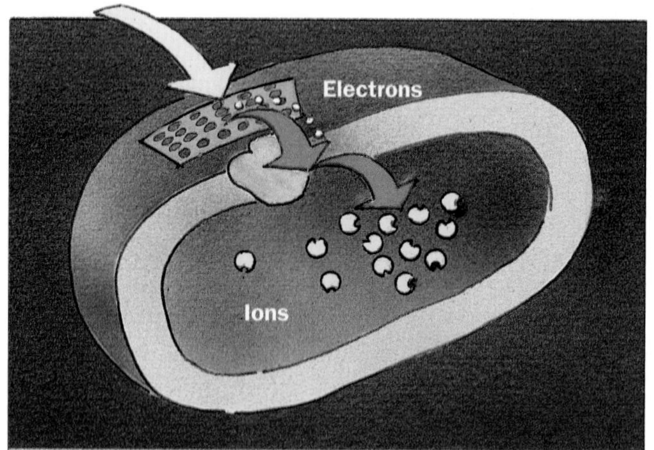

PUTTING IONS IN A SAC: Electrons, excited by sunlight, enable ions to collect in a sac.

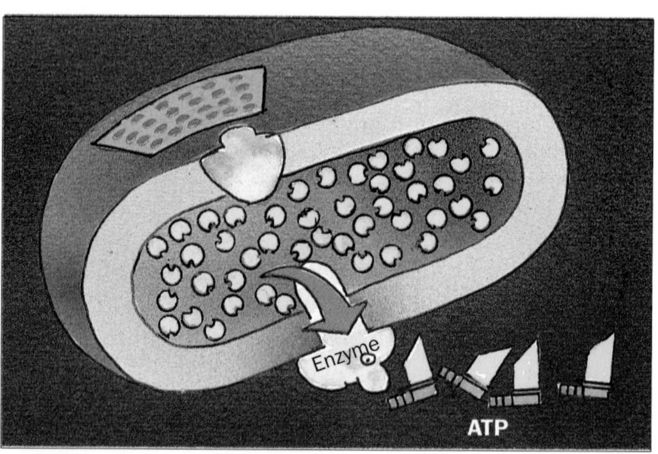

PRODUCING ATP: The escaping ions, channeled through an enzyme, power the making of ATP.

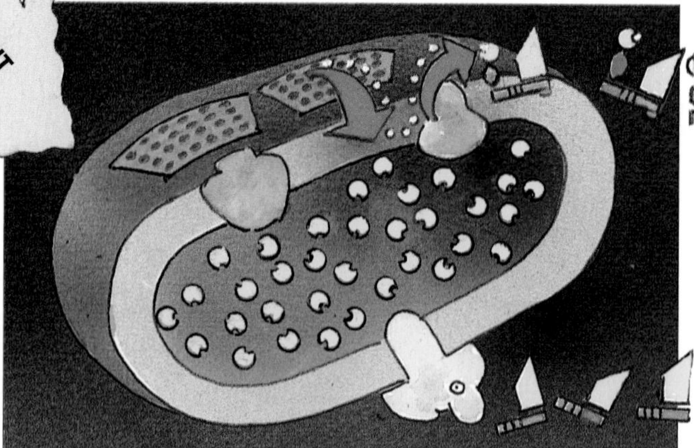

PRODUCING "HOT" HYDROGENS: The electrons, re-excited by more sunlight, are taken up by hydrogen ions, and attached to a molecule called NADP (nicotinamide adenine dinucleotide phosphate). Because their electrons are excited, these hydrogens are "hot."

MAKING SUGAR: Using the energy of ATP, a circle of enzymes combine the "hot" hydrogens with carbon dioxide to make sugar.

The Details

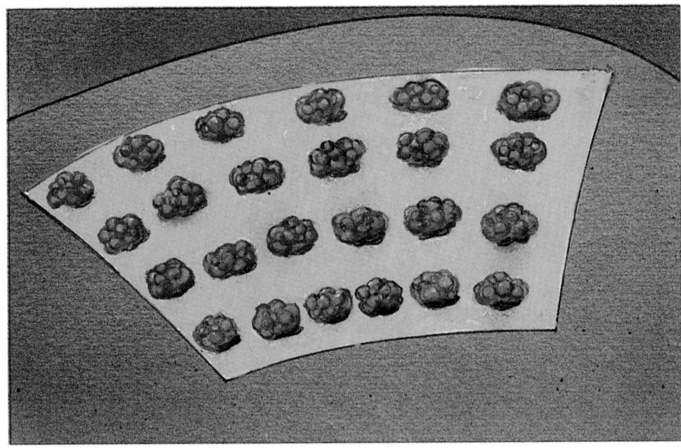

1. Chlorophyll molecules are arrayed in clusters as solar antennae.

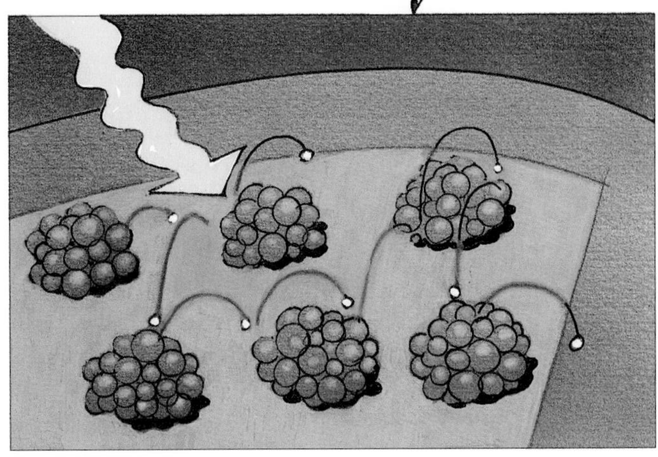

When sunlight strikes them, their electrons jump to higher-energy orbits, bouncing around until...

2. ...a special chlorophyll-enzyme combination transfers them to carrier molecules in the thylakoid membrane.

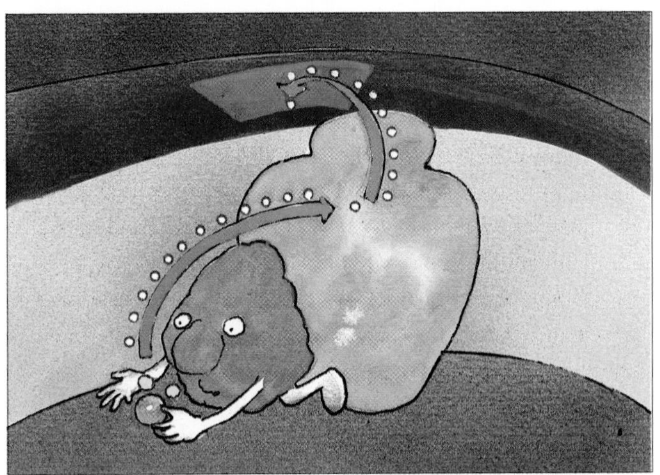

3. The electrons removed from chlorophyll are replaced by electrons from water (see the closeup — next box)...

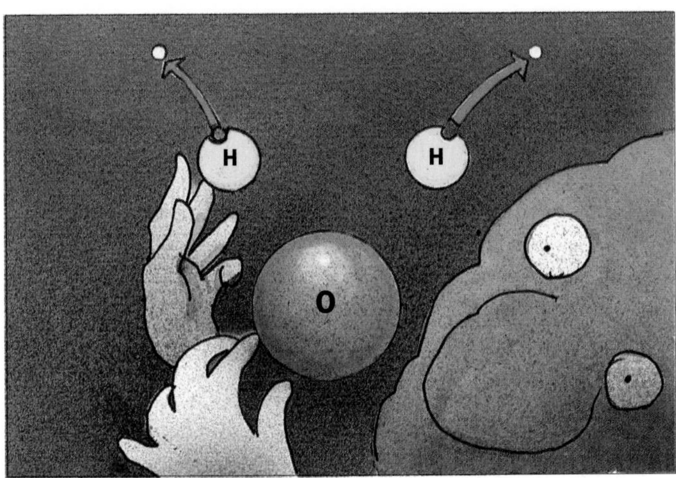

...provided by a water-spitting enzyme that divides water into two electrons, two hydrogen ions, and one oxygen atom.

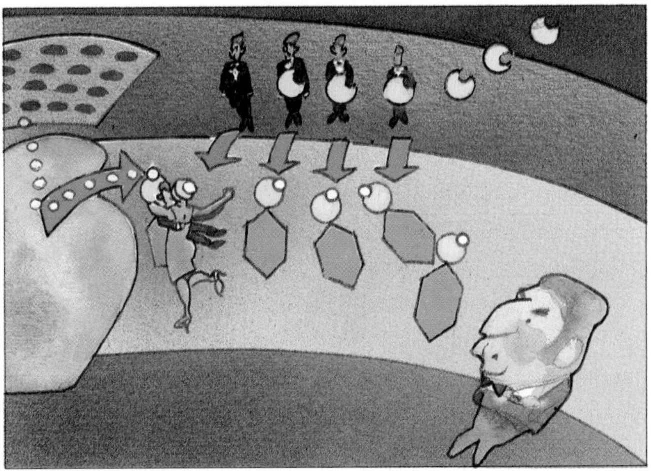

4. The electrons on the carriers attract hydrogen ions (H^+) from outside the sac, since opposites attract. Recall that a hydrogen ion plus an electron equals a hydrogen atom.

The Details

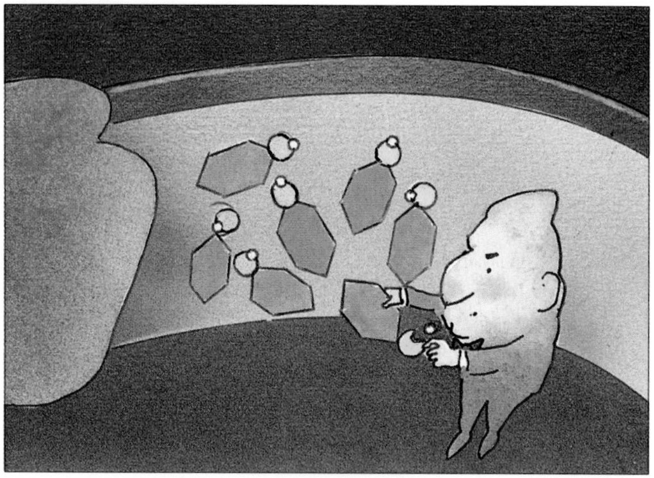

5. As the carriers reach the inner surface of the membrane, the bouncer enzyme grabs the ions...

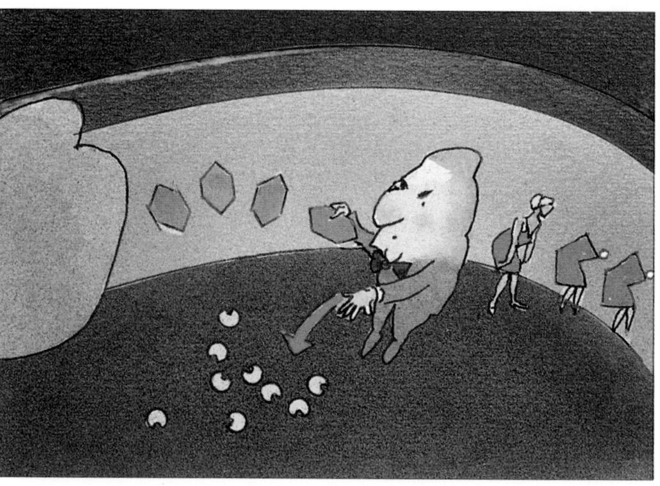

...and throws them into the sac, leaving the spent electrons to continue on, attached to a new carrier.

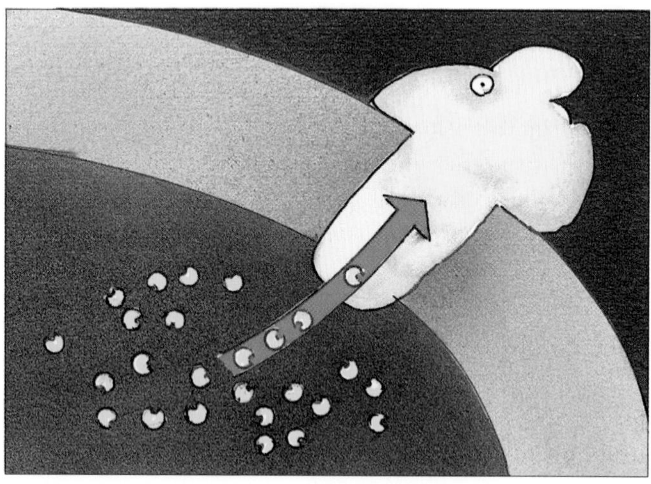

6. The only way out is via a channel through an ATP-making enzyme. The movement through this enzyme...

...supplies the energy to reattach phosphates to spent ATP molecules.

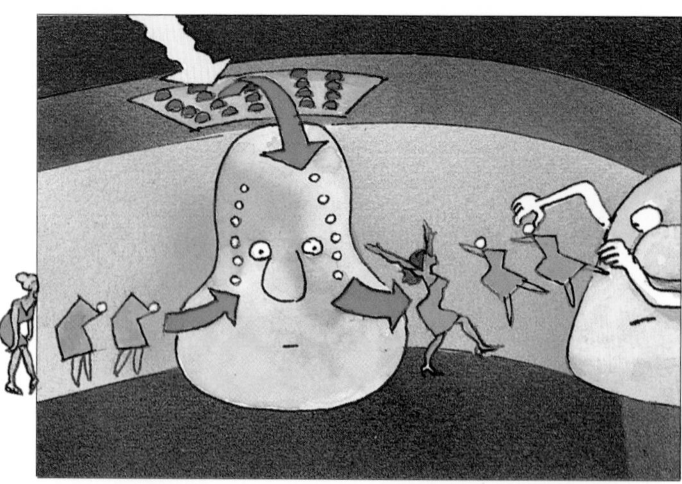

7. The spent electrons, stripped from their ions, replace electrons bounding off a new set of chlorophylls energized by sunlight. Restored chlorophylls are ready to do more work.

8. A final "hot" hydrogen enzyme transfers each energized electron, with a hydrogen ion, to a final carrier, NADP.

The Details

9. The action now moves to the stroma — the space in a chloroplast outside the sac. Here begins the Calvin Cycle, in which five enzymes cooperate to assemble half-molecules of sugar, using the ATP and the "hot" hydrogens the chloroplasts just made.

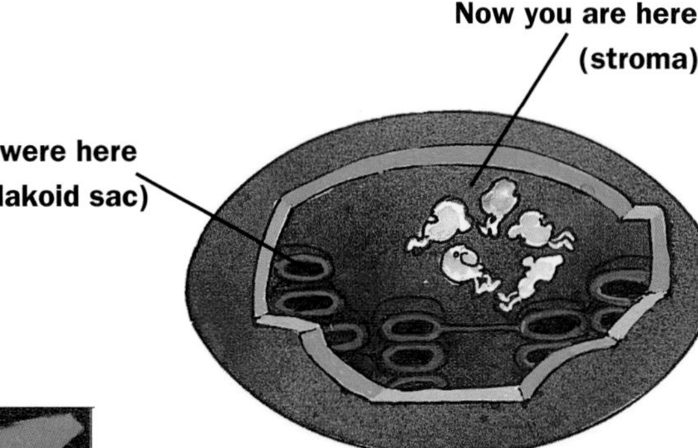

Chloroplast

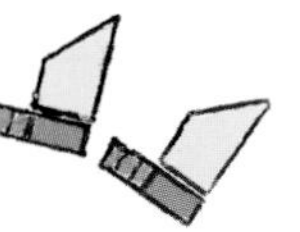

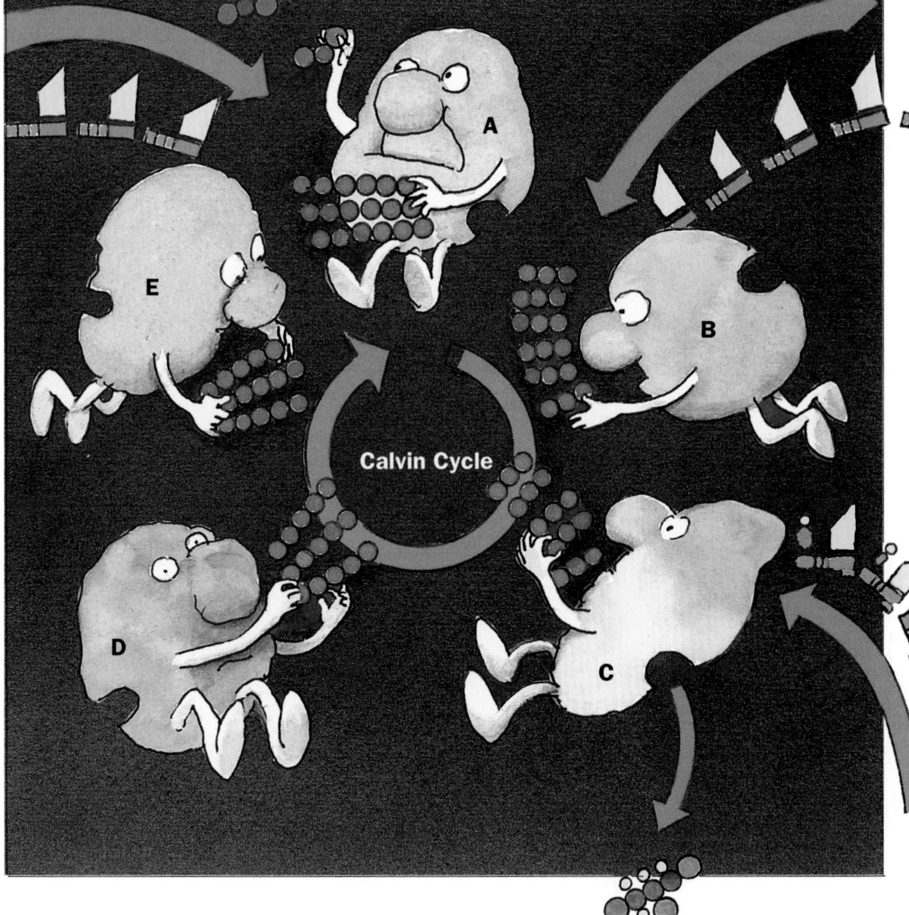

More sub-robots at work

Enzyme A attaches three carbon dioxides to each of three five-carbon sugars. (The oxygens are not shown in the cycle.)
The resulting three six-carbon sugars break into six three-carbon sugars.
Enzyme B energizes the three-carbon sugar fragments with ATP.
Enzyme C attaches hydrogens to the six sugars and kicks one off the assembly line.
Enzyme D rearranges the five remaining three-carbon sugars to make three five-carbon sugars.
Enzyme E energizes these with ATP...
...and the cycle is ready to repeat.

Making Sugar Out of Thin Air

Perhaps the most startling thing about life is that it turns air, energy, and water into living substance. A team of five enzymes initiates this feat by transforming carbon dioxide into sugar. The enzymes each make small changes in the product molecules as they pass them around. At several key points, they use energy supplied by ATP. The "hot" hydrogens they need are on NADP. An interesting feature of this kind of enzyme-catalyzed cycle is that it always needs some of the product it's making in order to make more of that product. In this case, for every six sugar fragments that travel around one turn of the Calvin Cycle, only one actually rolls off the assembly line as a finished product. The other five get recycled because they are needed to initiate the first step of the cycle. A production line that makes six products only to send five of them back into the process might seem inefficient. But the enzymes work incredibly fast, producing thousands of product molecules (energized sugars) per second.

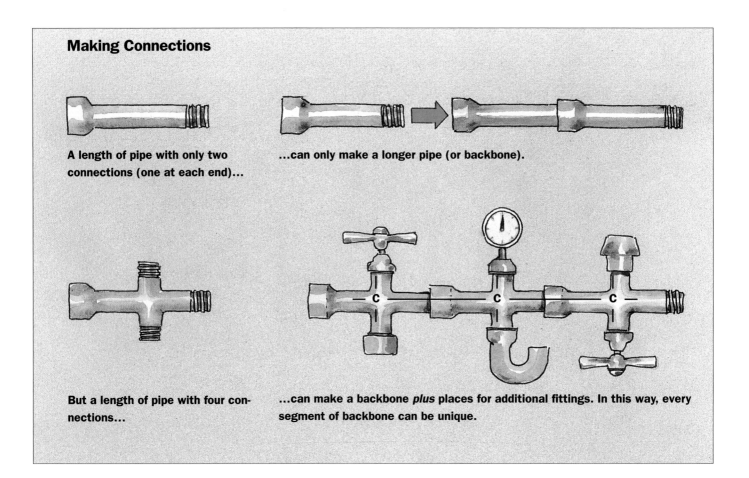

Making Connections

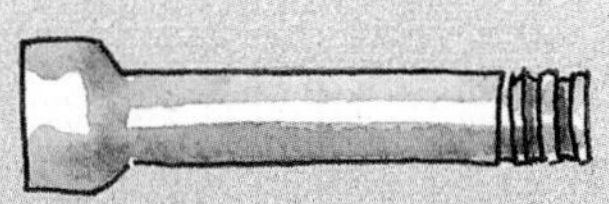

A length of pipe with only two connections (one at each end)...

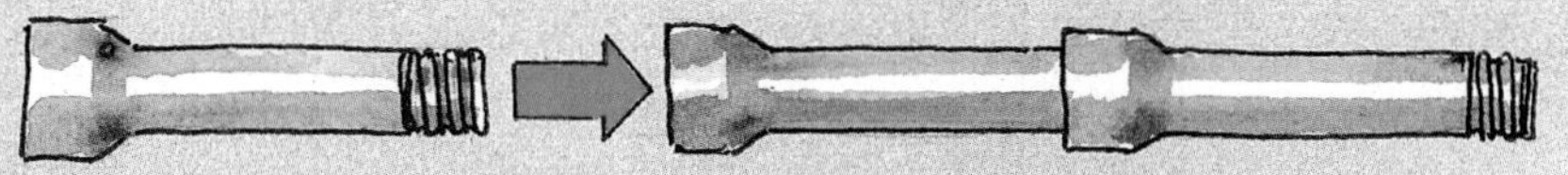

...can only make a longer pipe (or backbone).

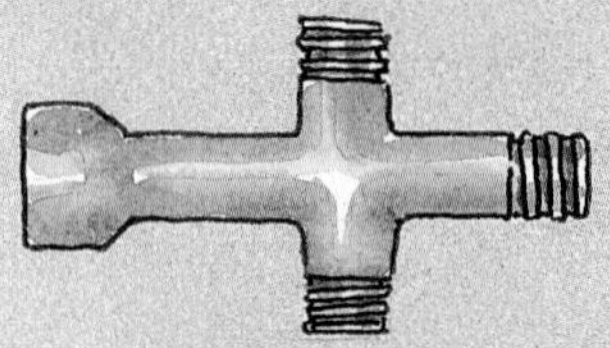

But a length of pipe with four connections...

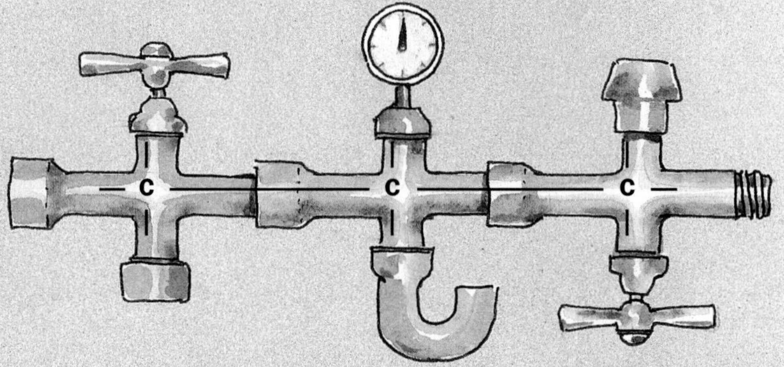

...can make a backbone *plus* places for additional fittings. In this way, every segment of backbone can be unique.

3.11 Three Cheers for Carbon

Carbon plays so central a role in life that we say life is "carbon based." Carbon's place of honor stems from its unique ability to make four separate bonds with other atoms (i.e., share four of its electrons with other atoms). Oxygen can make only two bonds, and hydrogen one. You can see the value of carbon's extra bond-making capacity through the above plumbing analogy. Pipe lengths that have two connections, one at either end, can fit together to make a longer pipe but nothing else. Pipe lengths that have three or four connections can be joined to make a longer pipe with joints for additional fittings sticking out at right angles. Carbon plays a similar role in life's long-chain molecules by enabling a molecule to add both length to its backbone and connections for side groups (though carbon's connections aren't at right angles). The backbone establishes the chain; the side groups give the chain its unique chemical character and informational value.

About Sugar

The eventual products of the Calvin Cycle are three-carbon sugars. This marks the end of photosynthesis; but it's not the end of the line for the sugar fragments. At this point, they are shuttled through the chloroplast's outer membrane into the cell's interior, where enzymes bond them together in pairs to make the six-carbon sugar called glucose. Glucose travels through cells in a host of modified forms — in such disguises as sucrose, ribose, lactose, cellulose, starch, and glycogen. Glucose provides all the energy and almost all the building materials life needs.

Van Helmont's Experiment

DOING *Science*

Before the early 1600s, people assumed that the substance of plants — roots, trunk, branches, and leaves — came from the earth in which they grew. In 1630, Jean Baptista van Helmont, a Flemish physician, did a simple experiment: He planted a willow branch weighing 5 pounds in 200 pounds of soil. Five years later, after regular watering, the branch had gained 165 pounds and the soil had lost only 2 ounces! Van Helmont concluded reasonably that the material in the tree couldn't have come from the earth — that it must have come from the water. He was right in the first conclusion, but only half right in the second. It wasn't known at the time that much of the substance of life is carbon, and it didn't occur to van Helmont that the air might have been the source of the material in the tree. His experiment was notable, though, because he dealt with quantifiable information: By carefully weighing things, he'd at least been able to rule out earth as the source of the material. Truth comes in small pieces.

Later in life, van Helmont got interested in the gas that's produced when wood is burned. He called it "gas sylvestre" (gas from wood) but never realized it was the very carbon dioxide that had been used to make his willow tree.

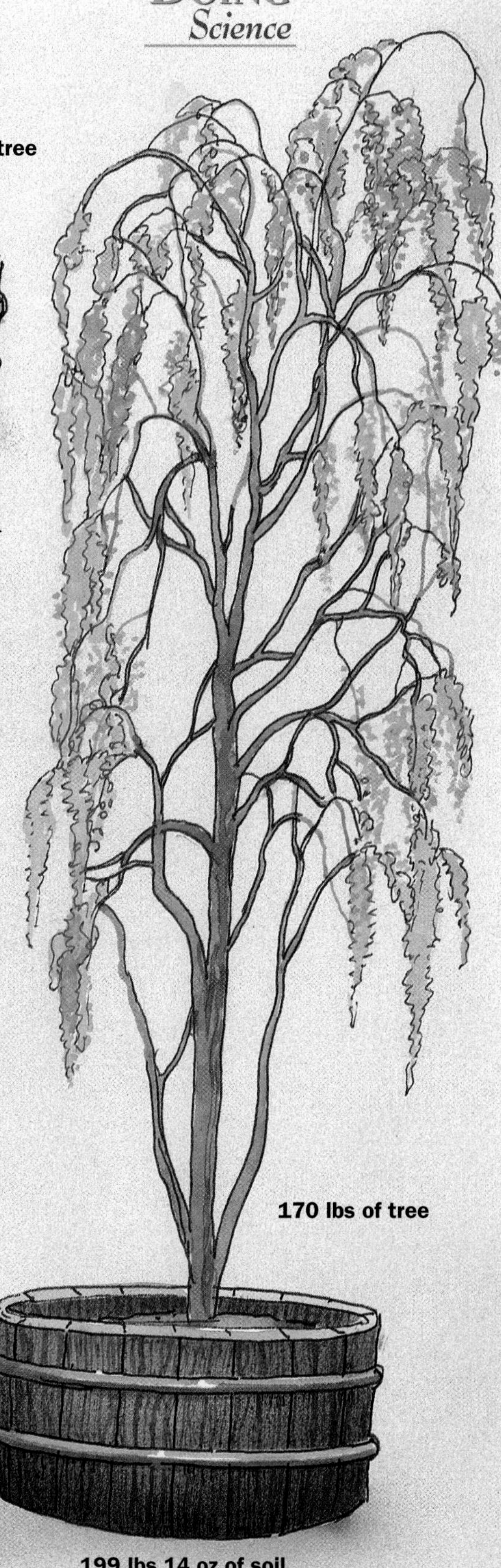

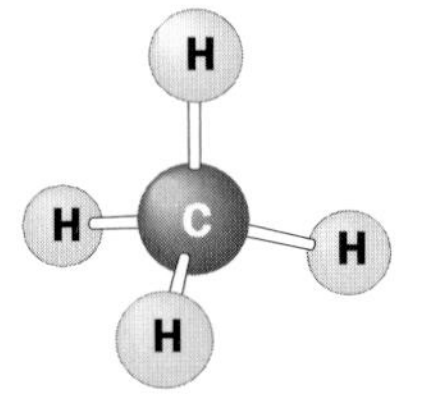

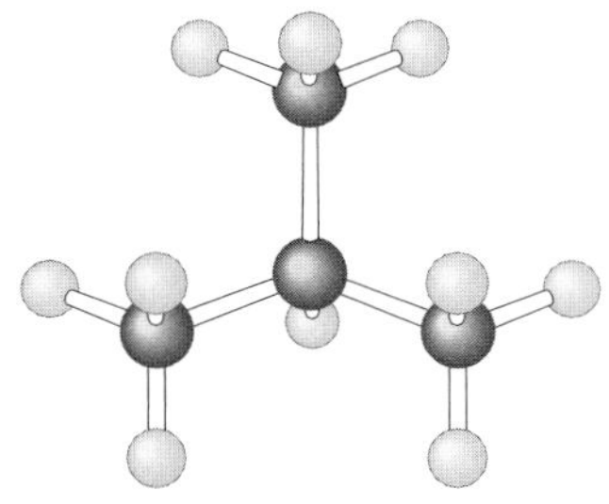

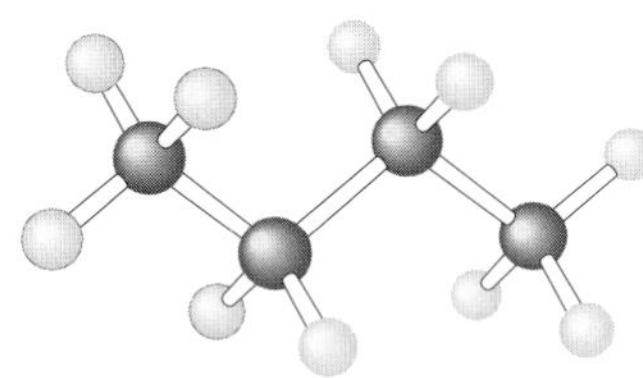

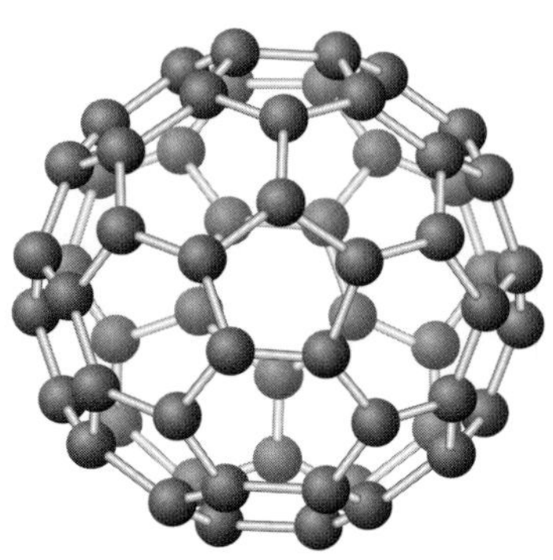

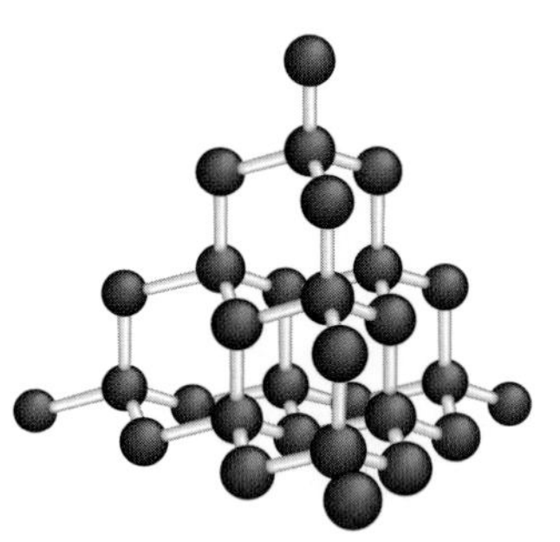

ATOMIC ARCHITECTURE

Take Carbon for example then
What shapely towers it constructs
To house the hopes of men!
What symbols it creates
For power and beauty in the world
Of patterned ring and hexagon —
Building ten thousand things
Of earth and air and water!
Pride searches in the flues of earth,
For the diamond and its furious sun,
Love holds its palms before the glow
Of anthracite and purrs.
Five senses take their fill
Of raiment, rainbows and perfume,
Of sweetness and of monstrous pain.

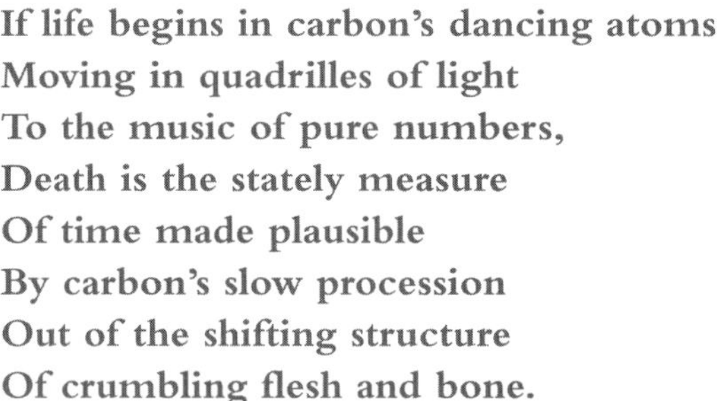

If life begins in carbon's dancing atoms
Moving in quadrilles of light
To the music of pure numbers,
Death is the stately measure
Of time made plausible
By carbon's slow procession
Out of the shifting structure
Of crumbling flesh and bone.

A. M. Sullivan, *Stars and Atoms Have No Size*, 1946

More About Carbon: The Modular Components of Atomic Architecture

Each carbon atom's four possible bonds to other atoms allow an enormous variety of molecular shapes: rings, spheres, lattices, straight or branching chains. When you add the possibility of another level of structure, such as chains made of connected rings, and remember that every bond between atoms is a storehouse of energy, you can see why carbon-based molecules have such a huge potential for energy storage.

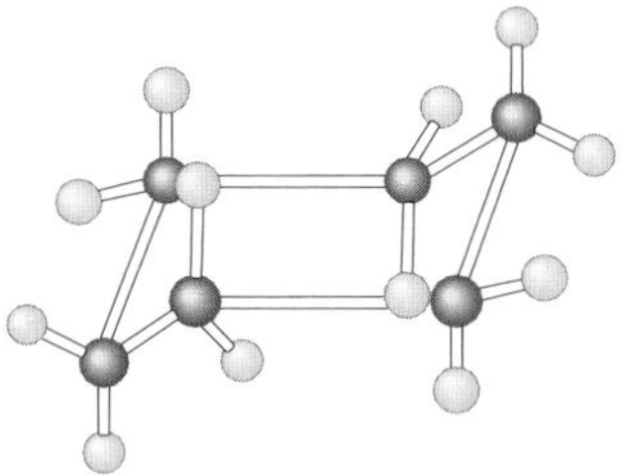

Hexane

Photosynthesis makes the two halves of hexagonal glucose molecules. A hexagonal ring of carbons with only hydrogens attached is hexane, shown above left. Add a few oxygens and rearrange things a bit, and you get a glucose molecule (left, with the atoms now represented by letters: C = carbon, H = Hydrogen, O = Oxygen). From glucose molecules, plant cells construct *chains* — starch (below) for efficient energy storage, or cellulose for structural use. When you light a wood fire, you're jump-starting the release of energy from the wood's cellulose chains, and you warm yourself with the heat released as that stored energy dissipates.

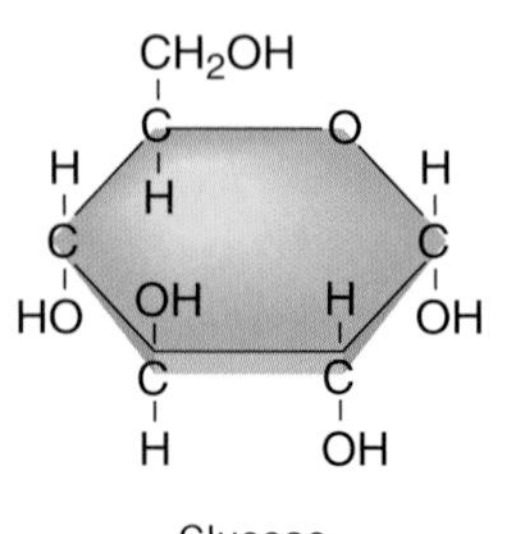

Glucose

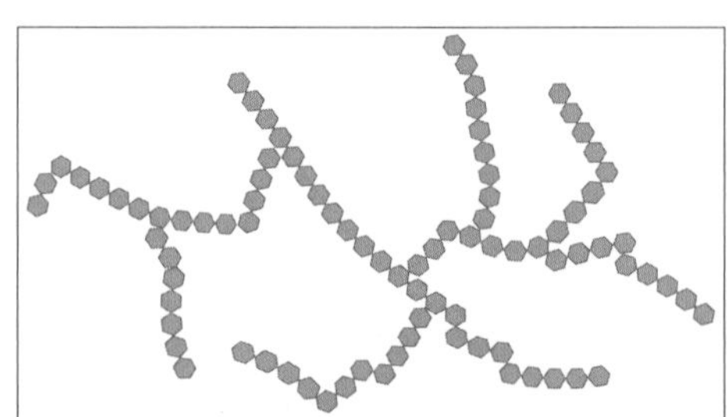

Starch

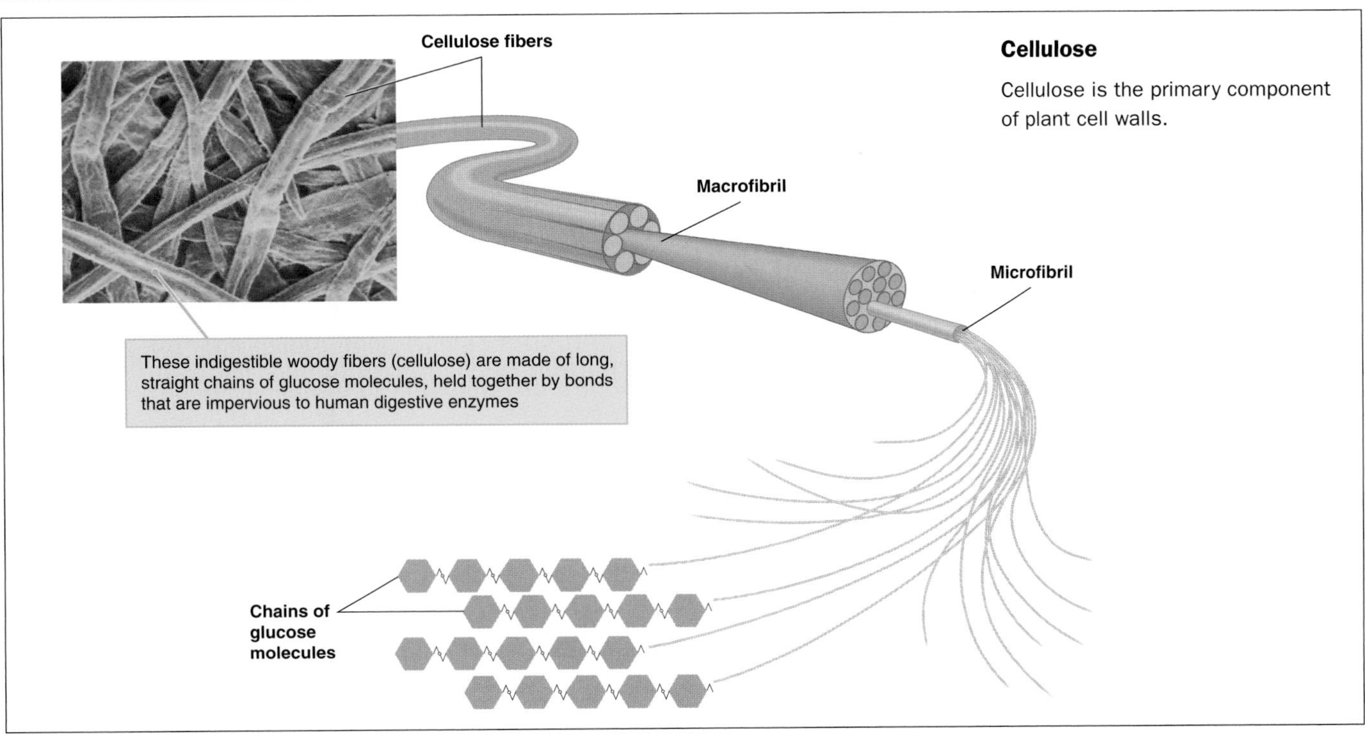

Cellulose

Cellulose is the primary component of plant cell walls.

Starch can be converted into other energy-storage molecules (fats and glycogen) or into functional molecules (amino acids and nucleotides). Fats are highly efficient energy-storage molecules made of carbon chains. Glycogen, at right, is an elaborately branched chain of glucose molecules used by animals to store energy for rapid supply to muscle tissue.

Carbon's importance as a unit of life's molecular modules is demonstrated by the fact that it makes up only 0.03 percent of the earth's crust but 18.5 percent of the human body mass. The only element more prevalent in the human body is oxygen, which is found there mainly as a component of water.

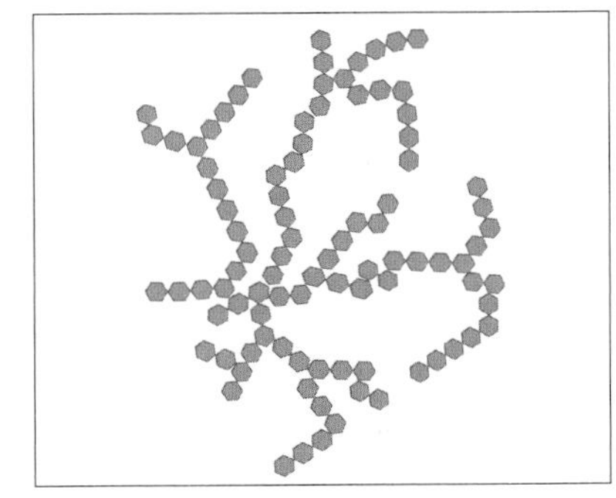

Glycogen

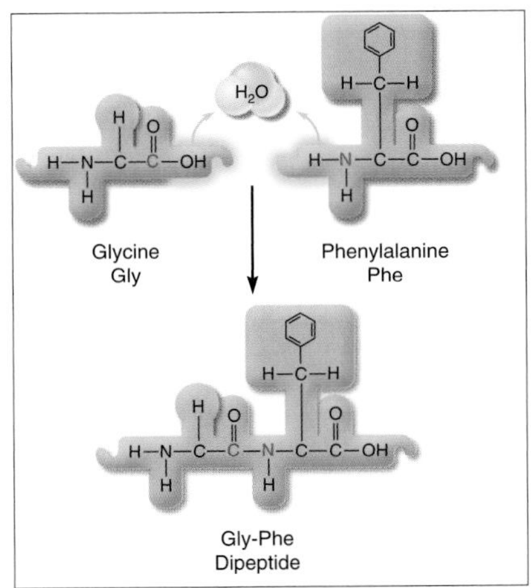

▲ Building a backbone

When amino acids link up to form a protein each link is a carbon-nitrogen bond. Amino acids are the protein's backbone. Similarly, in DNA and RNA, the phosphate group of one nucleotide links to the sugar of the next nucleotide to form a backbone.

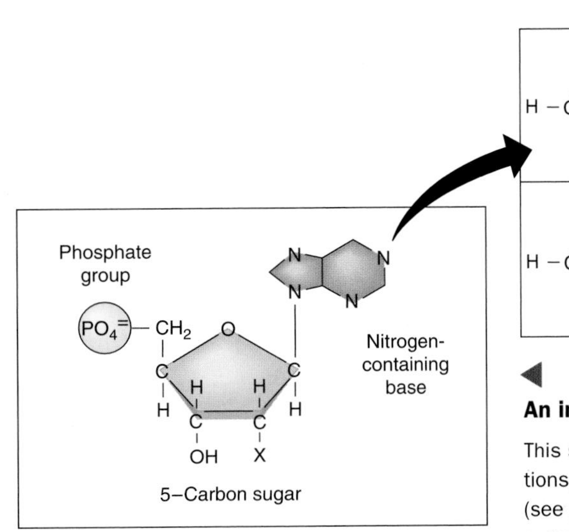

Sugar phosphate DNA or RNA backbone

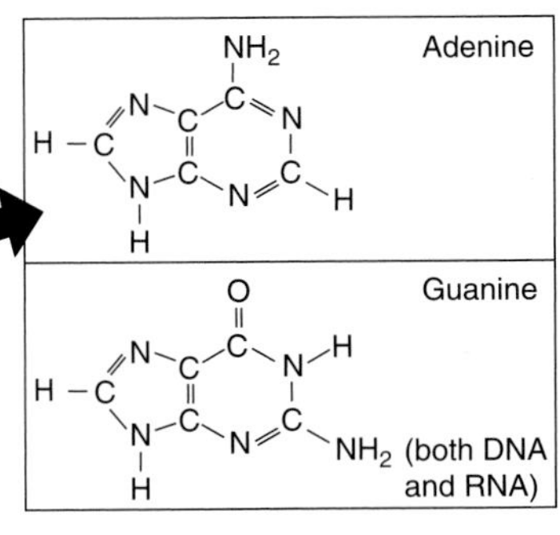

◀ An important building block

This structure, with very slight variations, is the module for both ATP (see page 43) and for the repeating sugar-nucleotide units that make up DNA and RNA.

3.12 Respiration — Breaking Down Sugar to Make ATP

Slow Burning

Making ATP is like burning wood. When you burn real wood, you make hydrogen- and carbon-rich material, break its bonds, and combine the pieces with oxygen to produce carbon dioxide, water, and heat. When sugar is burned in the mitochondria of all cells with nuclei, its bonds are broken down and combined with oxygen to make carbon dioxide and water. But, in the mitochondria of plants and animals, half of the energy is released as heat, and the other half ends up in the bonds of ATP. To accomplish this, enzymes "manhandle" the sugar to extract its hydrogen atoms. The electrons stripped from these hydrogens then flow along the mitochondrion's membranes, ultimately producing ATP.

So here's the sequence: Enzymes manipulate food (fragments of sugar molecules), extracting energetic hydrogens (1). They pass the electrons from these hydrogens along a series of carriers in a membrane inside the mitochondrion, picking up hydrogen ions as they go (2). Enzymes along the way separate the ions from the carriers and eject them into the cristae (intermembrane spaces) (3). The accumulating ions force their way through an ATP-making enzyme (4). Finally the spent electrons combine with hydrogen ions and oxygen to produce water (5).

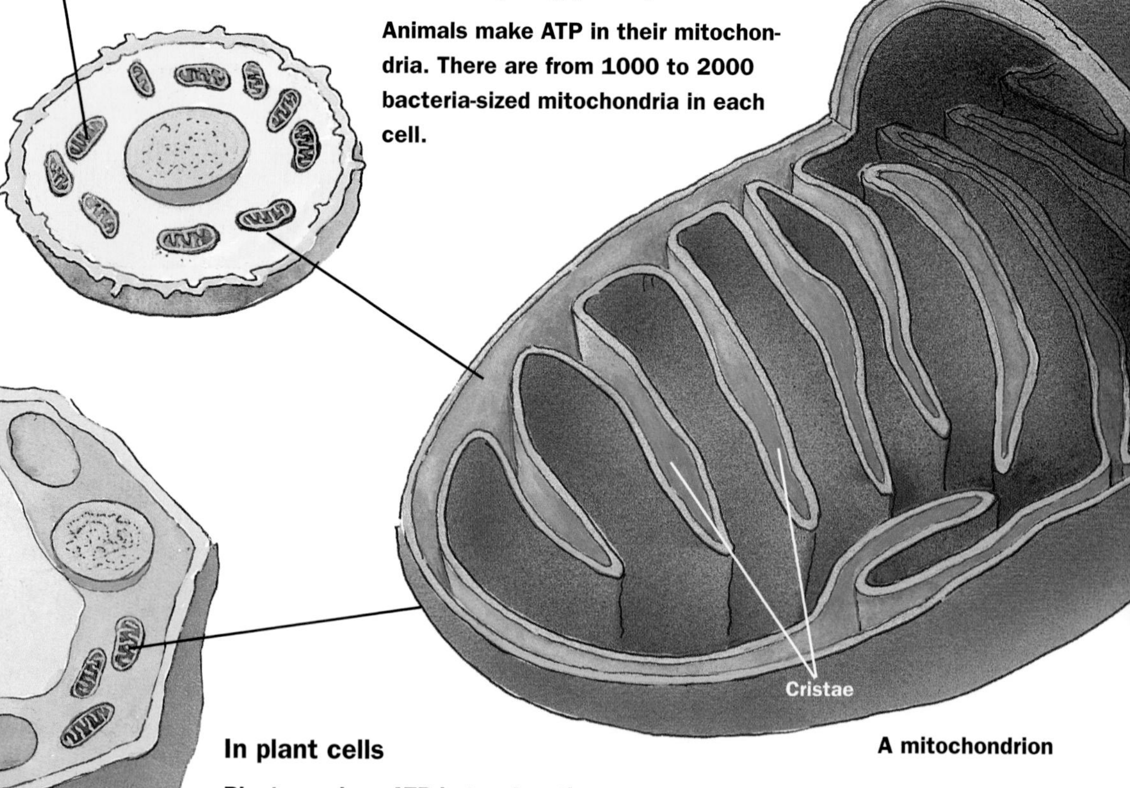

In animal, fungi, and protist cells

Animals make ATP in their mitochondria. There are from 1000 to 2000 bacteria-sized mitochondria in each cell.

A mitochondrion

In plant cells

Plants produce ATP in two locations: in chloroplasts for sugar-making and in mitochondria for everything else.

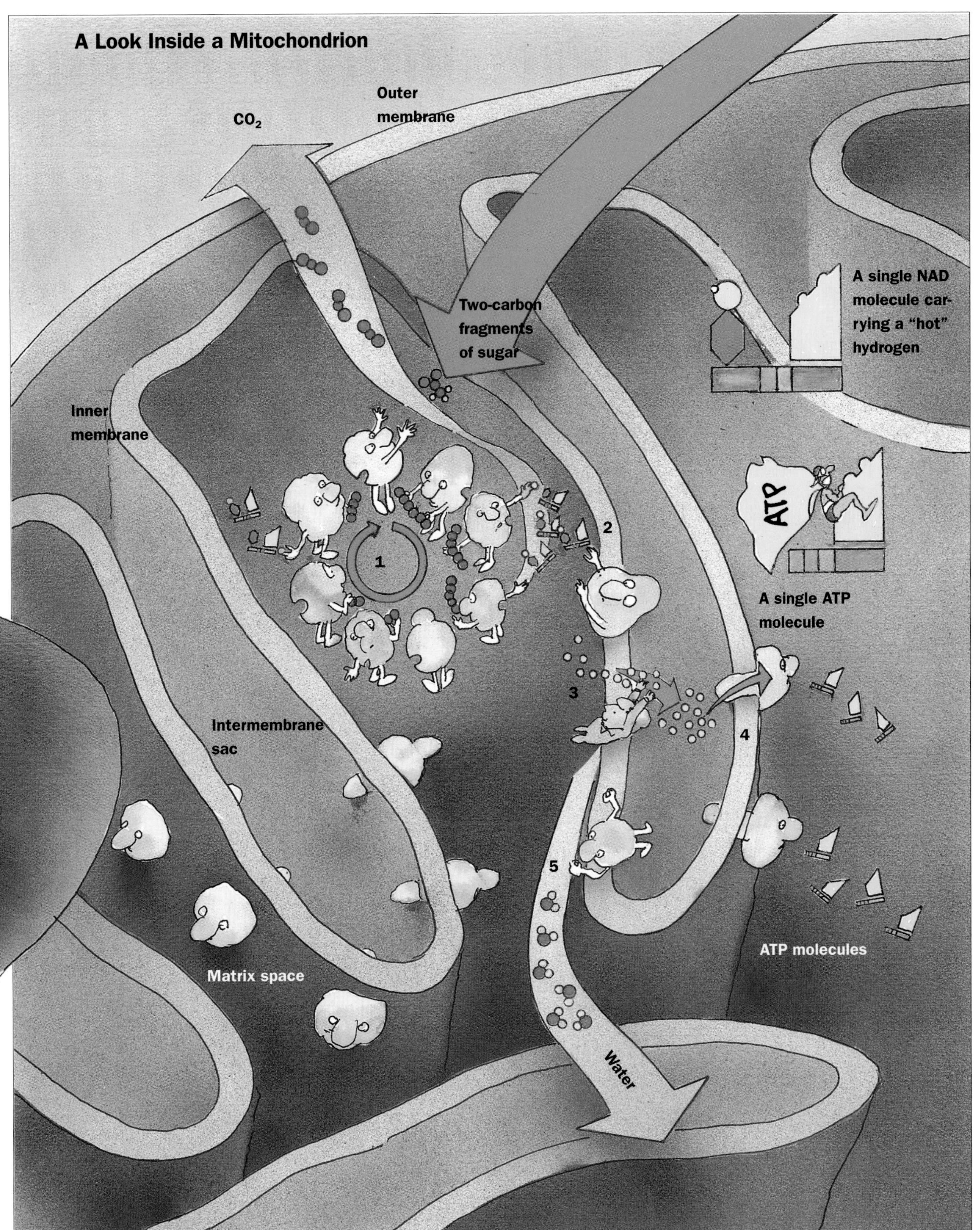
A Look Inside a Mitochondrion
CO_2
Outer membrane
Two-carbon fragments of sugar
A single NAD molecule carrying a "hot" hydrogen
Inner membrane
ATP
A single ATP molecule
1
2
3
4
5
Intermembrane sac
Matrix space
ATP molecules
Water

STEP BY STEP

Respiration in a mitochondrion is similar to photosynthesis in a chloroplast — but almost in reverse. Both processes involve a cycle of enzymes manipulating molecules, and a flow of electrons in a membrane. The numbers on the map at the right correspond to the numbered illustrations on pages 125 and 126.

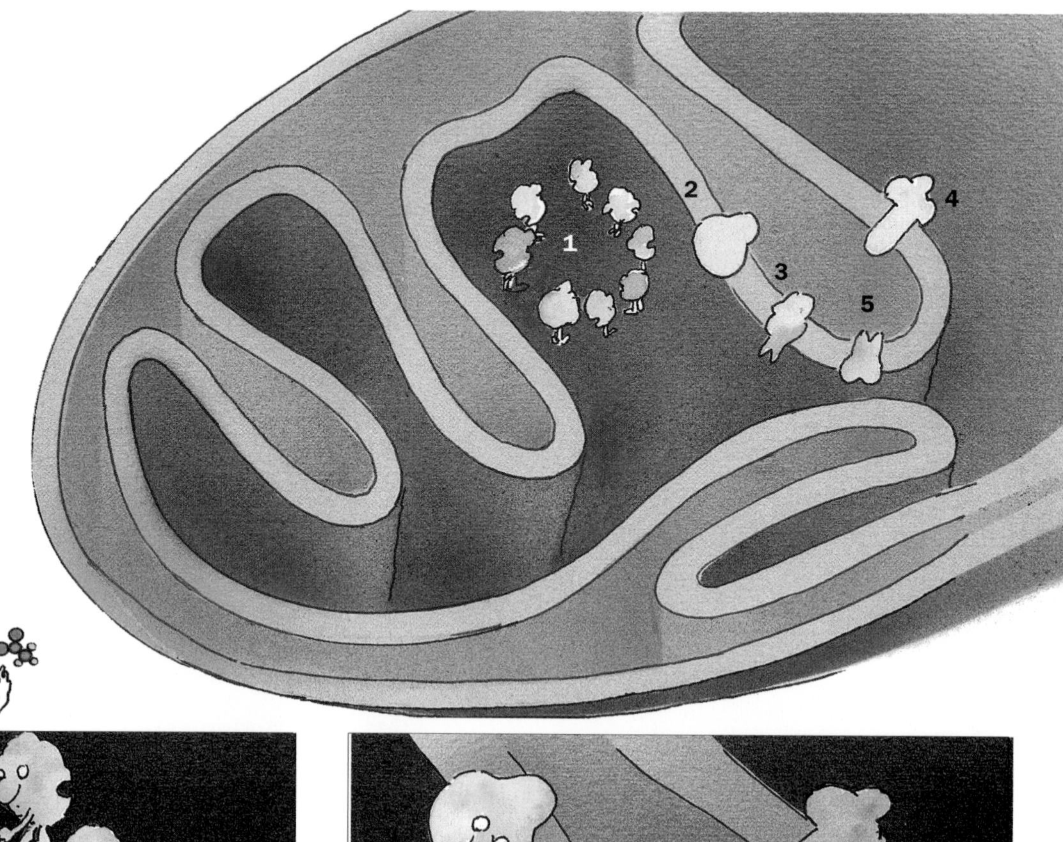

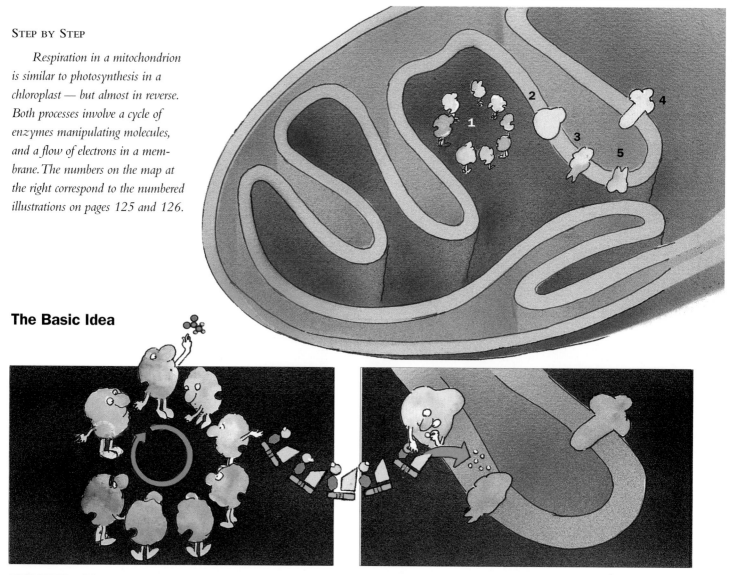

The Basic Idea

REMOVING "HOT" HYDROGENS: Enzymes remove hydrogens from sugar, putting them on carriers (NAD, which thus becomes NADH).

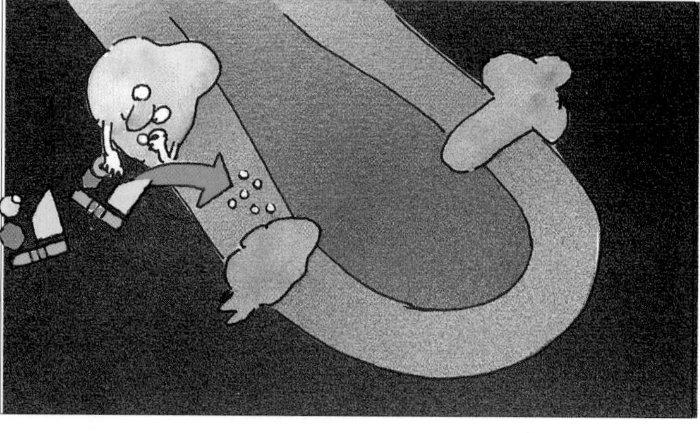

STARTING AN ELECTRON FLOW: Another enzyme strips electrons from the "hot" hydrogens.

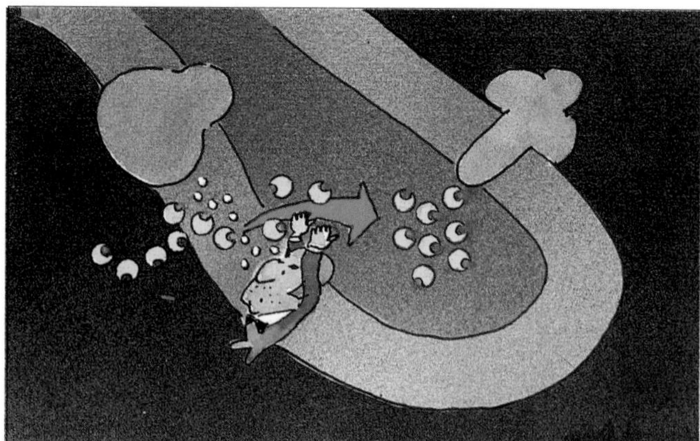

PUTTING IONS IN A SAC: The flowing electrons attract hydrogen ions, which get pushed into the intermembrane sac in the cristae.

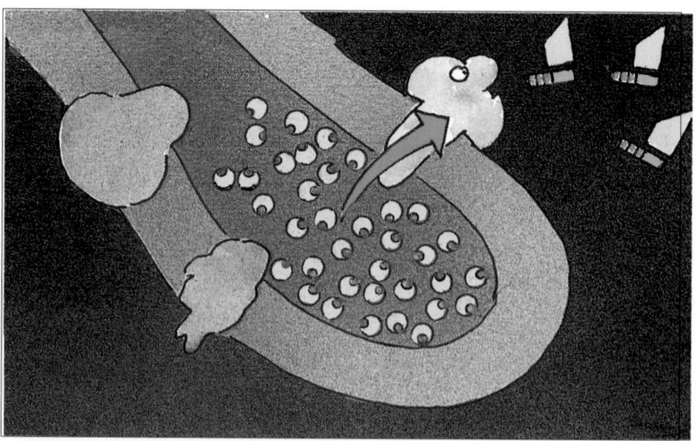

PRODUCING ATP: The force of escaping ions channeling through an enzyme produces ATP.

The Details

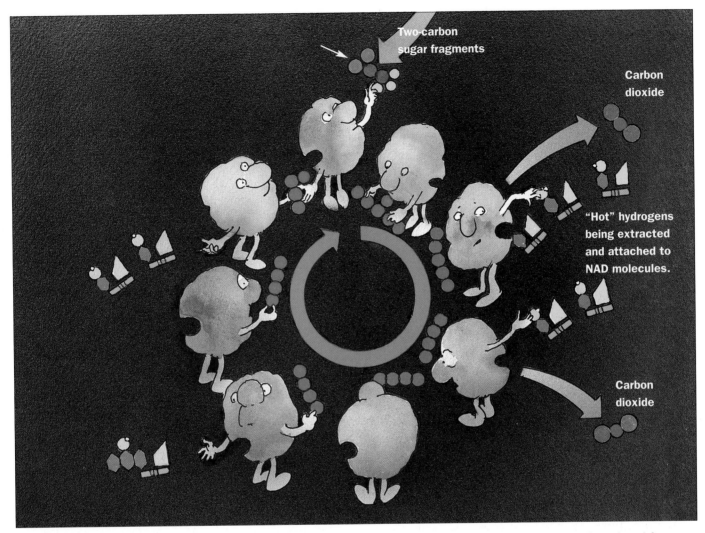

Two-carbon sugar fragments

Carbon dioxide

"Hot" hydrogens being extracted and attached to NAD molecules.

Carbon dioxide

1. Sugar enters the mitochondrion as two-carbon fragments (top), having been broken off from glucose (six carbons) in a process called glycolysis (see page 130). Several enzymes then manipulate these fragments to extract "hot" hydrogens at several points in the cycle. Carbons and oxygen combine and are discarded as carbon dioxide — which animals exhale.

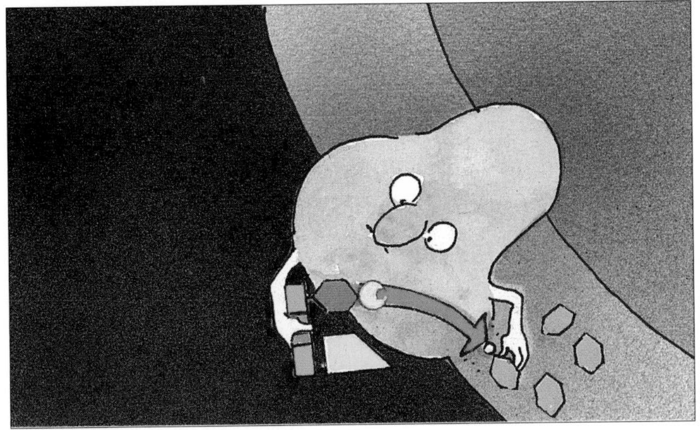

2. "Hot" hydrogens, passing in an endless stream, give up their electrons to an enzyme in the inner membrane...

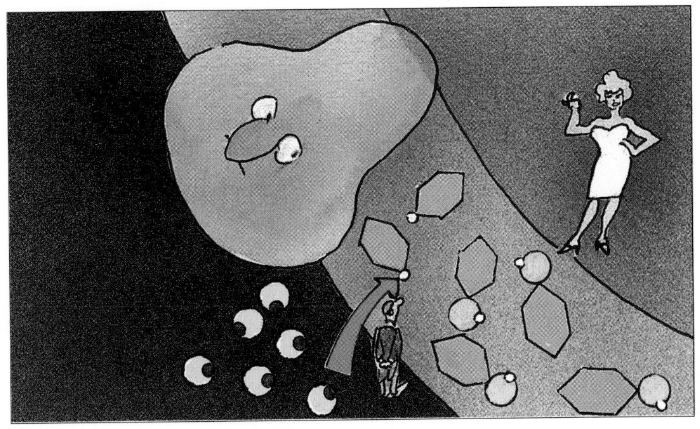

...which passes the electrons to carrier molecules floating in the membrane. Each electron is picked up by a hydrogen ion (making it a hydrogen atom on the carrier).

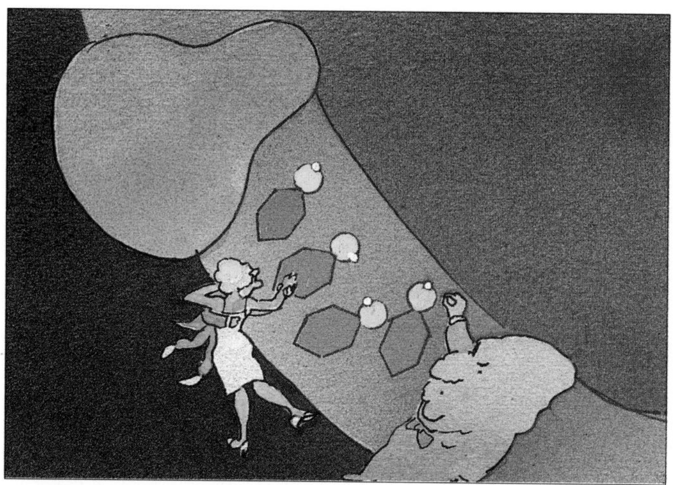

3. The carriers randomly "dance" the energized hydrogens over to the next enzyme...

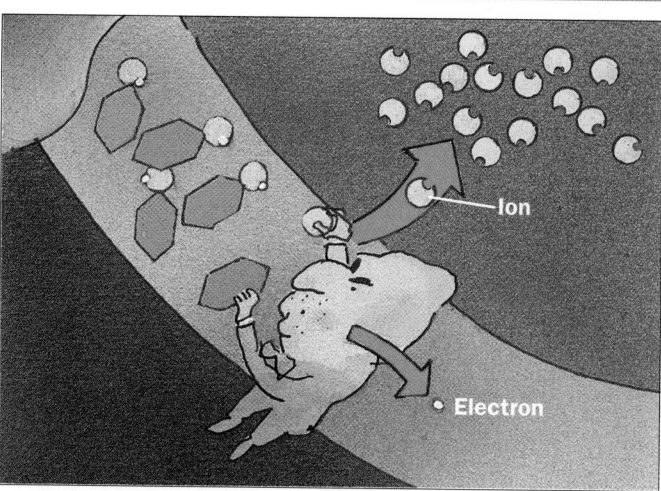

...which takes the electrons off the carriers and passes the ions into the intermembrane sac.

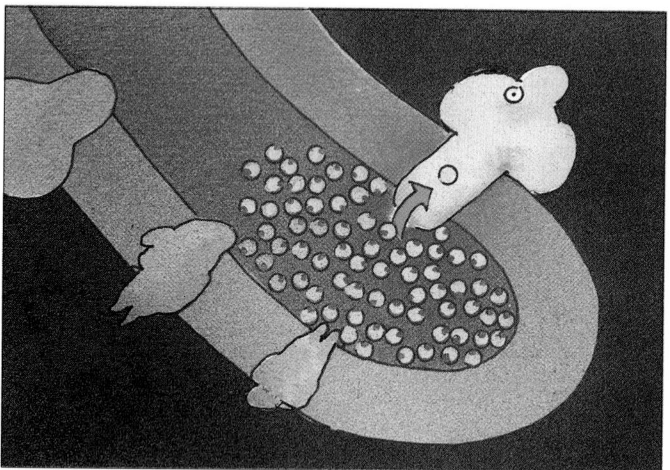

4. As the ions accumulate in the intermembrane sac, some of them escape through ATP-making enzymes.

The ions, moving through the enzyme, energize it, and it can then reattach phosphates to spent ATP molecules...

...so a steady stream of rejuvenated ATP molecules emerges, ready to supply the energy for the work of the cell.

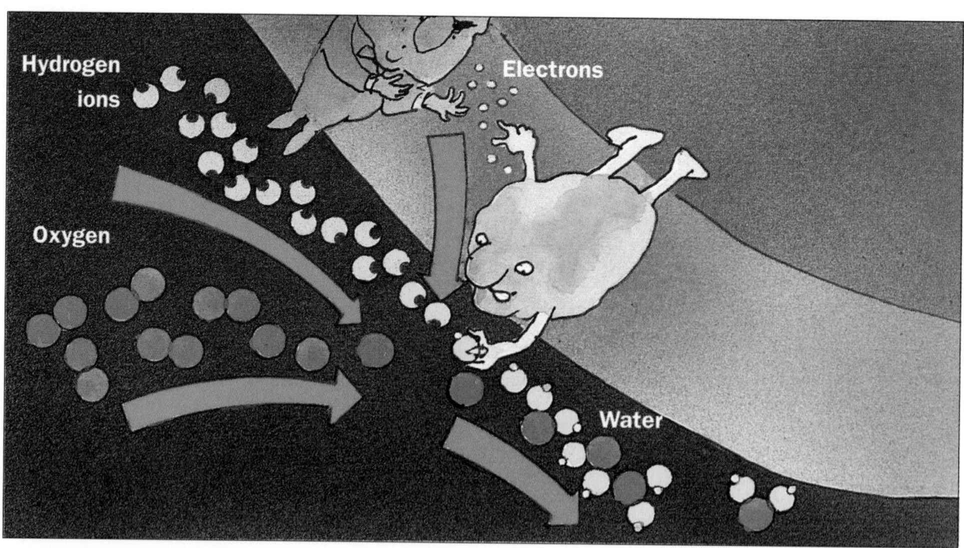

5. Finally, a third enzyme combines the spent hydrogen ions and oxygens to form water — a byproduct of the process.

Let's Hear It for Oxygen

As oxygen slowly increased in the Earth's atmosphere, it was poisonous to most organisms because it produced ionized molecules that could damage DNA. But, as often happens in evolution, adverse conditions create new opportunities. The "respirers" — simple bacteria that adapted to oxygen by evolving various ways to neutralize the unwelcome ions — thrived spectacularly.

Oxygen greatly expands an organism's ability to produce energy. A "fermenter" — a microorganism that doesn't use oxygen (see page 130) — can get only two ATPs out of one glucose molecule. A respirer can get at least twenty! With this advantage, the respirers, which include many bacteria and virtually all multicellular organisms, have come to dominate the living world.

There's one odd thing about oxygen's role in respiration: It is *hydrogen* — actually its electrons and ions — not oxygen, that an organism needs to produce energy. But the organism must have a way to dispose of its steady stream of spent hydrogens. That's where oxygen comes in, combining with the leftover hydrogen to make water (H_2O). So oxygen, on which our lives so totally depend, is not really in the show: it arrives at the stage door when the play's over, to pick up the exhausted performers.

DOING Science

Oxygen Discovered

Since earliest times, flames arising from burning material were seen as evidence that something essential was being released. This something became known as "phlogiston" in the 1700s. Scientists found that if they burned material inside an enclosed space, the flames would soon gutter out. What's more, the air inside the enclosure could no longer sustain an animal's life. It appeared that accumulated phlogiston inhibited both fire and life. Scientists found they could rejuvenate this "phlogisticated" air by putting plants in it — if the plants were exposed to sunlight. The plants somehow counteracted the effects of phlogiston.

The great French chemist Antoine Lavoisier decided to investigate the nature of phlogiston. In the 1780s, he carefully measured the amounts of everything involved in burning a given material. He showed, for instance, that when he set fire to a piece of metal it melted but actually gained weight; moreover, this increase in weight was exactly equal to the loss in weight of the surrounding air. (When wood is burned the ash is, of course, lighter than the original log. This is because, unlike metal, wood's cellulose combines with oxygen to form carbon dioxide and water which escape as part of the smoke. Add up the weights of the ash and the escaped gases, and they, too, would exceed the weight of the wood.)

Lavoisier later named the gas in the air the combined with the metal: oxygen. It all became clear. Air, depleted of oxygen by fire, can no longer sustain fire or life because these processes require oxygen. The photosynthesizing plant rejuvenates the air by releasing oxygen into it.

Keeping Those Fleas Jumping from Left to Right

In life, oxygen is what keeps the fleas jumping from dog to dog, so to speak. Without oxygen to pick up the leftover hydrogens from ATP-making reactions, the whole process of respiration would grind to a halt — and so would the living world.

Oxygen originates in water. It's the O in H_2O. Photosynthesis splits H_2O to 2Hs and an O, uses the Hs to make sugar and discards the O. Respirers strip Hs from sugar, use them to make ATP, then combine the Hs with O to make H_2O. Thus the H_2O to H_2O cycle comes full turn.

Call me oxygen

As we breathe, oxygen in the air diffuses across the cell membranes within our lungs and from there into invisibly small capillaries. Once in the bloodstream, oxygen crosses the membrane of a red blood cell, binds to one of the hemoglobin molecules inside, and is carried to every cell in the body. As the oxygen picks up the spent hydrogen ions that energized ATP production, water re-forms. At the cellular level, plants respire exactly the way animals do, but since they don't have to make the same large amounts of ATP, they don't use up all the oxygen they produce. That leftover oxygen is the breath of life for the rest of us.

Opening up a wider view of molecular exchanges, the illustration at right depicts the cycle of sugar, water, oxygen, and carbon dioxide through producers, consumers, and decomposers in land-based ecosystems. In general, this cycle of materials is the same in oceans, lakes, and streams, except that the producers there are mainly algae, a group of protists that includes species as large as hundred-foot-long kelp and as small as the microscopic green algae and photosynthetic bacteria called phytoplankton. It is estimated that phytoplankton produces 70 percent or more of Earth's oxygen supply.

Phytoplankton is also the major food source for tiny consumers called zooplankton. In general, planktonic organisms (zooplankton and phytoplankton) are the food base for virtually all other aquatic creatures. The reproductive success of phytoplankton is, therefore, of vital importance not only to aquatic life but to terrestrial consumers of oxygen, like us.

Question.

This photograph of a branch of the common waterweed *Elodea canadensis* was taken in bright sunlight. What are the hundreds of bubbles surrounding the plant?

Answer...

The plant is photosynthesizing, using water and dissolved carbon dioxide and generating little bubbles of oxygen (as well as glucose molecules, which we can't see).

Magnified 3 times

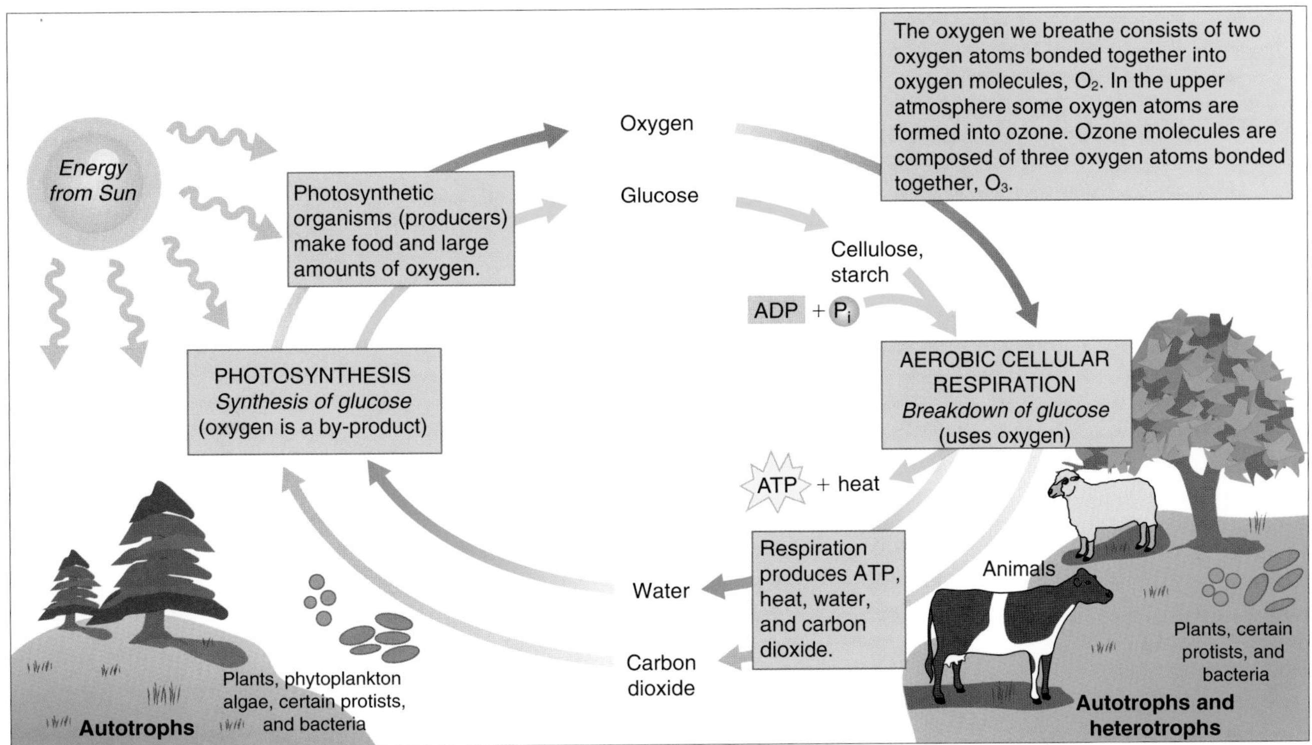
Energy from Sun
Photosynthetic organisms (producers) make food and large amounts of oxygen.
Oxygen
Glucose
The oxygen we breathe consists of two oxygen atoms bonded together into oxygen molecules, O_2. In the upper atmosphere some oxygen atoms are formed into ozone. Ozone molecules are composed of three oxygen atoms bonded together, O_3.
Cellulose, starch
ADP + P_i
PHOTOSYNTHESIS
Synthesis of glucose
(oxygen is a by-product)
AEROBIC CELLULAR RESPIRATION
Breakdown of glucose
(uses oxygen)
ATP + heat
Respiration produces ATP, heat, water, and carbon dioxide.
Animals
Water
Carbon dioxide
Plants, phytoplankton algae, certain protists, and bacteria
Autotrophs
Plants, certain protists, and bacteria
Autotrophs and heterotrophs

Recently, scientists have made the disturbing observation that the reproductive cells in some species of algae that live in Arctic and Antarctic seas are many times more vulnerable to ultraviolet radiation than the other algal cells. The ozone layer screens the Earth's surface from most of the ultraviolet radiation that reaches the upper levels of the atmosphere from the Sun. But, since we've been releasing chlorinated fluorocarbons (CFCs) into the atmosphere, holes have appeared in the ozone layer over the North Pole and Antarctica. These holes allow more ultraviolet rays to penetrate to the ocean surface. During the spring months, up to two-thirds of the ozone shield over Antarctica is destroyed. For algae, this is also the period of most active reproduction. So, the timing of the most intense penetration of ultraviolet radiation coincides with the period of maximum algal vulnerability – a highly destructive combination!

When ultraviolet penetration is high, some studies have shown a 6 to 12 percent loss of phytoplankton. This could well translate to a 4 to 8 percent loss of atmospheric oxygen. Look at the cycle of materials above to get an idea of the potential impact of such a loss.

3.13 Glycolysis

Energy Without Oxygen

In the Earth's ancient oceans, before anything could perform photosynthesis, organisms developed ways of getting energy from sugar-like materials *without the need for oxygen*. This operation was a primitive form of what we recognize today as glycolysis in animal cells or fermentation in microorganisms. In these processes, each molecule of glucose gets broken into smaller pieces by a series of enzymes, generating two ATP molecules. While this amount of ATP is much less than the twenty molecules that cells can make in their mitochondria by further breaking down the fragments of a molecule of glucose, it is an important emergency energy source for animals. For instance, when there's a sudden demand for heavy muscular work (as in running the 100-yard dash) and there's not enough time for oxygen to be delivered to the muscle cells by the blood, glycolysis supplies the necessary ATP.

A primitive form of glycolysis, consuming plentiful sugar-like materials in the ancient oceans before photosynthesis and oxygen became available, was probably one of life's earliest ways of producing usable energy.

In the plant cell ▶

The half-molecules of glucose made by chloroplasts are paired to make glucose in the cytoplasm, and stored in other forms such as sucrose and starch. The glucose molecules are broken down as needed into two-carbon pieces by glycolysis. The pieces are further broken down in plant mitochondria to make ATP, which supplies the energy for the plant's activities.

In the animal cell ▶

When organisms eat plants, liberated glucose molecules are broken down into two-carbon pieces by glycolysis, and further broken down in mitochondria to make ATP, which supplies the energy for the organism's activities.

Pasteur's Wine

DOING *Science*

Interest in fermentation — the breakdown of sugar to alcohol — goes far back in human history. Until 1860, it was believed to be a purely chemical process, having nothing to do with life. Then Louis Pasteur showed that fermentation was a living process carried out by yeast and bacteria. He found he could prevent spoilage of wine and beer by pasteurization — the process of using heat to destroy bacteria that produced vinegary, unpalatable acids. His achievements not only benefitted the French wine and beer industries, but led to discoveries in the early 1900s that showed that the breakdown of sugar also occurred in animals and plants. The general process was called glycolysis, meaning sugar breakdown, and was found to be accompanied by the formation of ATP. Soon it was shown that contraction of muscle tissue goes hand-in-hand with glycolysis and ATP production, and that ATP is quickly consumed as the muscle continues to work. These discoveries led to our present understanding: Glycolysis, which needs no oxygen, and respiration, which needs oxygen, both synthesize ATP, which supplies the energy for all cell work.

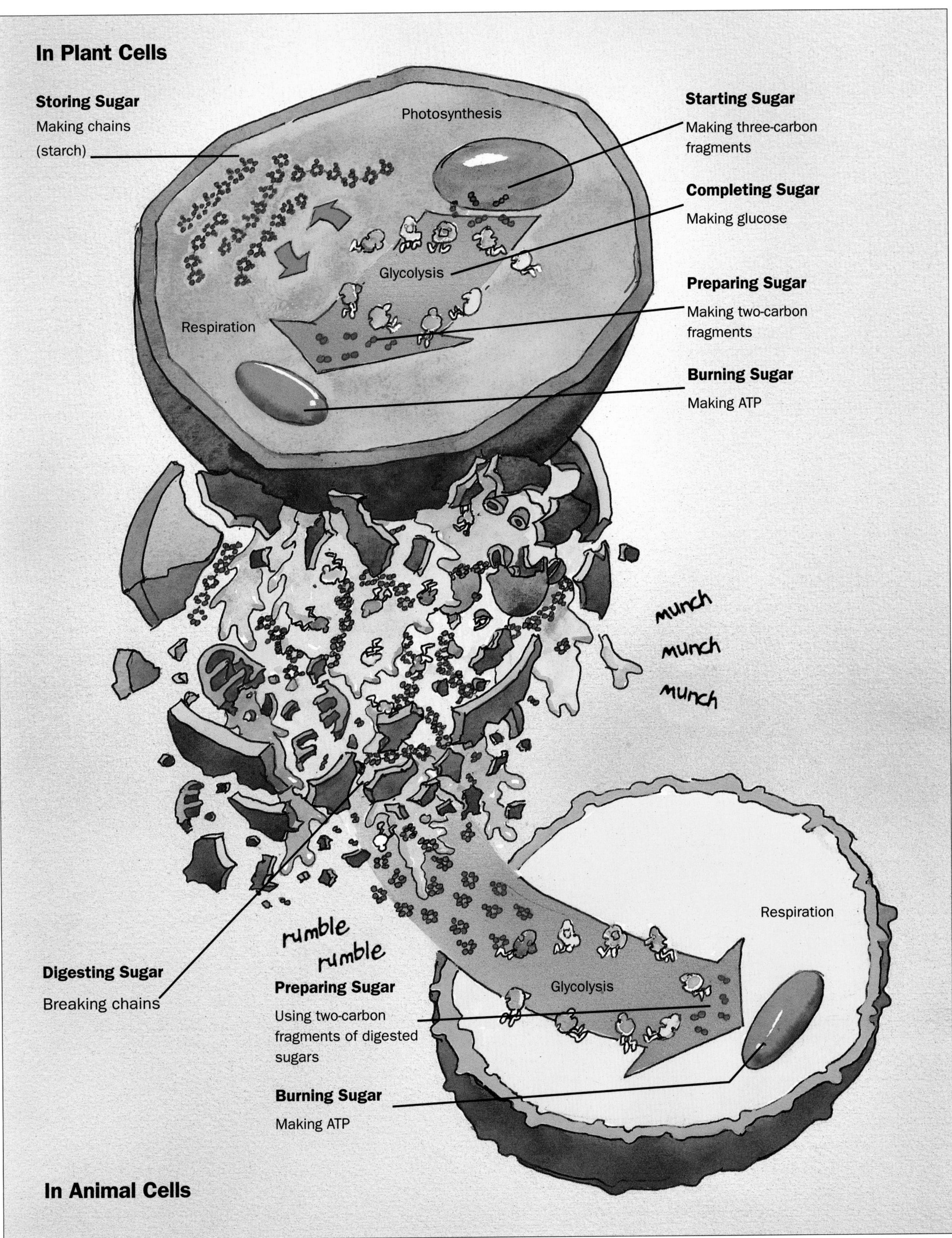
In Plant Cells
Storing Sugar
Making chains (starch)
Photosynthesis
Starting Sugar
Making three-carbon fragments
Completing Sugar
Making glucose
Glycolysis
Preparing Sugar
Making two-carbon fragments
Respiration
Burning Sugar
Making ATP
munch
munch
munch
rumble
rumble
Respiration
Digesting Sugar
Breaking chains
Preparing Sugar
Using two-carbon fragments of digested sugars
Glycolysis
Burning Sugar
Making ATP
In Animal Cells

3.14 Getting Energy to the Community

Ham and Cheese on Whole Wheat, Please

Unicellular organisms get their nutrients pretty directly. A slime mold absorbs energy-rich molecules from decaying plant matter on the forest floor right through its cell membrane. But what about larger creatures — cellular communities? How do they spread the wealth around, so to speak?

The food we encounter as consumers usually consists of big chunks of diverse materials whose molecules and molecular structures are much larger than those of sugar or the other basic materials of life. What's on our plate is not nearly ready for our cells to use as fuel and building blocks. If you could present a ham-and-cheese sandwich with a pickle on the side to a human cell, it wouldn't recognize that as food. Ham, cheese, lettuce, bread, and pickle must be converted into forms that can enter individual cells.

That conversion is what the digestive system does. Digestion is the link between food and the millions of small molecules that can cross cell membranes in the intestinal lining and move into the bloodstream for transport to the rest of the cells in our bodies. There, the deconstructed molecules are converted either into the energy that drives cellular function or into the materials that enable cellular reconstruction (see page 67).

First bite

Digestion starts as soon as you take your first bite of that sandwich. Saliva appeared as soon as you smelled — or even thought about — the food, and its enzymes start to break starch's molecular bonds as you chew, splitting its very long chains of glucose units into shorter ones.

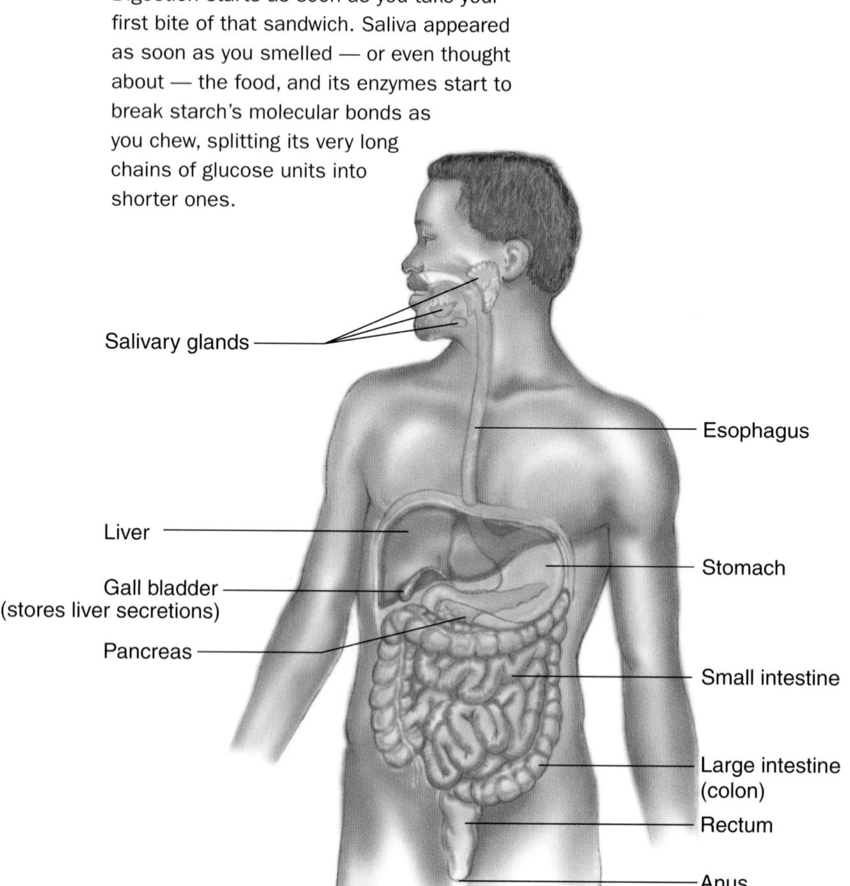

Swallowing

Once chewed, food passes through the tubular esophagus to the stomach.

In the stomach

The stomach churns up the food even more, exposing it to enzymes and acids that further break down starch chains and other large molecules. This semiprocessed food, now about the consistency of pea soup, is released little by little into the small intestine, where the final breakdown takes place.

Final breakdown

The small intestine produces still more digestive enzymes, which are aided by bile and fluids released by the pancreas and liver. Eventually, all that's left of the ham-and-cheese sandwich are minuscule fat droplets and molecules small enough to cross the membranes of the cells lining the intestine, and through those cells into the circulatory system.

Elimination

The large intestine reabsorbs water from the remaining intestinal contents. Indigestible wastes, including cellulose chains, pass into the rectum and are expelled through the anus.

A Closer Look at the Small Intestine

A close-up view of the small intestine (you saw a photograph of it on page 68) shows what makes its absorptive surface area so big. It is a deeply furrowed community of cells arranged as fingerlike villi. These villi contain threadlike capillaries of the circulatory system. The small molecules resulting from digestion can pass into and through the villi to get to the bloodstream.

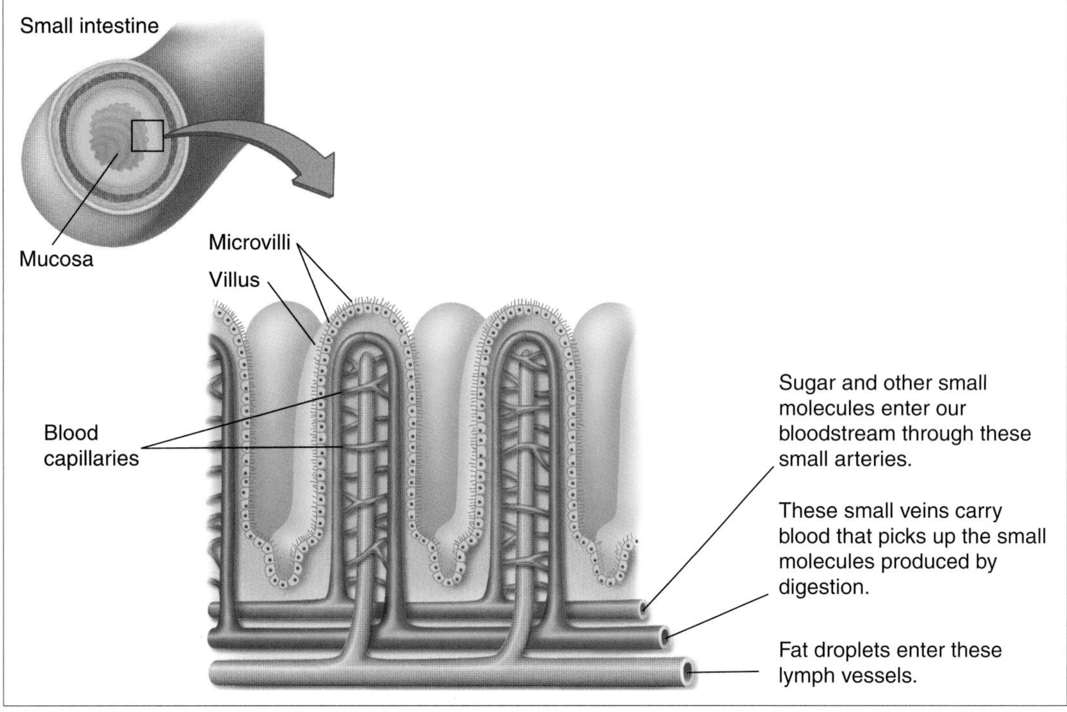

Small intestine
Mucosa
Microvilli
Villus
Blood capillaries
Sugar and other small molecules enter our bloodstream through these small arteries.
These small veins carry blood that picks up the small molecules produced by digestion.
Fat droplets enter these lymph vessels.

Sharing the Wealth

The heart pumps blood that carries oxygen and digested molecules to every cell in the body, where, reabsorbed from the blood, they become part of the cell's life processes. The simple rule that drives much of this cross-membrane absorption and reabsorption is that the digested molecules move from a more concentrated state to a less concentrated one (just like the milk in the coffee on page 96). They go first from a high concentration in the villi to a lower one in the blood, and later from a higher concentration in the blood to a lower concentration in body cells.

Living Light

Over the past 4 billion years, cells that were once free-living have joined together in cooperating communities that we call multicellular organisms. Within these organisms, groups of cells have taken on special roles — becoming muscle, brain, bone, skin, etc. These extra duties often require the additional production and consumption of energy as ATP. Instead of spending energy as lone hunter-gatherers, specialized cells divert some of their ATP to the performance of "civic" duties that benefit the whole community. For example, individual cells in a firefly's tail have no need to light up. However, since they are members of a larger community that must mate and reproduce, their light-making becomes a vital, ATP-requiring activity.

From a Cellular Perspective

Every cell in a firefly must produce ATP for its own needs.

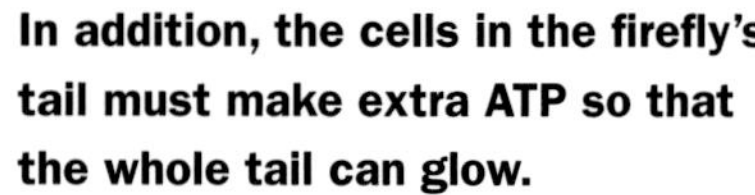

In addition, the cells in the firefly's tail must make extra ATP so that the whole tail can glow.

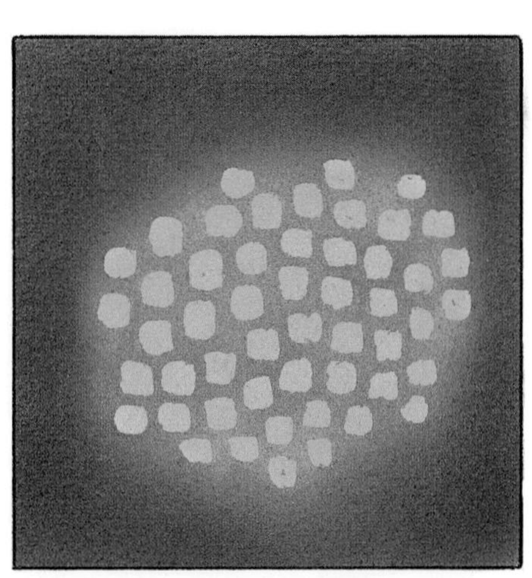

When millions of tail cells glow, the firefly has a good chance of mating and producing more fireflies.

From a Molecular Perspective

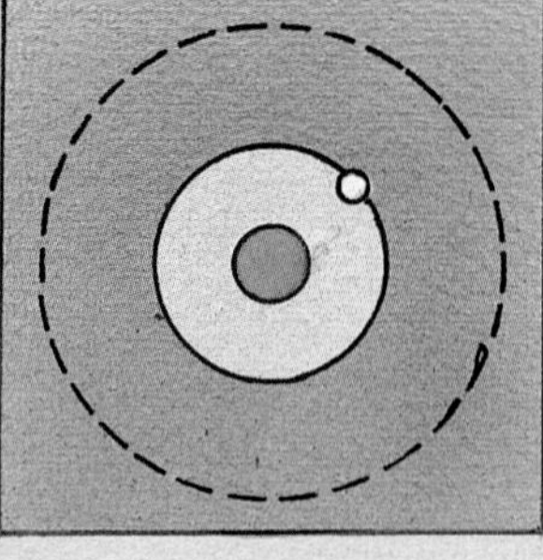

Electron in normal orbit

In the cells of the firefly tail, an enzyme attaches a part of ATP to a molecule called luciferin. This energizes luciferin, allowing oxygen (O_2) to bond to one of its carbon atoms and boosting an electron into a higher orbit. Luciferin then releases the oxygen and carbon as carbon dioxide (CO_2). As the electron drops back to its customary orbit, the spent energy is released as a tiny flash of light. This process, which uses ATP and oxygen to produce light and carbon dioxide, is the exact opposite of photosynthesis!

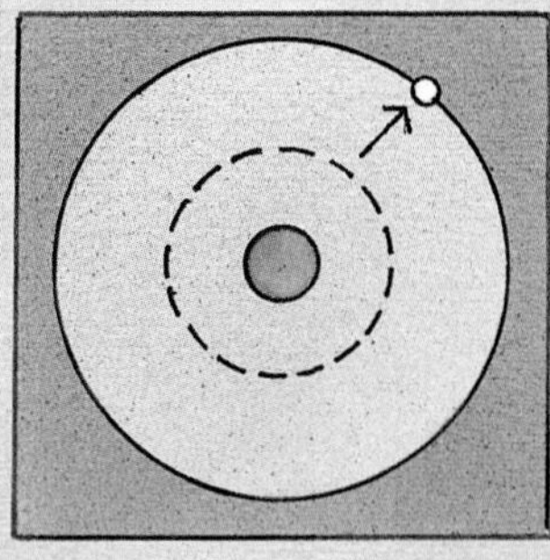

Electron boosted to a higher orbit

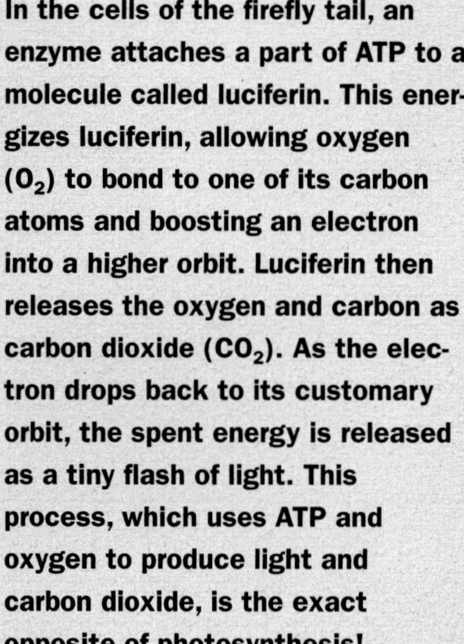

Electron falling back to a lower orbit, releasing energy as light

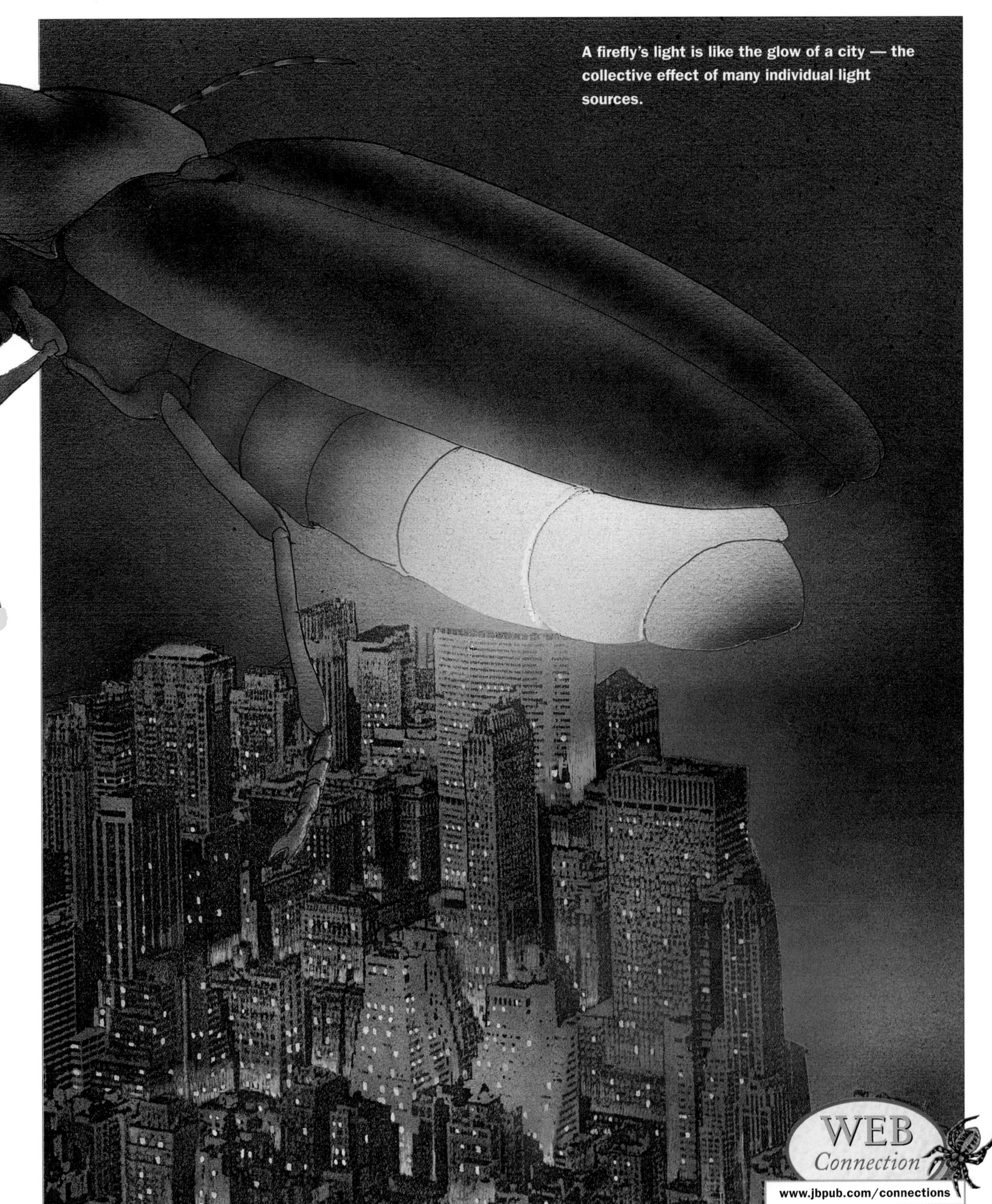

A firefly's light is like the glow of a city — the collective effect of many individual light sources.

Some of the Things You Learned About in Chapter 3

Questions About the Ideas in Chapter 3

1. Two simple molecular events enable living systems to utilize energy. What are they?
2. Energy flows from concentrated states (as in stars) to dispersed states (as in outer space). Why should we be glad it does?
3. Equilibrium is: (a) something life processes must achieve or (b) something life processes must avoid at all costs. Explain your answer.
4. Imagine that you are a cell and you want to join two low-energy molecules together. They resist joining. What two special types of molecules would you need to supply to overcome this obstacle?
5. If someone called you a parasite, you would probably be insulted. In some sense, they would be right. Why?
6. What does photosynthesis produce? How about respiration?
7. Interpret Benjamin Franklin's speculations on page 112 about fluid fire and fluid air in terms of what you now know about energy and photosynthesis.
8. The newly emerging skunk cabbage (upper right) is melting the spring snow that surrounds it. Where does the heat come from?
9. How do the two life forms you see in this photograph (lower right) get the energy they need to live, build, and recycle themselves? Explain their differences and similarities.
10. What's wrong with the project Jonathan Swift describes in the quotation on page 108? How does it resemble a firefly's light?

8

9

11. The NASA satellite images below show the islands of Indonesia and the surrounding ocean in (a) early spring and (b) summer. The red-orange through green colors indicate the presence of chlorophyll contained in ocean-dwelling algae. What is the reason for the difference between the two photographs? What could you predict about the relative amounts of oxygen and carbon dioxide at the ocean's surface in (a) and (b)?

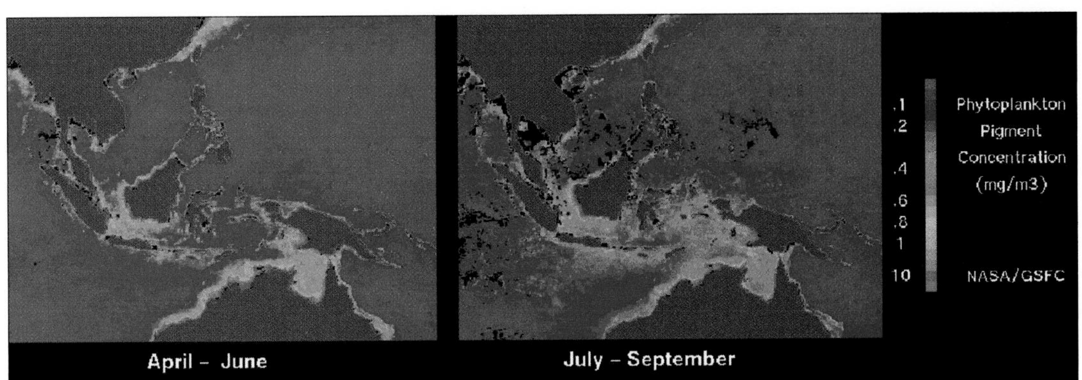

12. Trace the flow of energy from a fast-flowing stream to a millwheel (of the type seen in re-creations of early colonial settlements such as Sturbridge Village or Colonial Williamsburg) and then to the grinding of corn or the turning of a sawmill blade to cut lumber or the movement of a shuttlecock in a loom of a fabric mill. Is all the energy of the moving water transferred to the wheels, gears, pulleys etc.? Can the mill be run "backwards"? Compare this to a ray of sunlight hitting a leaf and eventually becoming cell movement or DNA replication. Be as specific as you can.

References and Great Reading

Anderson, Nancy K. and Linda S. Ferber. 1990. *Albert Bierstadt: Art and Enterprise.* New York: The Brooklyn Museum in association with Hudson Hills Press.

Calvin, William H. 1986. *The River That Flows Uphill: A Journey from the Big Bang to the Big Brain.* San Francisco: Sierra Club Books.

Dillard, Annie. 1999. *Pilgrim at Tinker Creek.* New York: HarperPerennial.

Franklin, Benjamin. 1931. *The Ingenious Dr. Franklin: Selected Scientific Letters of Benjamin Franklin,* Edited by Nathan G. Goodman. Philadelphia: University of Pennsylvania Press.

Johnson, A. William. 1999. *Invitation to Organic Chemistry.* Sudbury, MA: Jones and Bartlett Publishers.

Jukes, Thomas H. 1966. *Molecules and Evolution.* New York: Columbia University Press.

Pagels, Heinz R. 1982. *The Cosmic Code: Quantum Physics as the Language of Nature.* New York: Bantam Books.

Pearce, Fred. 1998. Algal Gloom. *New Scientist. August 8.*

Postgate, John. 1994. *The Outer Reaches of Life.* Cambridge: Cambridge University Press.

Shapiro, Robert. 1986. *Origins, a Skeptics Guide to the Creation of Life on Earth.* Arlington: Summit Books.

Schrodinger, Erwin. 1944. *What Is Life?* Cambridge: Cambridge University Press.

Scott, Andrew. 1991. *Basic Nature.* Cambridge: Basil Blackwell Ltd.

Sullivan, A. M. 1946. "Atomic architecture," in *Stars and Atoms Have No Size.* New York: Dutton.

For more questions and links to web resources, go to

CHAPTER 4

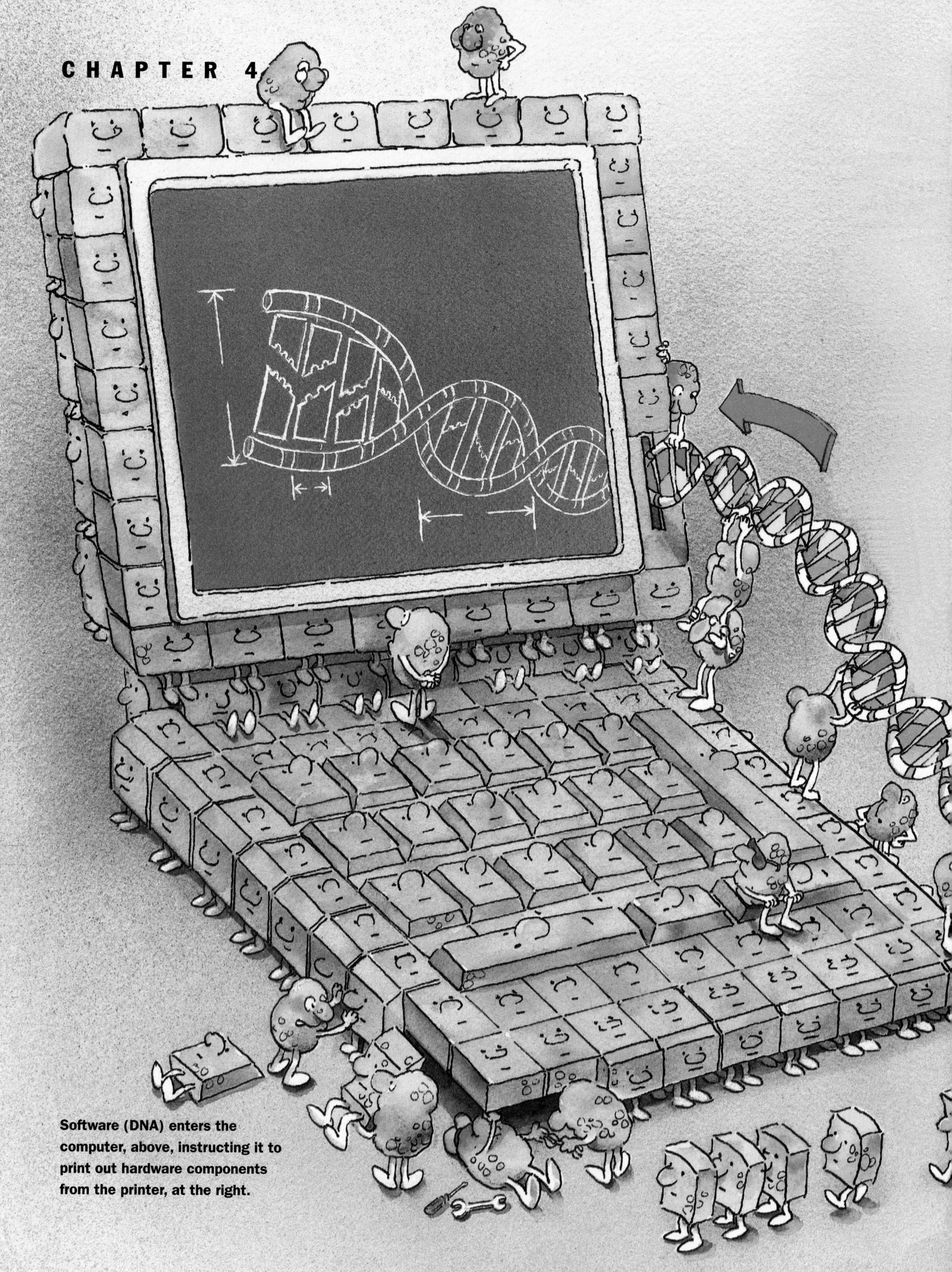

Software (DNA) enters the computer, above, instructing it to print out hardware components from the printer, at the right.

INFORMATION

The Storehouse of Know-How

IMAGINE A COMPUTER THAT CAN BUILD ITSELF — A COMPUTER THAT CONTAINS WITHIN its software the recipe for its own assembly. Its hardware, made from the software's instructions, handles the building, maintenance, and repair tasks. It also reads the software's instructions. Before the computer reaches the end of its workable lifetime, the hardware makes an exact copy of the software. This begins an entirely new copy of the software. This begins an entirely new and duplicate computer. And the cycle repeats itself indefinitely.

A living cell is such a self-organizing system. As such, it embodies a paradox: If hardware depends on software, and software depends on hardware, how could such a process ever get started in the first place? You'll recognize this as another version of the classic question, "Which came first, the chicken or the egg?" (see page 176).

More puzzles center on the nature of the software that life uses. For example, where does a young sapling store the information needed to build another oak tree? And how does this information interact with the rest of the tree's hardware? In this and the next chapter, *Machinery*, we explore the relationship between software and hardware, between the "information" and "machinery" of life.

4.1 Why Life Must Come From Life

It is no more likely that life could suddenly arise from non-life — for example, that flies could be created by decaying meat — than that a 747 could accidentally be assembled by a tornado blowing through a junkyard.

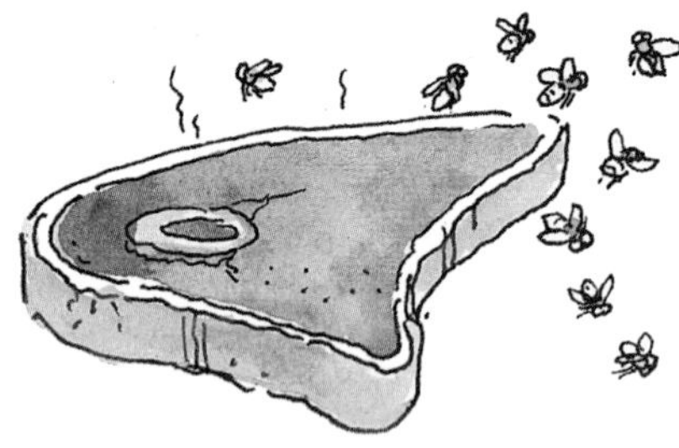

Flies pose an assembly problem much tougher than a 747. They are the result of billions of years' worth of accumulated "research and development," i.e., information built up by trial and error. Flies, like airplanes, can't be built without lengthy assembly instructions.

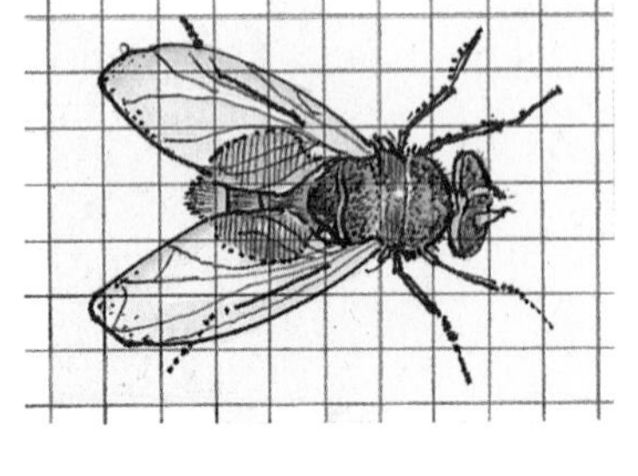

An Unbroken Chain

Throughout most of history, people believed life was controlled by mysterious, supernatural forces. For instance, seeing worms, maggots, flies, or even mice squirming around in decaying grains, mud, or rotting meat convinced early scientists that life arose spontaneously from non-life. Experiments later refuted this notion, but it still took a long time to appreciate why "spontaneous generation" is impossible: Life took 4 billion years to reach its present level of complexity. To maintain it, life must always come from life, flowing from generation to generation in an unbroken chain (except at its beginning on Earth four billion years ago). This inevitable conclusion comes from our modern understanding of the key role played by the information living things store within themselves.

The Death of "Spontaneous Generation"

Doing Science

In 1668, in one of the earliest carefully controlled biological experiments, Francesco Redi hypothesized that the life that was apparently generated from decaying meat didn't arise from non-life, but actually came from eggs laid on the meat. He took eight flasks with meat in them, left four open to the air, and sealed the others. After a while, maggots (fly larvae) infested the four open flasks, but none appeared in the sealed flasks. Redi then tried covering the formerly sealed flasks with gauze: no maggots. He correctly concluded that the maggots had come from eggs deposited on the meat by flies. This experiment disproved the notion that visible creatures, at least, were spontaneously generated from some "vital principle" in decaying matter.

Yet people still believed that microorganisms such as bacteria and yeast sprang spontaneously from decaying matter. The controversy raged on until 1864, when Louis Pasteur determined to end it. First, he tested air and dust and showed that they contained living organisms (he called these "ferments"). Next, he added air and dust to thoroughly sterilized materials, sealed them in flasks, and noted that life rapidly multiplied within. Then he placed sterilized matter inside a flask with a long S-shaped neck stopped up with a cotton plug; no life arose inside the flask. If Pasteur tipped the flask so that its contents touched the cotton plug, thereby contaminating them with microorganisms trapped in the plug, life sprang forth inside the flask within 48 hours. And, of course, if he broke the neck and let air in, growth inside the flask rapidly ensued. Said Pasteur, "Never will the doctrine of spontaneous generation recover from the mortal blow that a simple experiment has dealt it." It hasn't.

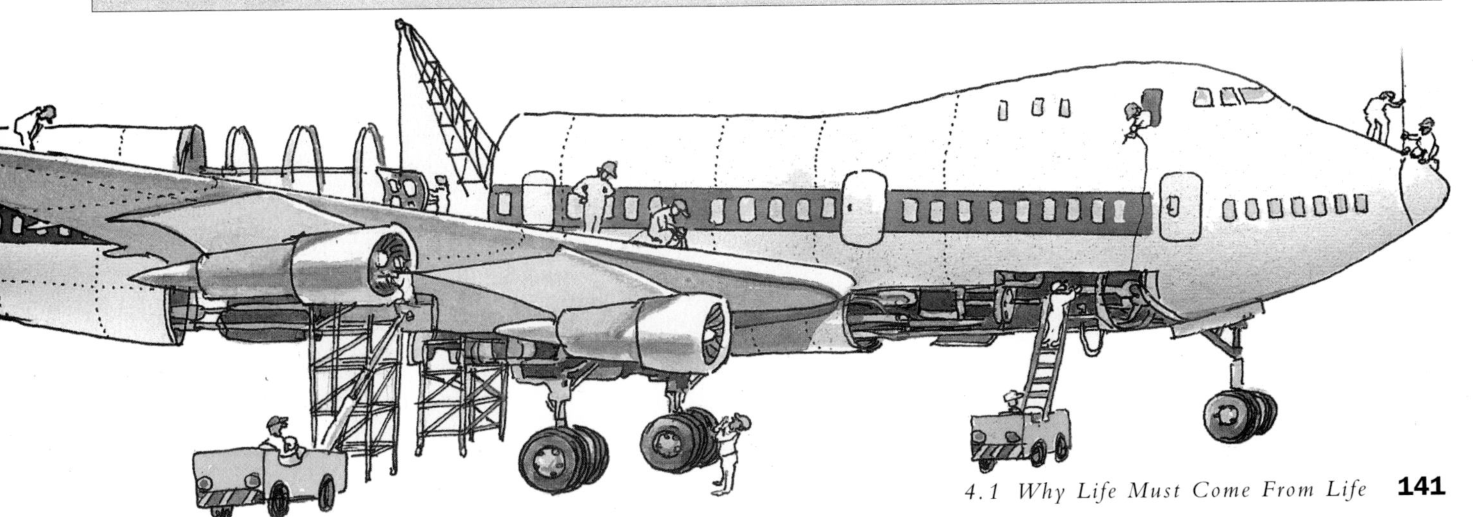

People who held strongly to the idea of spontaneous generation were hard to convince, though. They were sure that Pasteur was not accounting for a mysterious "life force" in his experiments.

The illustrations below show the different experimental methods Pasteur used to accommodate and disprove the "life force" hypothesis.

The sequence of Pasteur's experiments

Pasteur's and his critics' interpretations of the results follow the sequence.

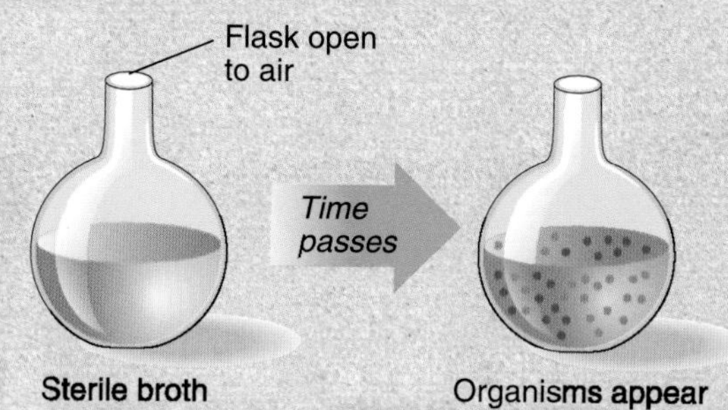

Pasteur: The broth provides a nutrient medium for the growth of unseen organisms in the air: life comes from other life.

His critics: A sterilized broth gives rise to life: spontaneous generation.

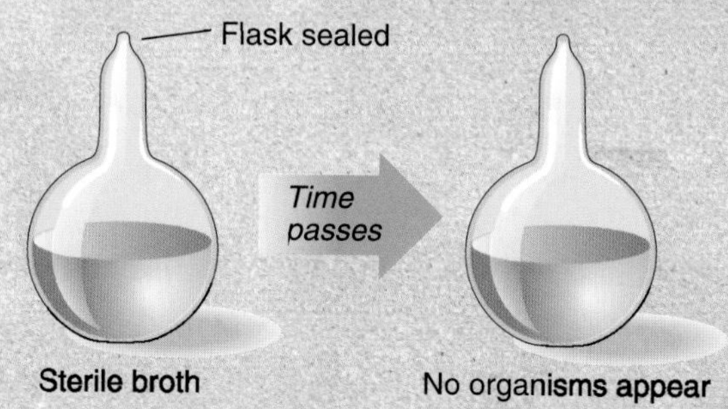

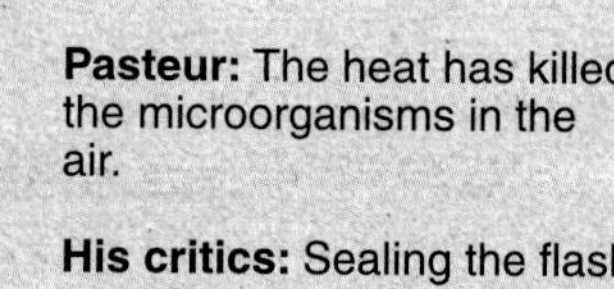

Pasteur: The heat has killed the microorganisms in the air.

His critics: Sealing the flask prevents entry of the "life force."

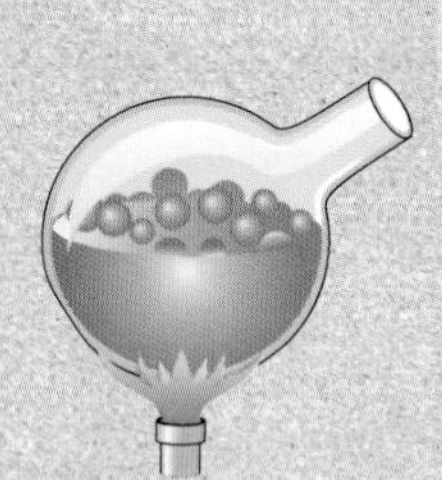

Each experiment begins with sterilized broth. Any living things the broth may have contained are destroyed by heat.

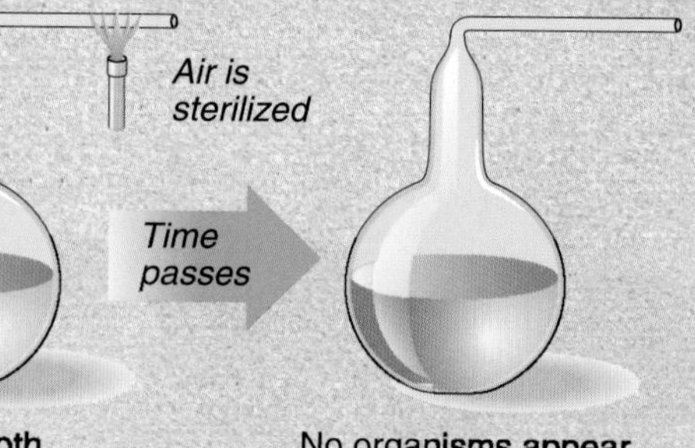
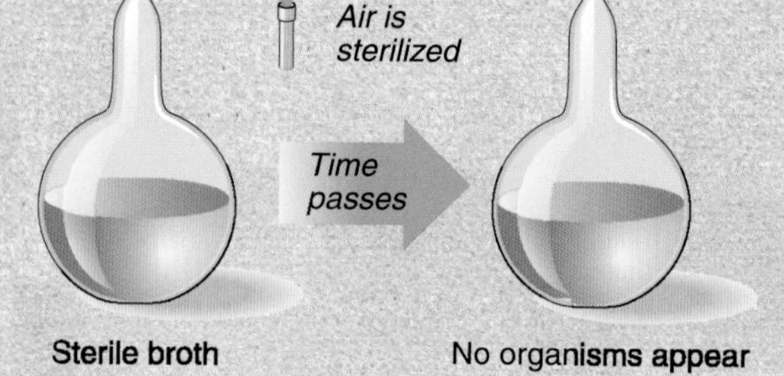

Pasteur: The heat has killed the microorganisms in the air.

His critics: Sterilizing the air kills the "life force."

(d) Sterile broth — Swan-neck flask — *Time passes* — *Air enters* — Microorganisms are trapped — No organisms appear

Pasteur: No living thing will appear in the flask because microorganisms will not be able to reach the broth.

His critics: If the "life force" has free access to the flask, life will appear, given enough time.

Some days later the flask is still free of any living thing. Pasteur has disproved the doctrine of spontaneous generation.

The Germ Theory and Its Applications to Medicine and Surgery
Read by Louis Pasteur before the French Academy of Sciences, April 29th, 1878.
Published in *Comptes rendus de l'Academie des Sciences,* lxxxvi, pp. 1037-1043.

The Sciences gain by mutual support. When, as a result of my first communications on the fermentations in 1857-1858, it appeared the ferments, properly so-called, are living beings, that the germs of microscopic organisms abound in the surface of all objects, the theory of spontaneous generation is chimerical; that wines, beer, vinegar, the blood, urine, and all the fluids of the body undergo none of their usual changes in pure air, both Medicine and Surgery received fresh stimulation. . . .

Our researches of the last year left the etiology of the putrid disease, or septicemia, in a much less advanced condition than that of anthrax. We had demonstrated the probability that septicemia depends upon the presence and growth of a microscopic body, but the absolute proof of this important observation was not reached. To demonstrate experimentally that a microscopic organism actually is the cause of a disease and the agent of contagion, I know no other way in the present state of Science than to subject the *microbe* . . . to the method of cultivation out of the body.

Our researches concerning the septic vibrio had not so far been convincing, and it was to fill up this gap that we resumed our experiments. To this end, we attempted the cultivation of the septic vibrio [a type of microbe] from an animal dead of septicemia. It is worth noting that all of our first experiments failed, despite the variety of culture media that we employed. Our cultural media were not sterile, but we found . . . a microscopic organism showing no relationship to the septic vibrio . . . an impurity introduced unknown to us . . . into the abdominal fluids from which we took our original cultures of the septic vibrio. [We found] a pure culture of the septic vibrio in the heart's blood of an animal recently dead of septicemia but . . . all our cultures remained sterile.

It occurred to us that the septic vibrio might be an obligatory anaerobe [an organism that cannot survive in the presence of oxygen] and that the sterility of our inoculated culture fluids might be due to the destruction of the septic vibrio by the atmospheric oxygen dissolved in the fluids. . . .

It was necessary therefore to attempt to cultivate the septic vibrio in a vacuum or in the presence of inert gases — such as carbonic acid.

Results justified our attempt; the septic vibrio grew easily in a complete vacuum, and no less easily in the presence of pure carbonic acid.

. . . It is a terrifying thought that life is at the mercy of the multiplication of these minute bodies, it is a consoling hope that Science will not always remain powerless before such enemies. . . .

Question.

What three specific steps in the scientific process does Pasteur describe in the paper above? What personal attributes of the researchers contributed to their eventual success?

Answer...

Pasteur describes an observation (specific microscopic organisms are associated with certain diseases), a hypothesis (a microscopic organism is the agent of disease and contamination), and an experiment (the attempt to grow these organisms and, by implication, eventually to inoculate healthy animals with it and see whether they caught the disease). The curiosity, determination, and creative imagination of the researchers shine through the description of their work.

www.jbpub.com/connections

4.2 An Abbreviated History of Genetic Discoveries

Uncovering the Secrets of Heredity

Once scientists realized that life can only come from life, they began to look more closely at inheritance. Yes, our offspring look like us…but why?

This short survey takes us up to the 1940s.

1860s

"Factors" Determine Inheritance

Austrian monk Gregor Mendel discovers that something he dubs "factors" somehow determine inheritance in pea plants. Every trait appears to be controlled by a pair of these factors. Further, a trait may have "dominant" and "recessive" forms. For instance, if Mendel bred a tall plant with a short one, the offspring were mostly tall; tallness is dominant, and shortness recessive. However, the recessive trait isn't lost — it can reappear in a later generation; two tall pea plants bred together might produce a short one.

Mendel's work remained largely unread until around 1900.

1890s

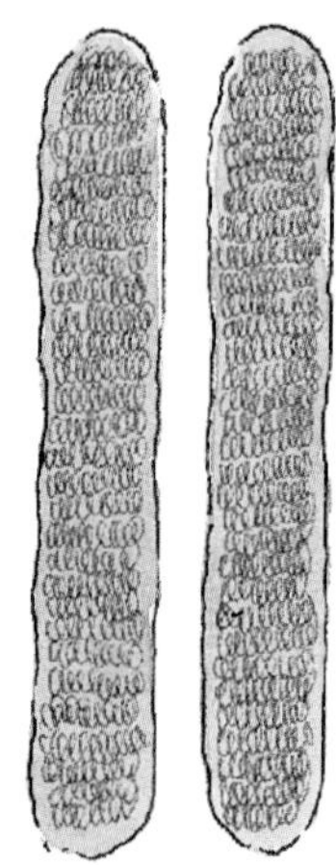

Chromosomes

Chromosomes, microscopic structures in the cell nucleus, are discovered by many researchers. They note that chromosomes, which come in pairs, double before cell division and are then shared between daughter cells. It is suspected that chromosomes are the carriers of heredity.

1903

"Factors" Are on Chromosomes

William Sutton makes the connection between Mendel's factors and chromosomes. One member of each pair of trait-determining factors is on one of a pair of chromosomes. One chromosome comes from the mother's egg, and the other from the father's sperm.

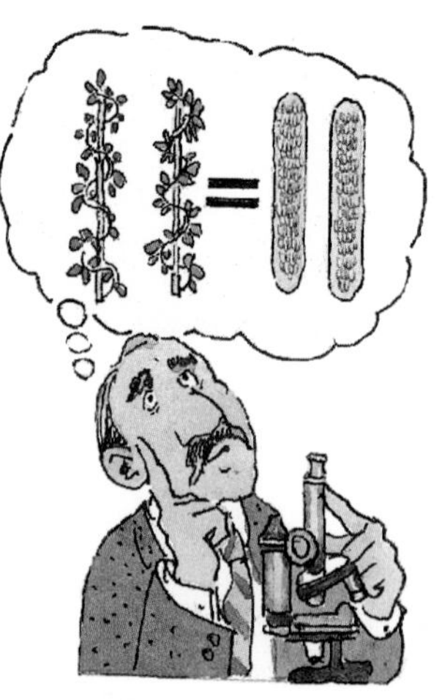

1905

Chromosomes Actually Determine Inheritance

Edmund B. Wilson and N. M. Stevens discover that a particular chromosome called the X chromosome, of which there are two in female cells and one in male cells, determines the sex of the offspring and explains why there are equal numbers of females and males: All eggs have an X, but only half of sperm do (the other half get a Y chromosome). This is the first evidence that a specific chromosome carries a specific hereditary property (sex).

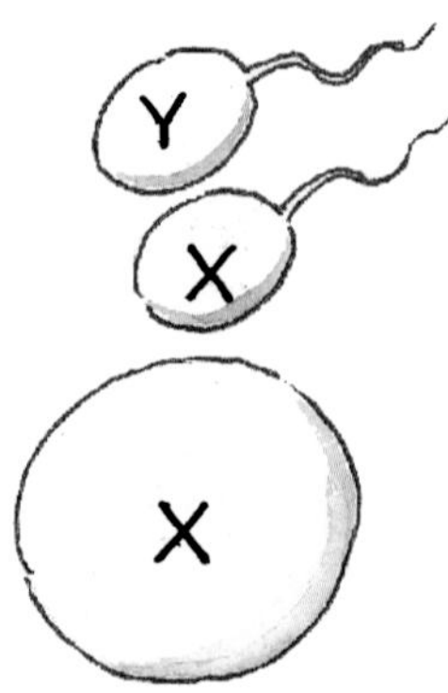

1906

Mendel's "Factors" Are Genes

Scientists coin the term "gene," meaning a piece of genetic information specifying a particular trait or characteristic. Genes are the factors Mendel discovered.

Genes Are Inherited Together

Thomas Hunt Morgan shows that many genes are inherited together, as would be expected if they are linked to each other in chromosomes. (The fruitfly has four chromosomes, and it has four groups of linked genes.) Chromosomes, then, are chains of genes.

1908

Genes Are Lined Up Along Chromosomes

Morgan observes that even though genes tend to be inherited together, this occurs more frequently with some pairs than with others. He infers that the farther apart genes are on a chromosome, the less likely they are to be inherited together. (This is because an actual physical exchange of genes takes place between chromosomes.) Morgan is able to "map" the relative positions of genes along fruitfly chromosomes.

1909

Hereditary Diseases May Be Caused by Defective Genes

Archibald Garrod postulates that certain inheritable human diseases result when particular proteins fail to perform their normal function.

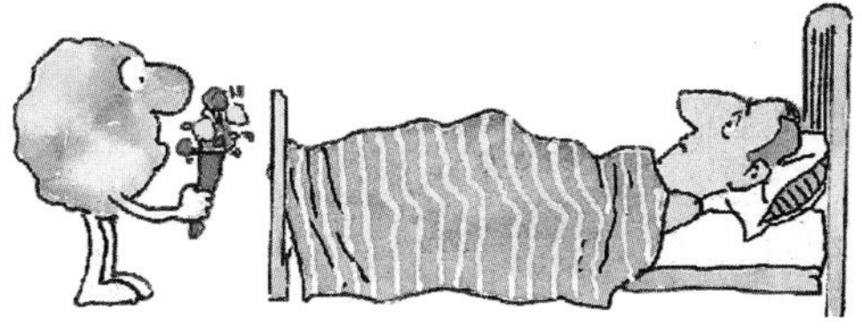

New Traits Are Caused by Mutations

Scientists realize that mutations — changes in genes — are what produce new genetic characteristics (as well as inherited diseases). They further realize that without mutations, there can be no evolution (see Chapter 8, *Evolution*). Hugo de Vries had discovered genetic mutations in 1886. Hermann Muller first produces mutations with X-rays in 1927.

1927

1942

One Gene — One Protein

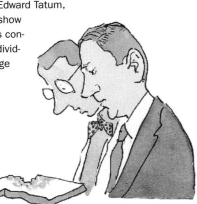

George Beadle and Edward Tatum, using bread molds, show that individual genes control production of individual proteins (see page 149).

1944

Natural Selection Operates on All Living Things

Salvador Luria shows that bacteria are subject to the same genetic and evolutionary forces that operate on plants and animals. Bacteria, because they reproduce so rapidly, become the main experimental subject of molecular genetics.

Genes Are Made of DNA

Oswald Avery and associates show that genes are made of deoxyribonucleic acid — DNA.

4.3 Life's Coding and Decoding Systems

The Recipe and The Cake

Here we've listed some instances of the relationship between encoded "ideas" and their decoded "products."

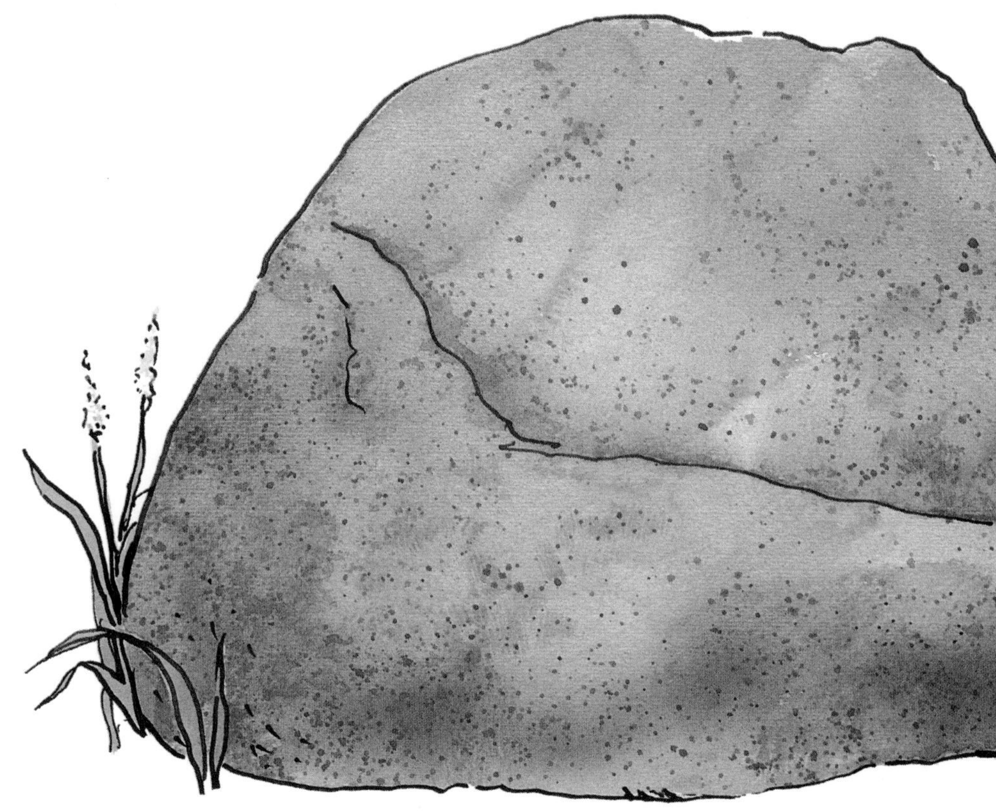

Idea	Product
Blueprint	Building
Recipe	Cake
Menu	Meal
Sheet Music	Symphony
Genes	Proteins

We can consider the items in the left-hand column as chunks of information corresponding to the actual products in the right-hand column. But, strictly speaking, coded information also exists in material form — ink, paper, molecules — so the items on the left are also products.

He's just full of information.
Yep…or grasshoppers.

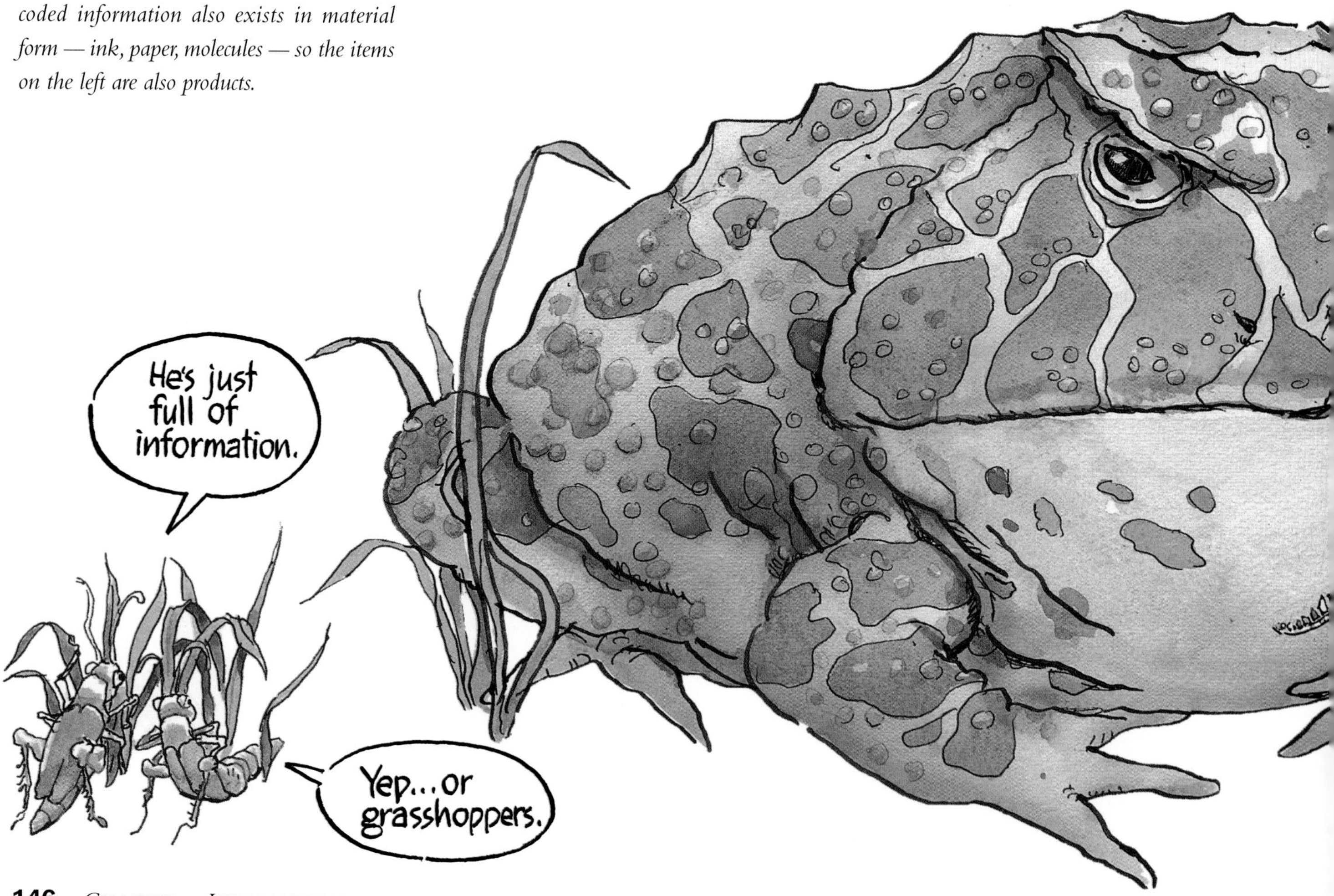

Rocks are simple, stable arrangements of molecules settled into low-energy states.

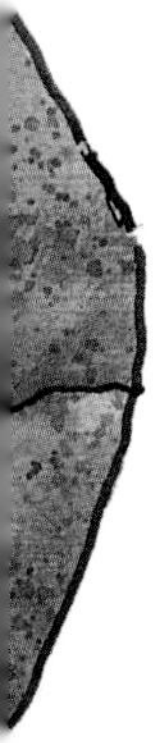

A toad's cells are complex arrangements of high-energy molecules, dynamically organized by information.

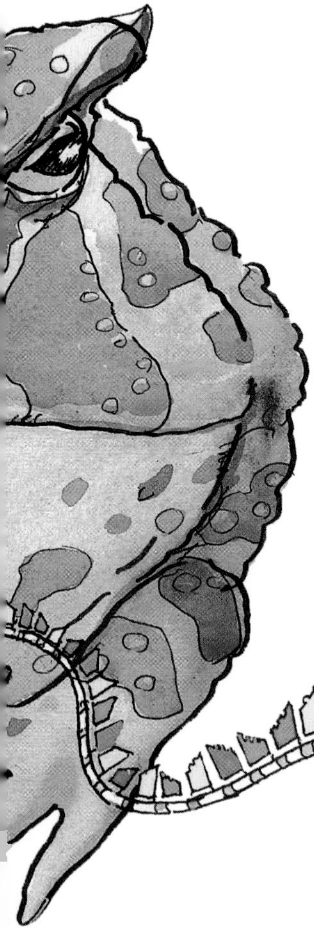

Information Is Embedded in Living Things

A fundamental difference between living and non-living things is that living things use information to create and maintain themselves. Rocks contain no instructions on how to be rocks. Toads contain instructions on how to be toads.

Information, like ideas, is dimensionless. It's simply a comparison between one thing and another, a registering of differences. Information becomes tangible when it is encoded in sequences of symbols: zeros and ones, dots and dashes, letters of the alphabet, musical notes, etc. Such sequences of symbols, in turn, are decoded — by machinery or by us — into computer output, Morse code messages, books, symphonies, etc. In order to be stored or transmitted, then, information needs to be put into some physical form, a process that requires energy. In this sense, you might say that "mind" and "matter" are inextricably linked.

Life's information — the "ideas" governing how it operates — is encoded in sequences of nucleotides (genes), which are, in turn, decoded by machinery (proteins) that manufactures parts that work together to make a living creature. Like the computer that builds itself, the process follows a loop: Information needs machinery, which needs information. This relationship can start simply and then, over many generations, build into something complex. Similarly, our deeper thoughts evolve out of simpler bits and pieces — hunches, ideas, memories.

This tells me nuthin'!

Information Needs Difference

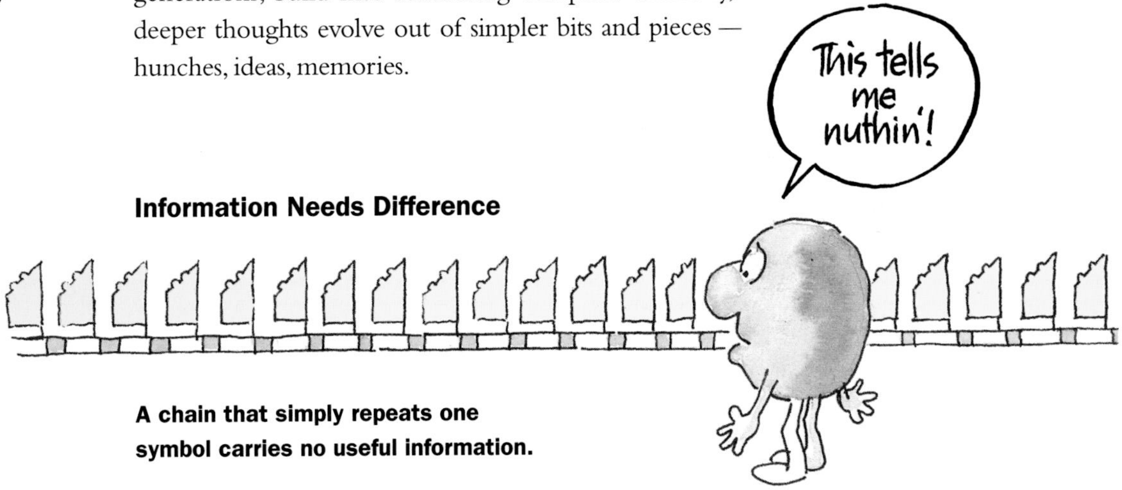

A chain that simply repeats one symbol carries no useful information.

Interesting...

But a chain made up of different symbols can encode information. All of life's genetic instructions are spelled out in combinations of four different "letters."

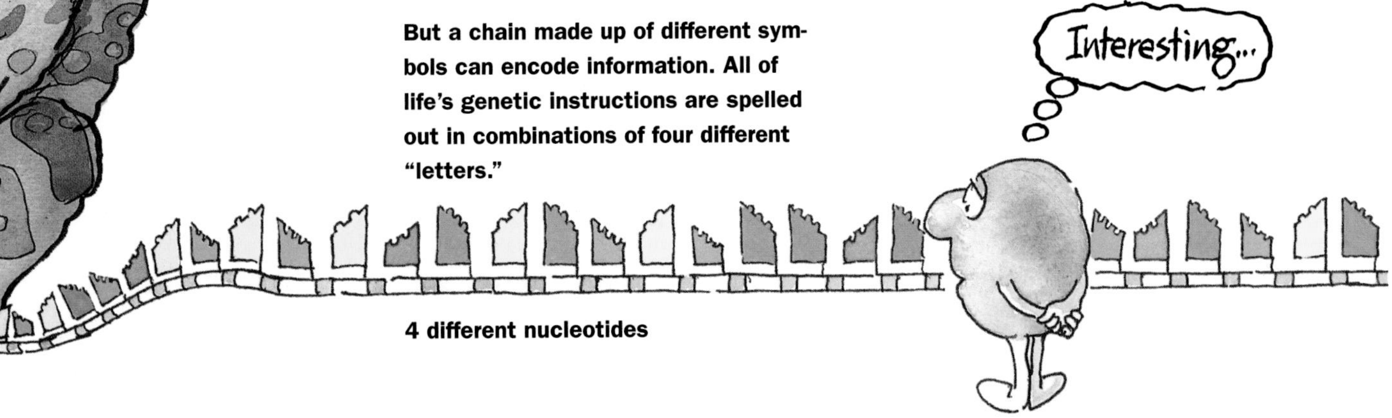

4 different nucleotides

4.4 DNA — What Does It Actually Say?

Not a Blueprint But a Recipe

We might never understand life's complexity were it not for the discovery that life is orchestrated by "intelligent" worker molecules called proteins. These proteins are various combinations of twenty, and only twenty, different amino acid molecules linked together into chains of various lengths. Every unique function of a protein is determined by the order of the amino acids in its chain.

Here we have a powerful insight into the way life works: One chain, DNA, carries information; a second chain, of amino acids linked into proteins, does life's work of growing, maintaining itself, and reproducing. DNA's sequence of units determines the sequence of amino acid units in proteins. Thus, DNA is not like a blueprint, which contains an image or a scale model of the final product; it is more like a recipe — a set of instructions to be followed in a particular order.

So life's complexity arises from a breathtaking simplicity: DNA's message says, "Take this, add this, then add this...stop here. Take this, add this, then add this,...etc." While the idea is simple, accomplishing it requires some ingenious machinery (see Chapter 5).

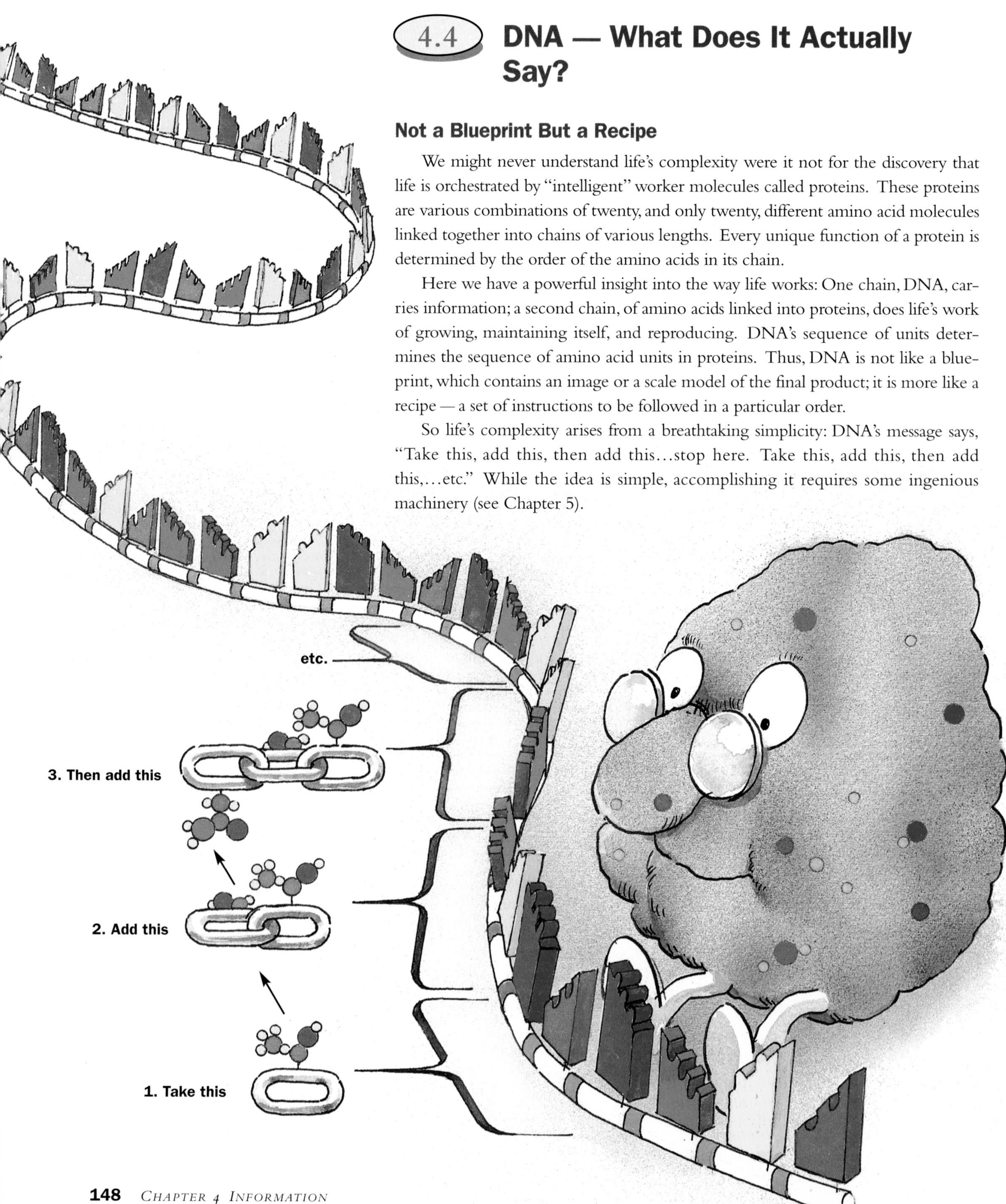

DOING *Science*

A Key Discovery: One Gene Makes One Protein — Beadle and Tatum and Bread Mold

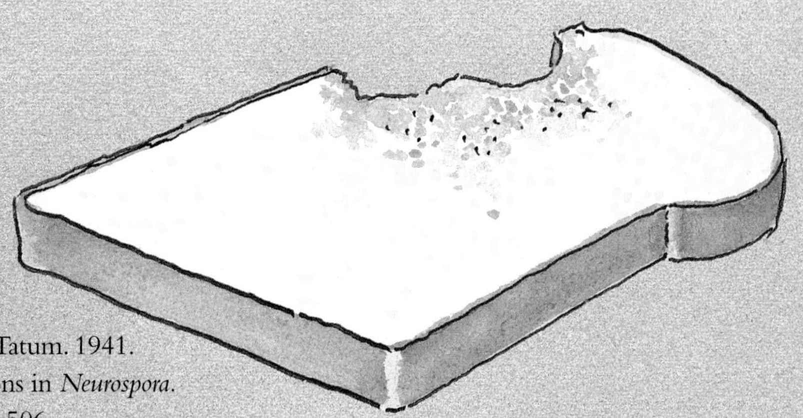

George W. Beadle and Edward L. Tatum. 1941.
Genetic control of biochemical reactions in *Neurospora*.
Proc. Nat. Acad. Sci. 27: 499-506.

Most of the inherited traits studied up to the early 1940s were complex functions: height of pea plants, fruitflies' wing shape or eye color, etc. These were probably controlled by many genes.

George Beadle realized he had to narrow the focus — to find one simple trait controlled by one specific gene. Inspired by Thomas Morgan, he started working with fruitflies but soon found a better subject — the common bread mold *Neurospora*. Here's the kind of experiment he and his associate, Ed Tatum, did. Normal molds can convert sugar, step-by-step, into all twenty amino acids. For instance, amino acid Z is made by converting molecule A into molecule B, then B into C, and finally C into Z. Beadle and Tatum exposed molds to X-rays, causing them to change — to mutate — so they and their progeny could no longer make certain amino acids. One mutant could no longer make Z unless Beadle and Tatum supplied it with molecule C. Giving it A or B didn't help. They concluded that the mutant had lost the ability to convert B to C — in other words, X-rays had damaged the protein (an enzyme) that converts B to C. Another mutant was unable to make Z unless it was supplied with molecule B. Beadle and Tatum concluded that this mutant had lost its ability to convert A to B — that is, X-rays had damaged the protein that converts A to B. Beadle and Tatum correctly surmised that each mutant had sustained X-ray damage to a specific gene that was responsible for making a specific protein. This simple idea — that one gene codes for one protein — opened the door to deeper understanding of how genes work.

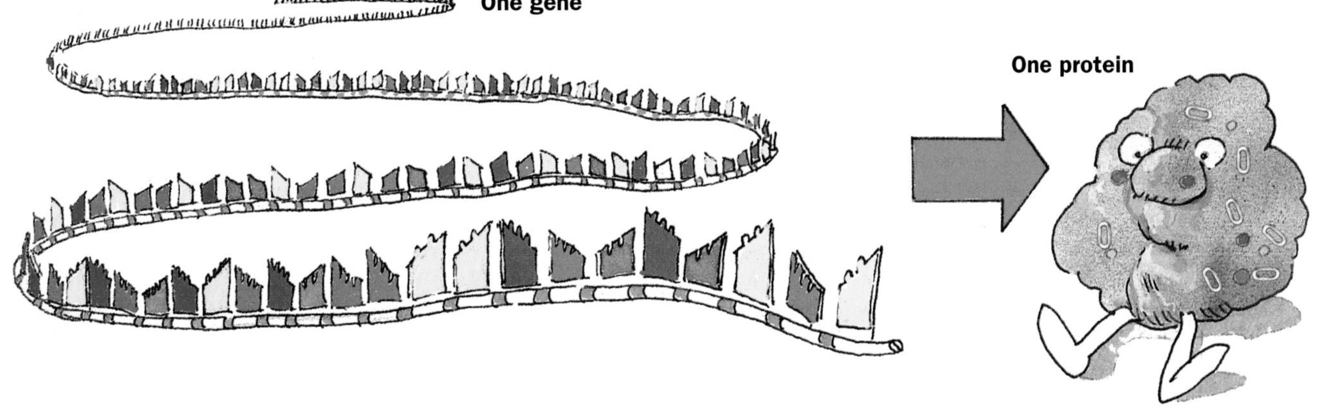

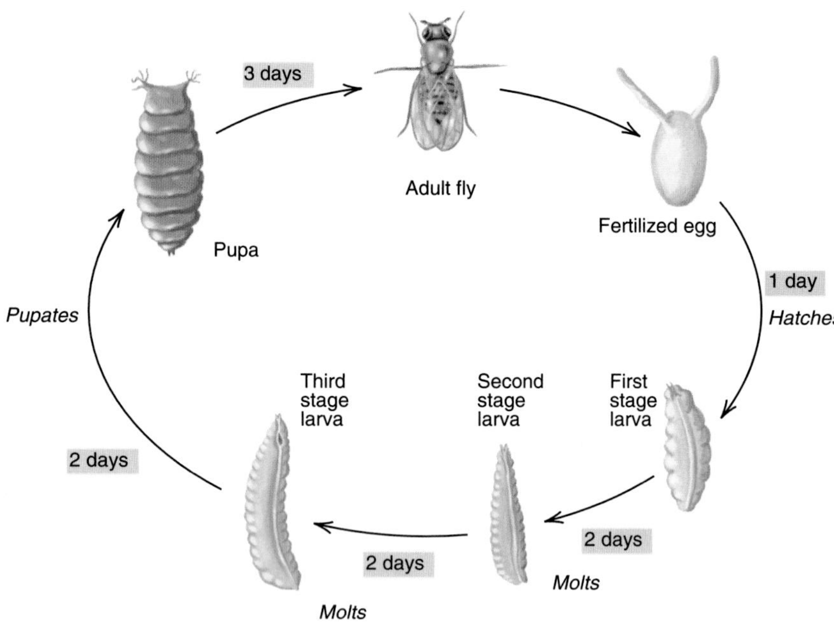

Fruitfly life cycle

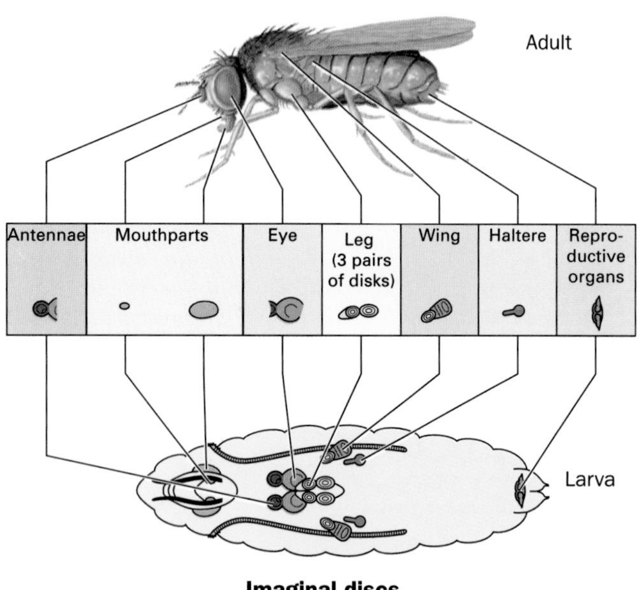

Imaginal discs

Cooking Up a Fruitfly

The recipe for a fruitfly *(Drosophila melanogaster)* produces an organism with an extraordinarily complex life cycle. As you can see in the illustration, the process of getting from a fertilized egg to an adult fly is no simple matter. The "stop here," "start there," "take this," "add this" instructions in a fruitfly's DNA are translated into machinery that builds several very different forms along the way to adulthood and egg-laying. During the fly's life cycle, a fertilized egg, essentially a sac with single nucleus in it, becomes a wormlike form, the larva, which is eventually transformed into a winged, six-legged adult capable of producing eggs or sperm that can combine to form fertilized eggs. Each fertilized egg contains all the instructions it needs to assemble the next generation.

The egg, about one-third the size of an adult fruitfly, carries in its DNA the genetic recipe needed to produce all the forms in the fly's life cycle. That means not just the instructions for making the parts but also those for making them at the right time and in the right sequence! Genes containing the instructions for making wings and legs can't be active in the larval stages. When one set of instructions is being implemented, the others are on hold, temporarily inactive.

As the egg develops, groups of genes turn on and off in response to chemical signals produced by other genes, and the first-stage larva is eventually formed. It grows and molts, becoming a second-stage larva. Then, after a second molt, the instructions that assemble larvae shut down, and those for making a pupa become active. While a larva is growing, it develops little patches of cells called imaginal discs, shown at left. These remain dormant until instructed to grow in the pupal stage, when they begin to develop into adult structures. In the laboratory, this entire cycle takes only 10–14 days from start to finish.

A Look at the Basic Ingredients

Let's take a closer look at part of the assembly process. The road to adulthood begins when the single fertilized nucleus in the egg forms a cluster of nuclei that subsequently migrate to the inner surface of the egg. Two basic types of cells form around these naked nuclei. One type consists of a single layer of cells that forms around the periphery of the egg. These will become body cells. The other group of cells, called pole cells, forms at one end of the egg. Pole cells are destined to become the cells that will pass on instructions to the next generation. Genetic changes (mutations) that occur in these cells will become part of the next generation's set of instructions — changing the recipe. If the change is advantageous to the offspring, the process of evolution will tend to keep the revised recipe in the cookbook. If the change is detrimental, as changes most often are, the offspring of the adult formed from this egg will not survive or will not reproduce as well as other individuals. The change will be lost over time. That particular version of the recipe will go out of print.

Find the fruitfly

This is the actual size of an adult fruitfly. You can imagine how small its egg is! Yet that egg contains all the instructions needed to make not only several larval stages and a pupa but also the adult's eyes, mouthparts, wings, antennae, reproductive organs, and hundreds of other parts.

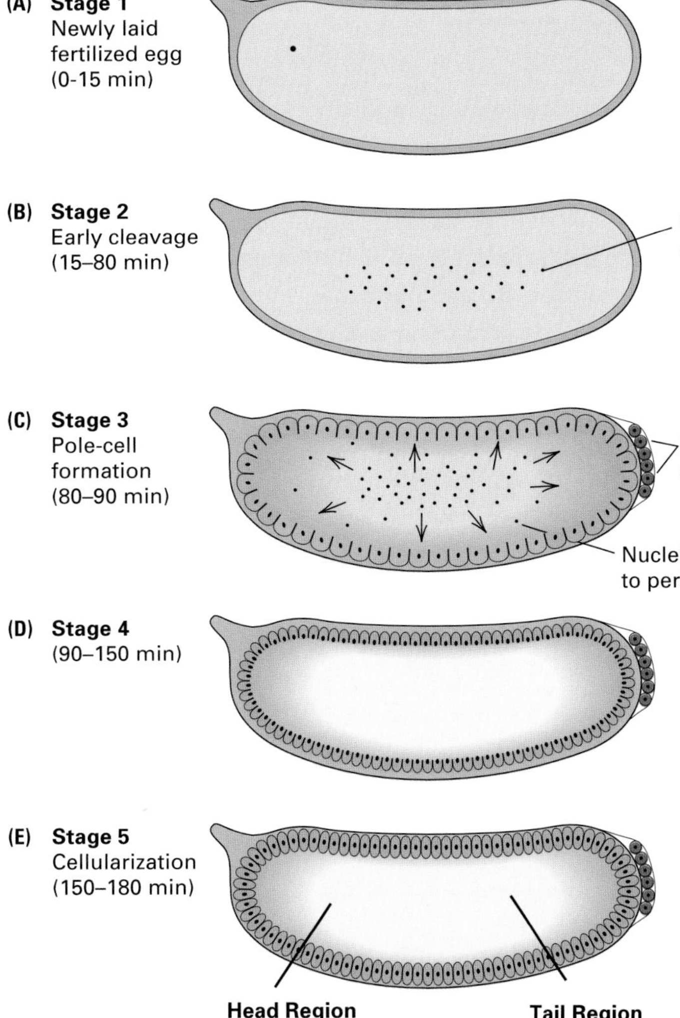

The fertilized nucleus

The egg cell's nucleus contains one set of genes from each parent — two copies of the same recipe. In biology, as in other things, variety is the spice of life, and biological variety arises in part from the fact that no two recipes are exactly the same. The overall instructions give similar results, but one recipe may yield a fly with red eyes and another a fly with white eyes. Differences in instructions are worked out during development. Variety allows evolution to test new genetic "ideas."

Nuclear migration

Nuclei migrate to the periphery and are surrounded either by pole cells or by the single layer of cells that will develop into the fruitfly's body parts.

ADENINE

THYMINE

GUANINE

CYTOSINE

SUGAR-PHOSPHATE BACKBONE

4.5 Nucleotides — Letters on a Backbone

Chemical Units of Information

Like the letters in the words of this paragraph, the four nucleotides of DNA comprise the letters of *its* language — the language of heredity. Each of the four nucleotides consists of a base, a sugar, and a phosphate. The base is either adenine (A), thymine (T), cytosine (C), or guanine (G) — each one a unique arrangement of carbon, nitrogen, oxygen, and hydrogen atoms. The base is bonded to a deoxyribose sugar, shown in all illustrations as a white cylinder, and to a phosphate. Like beads strung in a necklace, the repeating phosphate-sugar parts of the nucleotides link to each other in a continuous backbone that holds the sequence in order.

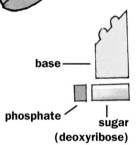

base

phosphate

sugar (deoxyribose)

From Nucleotide to Genome — A Hierarchy of Packaging

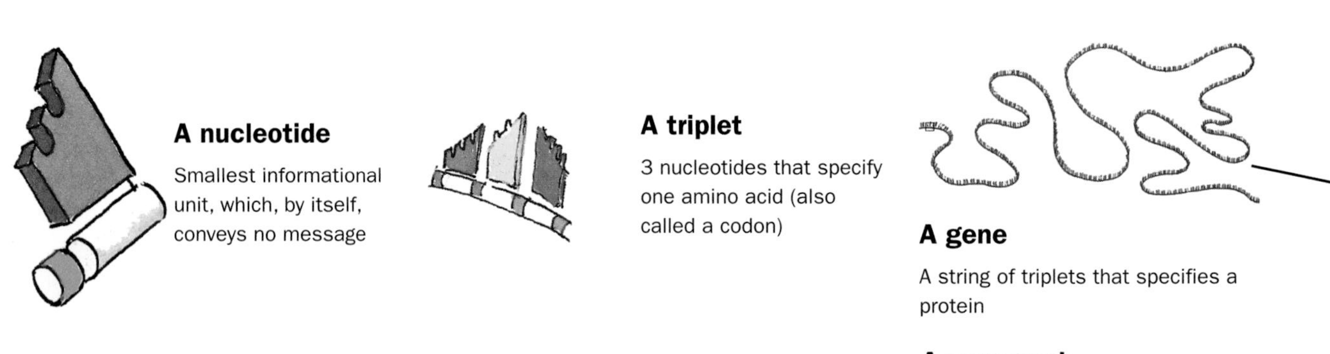

A nucleotide

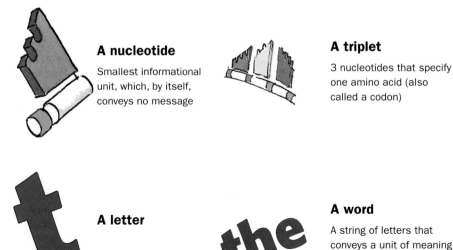

Smallest informational unit, which, by itself, conveys no message

A triplet

3 nucleotides that specify one amino acid (also called a codon)

A gene

A string of triplets that specifies a protein

A letter

the

A word

A string of letters that conveys a unit of meaning

A paragraph

A string of words that conveys an idea

Like the letters in the words of this paragraph, the four nucleotides of DNA comprise the letters of its language — the language of heredity. Each of the four nucleotides consists of a base, a sugar, and a phosphate. The base is either adenine (A), thymine (T), cytosine (C), or guanine (G) — each one a unique arrangement of carbon, nitrogen, oxygen, and hydrogen atoms. The base is bonded to a deoxyribose sugar, shown in all illustrations as a white cylinder, and to a phosphate. Like beads strung in a necklace, the repeating phosphate-sugar parts of the nucleotides link to each other in a continuous backbone that holds the sequence in order.

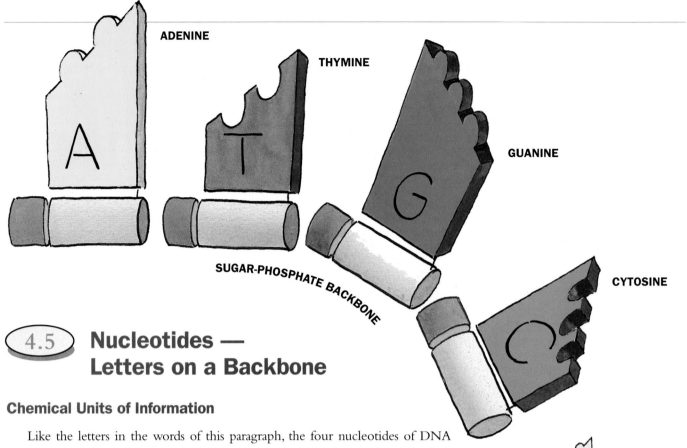

A Chemical Can Genetically Change Cells

In 1928, Frederick Griffith, a London medical officer, made a momentous discovery. At that time, the major cause of death worldwide was lobar pneumonia, caused by the pneumococcus bacterium. Scientists knew that certain mutant forms of these bacteria were benign; i.e., they didn't cause disease. Griffith discovered that if he mixed these *living* harmless pneumococci with dead disease-causing pneumococci and injected the mixture into mice, the mice all died of pneumonia. Moreover, their bodies were teeming with living, multiplying killer pneumococci! Something had been released from the dead killer cells and got inside the living benign cells and changed their inheritance; harmless cells had been permanently *transformed* into killer cells by engulfing the DNA of the dead killer cells. What was this transforming substance? Griffith was never to know — he died in the bombing of London in 1941.

It took many years of laborious chemical analysis and the painstaking development of methods for purifying and testing cell components before Oswald Avery, Colin MacLeod, and Maclyn McCarty at the Rockefeller Institute in New York announced, in 1944, that the transforming agent was DNA (until this time, protein was thought to be the hereditary material). Their work confirmed that DNA is the genetic molecule; genes are made of DNA.

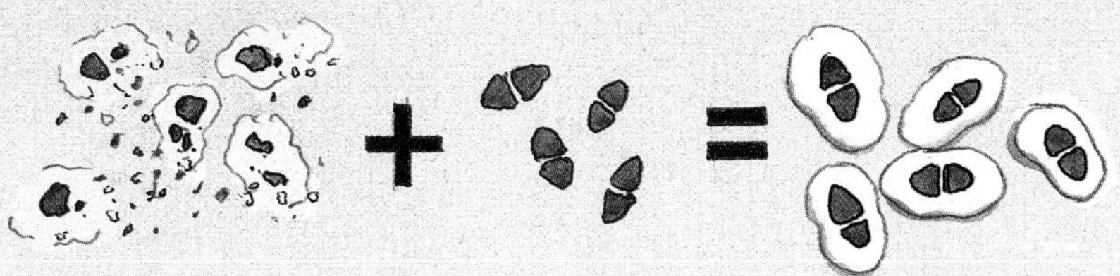

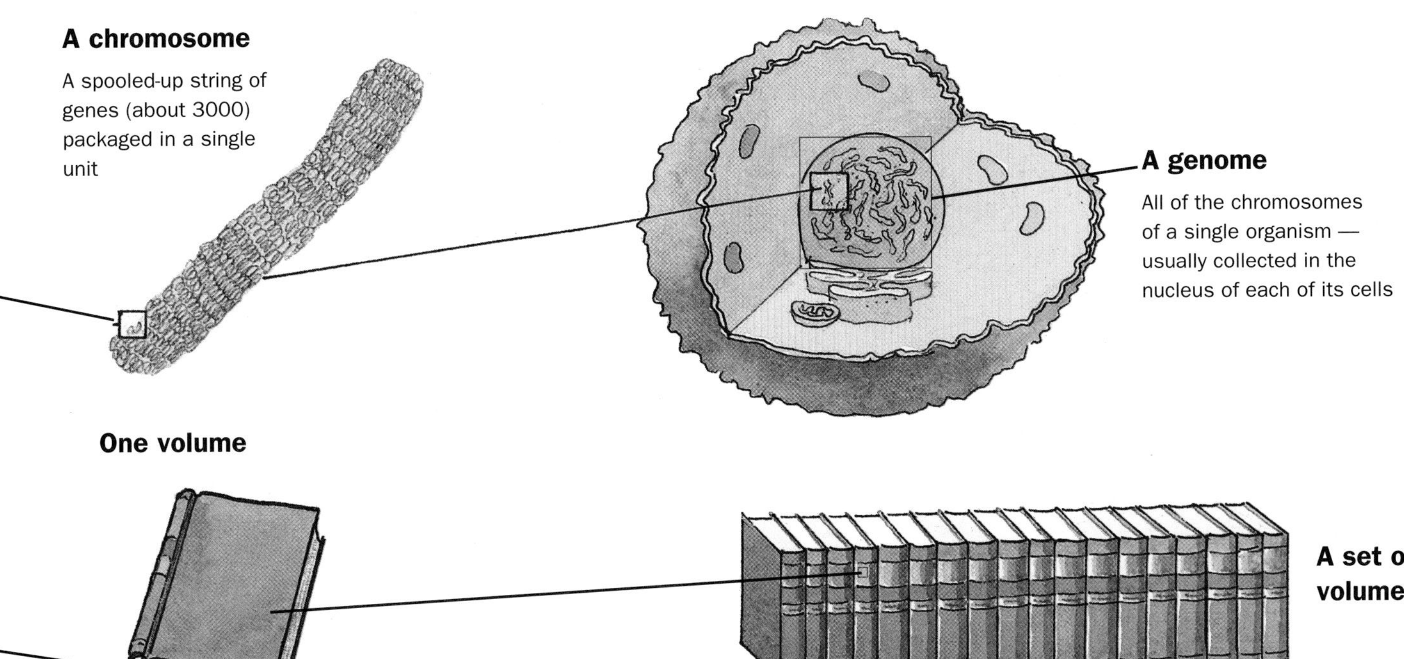

1. The nucleotides adenine and thymine...

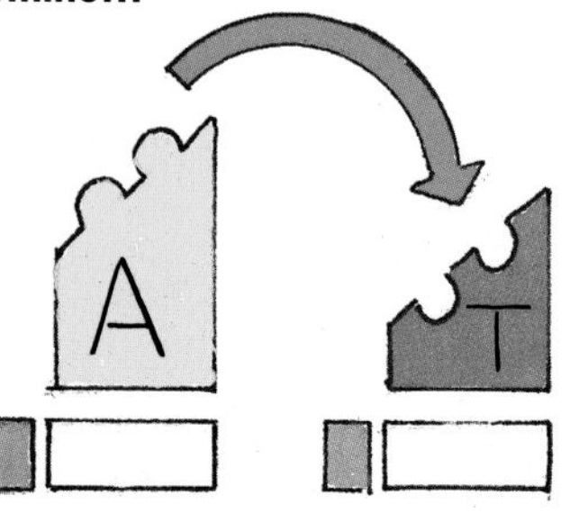

2. ...go together in a perfect fit.

4.6 DNA — Base Pairs and Weak Bonds

Nucleotide Pairs — A Key to Structure and Function

DNA is always found as a double chain, one lineup of different nucleotides paired with another. As you can see in the illustrations, the base parts of the four nucleotides (marked A, T, G, and C) match up in pairs. Their shapes and chemical make-up are such that A fits only with T and G fits only with C. These pairs, when fitted together, have exactly the same width (the distance from sugar to sugar). So the sequence of nucleotides in one chain of DNA will exactly match a complementary sequence in the other chain — and the backbones of the two chains will always be exactly the same distance apart. For example, if the sequence of nucleotides on one side is G-T-A-C-C, the sequence on the other side is C-A-T-G-G.

Moving In on the Structure of DNA

James Watson and Francis Crick, who began to work together in Cambridge, England, in 1951, believed that if they could visualize the form of a DNA molecule, they might see how it carried information and how it made copies of itself. They already knew a lot about DNA's chemistry timeline. DNA was first discovered by Johann Miescher in Switzerland back in 1869, and, over the years, many chemists had identified its four nucleotides and found out how they were linked in a chain. Furthermore, in 1949, Irwin Chargaff, a chemist at Columbia University in New York, had shown that while samples of DNA taken from different organisms — animals, plants, yeast, or bacteria — contained different amounts of the four nucleotides, the amount of adenine in each sample *always* equaled the amount of thymine, and the amount of guanine *always* equaled the amount of cytosine. At the time, no one knew why the quantities of these pairs of nucleotides showed this consistent relationship. What structure could account for this property?

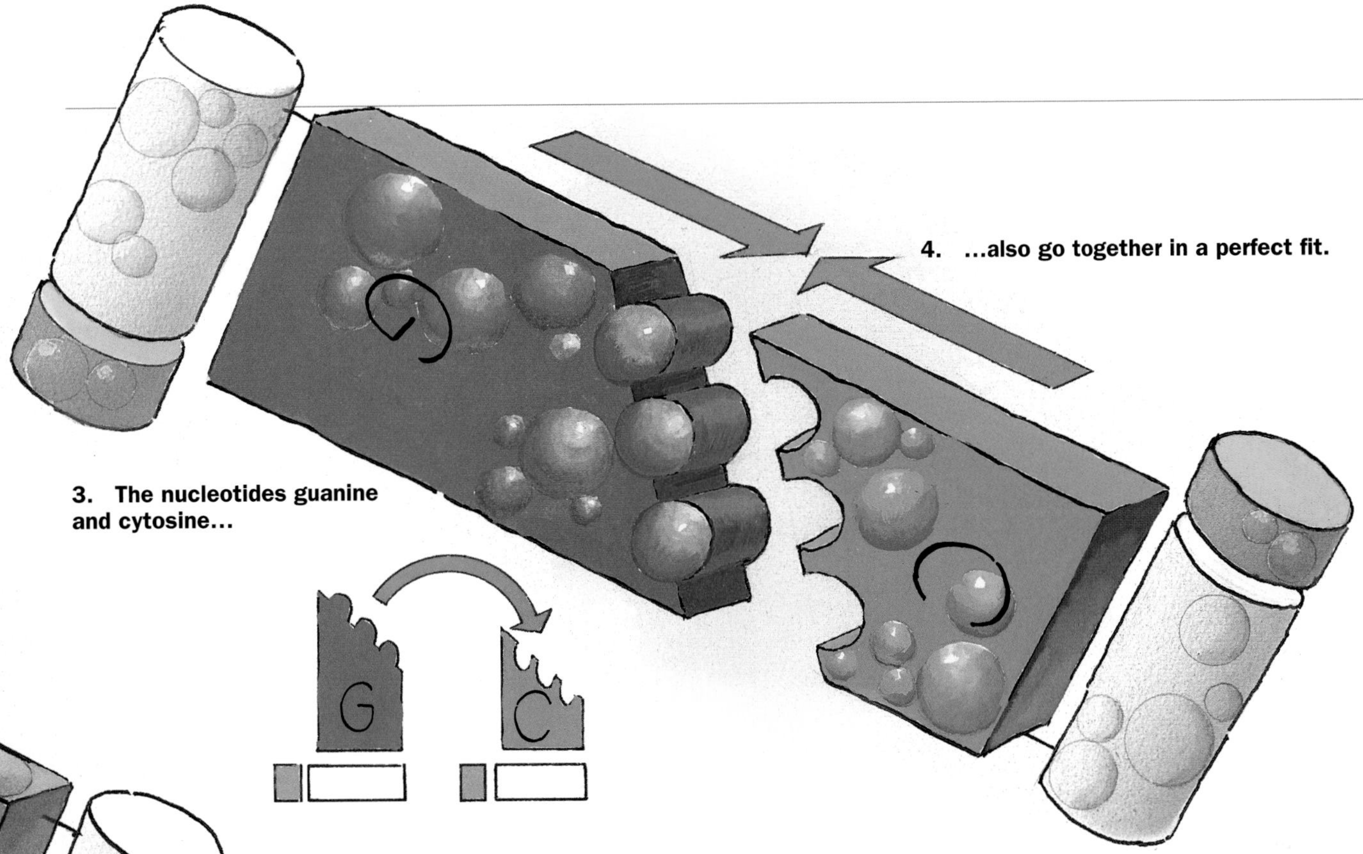

3. **The nucleotides guanine and cytosine...**

4. **...also go together in a perfect fit.**

G C

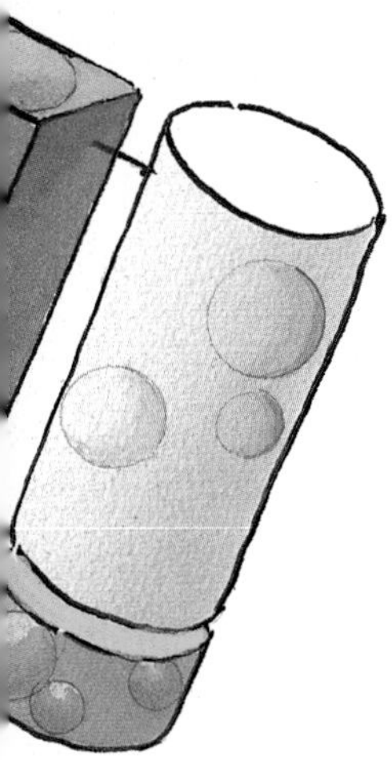

One chain is a counterpart, or complement, of the other. Note, too, that because of the way A and T or G and C must match up, the two chains must have opposite chemical directions — indicated at right by the opposing arrows.

Nucleotide pairing enables the two chains of DNA to fit together perfectly.

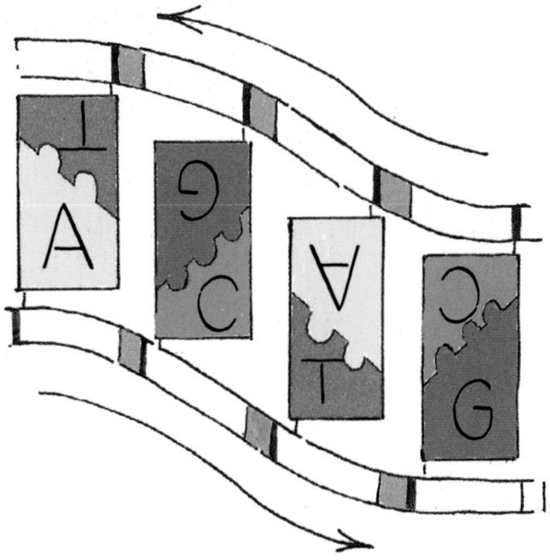

T A G C A T C G

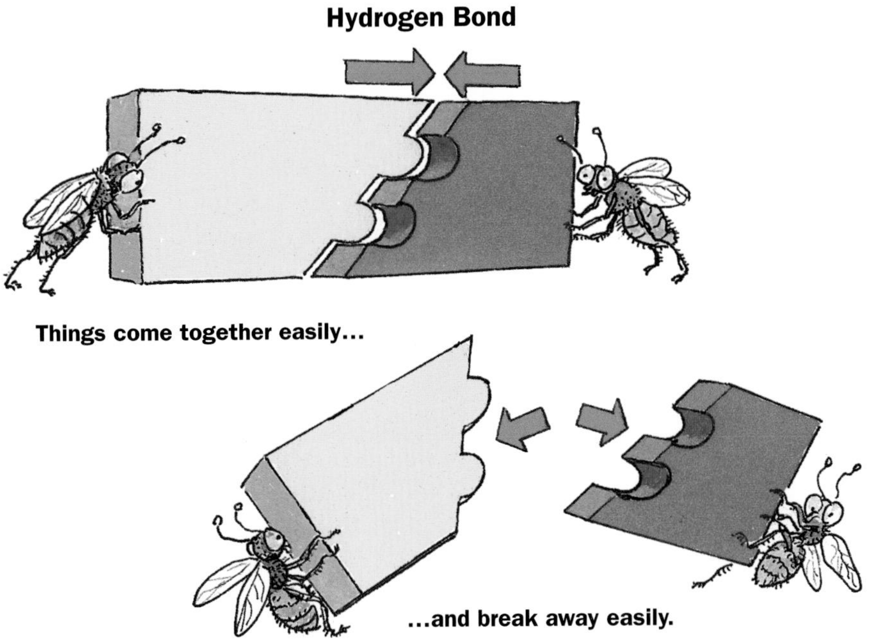

Hydrogen Bond

Things come together easily...

...and break away easily.

Weak Bonds

Weak bonds make it possible for big molecules to change shape or come apart and rejoin. Twenty times weaker than the covalent bonds that hold atoms together in molecules, these weak attractions between positive and negative charges can form only at very close range. Such weak bonds hold A to T and G to C in DNA and allow the two chains to separate readily, which they must do to replicate themselves.

4.7 DNA — The Double Helix

Information with a Twist

DNA resembles a spindly ladder that has been twisted so that its sides form spirals. Its exceeding thinness means it can easily be packed into small places. Its doubleness ensures that it won't get tangled up in itself; and it also protects the precious inward-facing nucleotide sequence — DNA's letters — from damage. And, as we shall see, this doubleness is what allows DNA to be copied.

Bacteria carry their DNA in one long double helix. In our cells, the DNA resides within the nucleus in 46 chromosomes — 46 double-stranded helices. The chains are stupefyingly long: If we think of the links in each DNA chain as letters, bacterial DNA represents about 60 average novels; human DNA about 1500! If all of the DNA in one of our cells was laid out end to end, it would be about 2 yards long. For a double chain that long to fit into a space as small as a cell nucleus, it must be incredibly thin and thus capable of folding and wrapping its strands up into dense, small structures. Since we have about 5 trillion cells, the *total* length of DNA in each of us would reach the 93 million miles from here to the Sun 30 times.

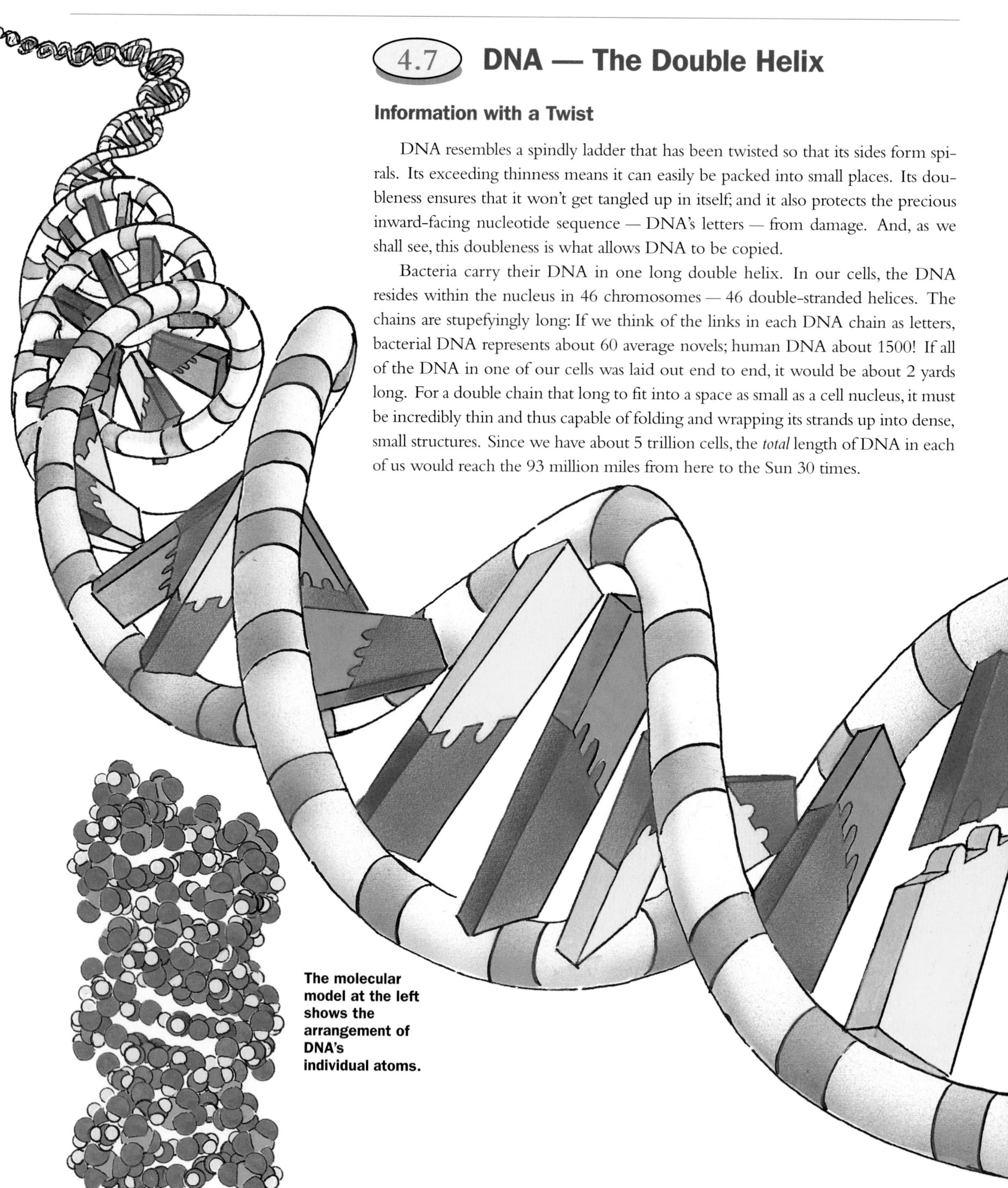

The molecular model at the left shows the arrangement of DNA's individual atoms.

Doing Science

Watson and Crick Discover the Structure of DNA

In London in 1952, Maurice Wilkins and Rosalind Franklin were using a process called X-ray diffraction (see page 13) to examine the shape of DNA. They shone X-ray beams through DNA and recorded on photographic film the pattern of scattering caused by the DNA molecules. Their work suggested that DNA was in the form of two or three chains whose bases somehow stacked near one another.

At Cambridge, Watson and Crick made cardboard and then sheet metal cut-outs of the nucleotides, based in part on knowledge obtained by Rosalind Franklin and Maurice Wilkins. This model-building approach was a key to their ultimate success.

A big eye-opener came when Watson and Crick learned that the *molecular* shapes of DNA's nucleotides were such that adenine fit *only* with thymine, and guanine fit *only* with cytosine. This made sense of Chargaff's discoveries (page 154). When Watson and Crick "mated" these base pairs inside DNA's sugar-phosphate backbones in a double helix, everything fit beautifully.

Watson and Crick triumphantly presented their model to the scientific world in 1953. Its acceptance was immediate, not only because of its intrinsic elegance but because it at once suggested how DNA could replicate itself: One strand was a complementary copy of the other; if the two strands were separated, new nucleotides could be laid down along each to form new strands (see page 159).

WEB Connection

www.jbpub.com/connections

DNA — Creating Its Own Future

Doubling of Information

Before a cell divides to become two, its DNA must be doubled so that each daughter cell will receive a perfect copy. This means the strands of DNA must first be separated, then complementary nucleotides must be linked along each of the separated strands.

DNA Replication — The Basic Idea

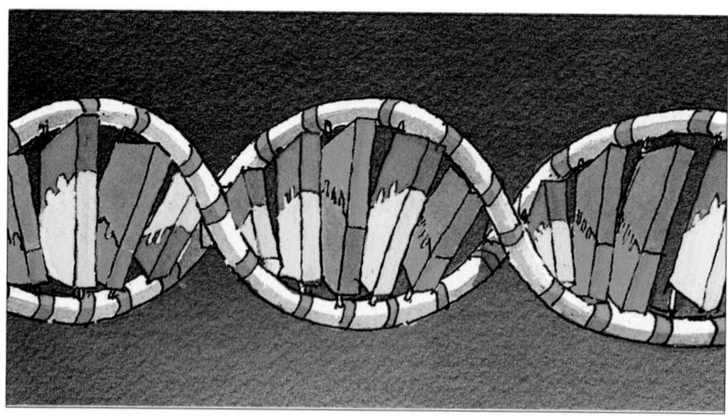

A double strand of DNA...

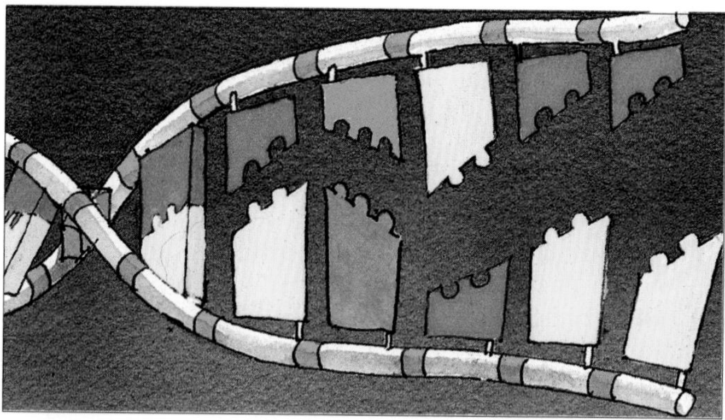

..."unzips" like a zipper, making its nucleotide bases accessible.

Free-floating nucleotides (made in other parts of the cell match with their complements...

...and link up, the original strand serving as a template (pattern) for the new.

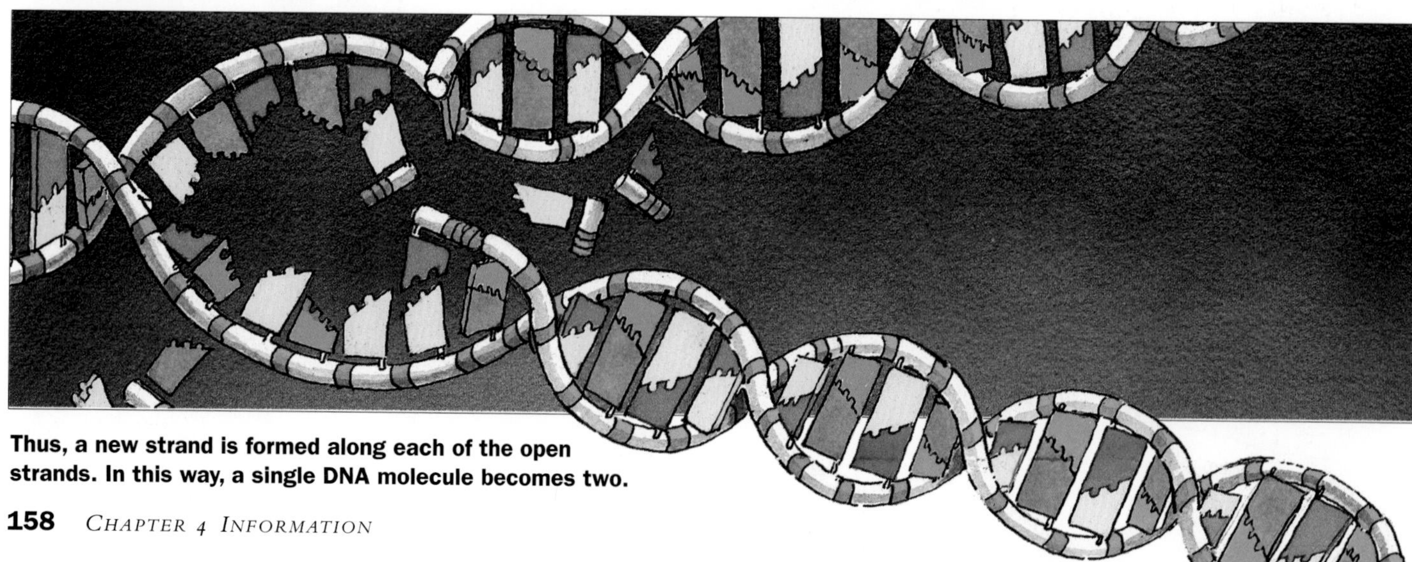

Thus, a new strand is formed along each of the open strands. In this way, a single DNA molecule becomes two.

4.9 How Enzymes Copy DNA

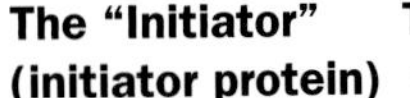

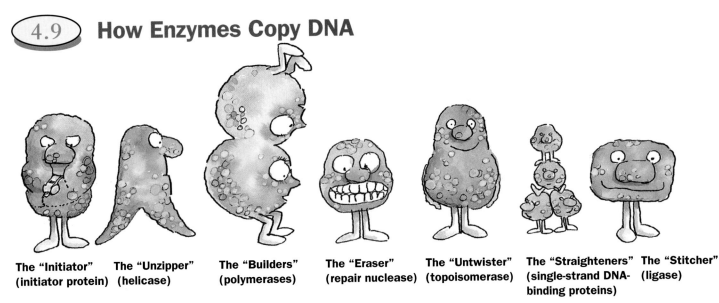

The "Initiator" (initiator protein) The "Unzipper" (helicase) The "Builders" (polymerases) The "Eraser" (repair nuclease) The "Untwister" (topoisomerase) The "Straighteners" (single-strand DNA-binding proteins) The "Stitcher" (ligase)

A Cast of Ingenious Characters

The sequence at the left oversimplifies. DNA doesn't copy itself any more than a recipe bakes a cake. DNA passively stores information. The team of proteins (enzymes) shown above and found in all self-replicating organisms does the actual copying, or replication. They do it with an accuracy of only one mistake in every hundred thousand or so nucleotides!

DNA Replication — The Details

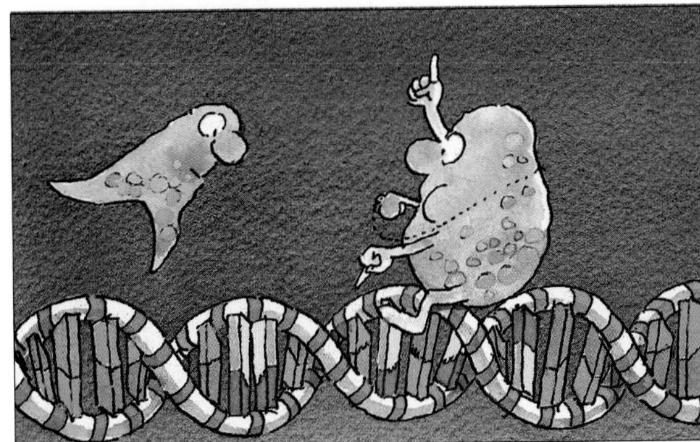

1. The initiator finds the place to begin copying and guides the unzipper to the correct position.

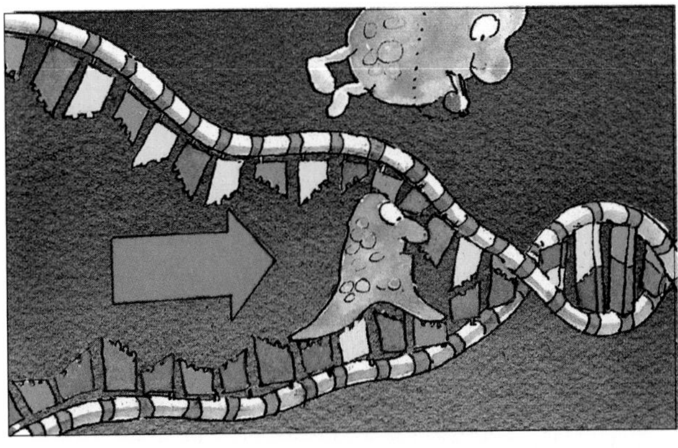

2. The unzipper separates the DNA strands by breaking the weak bonds between the nucleotides.

3. Then the builders arrive to assemble a new DNA strand along each of the exposed strands.

4. They build by joining individual nucleotides to their matching complements on the old strand.

DNA Replication — The Details

5. Free-floating nucleotides bring their own energy.

6. As each new nucleotide is added to the growing chain, its phosphate bond energy goes into making the new bond.

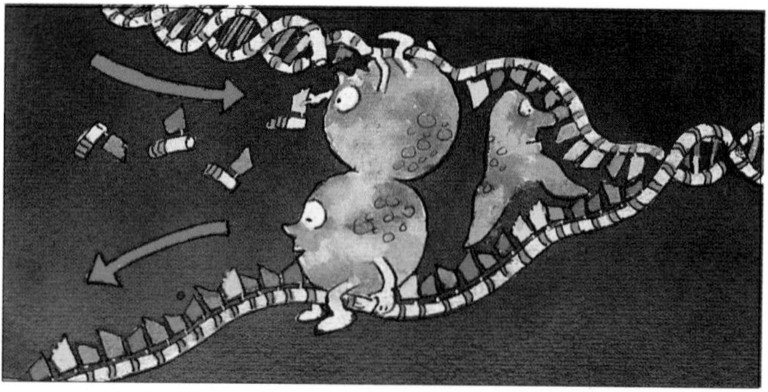

7. The upper builder follows behind the unzipper, but the lower strand runs the opposite way.

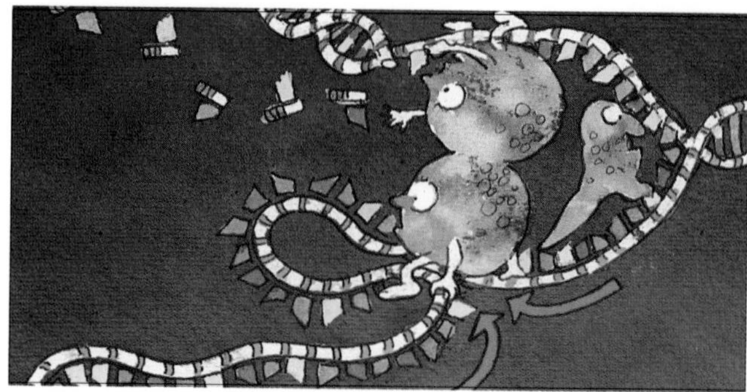

8. Yet the lower builder must build in the same chemical direction. She solves this by making a loop...

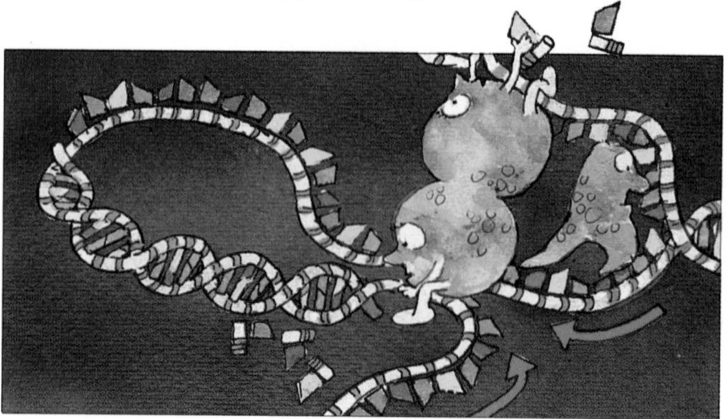

9. ...and building along the bottom half of it.

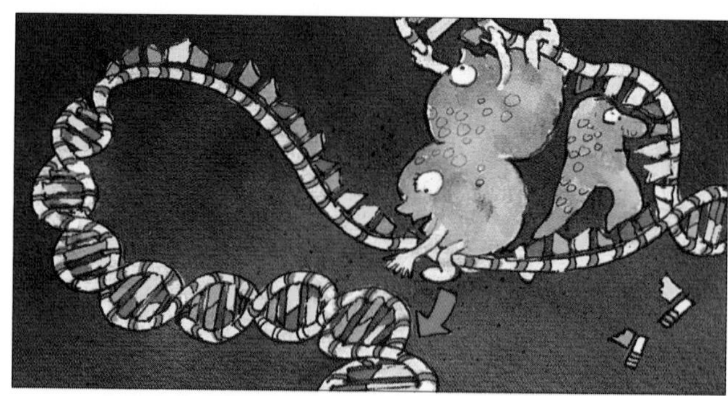

10. When she finishes a length, she lets go of the completed end...

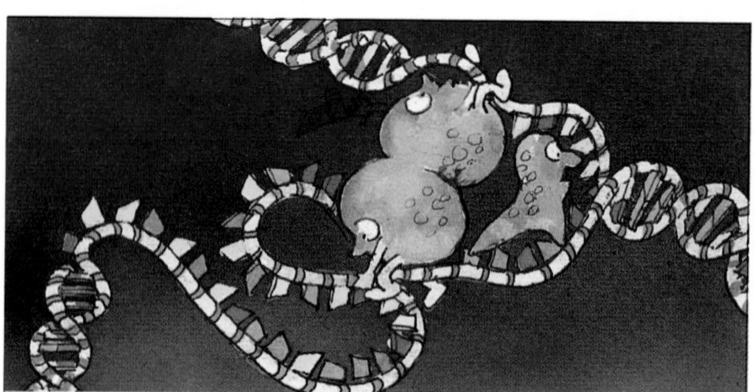

11. ...grabs a new loop, and continues linking nucleotides along a new stretch.

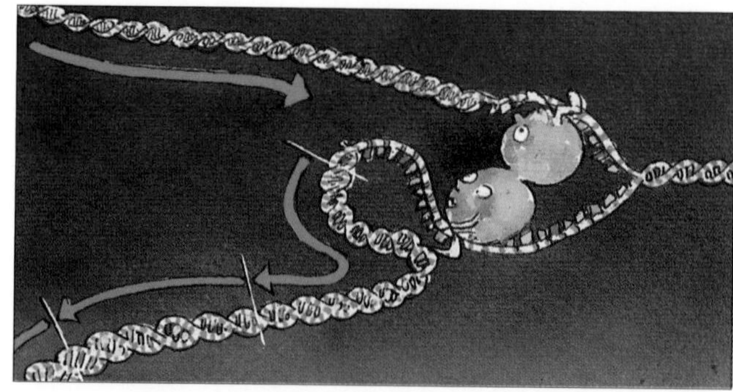

12. So, while the top new strand is built continuously, the bottom new strand is assembled in short lengths...

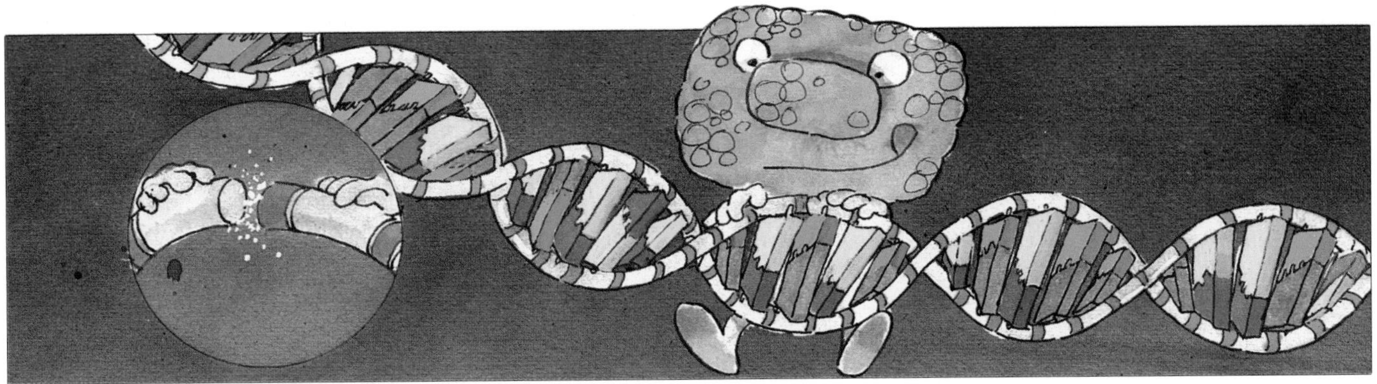

13. ...which are then spliced together by the stitcher. This reaction requires energy, supplied by ATP.

14. The straighteners keep the single DNA strands from getting tangled.

15. And the untwister unwinds the double helix in advance of the unzipper.

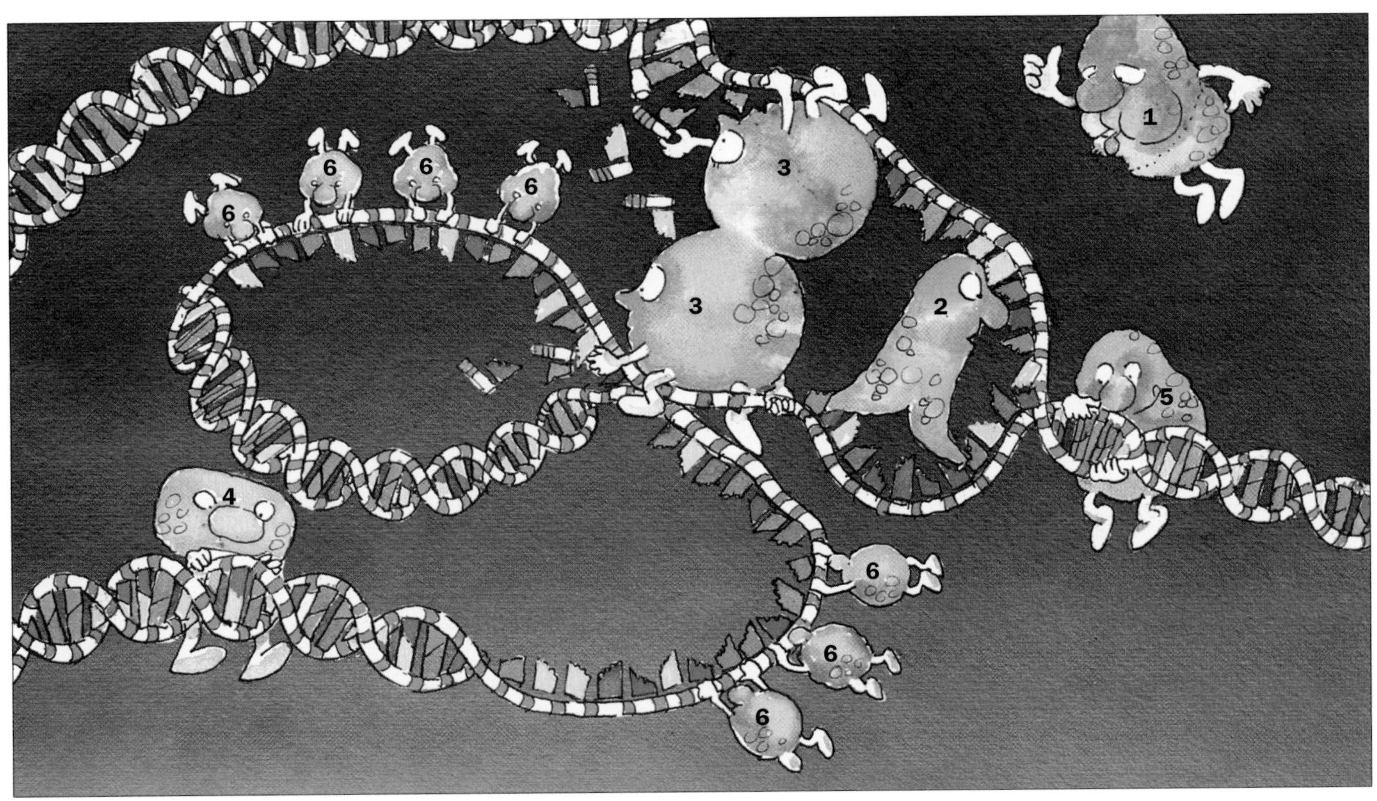

16. The initiator (1), the unzipper (2), the builders (3), the stitcher (4), the untwister (5), and the straighteners (6) work together in tight coordination, making near-perfect copies at the rate of fifty nucleotides per second!

4.10 Genomes

It has become increasingly clear that full disclosure of the language genes speak — the sequence of their nucleotide paragraphs — is essential to clarifying how their protein and RNA products interact in health and disease. We can now envision the eventual successful prevention or treatment of large numbers of purely genetic diseases, as well as major illness like cancer, cardiovascular disease, diabetes, multiple sclerosis, and Alzheimer's. And gene sequencing can be expected to improve the nutritional value and disease resistance of crops.

From a more basic research perspective, gene sequencing will continue to probe deeper into the evolutionary and present relationships of humans to all other living creatures and into how we grow, develop, and function. With the advent of techniques to be described in the following pages, and with ever increasing computer capacity, scientists have come to realize it would actually be possible to determine the full sequence of letters in the human genome — all three billion of them. It has long be recognized that each species has its own unique sequence of genes. Between any two humans, for example, there are many differences *within* their genes but there is only one *order* of genes common to all — only one "human genome."

The completion of the sequencing of the human genome is only the beginning. For once we know all the letters in the paragraphs and books, we must still find out what they mean — what are the protein products of those genes, how do they interact with and control each other, and how does their failure to function compromise our health.

Right now, we don't even know how many genes are in the human genome. Estimates range from 40 thousand to 120 thousand, and we're a long way from knowing what their functions are. Furthermore, only about 5% of our genome *is* genes. The function of the other 95% of our DNA is unknown.

Identifying DNA's Uniqueness

Simple viral genomes were the first to be sequenced. Then, in 1996, a yeast, *Saccharomyces cerevisiae,* which has fewer than 7000 genes (about 13 million nucleotides) was sequenced. In 1997, researchers announced that the sequence of the common intestinal bacterium *Escherichia coli,* which has about 4300 genes (more than 4 million nucleotides). Less than two years later, scientists reported that the first animal genome had been sequenced. *Caenorhabditis elegans,* a millimeter-long nematode (worm) that has nearly 1000 cells as an adult, (see the illustration, top right) and some 20,000 genes (97 million nucleotides). Researchers are currently working on the genomes of a number of other organisms.

C. elegans (magnified 130 times)

This minuscule worm's genome prescribes the types and arrangements of its thousand or so cells into all the body parts you see here.

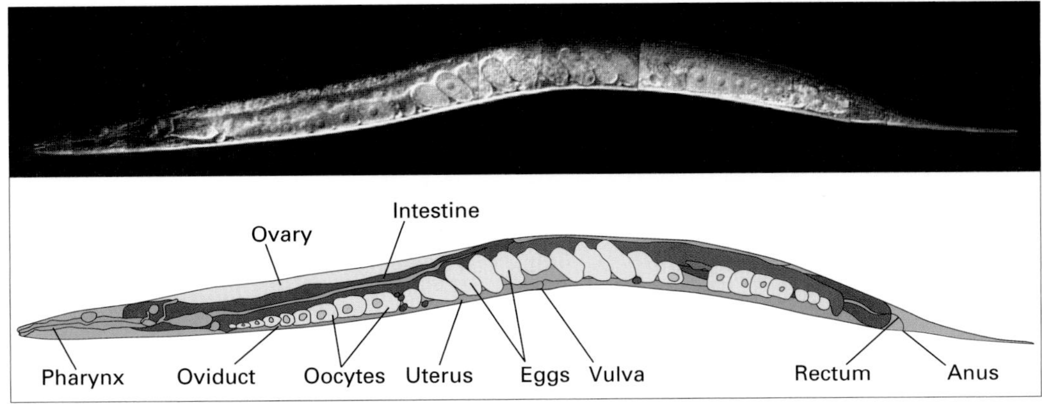

Success in moving on will continue to come from the study of simple models — simpler organisms. In the case of *C. elegans*, researchers have identified functions for less than half of the nearly 20,000 genes. They plan to selectively inactivate by mutation each of the remaining genes in an attempt to determine what it does. And sequencing of the mouse genome is nearing completion, which should speed up annotation of the human genome. Not only do the mouse genome and the human genome has approximately the same number of nucleotides (about 3 billion), but mouse genetics has been studied extensively for years and the functions of many genes have already been described.

Nearly every great scientific or technological advance carries the potential for both great good and great harm; genetic engineering, a tool already widely used in pharmaceutical manufacturing and plant science, is no exception. Critics worry, however, that genetically engineered organisms will have unforeseen effects on other organisms and the environment. And the potential for deliberate introduction of genetically engineered superpathogens can't be dismissed lightly. Genetic engineering isn't going to go away, however; in fact, its potential has only begun to be tapped.

Reading Genomes — The Technology: 1

A length of DNA from any chosen source can be inserted into the DNA of living bacteria and copied repeatedly as the bacteria multiply. The discovery of this technique, called recombinant DNA technology, opened the door in the mid-1970s to the modern era of genetic engineering. In practice, the subject DNA is first spliced into the DNA of a plasmid — a wandering virus-like piece of DNA that enters bacterial cells and replicates in synchrony with them (see pages 324–325). In a couple of days, millions of bacteria will have accumulated and their millions of copies of passenger DNA can be extracted and studied.

DNA polymerase

A new technique now in widespread use because it is faster and readily automated is PCR — polymerase chain reaction. It makes use of *DNA polymerase**, the enzyme in all living cells that makes DNA. It will copy as little as a single molecule of DNA from any source: such as body tissues and fluids, disinterred bodies, preserved prehistoric specimens, etc. The process involves the following steps:

1. The subject DNA to be copied (up to some 5000 nucleotides in length) is heated to separate the two strands.
2. Short stretches of RNA (5–20 nucleotides in length) called *primers*** synthesized to be complementary to short DNA stretches at either end of the subject DNA, are added.
3. Polymerase and lots of the four nucleoside triphosphates complete the mix.
4. Synthesis of new DNA proceeds to completion along each of the separated subject strands, starting from the primers at either end. This results in two double strands. The cycle is repeated: reheating to separate the strands and adding more nucleoside triphosphates (free nucleotides) and primers.

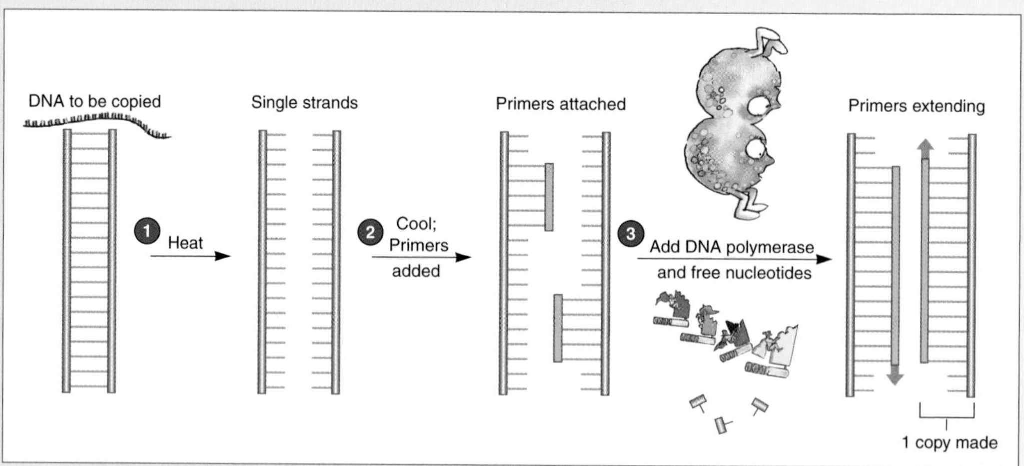

*Ordinary DNA builders - DNA ploymerases, like almost all proteins, are heat-sensitive — i.e., their action is crippled by the heat needed to separate the DNA strands in the first step. This problem is ingeniously avoided by using a heat-stable polymerase from a species of bacterium whose natural habitat is hot springs (page 79) where the temperature is near the boiling point of water! The particular bacterium used is *Thermus aquaticus* and its builder is called *Taq* polymerase.

So we start with a test tube containing the four kinds of nucleotides, a couple of primers, some *Taq* polymerase, and a sample of DNA, and we subject it to several dozen PCR cycles. What have we got? We have millions of copies of the targeted section of the sample DNA. The targeted section could be a gene or part of a gene, a mutation or infective viral or bacterial material. All nontargeted parts of the original DNA are essentially invisible after such extensive amplification. PCR is so sensitive that it can make copies from degraded DNA of miniscule cell fragments.

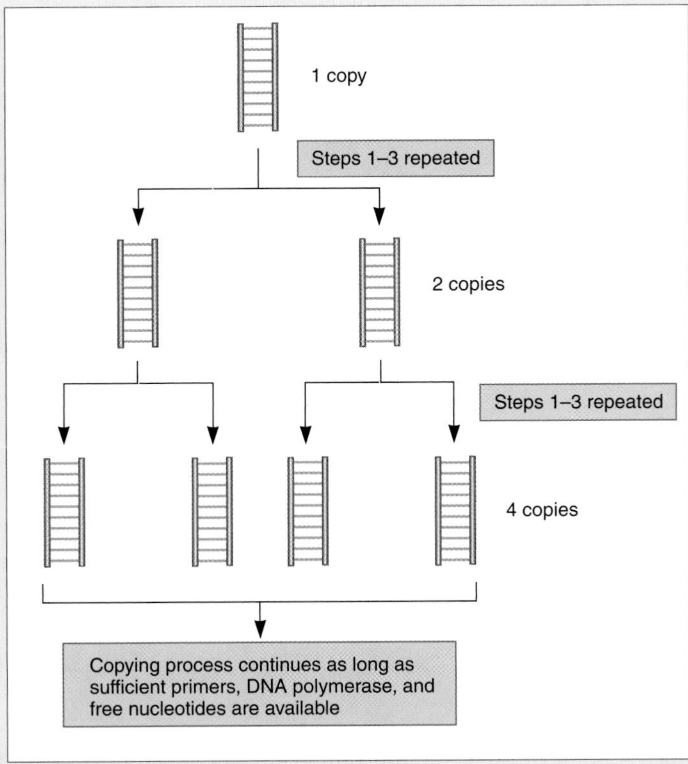

PCR has been used to study the DNA of mummies, extinct animals, and even an amber-encased termite from 3 million years ago. Researchers managed to extract some DNA from Neandertal bones, amplify it with PCR, and compare it with the DNA of modern humans. Their conclusion? Neandertals were not our ancestors; they diverged genetically from our ancestors 500,000 to 600,000 years ago. PCR is now used every day for many other laboratory applications, including, for example, routine forensic investigations, HIV detection when infection levels are very low, and amplification of unexpressed DNA molecules for cloning or genetic engineering.

**RNA primers are essential for DNA synthesis by DNA polymerase in nature and in the test tube. An enzyme called RAN polymerase first builds a short primer complementary RNA onto a chain to be copied. The DNA polymerase then initiates synthesis from the primer. The primer is later removed and replaced with DNA nucleotides. This step was omitted for simplification from the description of DNA synthesis on pages 158–161. The need for primers poses a limitation to PCR. To make them, the chemist must know the sequence of bases at either end of the subject DNA.

DNA Fingerprinting — The Technology: 2

All living creatures, and even members of the same species, have their own, unique, DNA text. Identifying an individual's DNA depends on demonstrating differences in nucleotide sequences. Fingerprinting human DNA focuses on stretches of nucleotides that are particularly variable from one person to another — called VNTRs: variable numbers of tandem repeats. (Five to ten percent of the human genome consists of these stretches of as few as two to as many as several thousand nucleotide sequences repeated over and over. Their function is unknown.)

VNTRs are snipped into smaller pieces by restriction enzymes. These are enzymes from bacteria that cut DNA double strands at particular short sequences of nucleotides — each enzyme recognizing and cutting at one and only one short sequence.

If several restriction enzymes are used to snip DNA in several different locations, a number of fragments will be produced of varying lengths. The number and lengths of the fragments depend on which restriction enzymes are used.

The fragments need next to be separated and visualized. This is done by gel electrophoresis. The chopped-up DNA samples are placed on the top of a gel — a sheet of jello-like material through which the fragments migrate. (Movement of the fragments, which are negatively changed, is in response to an electrical field through the gel — positive at the bottom.) The shorter the fragments, the faster and farther they move.

The gel sheet with the fragments now separated is treated to convert the fragments' double strands to single. It is then pressed against — blotted onto — a sheet of nylon to which the single strands stick. The fragments are made visible by annealing them with single-strand radioactive DNA fragments (called DNA probes). These bind to their

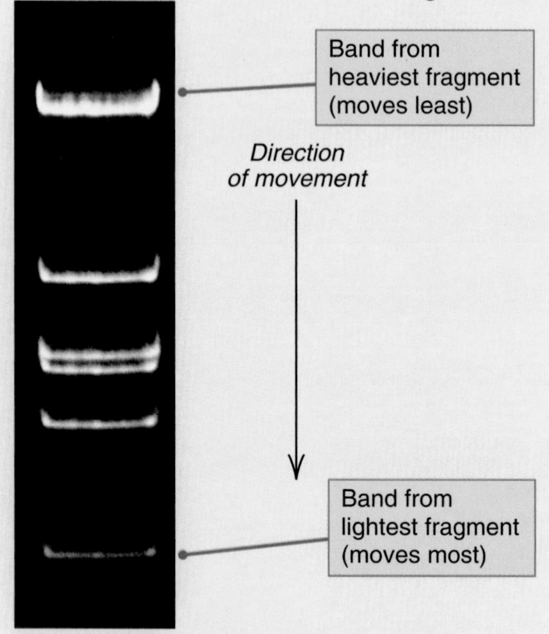

The apparatus

The gel acts a bit like a fine sieve. It's easier for small fragments to move through the sieve than it is for large ones, so the larger DNA fragments lag behind smaller ones as they progress through the gel.

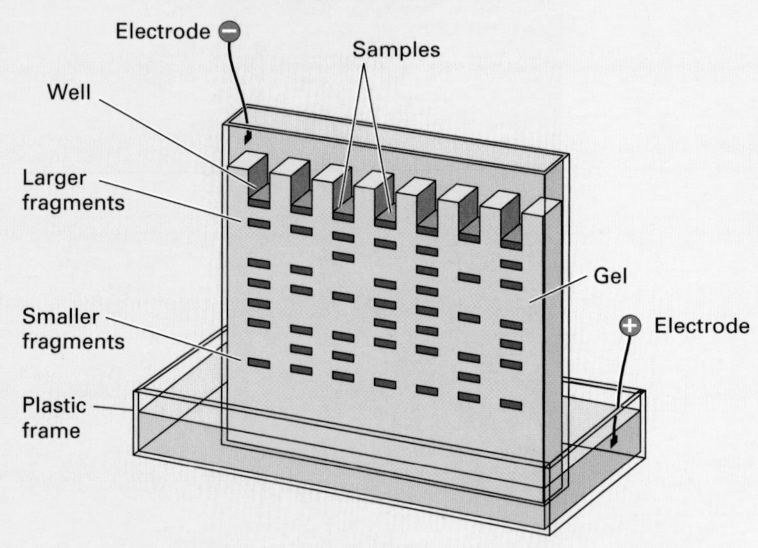

The outcome track

The dye binds to the DNA fragments, whose sizes are then measured by their positions along the gel track. Fragments of the same DNA line up with one another; fragments of differing DNA line up differently.

complements on the nylon sheet and, when the sheet is laid onto photographic film, show up on the film as dark bands.

The human genome yields, by both these methods, distinctive patterns of bands — distinctive fingerprints.

Using DNA Fingerprinting

Aside from its obvious uses in forensic analysis and legal proceedings, DNA fingerprinting is used to determine parentage (of animals and plants, as well as humans), to assess donor-recipient compatibility, and to optimize mate selection for captive endangered species. Patent cases involving genetically engineered or selectively bred organisms have been settled by DNA fingerprinting. The technique has also been used to track the source of Caspian caviar in an effort to protect nearly extinct species of sturgeon, and to identify the origin of hides, tusks, and meat in cases of suspected poaching. DNA from ancient plant, animal, and human remains has been analyzed for information on species and population evolution.

In December 1999, French scientists began DNA testing on the preserved heart of the presumed Dauphin Louis XVII of France, whose parents (the king and queen, Louis XVI and Marie-Antoinette) were guillotined in 1793 at the height of the French Revolution. The child was reported to have died in prison two years after his parents were executed, but numerous rumors of his escape surfaced over the years, and a number of persons claimed to be him or one of his descendants. The heart of the boy who died in prison, which had been removed and preserved at autopsy, changed hands a number of times but ended up at the royal crypt of the Saint-Denis cathedral outside Paris. DNA from Marie-Antoinette's hair follicle cells and from those of two of her sisters were compared with DNA from the preserved heart and from two of the sisters' known living descendants. The comparison confirmed that the heart was indeed that of the unlucky ten-year-old.

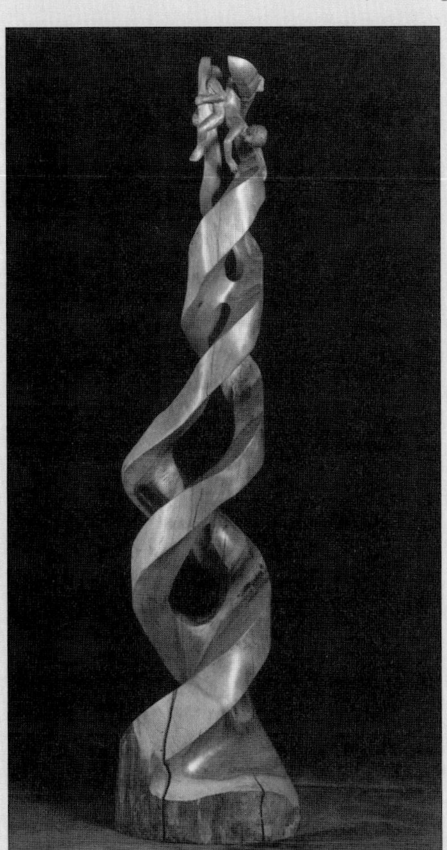

Mahlon Hoagland, *Exploring DNA,* 1990

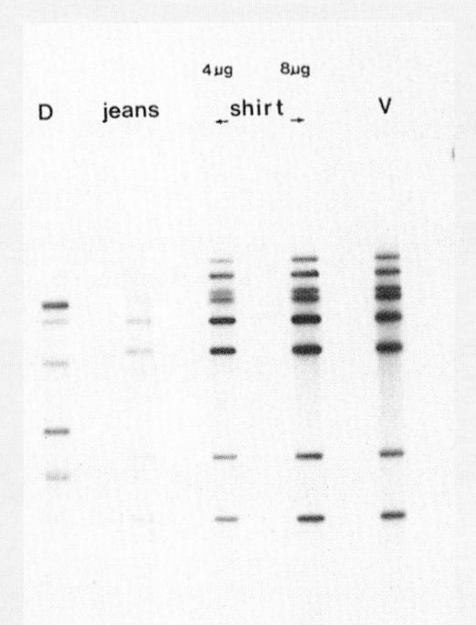

Comparing the "fingerprints"

Here you see DNA fragments from one person compared to those from another. Column V shows the DNA of a stabbing victim. Column D shows the DNA of a suspect in the stabbing. The columns labeled "jeans" and "shirt" are DNA from blood cells found on the suspect's clothing. What can you infer from this comparison?

The discoveries of Science, the works of art, are explorations — more, are explosions, of a hidden likeness.

Jacob Bronowski, *Science and Human Values,* 1956

Reading a Gene Sequence — The Technology: 3

Here is one way DNA sequencing is performed.

DNA is cut into lengths of a few hundred nucleotides by restriction enzymes. These are separated into single strands. Four reaction tubes are set up by containing:

1. The single-strand DNA.
2. The four nucleotides ATP, GTP, CTP, and TTP.
3. A primer complementary to the first few nucleotides of the DNA. This is made radioactive so that the sequence of nucleotides added to it may be detected at the end of the reaction.
4. The enzyme DNA polymerase.
5. And, in each tube, a small amount of a chemically altered form of either ATP or GTP or CTP or TTP. These are chain stoppers.

DNA synthesis proceeds. The enzymes add nucleotides, one by one, starting from the primer, each new nucleotide complementary to a nucleotide on the DNA strand being copied. Whenever one of the chain-stopper nucleotides enters the chain synthesis stops. This produces newly made single strands whose lengths depend on how many nucleotides got into the chain before lengthening stopped.

The lengths of chains is determined by gel electrophoresis as in fingerprinting. This method is so exquisitely sensitive it can separate sequences of 100 or more nucleotides that differ in length by only one nucleotide.

For example, using a primer of 10 nucleotides, in the tube containing chain-stopper GTP, sequences of length 11, 14, 17, and 19 are found. We conclude that the new sequence added to the 10 nucleotide primer had G in positions 1, 4, 7, and 9. In the tube containing chain-stopper CTP, sequences of length 13, 15, and 18 are found. This means C must occupy positions 3, 5, and 8. Similarly, the tubes containing chain stoppers ATP and TTP show A to be in position 2 and 10 and T to be in position 6.

We conclude that the sequence of the copy made in this run is:
GACGCTGCGA.

Once this sequence is revealed, the complementary sequence of the input DNA is known.

There are now ways to automate this whole process so that thousands of nucleotides can be sequenced daily!

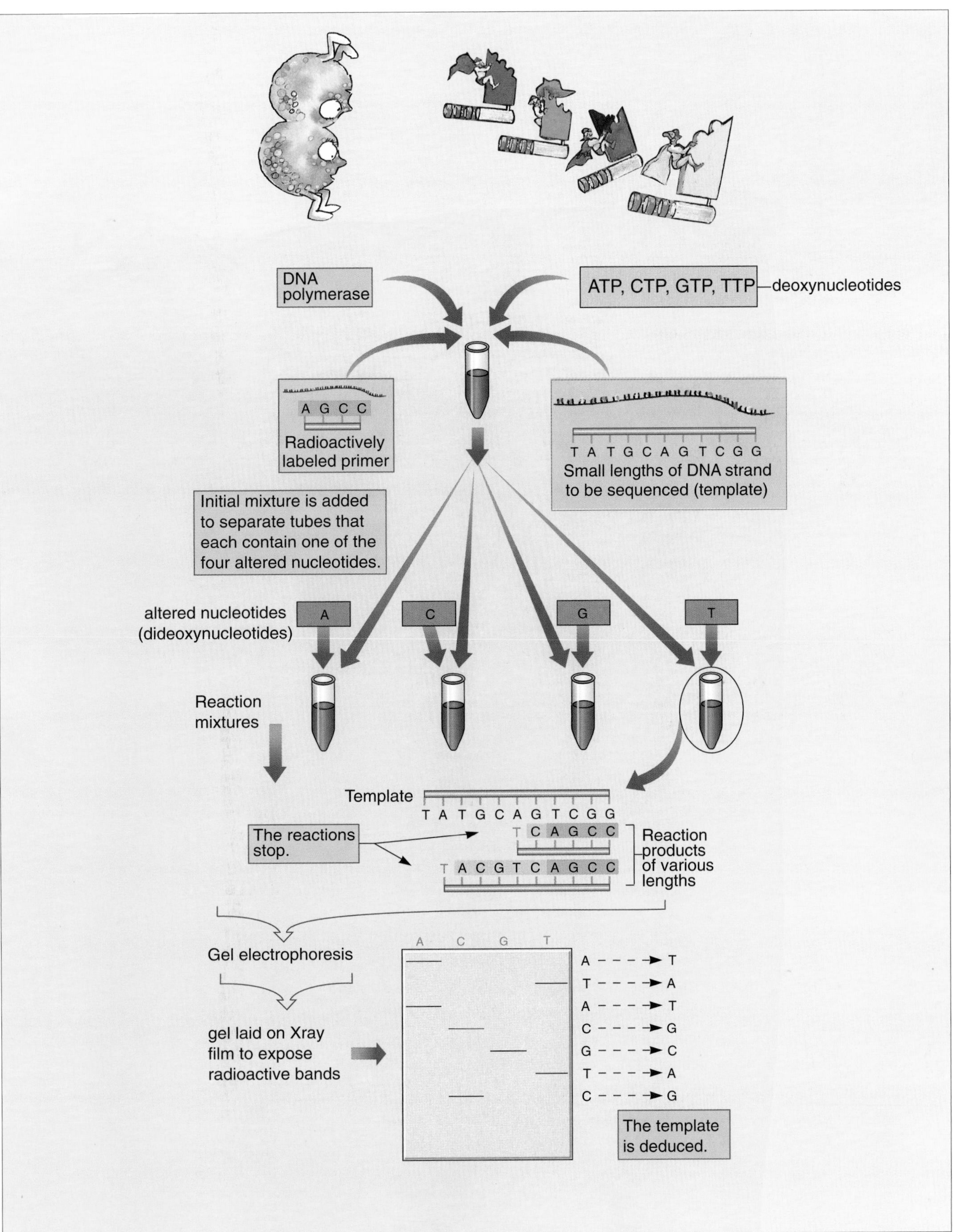
DNA polymerase
ATP, CTP, GTP, TTP
deoxynucleotides
A G C C
Radioactively labeled primer
T A T G C A G T C G G
Small lengths of DNA strand to be sequenced (template)
Initial mixture is added to separate tubes that each contain one of the four altered nucleotides.
altered nucleotides (dideoxynucleotides)
A
C
G
T
Reaction mixtures
Template
T A T G C A G T C G G
The reactions stop.
T C A G C C
T A C G T C A G C C
Reaction products of various lengths
Gel electrophoresis
gel laid on Xray film to expose radioactive bands
A C G T
A - - - -> T
T - - - -> A
A - - - -> T
C - - - -> G
G - - - -> C
T - - - -> A
C - - - -> G
The template is deduced.

DNA Repair

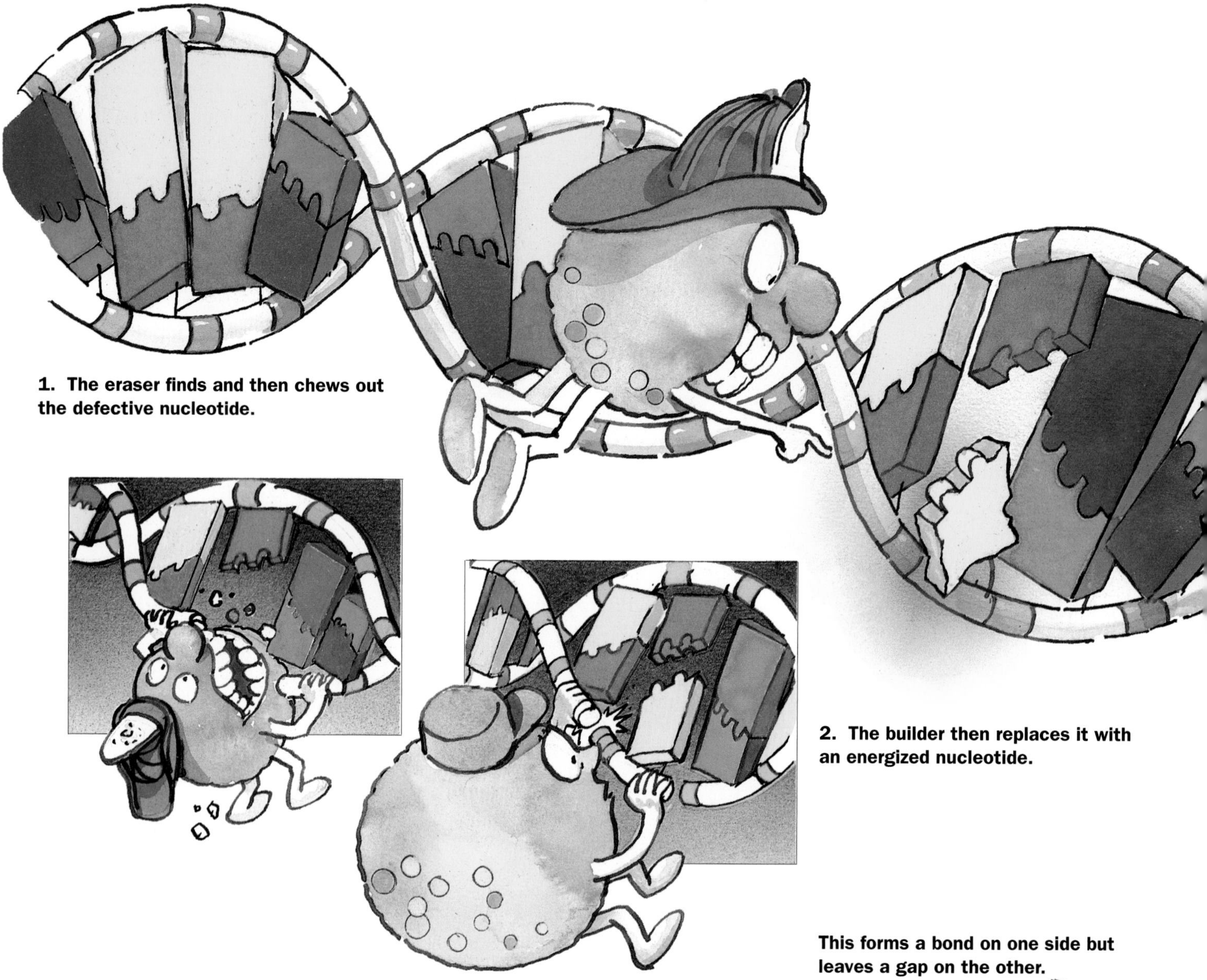

1. The eraser finds and then chews out the defective nucleotide.

2. The builder then replaces it with an energized nucleotide.

This forms a bond on one side but leaves a gap on the other.

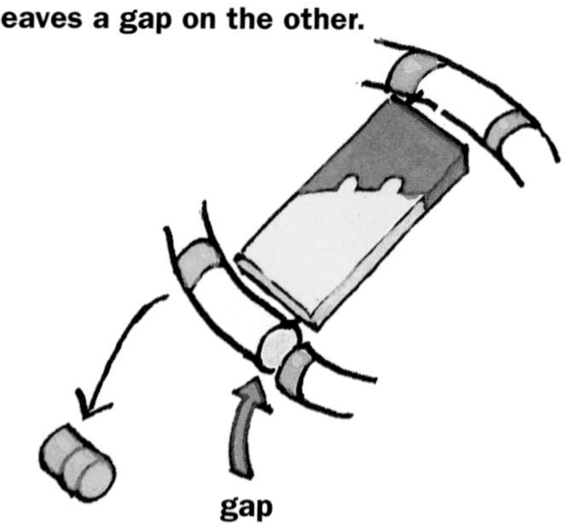

A Precise, Self-Correcting System

Although the system for copying DNA is extremely accurate, mistakes do happen; sometimes these mistakes can be devastating. Other threats to the integrity of DNA, which regularly damage nucleotides, include chemical events inside cells and ultraviolet light. The cell recruits an army of repair enzymes to handle these problems. Three kinds of repair enzymes regularly patrol DNA and repair any errors they find. First, erasers find poorly matching or damaged nucleotides and snip them out. Second, builders follow close behind to fill the gaps, using the other strand as a guide. Finally, stitchers restore the continuity of the backbone of the repaired strand.

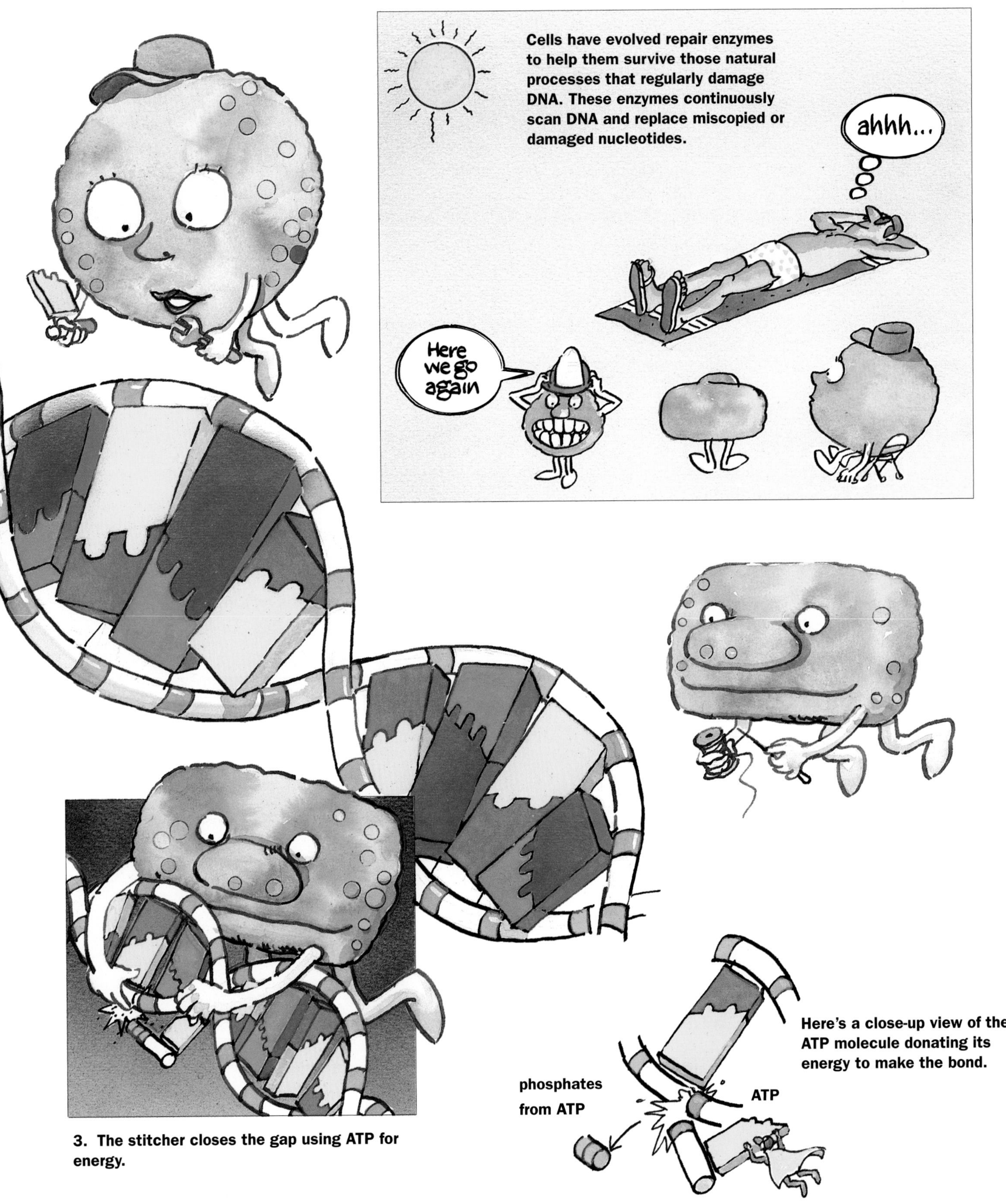

Cells have evolved repair enzymes to help them survive those natural processes that regularly damage DNA. These enzymes continuously scan DNA and replace miscopied or damaged nucleotides.
ahhh...
Here we go again
Here's a close-up view of the ATP molecule donating its energy to make the bond.
phosphates from ATP
ATP

3. The stitcher closes the gap using ATP for energy.

4.12 Permanent Changes in DNA

DNA has its own efficient repair mechanisms; however, sometimes even those mechanics can't do the job. The result is a mutation — a permanent change in the genetic material. Mutations can range from inconsequential to advantageous, inconvenient to fatal. The type of cell affected (germ cells, like eggs or sperm, versus all other somatic cells), the stage of development, and the recessiveness or dominance of the mutated gene are the determining factors.

Some mutations are spontaneous (which means they generally have unknown causes, but are often the result of replication errors during cell division); others are induced by exposure to some agent (such as certain chemicals and types of radiation) known to be a mutagen — a "hacker" that inserts, deletes, or rearranges a section of DNA code. If this occurs in a gene (instead of in the long "junk" sequences between genes), the substitution of even a single base pair can have serious consequences; sickle-cell anemia is a classic example (see page 51). You'll recall that each gene is the recipe for a specific protein. Proteins are made up of amino acids, and each of the 20 amino acids is specified by three nucleotides. In effect, a series of three-letter words (codons) makes up the paragraph that is the recipe for a particular protein. An insertion or a deletion, unless it is a multiple of three nucleotides, changes every codon from that point on to the end of the gene. Fortunately, the ratio of "junk" to protein-coding DNA segments is very large, about 95 to 1 in mammals, and most mutations have no detectable effects.

Mutations, as you recall from Chapter 2, play both advantageous and damaging roles, depending on their environment (Life Creates With Mistakes, see pages 47–49). An example given there is of various snow-dwelling species whose protective white coloration is a mutation that provides an advantage in a snowy landscape.

Spontaneous mutations are at the root of much of the selective breeding done by animal (and plant) fanciers. Witness the many breeds of domestic dogs — all belong to the same species *(Canis familiaris)*, but they vary amazingly in size, shape, color, and coat. Some of this variability is just part of the genome, of course. But over the years, certain mutations popped up (an unusual color, perhaps, or a longer or shorter coat) and were selectively bred for.

Standing out

You might wonder why this penguin, white among his darker brethren, stands alone on the Antarctic ice. Shouldn't his white coloration protect him and show up in many of his descendants? Alas, no. Antarctic penguins have no predators on land, where this one blends into the background. Underwater, though, he stands out like a McDonald's sign for the whales, sharks, and seals looking for fast food.

Life's few really bad mistakes tend not to live at all, or to die very soon, and thus don't get incorporated into the organism's gene pool. Somatic mutations although they can seriously affect an individual organism, don't ever make it into the gene pool (except through cloning). Germ-cell mutations are the ones that are passed along to the next generation. In cells that have DNA from two parents, a mutated gene can be either dominant or recessive. Dominant genes are always expressed whether the parental genes are the same or different, but recessive genes are expressed only when genes from both parents are the same. Many inherited diseases — cystic fibrosis is one — are due to mutated recessive genes that can be passed along for generations with no expression. Only when a carrier of such a gene mates with another carrier of that gene does the possibility for expression of disease exist, and even then an individual offspring has only a one in four chance of inheriting the recessive gene from both parents.

Just such a chance occurrence provided a means for studying immune system biology. In 1980, four laboratory mice came to the attention of an immunologist when their blood tests suggested that they had no immune reactions. As a result of a spontaneous mutation in one of their parents, the four littermates lacked the ability to make T and B cells, the white blood cells that fight disease and reject transplants. Today, the *scid* (severe combined immunodeficiency) mouse is essential to AIDS studies and donor/host tissue rejection research. The *scid* mouse can be implanted with human tissue, and thus human diseases, and will not reject the implants. Researchers can then experiment with various drug regimens and even gene therapy.

Sensitive cats

Siamese cats, bred for their coloration, have a "conditional mutation"; its expression depends on temperature. The enzyme responsible for black coloration is temperature-sensitive and inactive at normal body temperature. Black pigment is deposited in the hairs only on the cooler regions of the body: face, ears, paws, and tail. The mutation gave rise to the breed.

Like mice, fruitflies have been used for years in laboratory research. One of the first species to have its embryonic genes manipulated extensively, the fruitfly was mutated in the lab to produce, for example, myriad eye and wing variations as well as body part rearrangements. But the fruitfly is an invertebrate. The new kid on the block, a vertebrate, is the zebrafish *(Danio rerio)*.

The field is developmental biology, and the goal is to find the genes responsible for building the vertebrate embryo. A freshwater aquarium fish, the zebrafish (see page 257) is very hardy and has a 3-month life span. Its embryo is transparent, which allows researchers to observe every step of embryogenesis (which takes only 5 days), including nervous system development. Mutations are easily induced by exposing breeding adults to chemicals, viruses, or radiation. Many of these mimic human diseases and defects. Locating the precise gene affected, then cloning it, is the next step. One researcher, Nancy Hopkins at MIT, discovered that using viruses to induce mutations provided a ready-made label on the mutated gene. The potential benefits of fully understanding how vertebrate embryos develop are not lost on biotechnology firms, which are vying to fund zebrafish research.

Doing Science

A single gene can have a much larger effect than you might think, when you're told that all a gene does is to make a protein — as this study shows.

Ferguson, J. N., Young, L. J., Hearn, E. F., Marzuk, M. M., Insel, T. R., and Winslow, J. T. 2000. Social amnesia in mice lacking the oxytocin gene. *Nature Genetics,* 25.

For mice and other rodents, smell is the primary "social sense." Unlike humans, for whom visual cues are primary guides in social behavior, mice depend on olfactory cues to tell them whether other mice are familiar to them, interested in them, or hostile to them. This paper studies the effect of a single gene, the one that makes the brain protein oxytocin, on the social memory of mice.

Mice without the oxytocin gene are perfectly capable of responding to nonsocial olfactory stimuli, such as the smell of food or of a familiar place, but they don't respond to familiar mice. They suffer from social amnesia. When these mutant, oxytocin-deficient mice were treated with oxytocin, their social memory was "rescued," and when normal mice were treated with a substance that inhibited the action of oxytocin, they developed social amnesia.

4.13 DNA to RNA: Copying Genes into Messengers

Transcription: Preparing the Daily Work Orders

While replication of DNA is the grand event preceding a cell's division into two, DNA also regularly participates in the daily business of living. As in our imaginary self-assembling computer, DNA's software provides the instructions to its hardware. These instructions get sent out from DNA's central storehouse in the cell nucleus to the protein-making assembly plants in the cell's cytoplasm (see Chapter 5, *Machinery*), in the form of gene messengers, short stretches of information copied off the DNA. Messengers are made of a sort of throw-away version of DNA — good for limited work but not for long-term storage. Imagine going into a vault, taking out a set of precious instructions written on fine parchment, carefully copying the part you need on ordinary paper, returning the parchment to the vault, and then carrying the copy to the factory floor.

This process, called transcription, represents only the first stage of a larger operation dedicated to making proteins. You might notice (on the right-hand page) that transcription shares some of the mechanics of replication: DNA's double helix gets opened up, and a new nucleotide chain is built along a pre-existing strand that acts as a guide, or template. But the two processes differ. Transcription involves copying only one or a few genes at a time, not thousands. And the new, throw-away molecule that is produced is mRNA (ribonucleic acid), a close cousin to DNA.

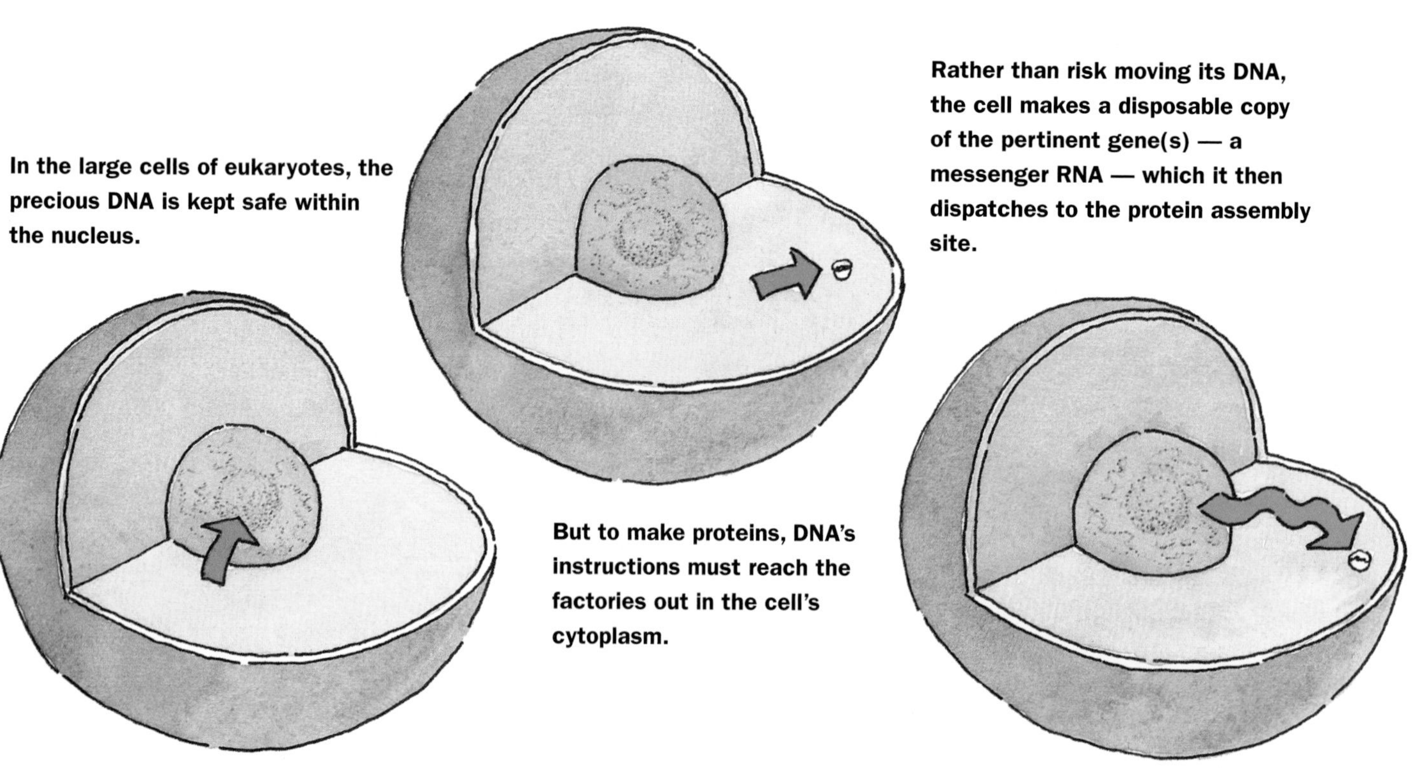

In the large cells of eukaryotes, the precious DNA is kept safe within the nucleus.
But to make proteins, DNA's instructions must reach the factories out in the cell's cytoplasm.
Rather than risk moving its DNA, the cell makes a disposable copy of the pertinent gene(s) — a messenger RNA — which it then dispatches to the protein assembly site.

Making a Messenger — The Basic Idea

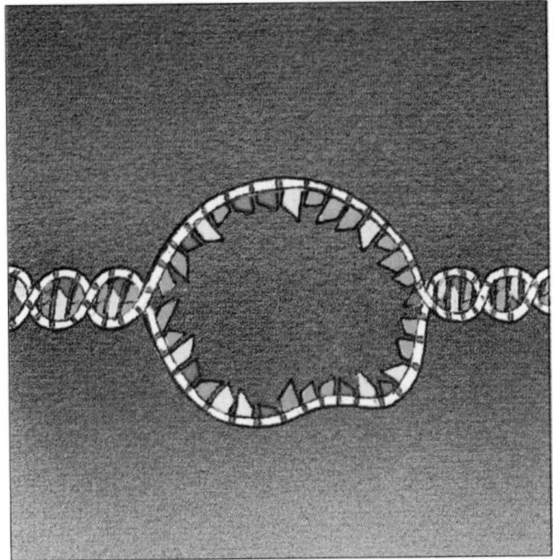

First, a small section of DNA is opened up.

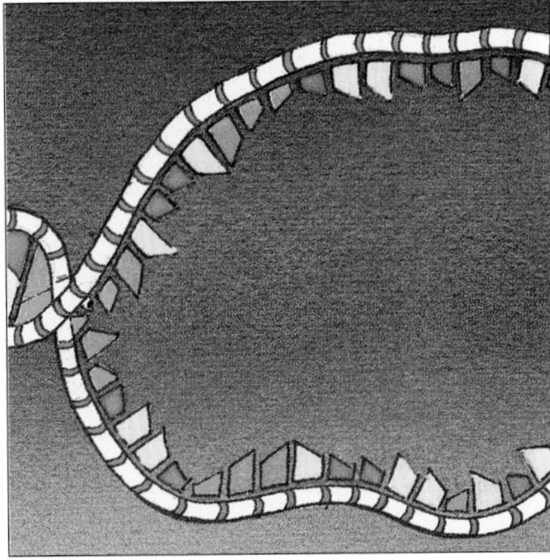

One strand conveys the actual message of the gene.

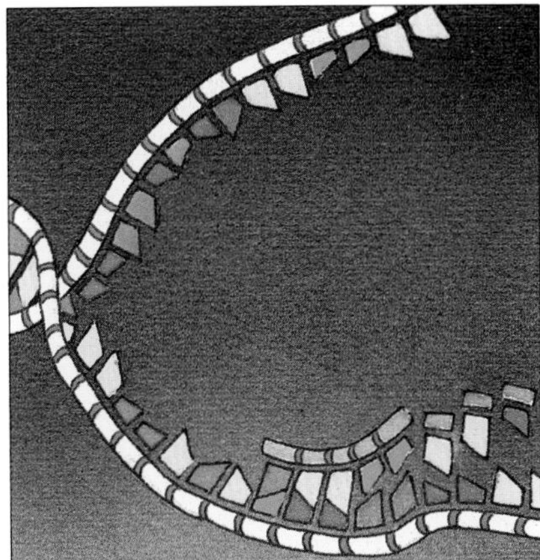

The other strand acts as a template on which the messenger is made.

The messenger is made of nucleotides, similar to how DNA is built.

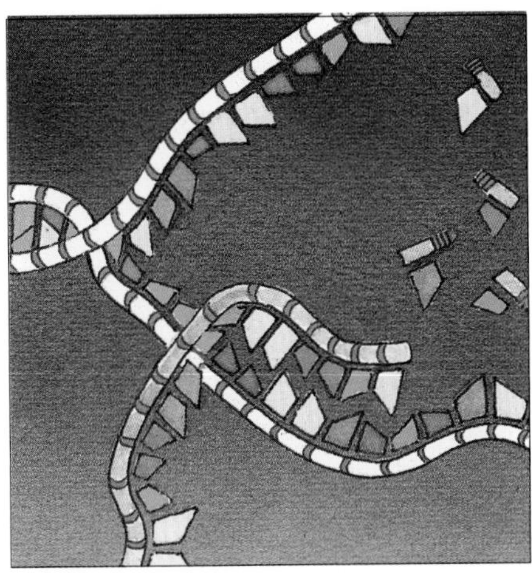

As the messenger is assembled, it separates from the template strand.

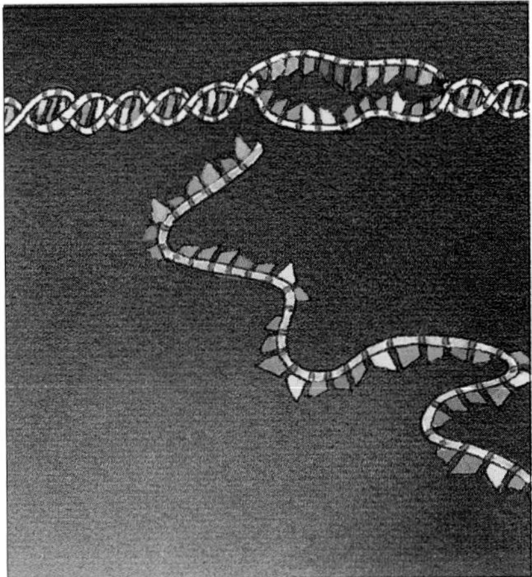

And when the entire gene is copied, the DNA releases the messenger.

Messenger-building requires the work of a single versatile enzyme.

It finds the starting point along DNA...

...copies the gene...

...and then closes the double helix.

The egg contains all the information needed to...
...make a new chicken...
...which grows up and...

4.14 The Chicken/Egg Problem

A New Way to Look at an Old Paradox

Untangling the chicken/egg problem ("Which came first?") produces some real insight into the way life works. The paradox plays a trick by seeming to ask a single question when in fact it asks two very different questions at the same time. The first deals with cycles, the second with evolution. We need to separate these two.

We begin with the simple observation that any true loop has no beginning and no end. Chicken produces egg, egg produces chicken — in an endlessly repeating cycle. So the answer to "Which came first?" must be "neither." If you want to understand the underlying mechanism, however, try looking at the chicken as machinery and the egg as information. Machinery makes information, which instructs machinery, etc. But this, too, oversimplifies. While it's true that the egg has all the information needed to make the chicken, information by itself cannot do anything without some decoding machinery, i.e., the proteins required to "unpack" that information.

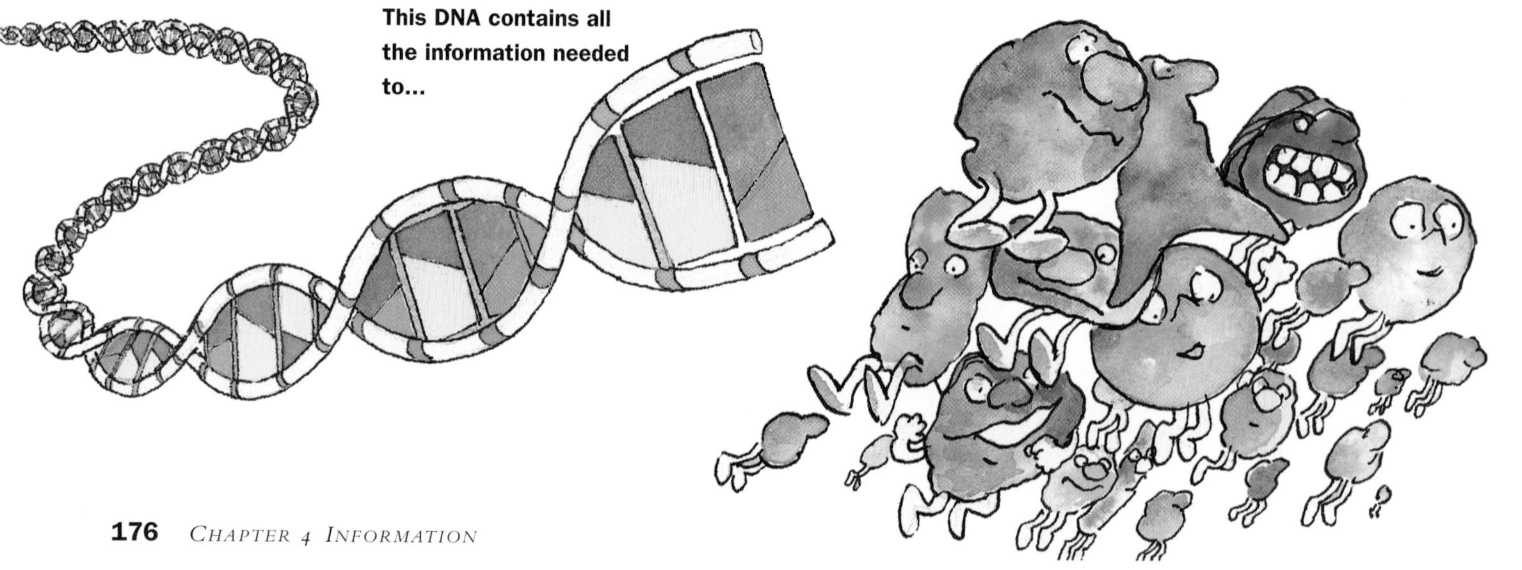

...make all the proteins for an individual chicken.

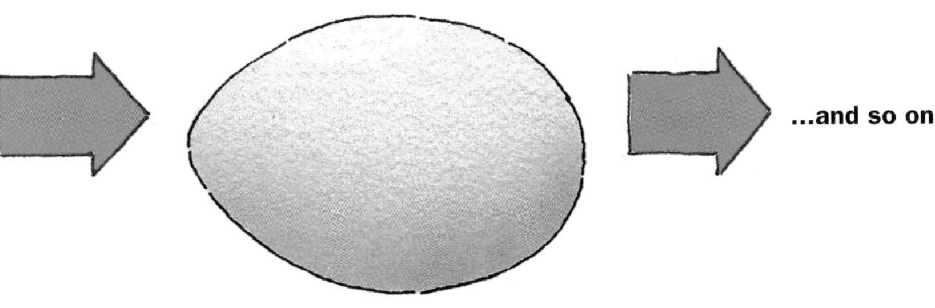
...makes a new egg...
...and so on.

"A hen is only an egg's way of making another egg."
—SAMUEL BUTLER

So we can say with more accuracy that the egg has all of the information plus just enough machinery to turn that information into living substance. In other words, every egg needs a little bit of chicken to go with it. The adult chicken, on the other hand, carries 100 percent of the information plus 100 percent of the machinery (that is, a complete chicken body); so producing a new egg is no problem.

The second question of the paradox might be rephrased as "Where did the chicken/egg cycle come from?" If we traced the ancestry of chicken and egg (both relatively recent "inventions") all the way back through billions of years, what would we find at the starting point? We can't be sure of the answer, but it may have been molecules that could function as both information and machinery (see page 292). From this beginning, chicken-ness would have arisen in steps, mostly tiny and gradual, over vast stretches of time.

...and so on.
Some of these proteins make more DNA...

4.15 DNA Packaging

Blowing in the Wind

DNA has found a wealth of ingenious ways to package itself; to create carriers that ensure that its message will get to the next generation: pollen, nuts, seeds, spores, sperm, egg, etc. These vehicles often carry food with them to sustain the early phases of new lives. They also contain enough of the necessary machinery for DNA to get a new foothold — to express itself in the form of the next generation's protein molecules.

Most of these vehicles for DNA will get lost before they find the proper environment in which to develop. Their substance will be broken down into simple molecules, and their message lost. To ensure that this won't be the fate of all, life, profligate with energy and materials, makes millions of DNA carriers so that a few will succeed in getting their message through. However, sometimes even large numbers aren't enough to get the job done. Over eons of trial and error, the DNA of some kinds of organisms has found ways of using *other* kinds of organisms to help it pass its message down the generations. A plant's DNA, for example, instructs the plant's flower to produce nectar to attract bees or birds, which, in the process of nourishing themselves, not only ensure the survival of their own DNA but also pick up the flower's DNA-containing pollen and carry it to new locations. Think of the DNA of one kind of organism, then, as having the power to enlist the help of the DNA of another kind of organism to accomplish the crucial job of re-creating itself in the next generation.

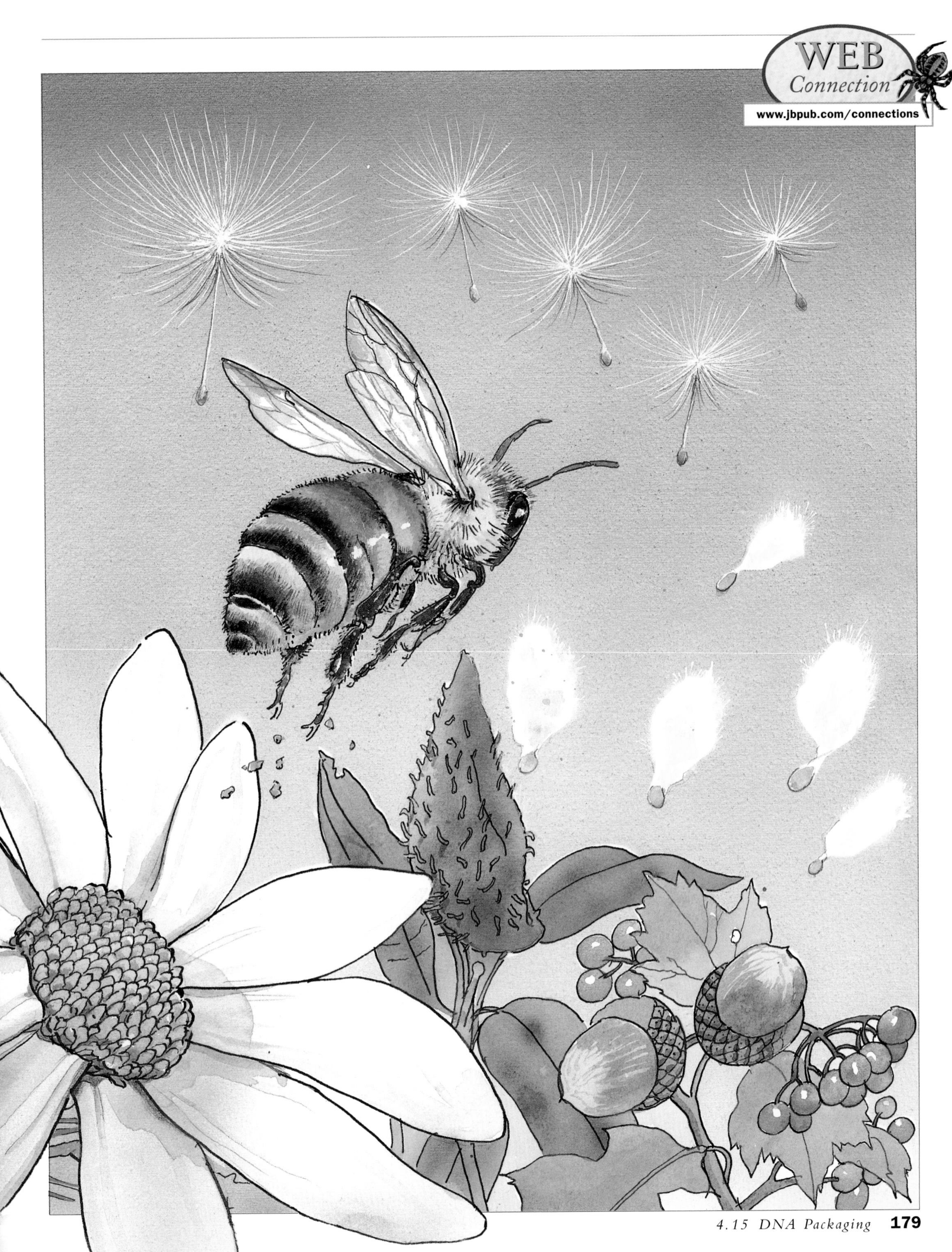
WEB
Connection
www.jbpub.com/connections

Some of the Things You Learned About in Chapter 4

Questions About the Ideas in Chapter 4

1. Information theorists tell us that all information, no matter what kind, depends on one simple feature. What is that feature?
2. People refer to genes for eye color, for height, and even for certain kinds of behaviors. But literally speaking, genes have only one function. What is it?
3. You may have heard DNA characterized as the blueprint for life. Why is this metaphor inaccurate? What's a better one?
4. The DNA in every cell copies itself before division. True or false?
5. Which came first, the chicken or the egg? Explain your answer.
6. Identify an everyday example where the shape, sequence, or arrangement of symbols transmits information for dynamic events or processes.
7. What might be the advantage of having complementary base sequences in the two strands of a DNA molecule?
8. If you were making a piece of furniture, you might want specific tools for each step of the job (i.e., a saw, hammer, drill press, mitre box, and router) rather than just one small, all-purpose hand tool. What is the advantage of having "dedicated" tools for a job? What are the "dedicated" tools of DNA replication?
9. If all life comes from life (there is no spontaneous generation) and the cell is the basic unit of life, then how could the first cell have arisen?

References and Great Reading

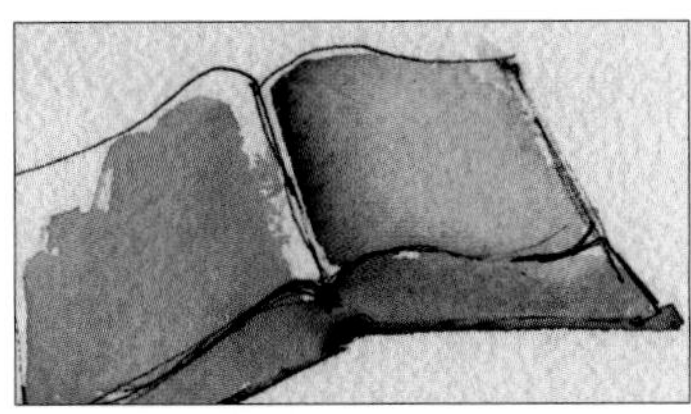

Beadle, G. W. and E. L. Tatum, 1941. Genetic control of biochemical reactions in *Neurospora*. *Proc. Nat. Acad. Sci.* 27: 499-506.

Dawkins, R. 1990. *The Selfish Gene.* Oxford: Oxford University Press.

Hartl, D. and Jones, E. 2001. *Genetics: Analysis of Genes and Genomes.* 5E. Sudbury, MA: Jones and Bartlett Publishers.

Krings, M., A. Stone, R. W. Schmitz, H. Krainitzki, M. Stoneking, and S. Paabo, 1997. Neandertal DNA sequences and the origin of modern humans. *Cell* 90:19.

Mullis, K.B. 1990. The unusual origin of the polymerase chain reaction. *Scientific American* 262(4):56-61, 64-65.

Pasteur, L. 1878. The Germ Theory and Its Applications to Medicine and Surgery. *Comptes rendus de l'Academie des Sciences,* lxxxvi, pp. 1037-43.

Watson, J. D. and F. H. C. Crick, 1953. Genetical implications of the structure of deoxyribonucleic acid. *Nature.* 171: 964-967.

Watson, J. D. and F. H. C. Crick, 1953 Molecular structure of nucleic acids – a structure for deoxyribose nucleic acids. *Nature.* 171: 737-738.

"The red fox can run far." Like nucleotide codons, the three-letter words in this sentence convey information. Describe the effect on information content of each of the following actions: (1) Delete a single letter from the first word and shift the following letters over by one so that the words continue to have three letters. (2) Delete a single letter from the last word and shift the remaining letters of the following words over by one so that the words continue to have three letters. (3) Substitute the letter m for the letter r. (4) Add the letter m to the start of the second word and shift the following letters over by one so that the remaining words continue to have three letters.

Question.

How does deleting letters or changing the groupings of the letters in this sentence (while keeping them in the same order) change the information content of the sentence? What does this have to do with DNA base sequences and the function of proteins?

Answer...

Changing the grouping of letters as well as the kinds and numbers of letters alters the amount of information transmitted. This is equally true for words in a sentence and for DNA base sequences. Altering the genetic code changes the kinds of proteins made and may alter or destroy protein function.

1. Her edf oxc anr unf ar. No information remains because the words are not understandable after the deletion.
2. The red fox can run fr. Most information is preserved up to the point of the deletion, so the sentence makes some sense even though it is incomplete.
3. The med fox can mun fam. A few words are understandable, but the entire sentence is not.
4. The mre dfo xca nru nfa r. No information is conveyed.

for more questions and links to web resources, go to
WEB
Connection
www.jbpub.com/connections

CHAPTER 5

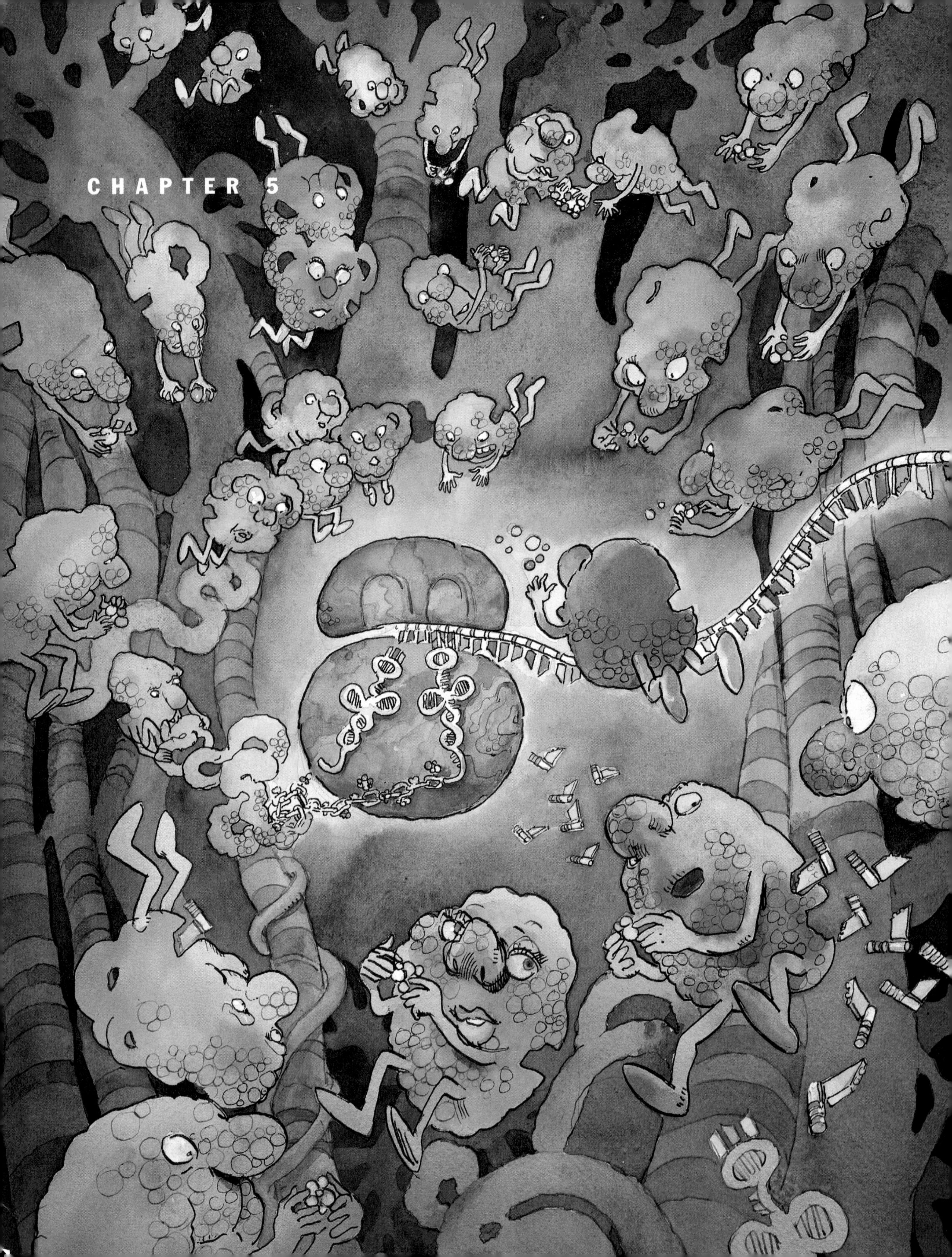

MACHINERY

Building Smart Parts

WHEN WE HUMANS BUILD A RADIO OR A CAR OR A COMPUTER, WE ASSEMBLE INANIMATE parts using the know-how we've accumulated over several hundred years. When our cells build us, they use information accumulated over four billion years — and they build know-how right into the parts. The parts are "smart." Instructions in DNA are translated into many thousands of ingenious devices, proteins, that do their tasks with astonishing fidelity, precision, and cooperation. Everything we do — think, laugh and cry, run and dance, conceive and give birth to children — emerges from the coordinated activities of a lively, intercommunicating society of protein molecules.

We call these proteins "machinery" because they do work that is accomplished by a simple movement. By making a subtle shift in its internal structure a protein can change its shape reversibly. If you watched one doing this all day, you'd likely be unimpressed by its I.Q. — for each protein knows only a single trick (or, occasionally, two). But movement can be put to all kinds of clever tasks (see the next page). And if you watched several proteins each performing its own task but working as a team, you'd begin to appreciate their cumulative "intelligence."

How does something as seemingly prosaic as DNA's long, monotonous sequence of only four different nucleotides get converted into the 20,000 or so different kinds of protein molecules that perform daily miracles in our bodies? That is the business of the cell's protein-making machinery.

About Proteins

What Proteins Do

Life's diversity can be traced to differences in the kinds and arrangements of protein molecules. More than half of the non-water weight of your cells is protein. Proteins do the daily business of living, giving cells their shapes and unique abilities. We've alluded to some of proteins' abilities earlier. Here's more about the key roles they play.

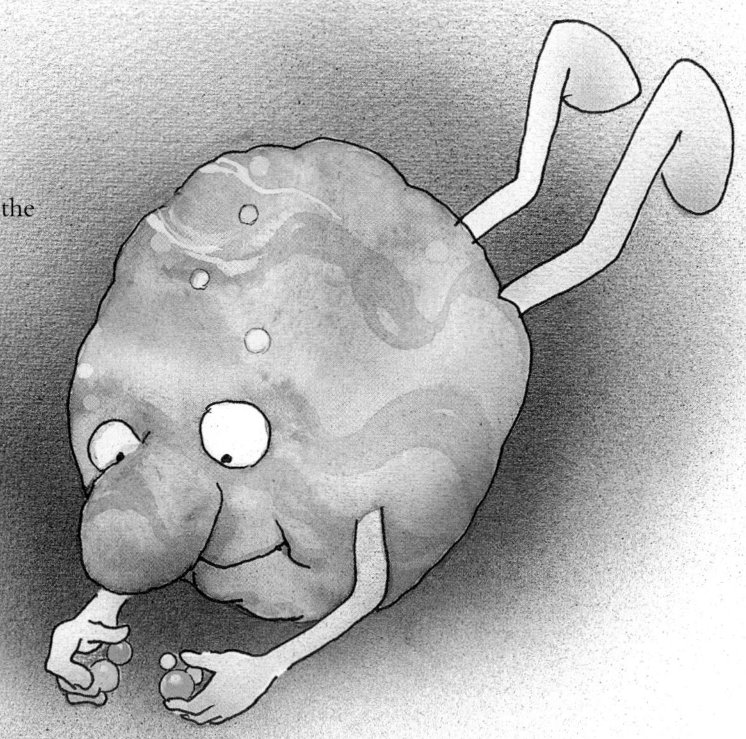

Enzymes

Enzymes are catalysts — they speed up the breaking apart and putting together of molecules. Their surfaces have special shapes that "recognize" specific molecules, similar to the way a lock accepts only a certain key. Enzymes themselves remain unchanged by the changes they bring about; they can be used over and over again.

Transporters

Special transporter proteins in cell membranes function as tunnels and pumps, allowing materials to pass in and out of the cell.

Movers

Because the shape of protein chains is mostly determined by weak, easily broken and remade chemical bonds, these chains can shorten, lengthen, and change shape in response to the input or withdrawal of energy. The energy molecule ATP can activate one part of a protein molecule, causing another part of the same molecule to slide or take a "step." Subsequent removal of ATP causes the protein to return to its original shape. Then the cycle can be repeated.

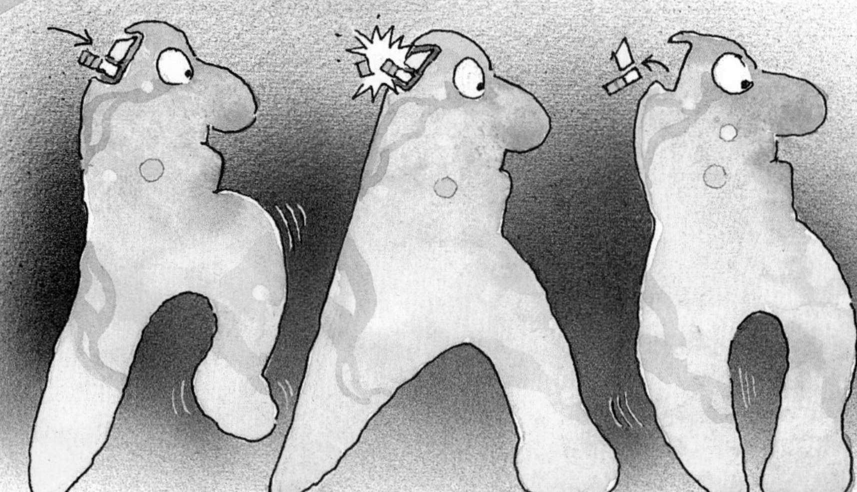

Supporters

Long chains of folded or coiled proteins can form sheets and tubes — the cell's equivalent of posts, beams, plywood, cement, and nails.

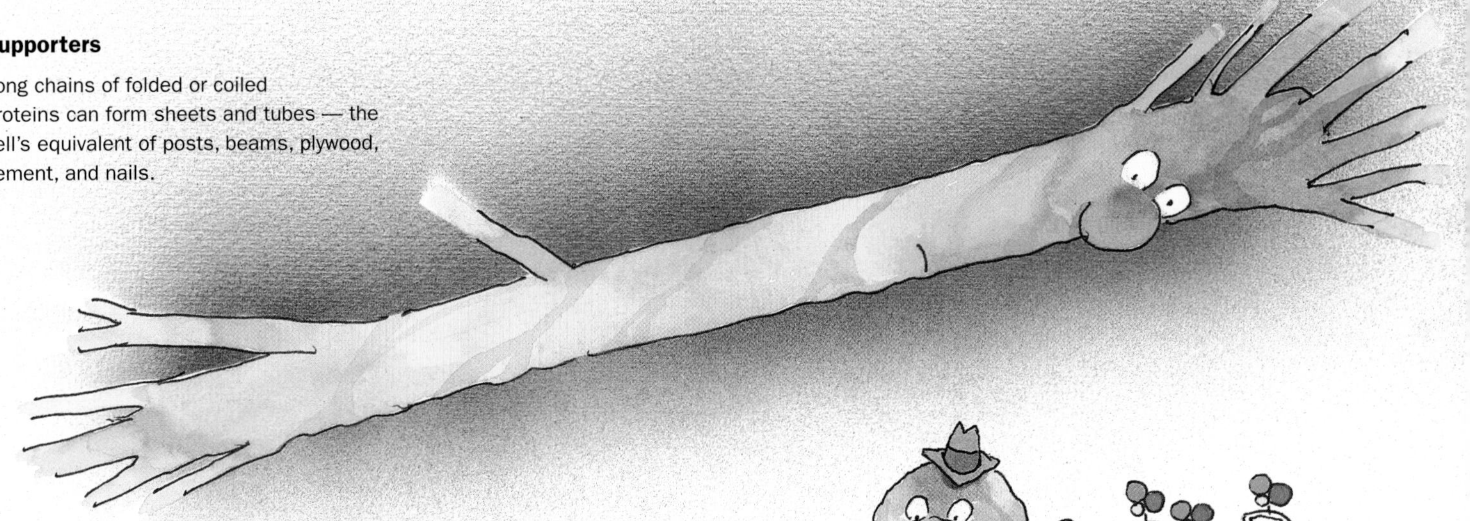

Regulators

Enzymes that convert one chemical to another must do so in several steps. The first enzyme in a cycle "notices" when enough of the final product builds up and shuts down the assembly line. This ability to respond to feedback is built into the regulator's structure (see Chapter 6).

Communicators

To work together in harmony, cells must be able to pass messages back and forth. Proteins can act as cells' chemical messengers. Hormones are examples. Communicator proteins sit on the surface of the receiving cell to gather the incoming signal.

Defenders

Antibodies are proteins with special shapes that recognize and bind to foreign substances, such as bacteria or viruses, surrounding them so that scavenger cells can destroy them and flush them out of the body.

Grrr...

WEB Connection

www.jbpub.com/connections

5.2 Multiplying Small Effects

Pumping Iron

Out of the 70,000 or more different kinds of proteins made in human cells, we have selected two — actin and myosin — to show how small molecular events can produce large effects. Actin and myosin are the proteins that make muscle work. Inside muscle cells, actin and myosin genes are translated into many millions of copies of each of these proteins. They line up to form a biochemical ratcheting device that uses ATP for energy to shorten and lengthen itself. This tiny molecular machine leads to the action of a bulging biceps through the simple means of scaling up. Millions of actin-myosin combinations are strung end-to-end in long fibers, and these fibers are bundled together into dense, parallel, elastic cables — the muscle cells. Each microscopic contraction of an actin-myosin combination is amplified into contraction of a cell. Collective cell contractions produce an overall grand contraction — the action of a muscle.

1. Actin molecules are long and thin; myosin molecules are thicker and have many "arms" and "hands" sticking out from their sides. The hands touch the actin molecules.

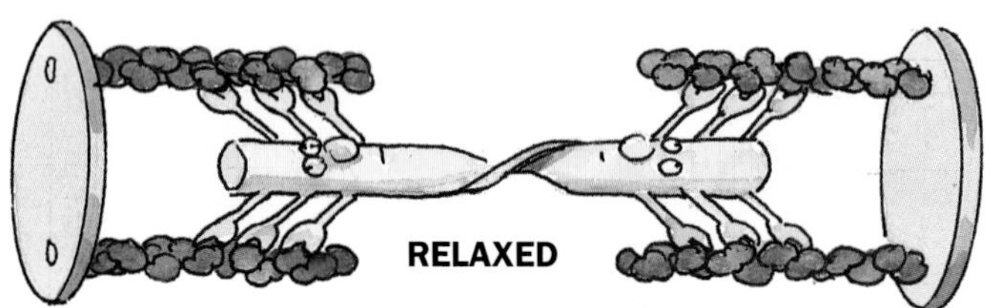

2. Each unit of contraction consists of two identical arrays of actins attached to discs and facing each other, connected by myosin. ATP binds to myosin's hands, releasing them from actin. The subsequent splitting of ATP to ADP and phosphate causes myosin's hands to grab actin, and its arms to draw back.

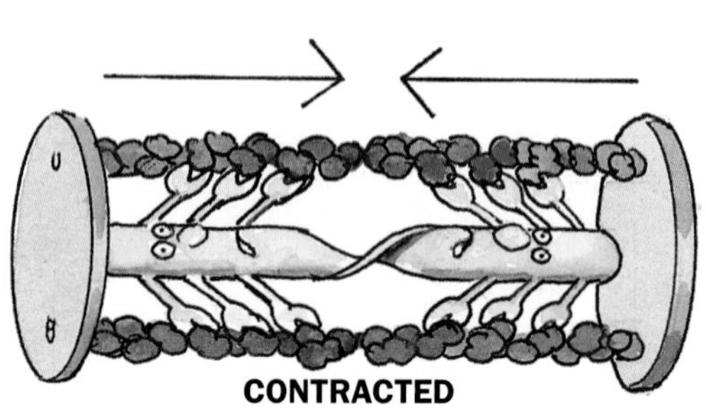

3. The release of ADP and phosphate from myosin causes the arms to make a stroke like an oar, pulling the actins with their attached discs toward each other; this causes contraction.

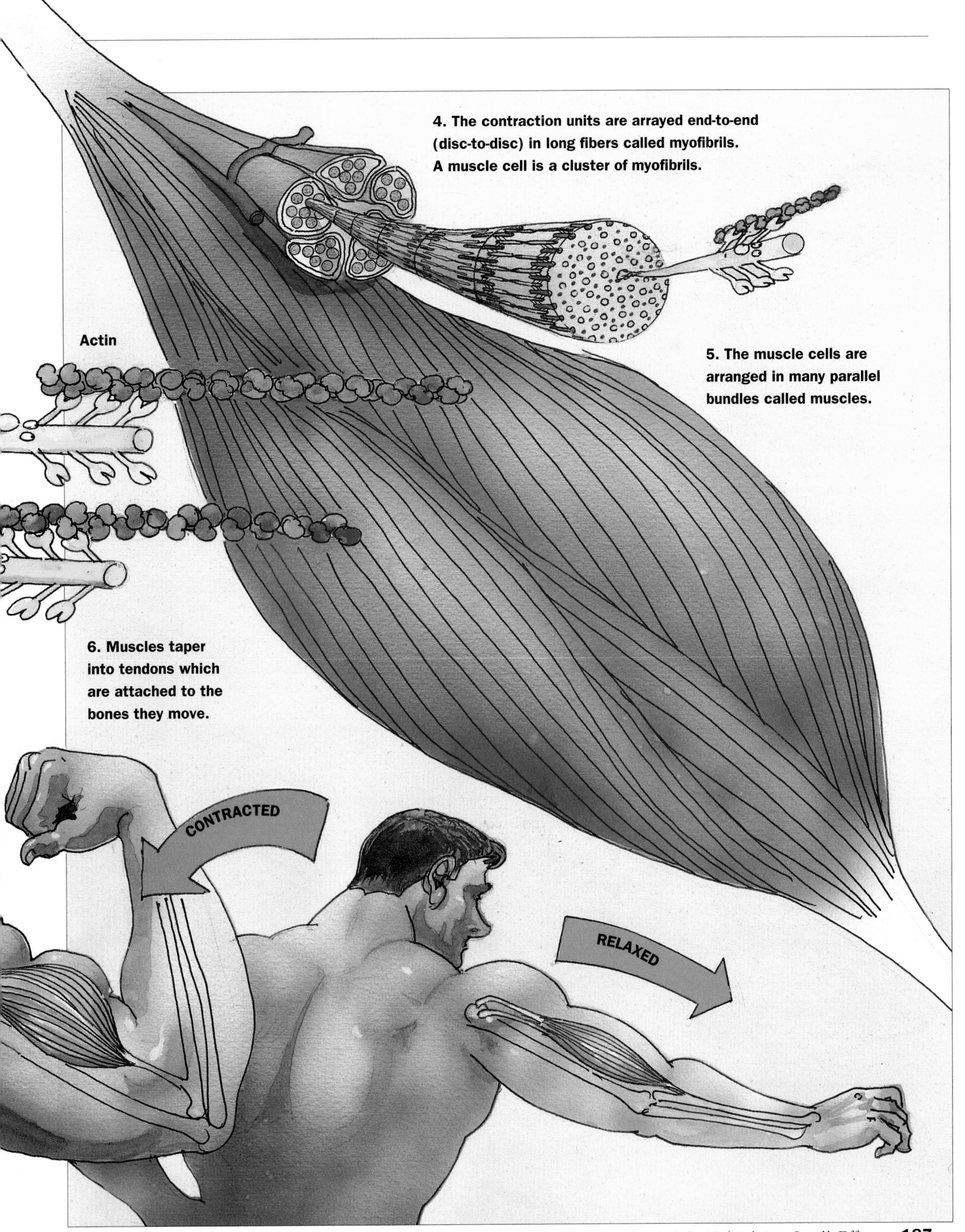
4. The contraction units are arrayed end-to-end
(disc-to-disc) in long fibers called myofibrils.
A muscle cell is a cluster of myofibrils.
Actin
5. The muscle cells are
arranged in many parallel
bundles called muscles.
6. Muscles taper
into tendons which
are attached to the
bones they move.
CONTRACTED
RELAXED

It's the Same Molecules Everywhere You Look

A WORLD OF SWARMING CHLOROPLASTS

I was in a laboratory, using a very expensive microscope. . . . In the circle of light formed by the two eyepieces trained on the translucent leaf . . . I could easily see what I had come to see; the streaming of chloroplasts.

. . . Around the inside perimeter of each gigantic cell trailed a continuous loop of these bright green dots. They spun like paramecia; they pulsed, pressed and thronged. A change of focus suddenly revealed the eddying currents of the river of transparent cytoplasm, a sort of "ether" to the chloroplasts, or "space-time," in which they have their tiny being . . . they swarmed in ever-shifting files around and around the edge of the cell; they wandered, they charged, they milled, raced and ran at the edge of apparent nothingness, the empty-looking inner cell; they flowed and trooped greenly, up against the vegetative wall.

Annie Dillard, *A Pilgrim at Tinker Creek,* 1999

The eddying cytoplasmic currents (described in the observation above) that carry *Elodea's* chloroplasts are driven by exactly the same minuscule actin-myosin motors that make muscles contract and extend, that drive the slime mold's motion, and that allow white blood cells to engulf invading germs.

Microfilaments composed of the proteins actin and myosin can be found in animal, plant, fungal and bacterial cells. When ATP binds to the myosin molecules in these long fibers, the myosin contacts the actin molecules, causing the microfilaments to contract at the same time and in the same direction. This causes movement of the fluid part of the cell's contents. Interior organelles, vesicles, and molecules float on these currents like surfers on a wave, traveling rapidly throughout the cell.

Moving materials through a plant's root cells

In this laser scanning fluorescence micrograph, the actin in actin-myosin microfilaments shows up as bright green fluorescent threads, and each cell's nuclear DNA fluoresces blue. You can see how the microfilaments permeate the cell, circling the nucleus and enhancing the intracellular movement of molecules and molecular structures.

Acetabularia

Acetabularia, whimsically called the mermaids wineglass, is a relatively enormous unicellular alga (it measures from 3 to 10 cm in length, and its "wineglass" cap is 1 to 3 cm in diameter). While its linear shape and filmy cap offer a large surface area for the diffusion of materials into and out of the cell, passive diffusion is not enough to move necessary molecules quickly throughout its volume. Cytoplasmic streaming, driven by actin-myosin interactions, is again the solution to this cell's traffic problem.

A forest of *Acetabularia* cells.

Allen, Nina S. 1974. Endoplasmic filaments generate the motive force for rotational streaming in *Nitella*. *Journal of Cell Biology* 63: 270-287.

This paper describes a clever experiment designed to show that the undulation of microfilaments in a large algal cell, *Nitella*, is the cause of the motion of particles throughout the fluid cell interior. *Nitella* were cultivated and collected. A window was cut into several cells using a mercury arc lamp, which allowed the experimenter to see into the cell and to film cytoplasmic streaming. The movement of particles in the cytosol was filmed by strobe light. The films showed particles moving in a serpentine pattern, which led to the conclusion that they were attached to unseen filaments. When a substance that inhibits actin-myosin interactions but does not affect other molecular structures in the cell was introduced, particle motion stopped. This led to the hypothesis that the filaments were made of actin.

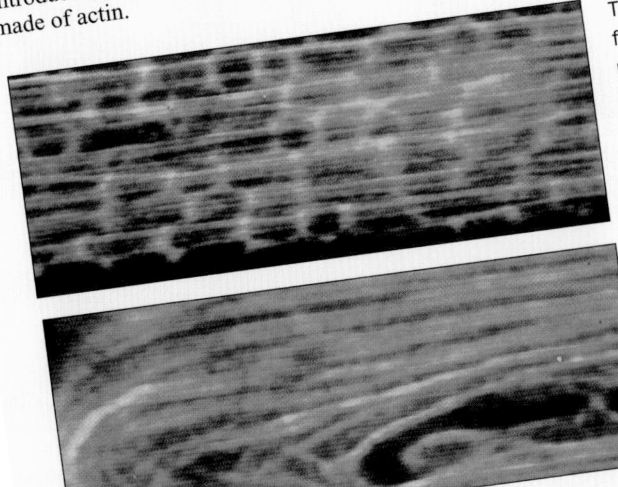

The actin structure in *Nitella* forms an endless belt that provides enough momentum to sweep the entire cell content in a circle. (Only the actin in the actin-myosin complex flouresces.)

Each amino acid has a different side group with a unique chemical character...

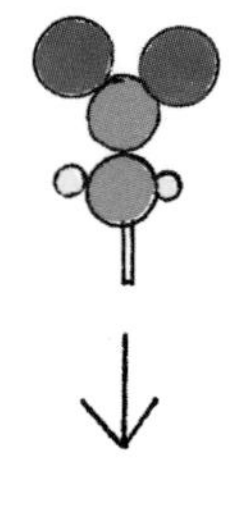

...attached to a backbone piece that's the same for every amino acid.

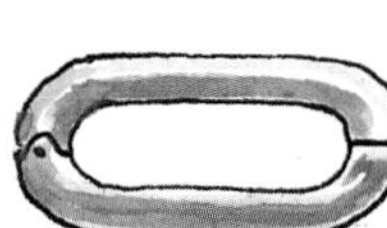

5.3 Proteins Are Chains Made from Twenty Amino Acids

Sequence Makes the Difference

Underlying the bewildering variety of protein shapes and sizes is a surprising simplicity. When proteins are unfolded and stretched out, they turn out to be chains of amino acids. The sole determinant of a protein's natural shape, and consequently its function, is the order of the amino acids in the chain.

There are twenty — and only twenty — amino acids (you can see all of them on page 195). Animals, plants, protists, fungi, and bacteria use some or all of these amino acids in their protein chains. All amino acids contain carbon, hydrogen, oxygen, and nitrogen atoms, and two of them have sulfur atoms as well. Ten of the amino acids have electrically charged side groups that are attracted to water. These cluster on the surface of the folded-up protein chain where it's easier for them to make contact with the surrounding water in the cell. The other ten amino acids have no electrical charge and so tend to cluster on the inside of the folded-up molecule where they'll stay dry. The amino acids are linked to each other by strong covalent bonds between their backbone pieces (what we show as chain links). Once a protein is assembled, its covalently linked amino acids form additional weak hydrogen bonds with each other. These easily broken and reformed weak bonds give protein molecules their remarkable ability to change shape, which is the key to their functioning. They also give proteins great flexibility and mobility.

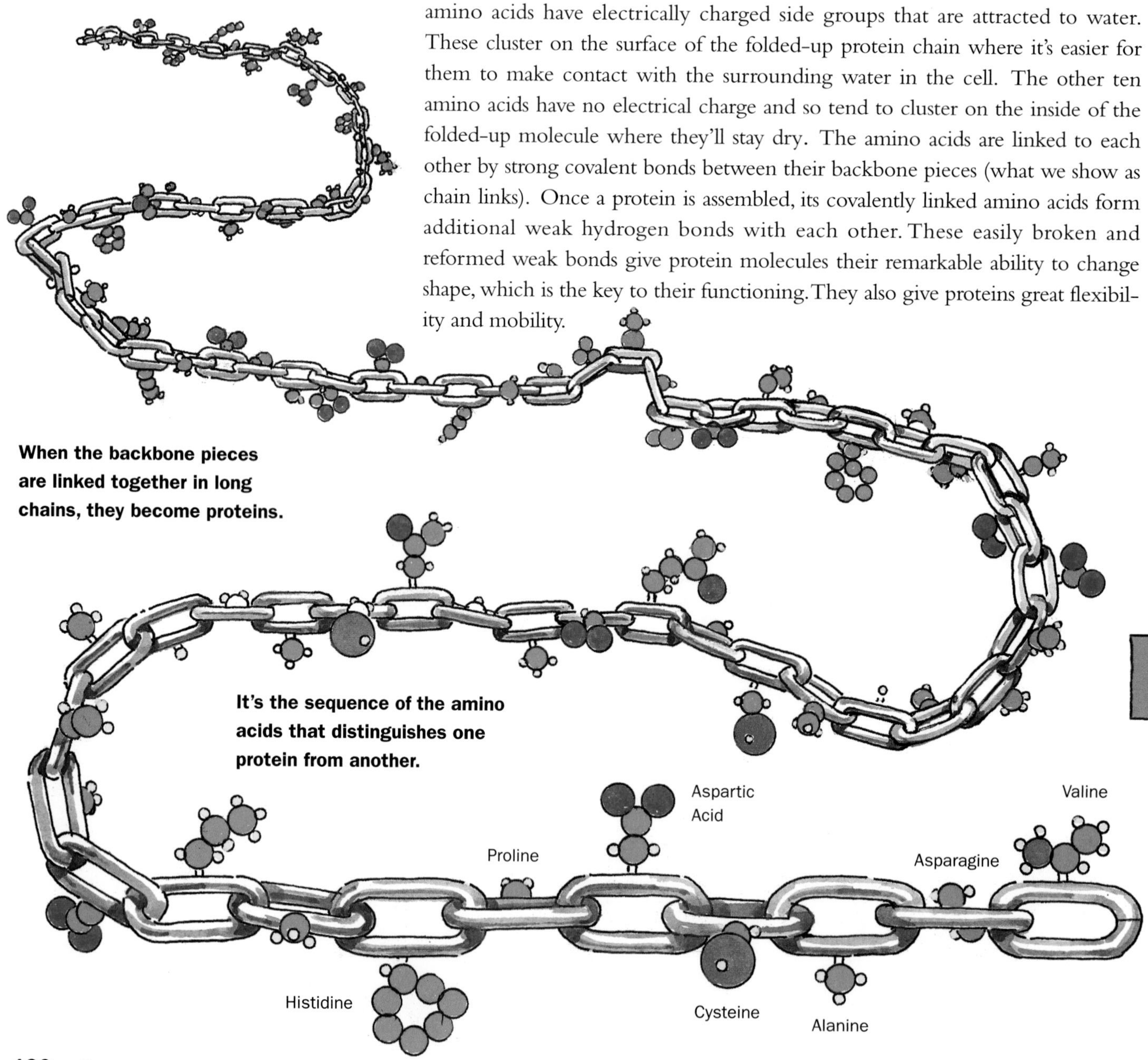

When the backbone pieces are linked together in long chains, they become proteins.

It's the sequence of the amino acids that distinguishes one protein from another.

Aspartic Acid

Valine

Proline

Asparagine

Histidine

Cysteine

Alanine

Protein Folding

Proteins find themselves mainly in one of two environments — water or fat. This explains why proteins fold the way they do. A protein in a watery environment folds its fat-liking amino acids tightly inside itself while its water-liking amino acids face the surrounding water. Proteins that reside in membranes, which are made of fat, do the opposite. Proteins can't do their work unless they're folded up correctly.

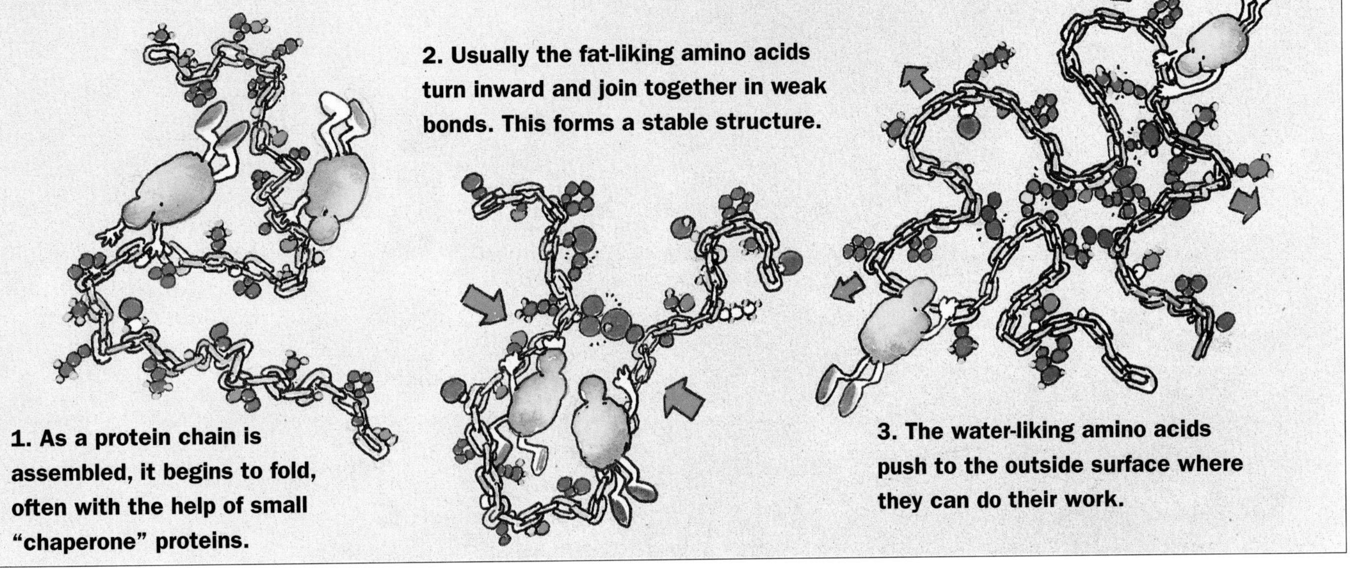

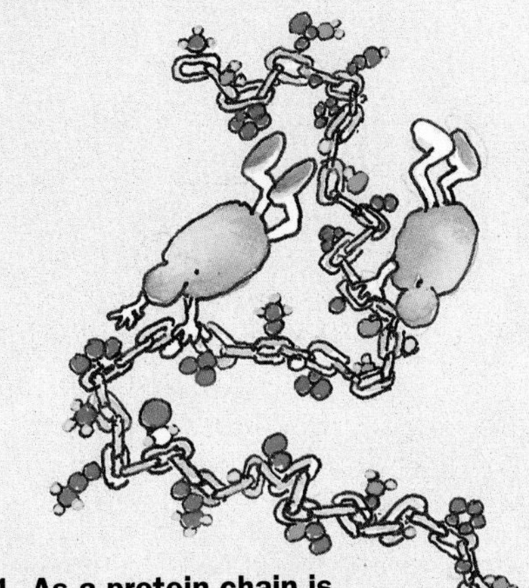

1. As a protein chain is assembled, it begins to fold, often with the help of small "chaperone" proteins.

2. Usually the fat-liking amino acids turn inward and join together in weak bonds. This forms a stable structure.

3. The water-liking amino acids push to the outside surface where they can do their work.

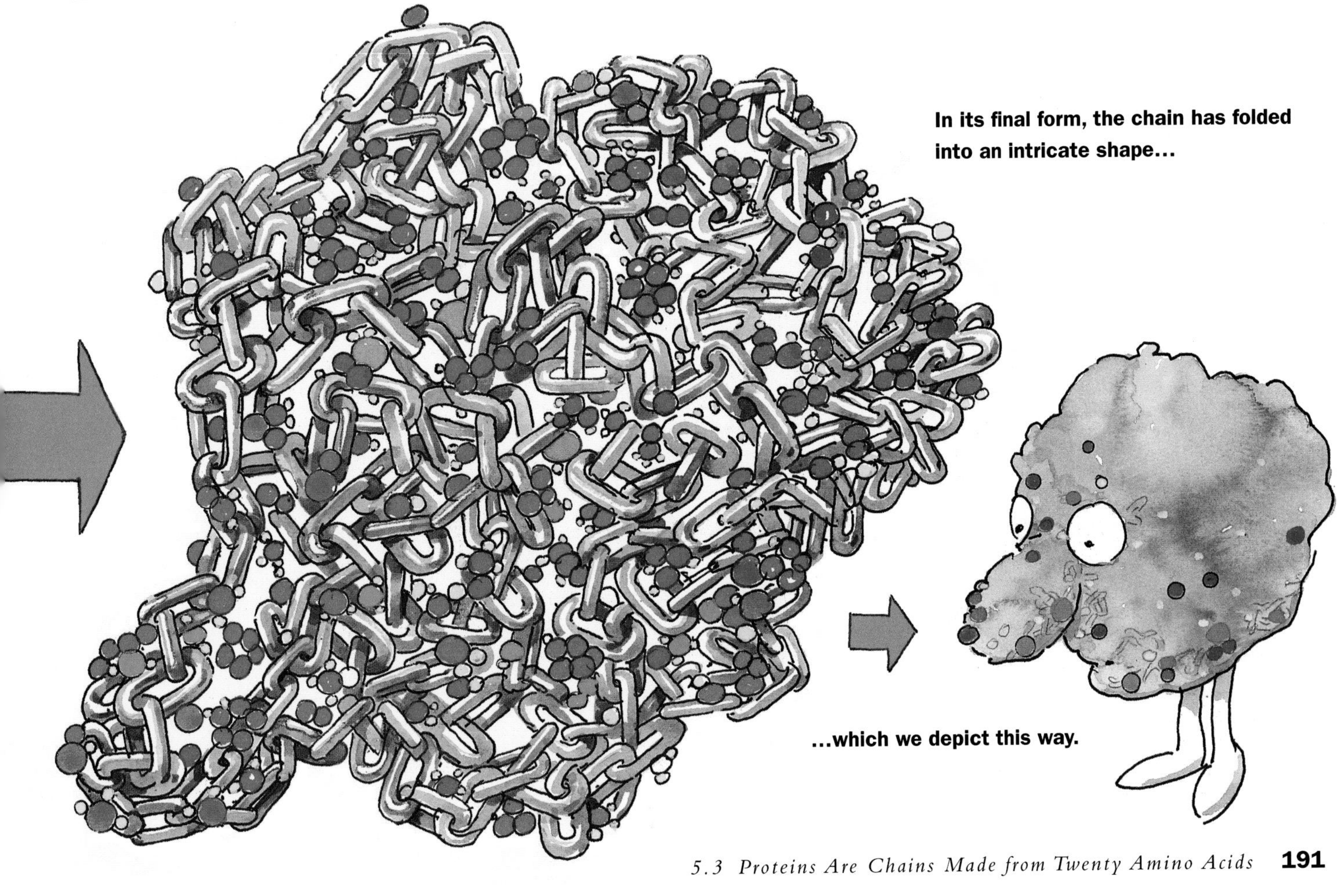

In its final form, the chain has folded into an intricate shape...

...which we depict this way.

How Orders Translate into Assembled Boxes of Donuts

One coconut, six chocolate, three glazed, one jelly...
Clerk

Clothespins and Donuts

DoNutArama, a popular donut shop, makes twenty kinds of donuts. The donuts are so good that people buy big boxes of them. And each customer is very particular about having exactly the right kinds of donuts in exactly the right order in the box.

At first, the clerk at the counter tried shouting the orders to the kitchen staff, but they made too many mistakes. Written orders were out because the employees couldn't read the clerk's handwriting. Then someone remembered the colored clothespins in the basement. Maybe the clerk could somehow use the clothespins to transmit orders for donuts to the kitchen.

The clothespins came in four colors. The donuts came in twenty varieties.

What's the most efficient way to use four units to represent twenty units? The clerk worked out a code.

He first tried using combinations of two colors of clothespins: i.e., red + blue = jelly; yellow + red = chocolate; etc. He soon realized that there weren't enough different two-color combinations to represent all twenty donuts. But a three-clothespin code could produce sixty-four (4 × 4 × 4) possible combinations — more than needed for twenty different donuts. So he and his staff worked out and memorized a three-color code: red + blue + yellow = jelly; yellow + red + green = chocolate; etc. As the clerk took the orders, he put the correct color sequence on the line. In the kitchen, the decoder read the code, then hung the proper donut on the hook next to it. The packager took the donuts off the hooks and put them in their proper sequence in the box. Counter orders were transcribed into clothespin sequences and decoded into boxes of donuts, and things worked sweetly ever after.

Jelly...

Decoder

www.jbpub.com/connections

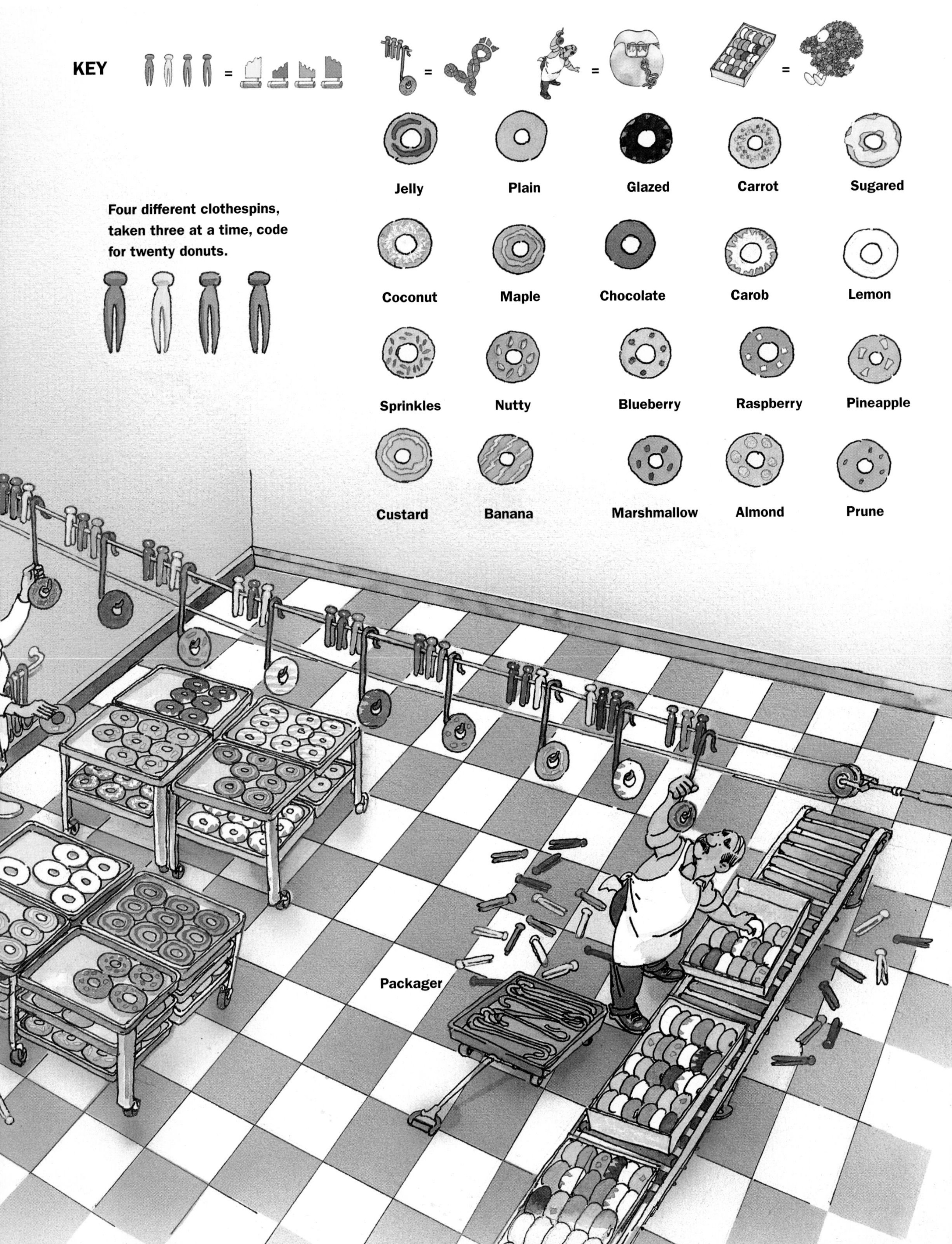
KEY
Four different clothespins, taken three at a time, code for twenty donuts.
Jelly
Plain
Glazed
Carrot
Sugared
Coconut
Maple
Chocolate
Carob
Lemon
Sprinkles
Nutty
Blueberry
Raspberry
Pineapple
Custard
Banana
Marshmallow
Almond
Prune
Packager

5.5 How DNA Information Translates into a Working Protein

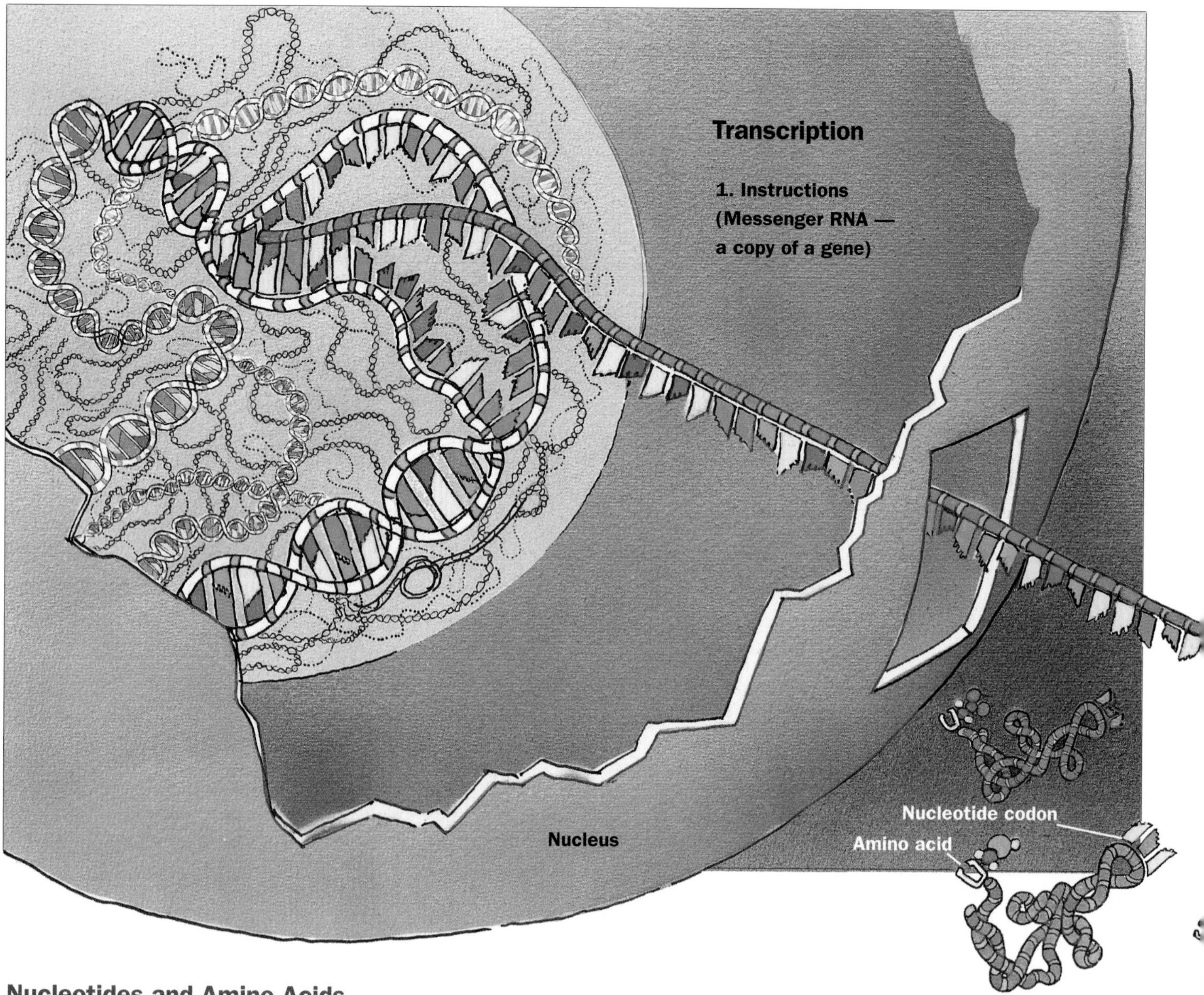

Transcription

1. Instructions (Messenger RNA — a copy of a gene)

Nucleus

Nucleotide codon

Amino acid

Nucleotides and Amino Acids

A DNA molecule is many, many nucleotides (clothespins) long. It is composed of genes, which are, on the average, some 1200 nucleotides long. Within each gene, the nucleotides are ordered in about 400 groups of three nucleotides apiece. Each nucleotide triplet (called a codon) gets translated into one of the twenty amino acids (donuts). The entire gene will be translated into a protein molecule that is about 400 amino acids long (the packaged donuts).

Here's how you make a protein. First, copy — transcribe — the sequence of nucleotides in a gene into a single strand of RNA (see Chapter 4, page 174) called messenger RNA (mRNA). Second, attach amino acids to small RNA molecules called transfer RNAs (tRNAs), or adaptors. These act like the decoder with her donut hook. Each adaptor recognizes a particular three-nucleotide codon. Third, bring the adaptors with their attached amino acids and the messenger RNA to a protein synthesis factory called a *ribosome* (the packager), which links up the amino acids to make the protein.

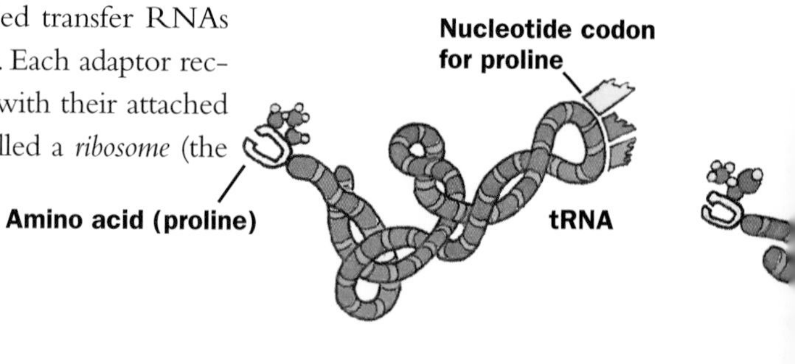

2. Adaptors (transfer RNA molecules with amino acids attached)

A transfer RNA is the key "decoding" unit between information and final protein product. Each has a three-letter codon at one end and an amino acid at the other end.

Nucleotide codon for proline

Amino acid (proline)

tRNA

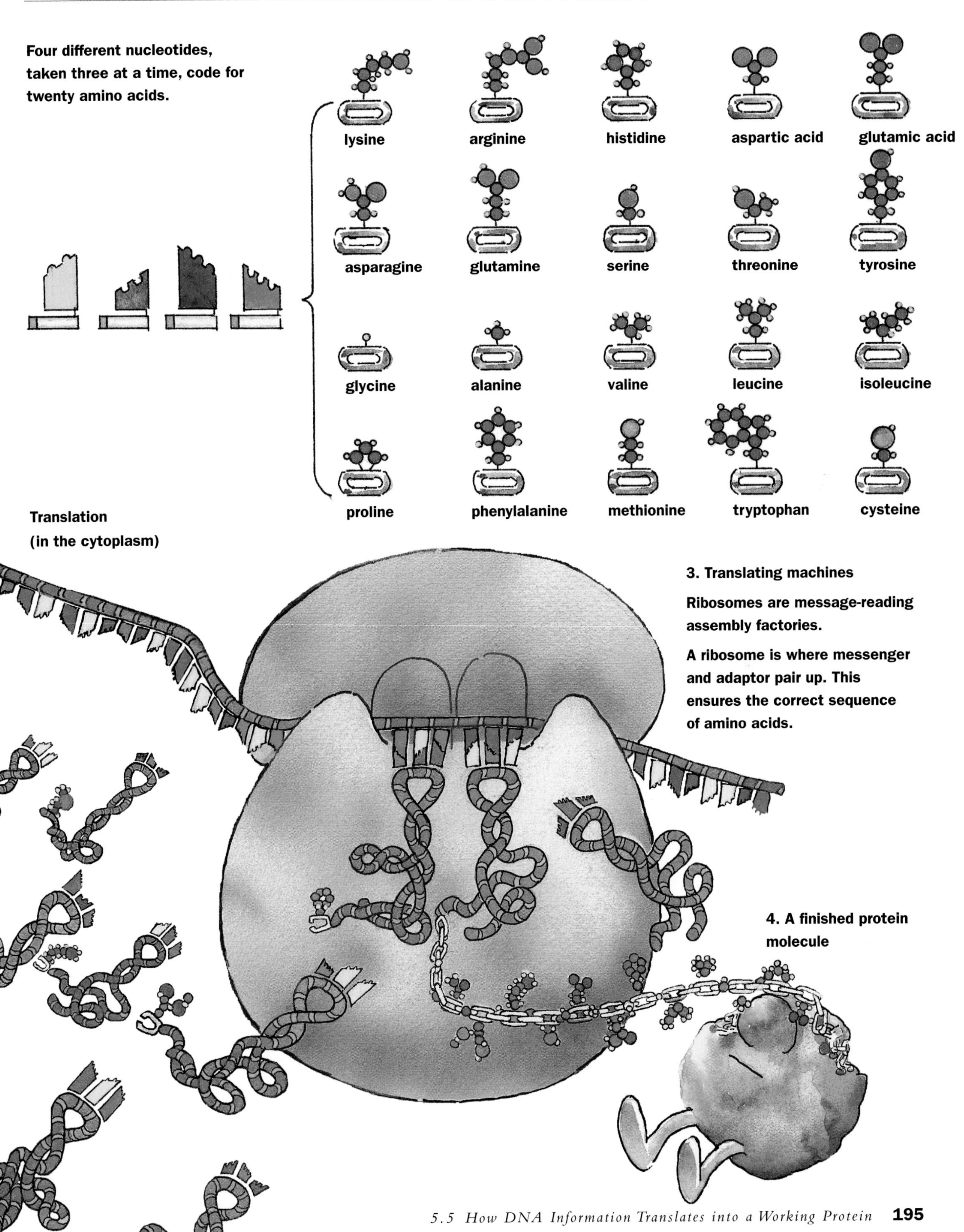
Four different nucleotides, taken three at a time, code for twenty amino acids.
lysine
arginine
histidine
aspartic acid
glutamic acid
asparagine
glutamine
serine
threonine
tyrosine
glycine
alanine
valine
leucine
isoleucine
proline
phenylalanine
methionine
tryptophan
cysteine
Translation (in the cytoplasm)
3. Translating machines
Ribosomes are message-reading assembly factories.
A ribosome is where messenger and adaptor pair up. This ensures the correct sequence of amino acids.
4. A finished protein molecule

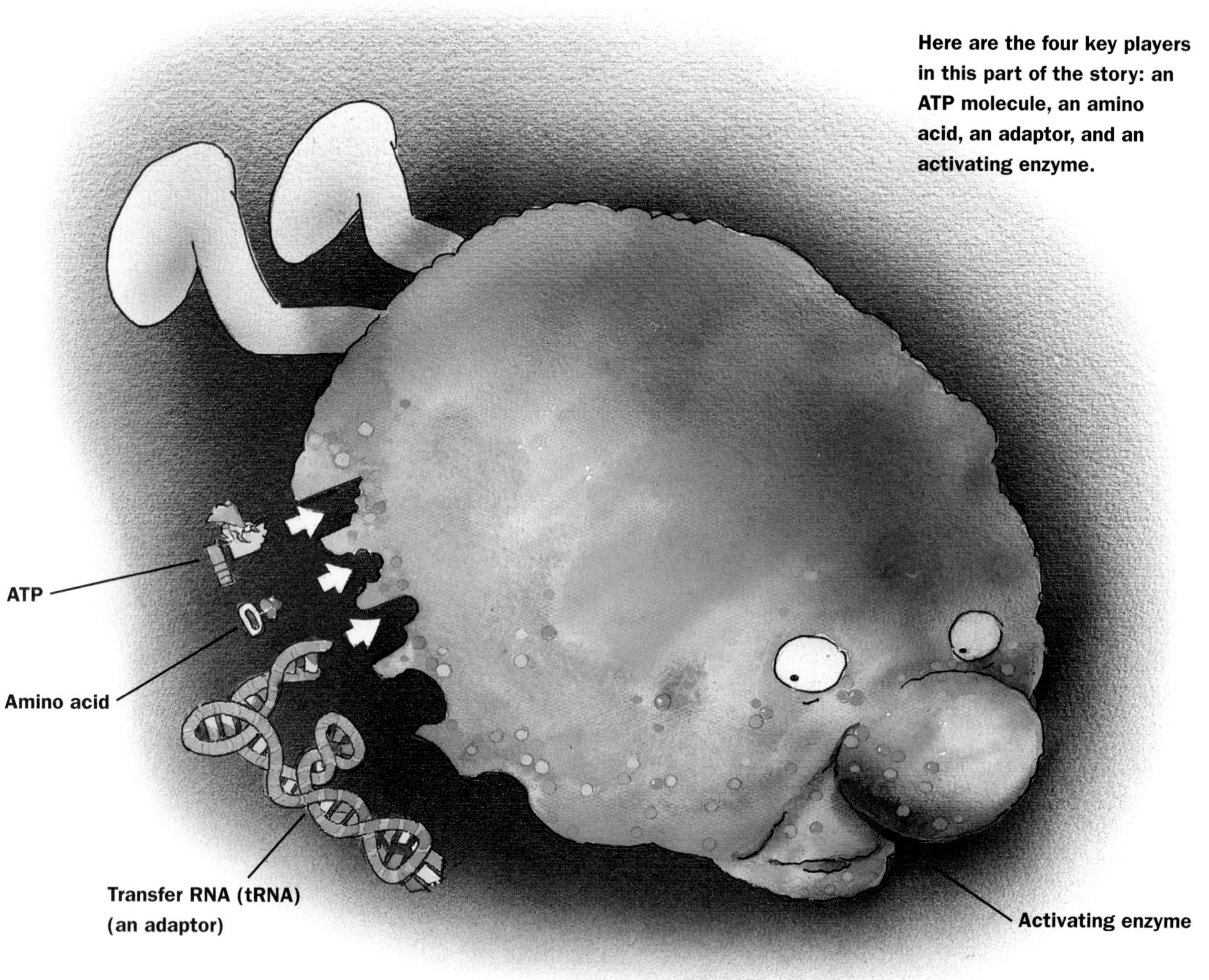

Here are the four key players in this part of the story: an ATP molecule, an amino acid, an adaptor, and an activating enzyme.

ATP

Amino acid

Transfer RNA (tRNA) (an adaptor)

Activating enzyme

5.6 From DNA to Protein — A Multistep Process

The Basic Idea

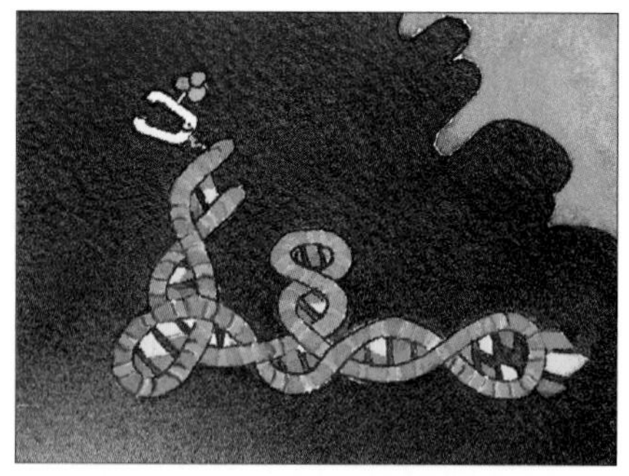

An energized amino acid gets put on an adaptor.

Charging the Adaptor

On the previous pages, we showed the process of transcription of DNA to mRNA in the nucleus, and the translation process in the cell's cytoplasm by which genes prescribe the order of amino acids in proteins. Now let's follow the key steps more closely. There has to be a chemical connection between each amino acid and each messenger RNA. Transfer RNA (tRNA) — the adaptor — makes that connection. One end of the adaptor carries a three-nucleotide code. This will match up with three complementary nucleotides on the messenger. A specific enzyme, called an amino acid activating enzyme, energizes each amino acid and then attaches it — just the right one — to the opposite end of the adaptor. Since there are twenty amino acids, there must be at least twenty different activating enzymes and twenty different tRNA adaptors. In the panels on the next page we show the first steps in the construction of a protein: energizing amino acids and linking them to their adaptors.

The Details

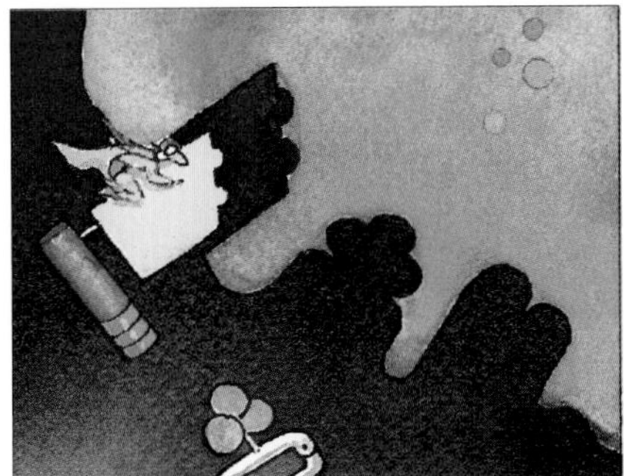

ATP floats near the enzyme and docks in a place tailor-made for it.

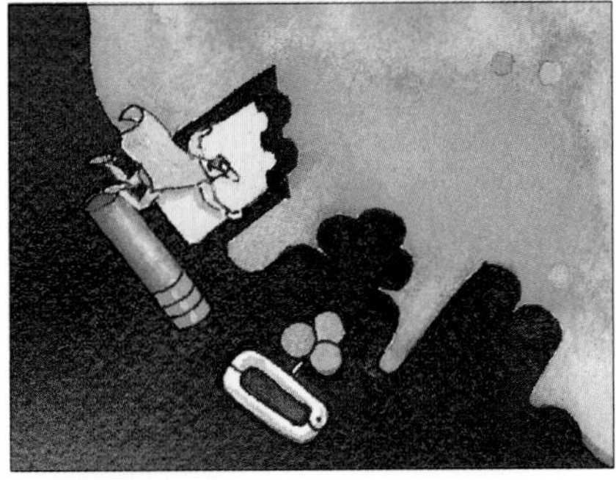

Meanwhile, an amino acid floats into a dock nearby.

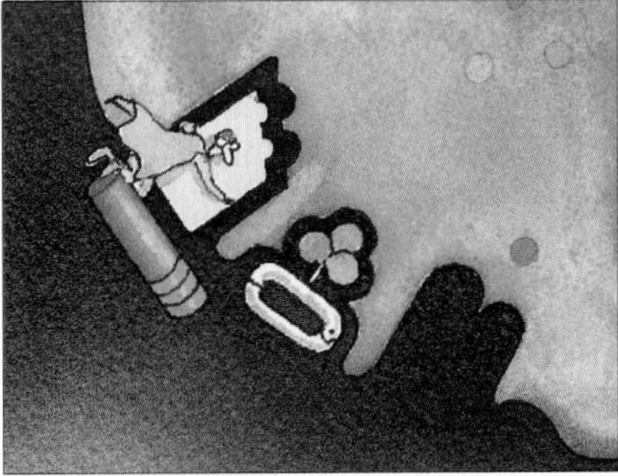

The two are brought closer together until...

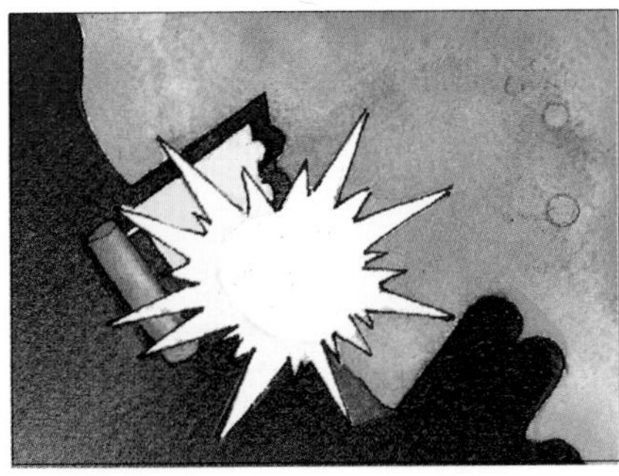

...they bond...

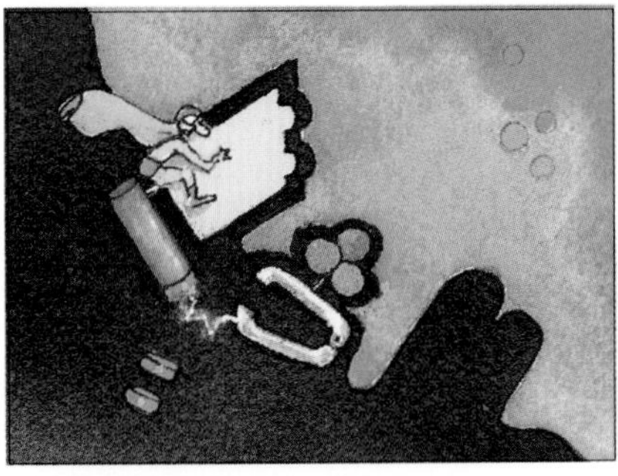

...ejecting two phosphates from ATP.

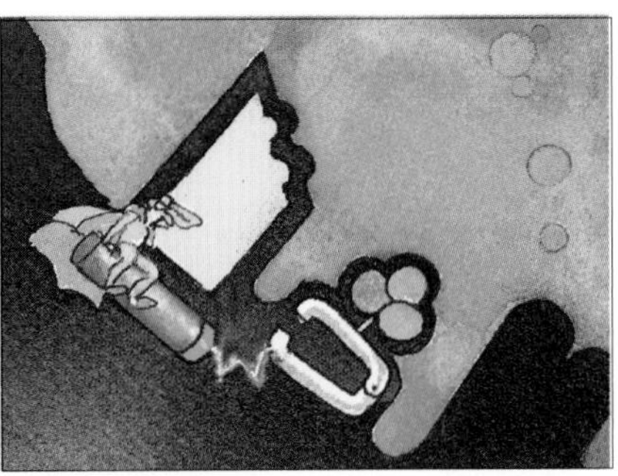

The amino acid is now *energized.* (Note how the link is now open.)

Next, the odd-shaped tRNA adaptor floats into view...

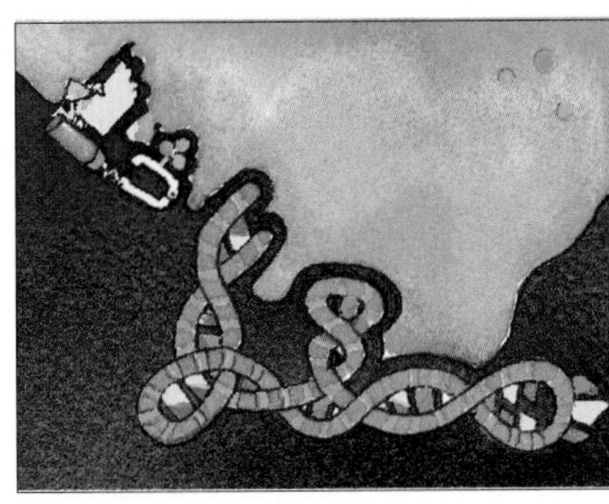

...and docks at another nearby site on the enzyme.

The end of the adaptor is brought closer to the amino acid until...

...the two are joined.

Energy flows into the new bond; the "spent" energy molecule is released.

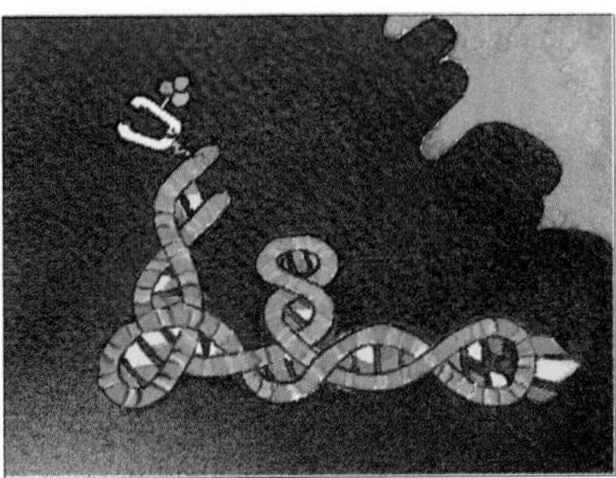

Then the adaptor is released, with its amino acid attached.

5.7 Translation

Assembling the Protein Chain

An energized amino acid has been attached to one end of a tRNA adaptor, which carries at its other end a three-nucleotide code specific for that amino acid. Now the amino acid needs to be linked into a chain with others, in a specific order, to create a specific protein. This next phase requires the help of special machinery that can use the adaptors to "read" the nucleotide triplet codons on the messenger and assemble the appropriate amino acid chains. That's the job of the ribosomes. A ribosome is made of a larger and a smaller piece, each composed of about equal amounts of ribosomal RNA and protein; it looks a bit like a designer telephone. The ribosome "reads" the tape-like mRNA message three units (one codon) at a time, linking amino acids together as it proceeds. When it gets to the last triplet, which signals "stop," it releases the finished amino acid chain (the packager closes the donut box).

The Three Key Elements

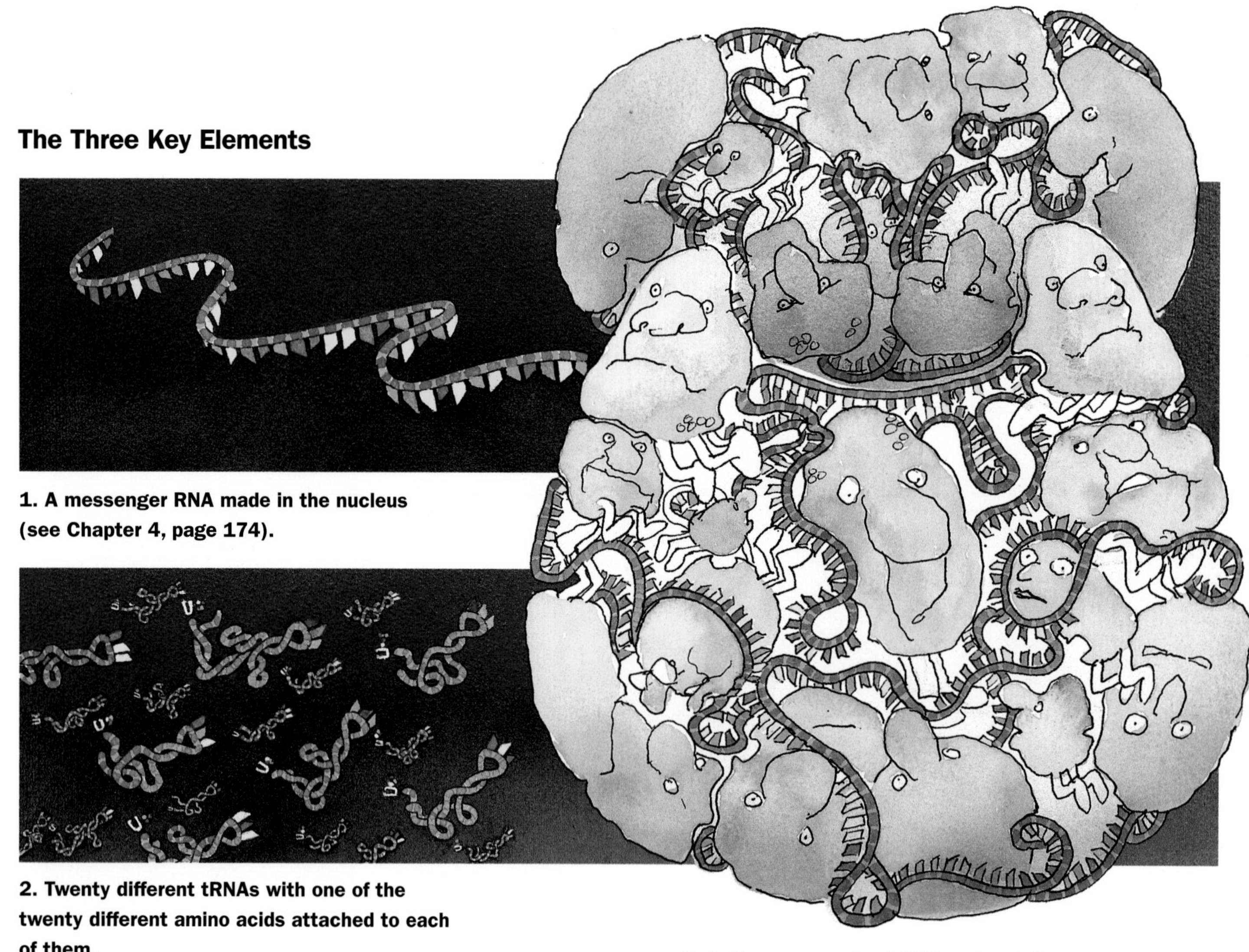

1. A messenger RNA made in the nucleus (see Chapter 4, page 174).

2. Twenty different tRNAs with one of the twenty different amino acids attached to each of them.

3. A ribosome made of RNA and proteins.

Assembling the Chain

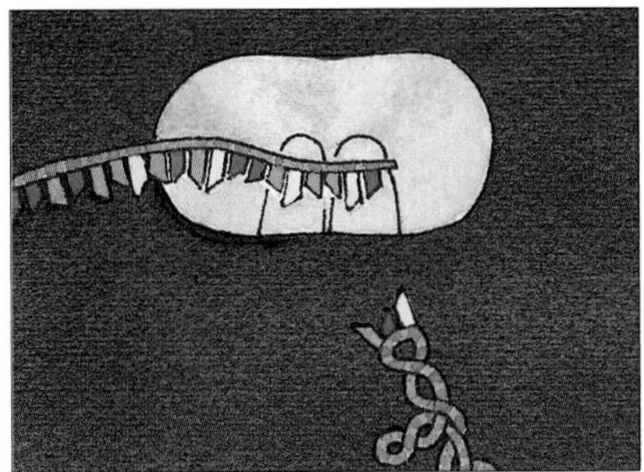

The messenger RNA attaches itself to the smaller subunit of the ribosome.

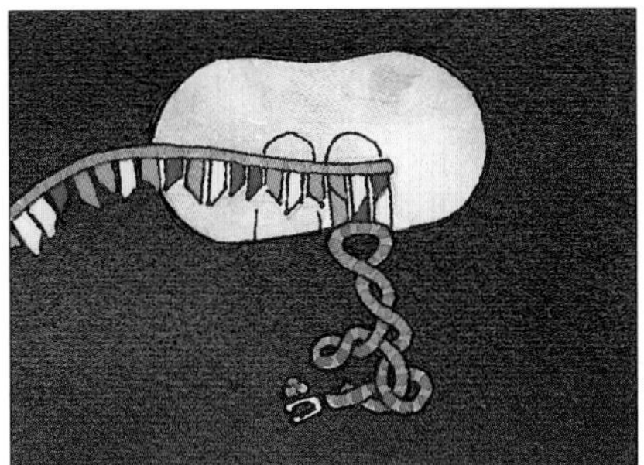

The first tRNA matches the messenger's first three nucleotides.

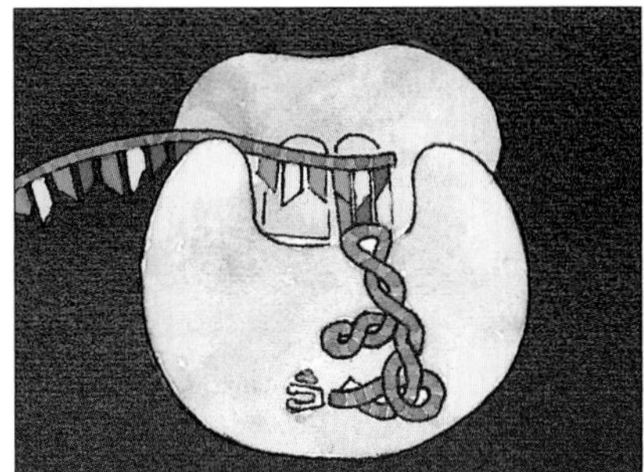

The larger subunit joins up with the smaller subunit.

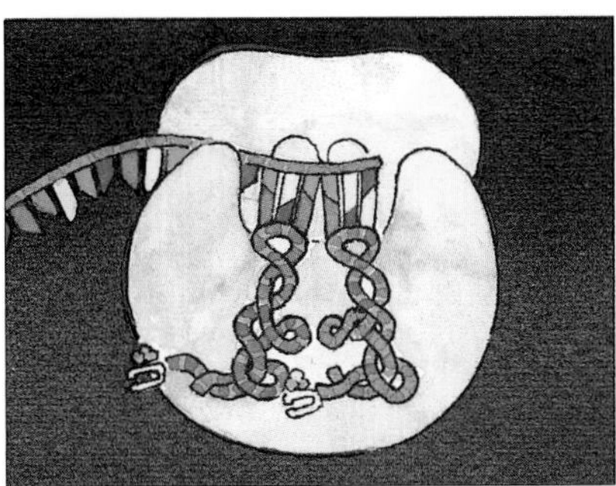

The second tRNA enters a second dock.

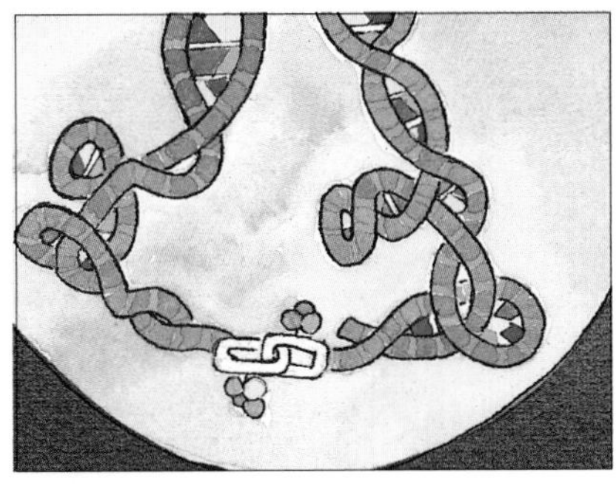

The backbone links of the first two amino acids join up.

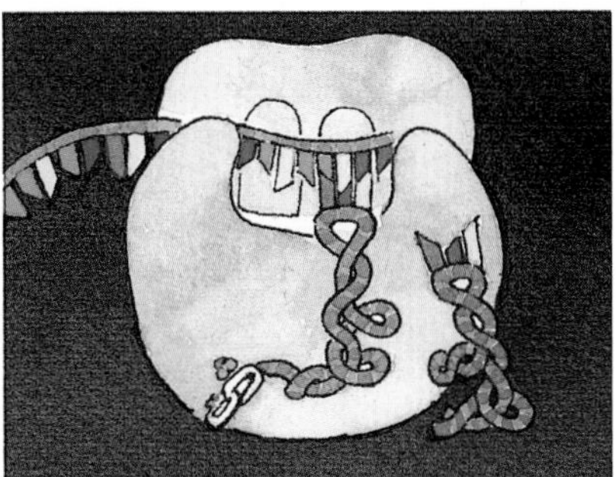

The messenger shifts to the right, and the first tRNA drops off.

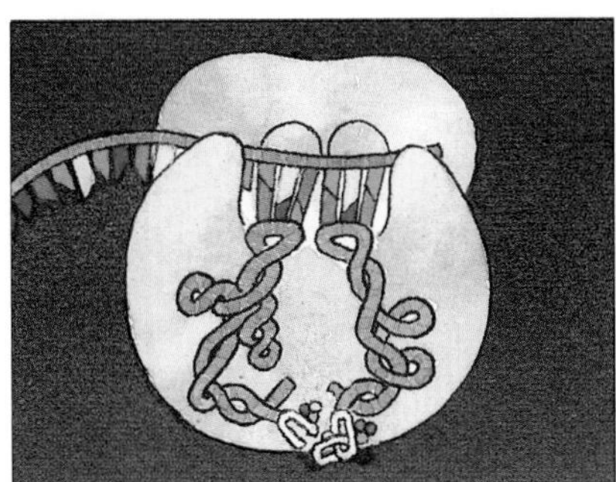

The next tRNA arrives at the second dock to add the next link.

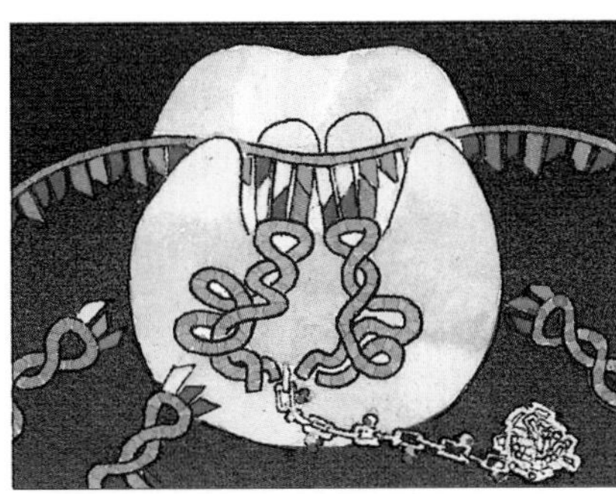

One by one, triplet codons are "read," and the protein chain grows.

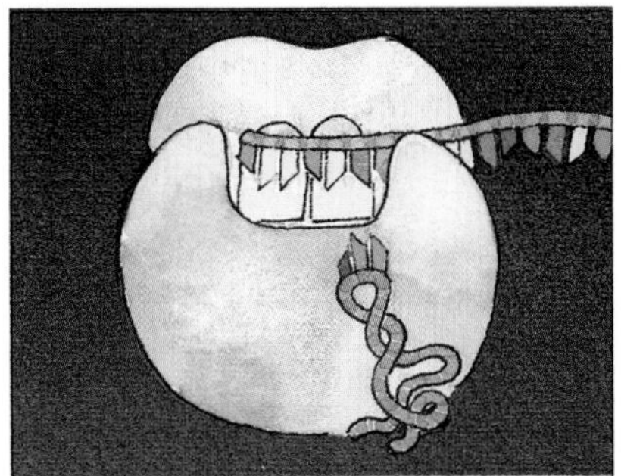

The final triplet codon signals "stop" — no adaptor fits it.

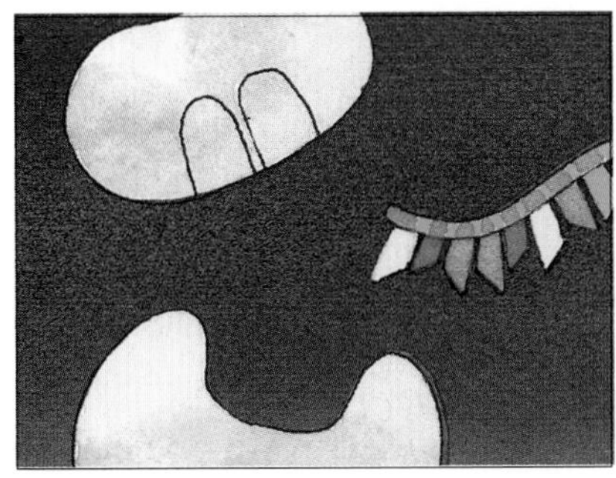

The ribosome separates and drops off the mRNA.

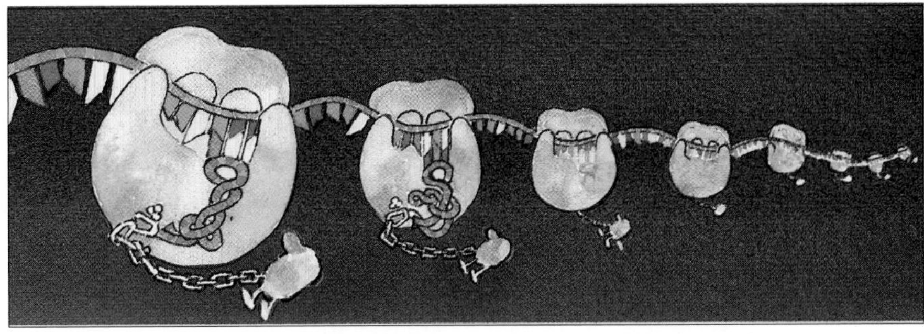

For efficiency in making multiple proteins, messenger RNA is read by more than one ribosome simultaneously.

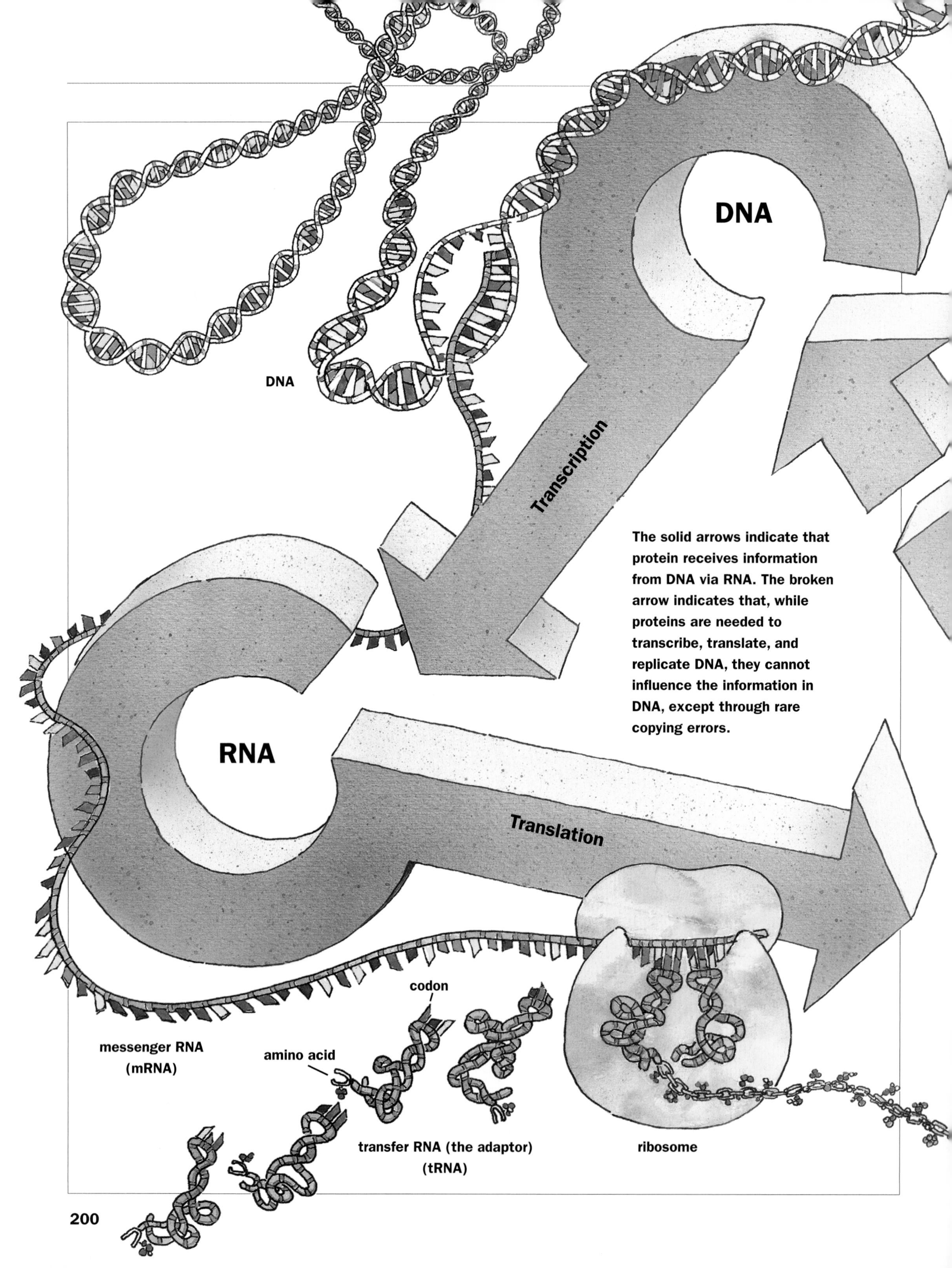

The solid arrows indicate that protein receives information from DNA via RNA. The broken arrow indicates that, while proteins are needed to transcribe, translate, and replicate DNA, they cannot influence the information in DNA, except through rare copying errors.

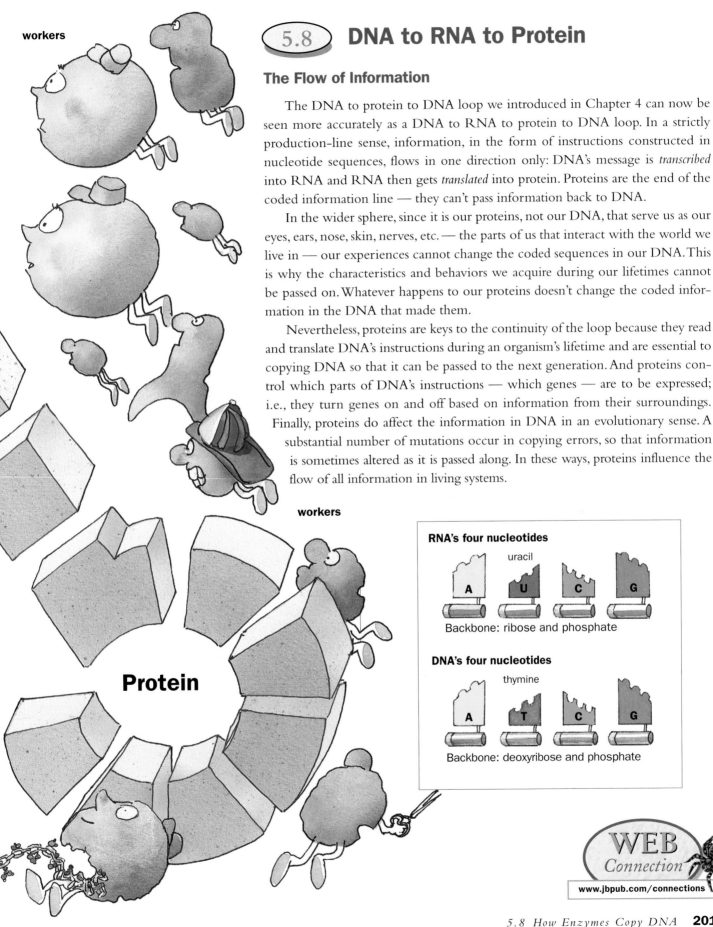

workers

5.8 DNA to RNA to Protein

The Flow of Information

The DNA to protein to DNA loop we introduced in Chapter 4 can now be seen more accurately as a DNA to RNA to protein to DNA loop. In a strictly production-line sense, information, in the form of instructions constructed in nucleotide sequences, flows in one direction only: DNA's message is *transcribed* into RNA and RNA then gets *translated* into protein. Proteins are the end of the coded information line — they can't pass information back to DNA.

In the wider sphere, since it is our proteins, not our DNA, that serve us as our eyes, ears, nose, skin, nerves, etc. — the parts of us that interact with the world we live in — our experiences cannot change the coded sequences in our DNA. This is why the characteristics and behaviors we acquire during our lifetimes cannot be passed on. Whatever happens to our proteins doesn't change the coded information in the DNA that made them.

Nevertheless, proteins are keys to the continuity of the loop because they read and translate DNA's instructions during an organism's lifetime and are essential to copying DNA so that it can be passed to the next generation. And proteins control which parts of DNA's instructions — which genes — are to be expressed; i.e., they turn genes on and off based on information from their surroundings. Finally, proteins do affect the information in DNA in an evolutionary sense. A substantial number of mutations occur in copying errors, so that information is sometimes altered as it is passed along. In these ways, proteins influence the flow of all information in living systems.

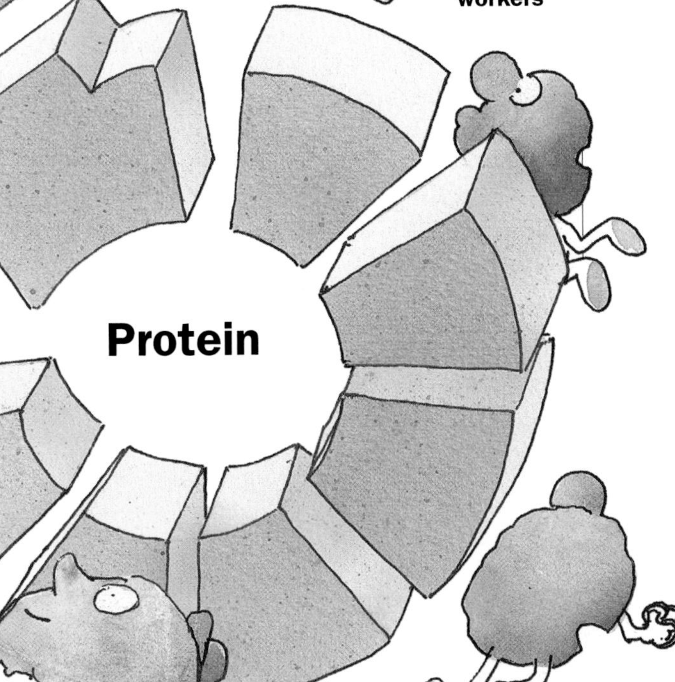

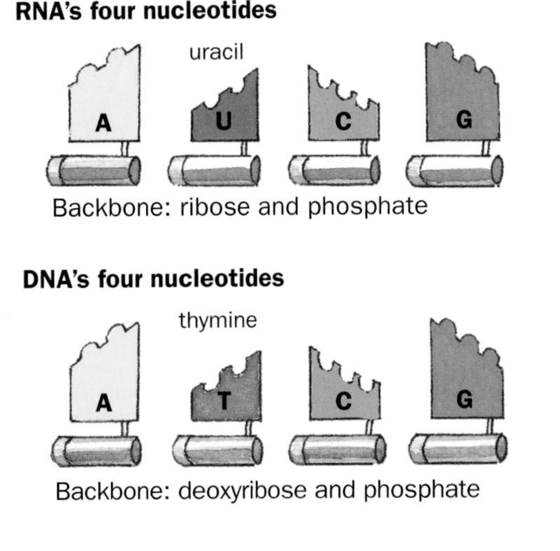

WEB Connection
www.jbpub.com/connections

5.9 Key Discovery

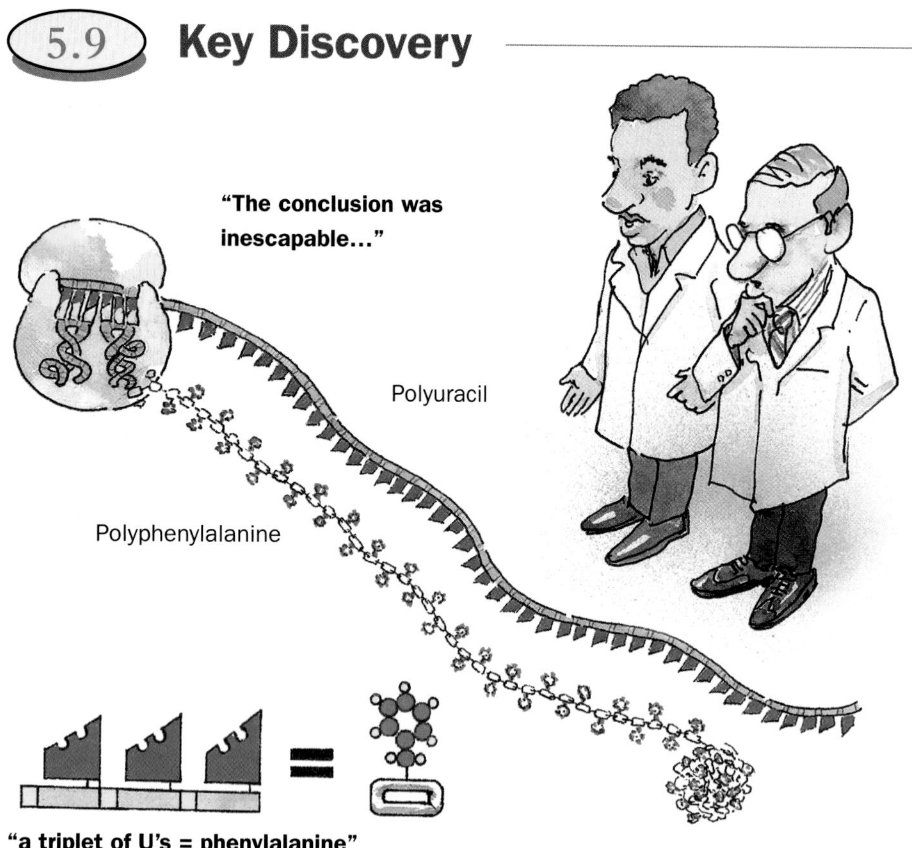

How to Read the Genetic Code

The chart on the right summarizes the genetic code. Read it like a map with coordinates. Three nucleotides code for one amino acid. Each triplet of nucleotides is called a codon. If you want to find the amino acid whose code is CAU, for example, find the box where C in the left-hand column meets A in the top row. This box contains histidine and glutamine. From this box look across to the right-hand column and find U. So histidine is represented by CAU. Note that it's also represented by CAC.

Life has used all but three of the sixty-four possible codons that can be made using four nucleotides. So most of the amino acids are represented by more than one triplet. The three triplets that don't code for an amino acid instead signal the protein-making machinery to stop; they are called stop codons.

Cracking the Genetic Code

In 1961, Marshall Nirenberg and Johann Matthaei, two young biochemists at the National Cancer Institute in Bethesda, Maryland, made an astonishing discovery. Not yet aware of the discovery of messenger RNA in Britain and France, they were searching for something like it: evidence that *some* type of RNA might program ribosomes to make protein. They took any samples of RNA they could lay their hands on and incubated them with ribosomes from bacteria, along with activating enzymes, ATP, transfer RNAs, and a mixture of amino acids. They looked to see if any of the RNAs simulated protein synthesis. The results were not particularly encouraging until, by chance, they added an artificial RNA — polyuridylic acid (U-U-U-U...) — chains made of one nucleotide, uracil, linked to each other as in natural RNA. Incredibly, the ribosomes obediently "read" the "poly U" chains into an artificial "protein," polyphenylalanine — long chains of the single amino acid phenylalanine! The conclusion was inescapable: The triplet code for phenylalanine must be UUU.

The exciting wider implication: If ribosomes can be induced to translate RNAs of any nucleotide sequence into protein, then RNAs of known nucleotide sequence could be incubated with ribosomes and watched to see what kind of amino acid sequence came out. Here lay the solution to the genetic code! Nirenberg and Matthaei pounced, as did others who learned of their discovery. A frenzy of experimentation ensued, with the result that all sixty-one of the triplet codes for the twenty amino acids were identified by 1965.

<table>
<tr><th rowspan="2">First Position</th><th colspan="4">Second Position</th><th rowspan="2">Third Position</th></tr>
<tr><th>U</th><th>C</th><th>A</th><th>G</th></tr>
<tr><td rowspan="4">U</td><td rowspan="2">phenylalanine</td><td rowspan="4">serine</td><td rowspan="2">tyrosine</td><td rowspan="2">cysteine</td><td>U</td></tr>
<tr><td>C</td></tr>
<tr><td rowspan="6">leucine</td><td rowspan="2">stop</td><td>stop</td><td>A</td></tr>
<tr><td>tryptophan</td><td>G</td></tr>
<tr><td rowspan="4">C</td><td rowspan="4">proline</td><td rowspan="2">histidine</td><td rowspan="4">arginine</td><td>U</td></tr>
<tr><td>C</td></tr>
<tr><td rowspan="2">glutamine</td><td>A</td></tr>
<tr><td>G</td></tr>
<tr><td rowspan="4">A</td><td rowspan="3">isoleucine</td><td rowspan="4">threonine</td><td rowspan="2">asparagine</td><td rowspan="2">serine</td><td>U</td></tr>
<tr><td>C</td></tr>
<tr><td rowspan="2">lysine</td><td rowspan="2">arginine</td><td>A</td></tr>
<tr><td>methionine</td><td>G</td></tr>
<tr><td rowspan="4">G</td><td rowspan="4">valine</td><td rowspan="4">alanine</td><td rowspan="2">aspartic acid</td><td rowspan="4">glycine</td><td>U</td></tr>
<tr><td>C</td></tr>
<tr><td rowspan="2">glutamic acid</td><td>A</td></tr>
<tr><td>G</td></tr>
</table>

5.10 The Unity of Biology

A Light That Blinked...

Life first strikes us with its diversity. Evolution has filled every niche: Bacteria thrive in hot springs; fish plumb the depths of the sea; birds soar skyward, defying gravity. But when we look below the surface at the ways molecules work in cells, we cannot but marvel at their unity. All living creatures use DNA and RNA to store and replicate information, building them from the same four nucleotides. They make those nucleotides using very similar pathways. They translate nucleotide chains into proteins using the same twenty amino acids and the same genetic code. They use very similar translation apparatus — ribosomes, tRNAs, mRNAs, activating enzymes. If we take ribosomes from bacteria and put them in a test tube, they'll translate human messenger RNAs into human proteins — and vice versa. And many of the proteins — supporters, movers, communicators, transporters, and catalyzers — when dissected into their primary amino acid sequences — are quite similar in most creatures throughout the living world.

The realization dawns on us that we all had common beginnings. Billions of years ago, a tiny light blinked on somewhere and has come to illuminate every nook and cranny of our Earth's surface.

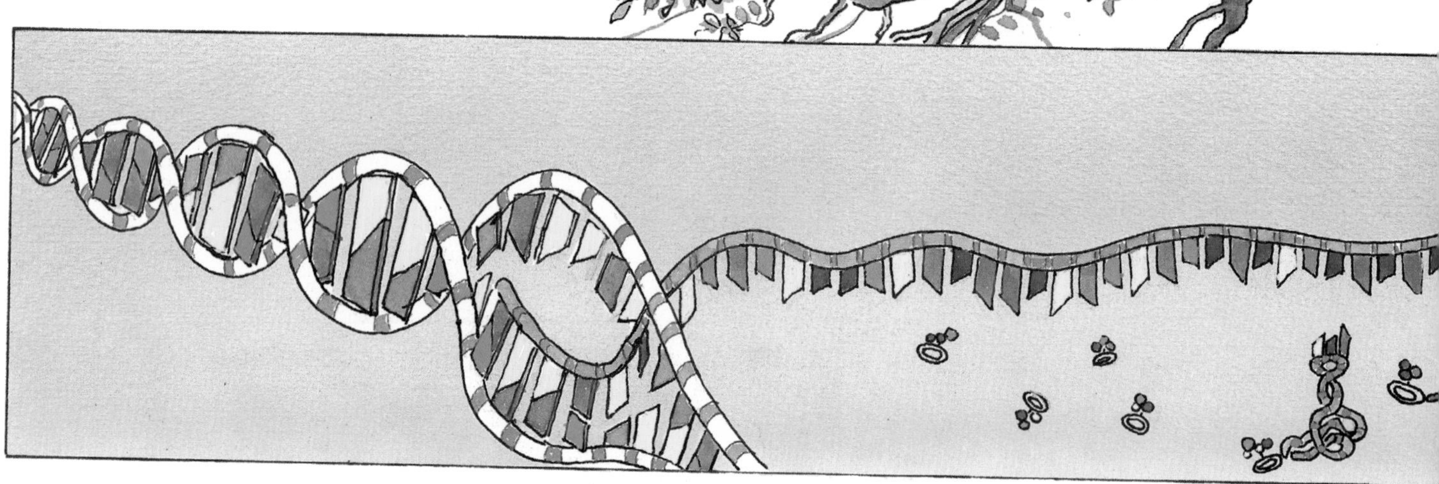

"We are educated to be amazed by the infinite variety of life forms in nature; we are, I believe, only at the beginning of being flabbergasted by its unity."

Lewis Thomas

WEB Connection
www.jbpub.com/connections

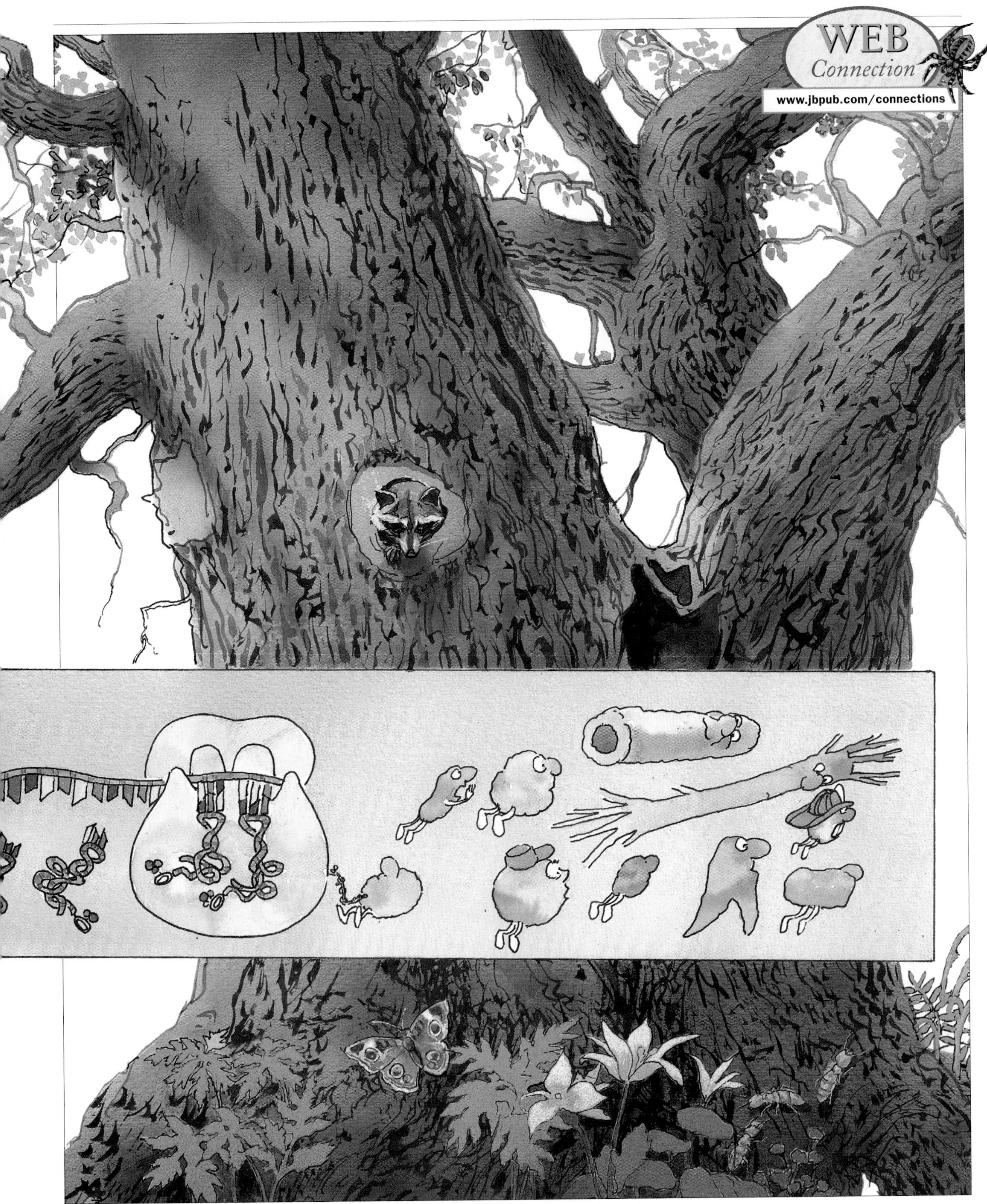

Some of the Things You Learned About in Chapter 5

Questions About the Ideas in Chapter 5

1. Name at least four of the key activities of proteins that make life possible.
2. What two things have to be done to amino acids before they can be linked together in a protein molecule?
3. Why is transcription necessary? Why don't cells make their proteins directly from DNA?
4. What are the key elements of the translation machinery?
5. How would you go about testing that a bacterium's protein-making machinery could make a human protein?
6. What is a triplet and what is its function?
7. How would each of the following mutations affect the final sequence of a specific protein (or even its production)?
   (a) a change in the DNA sequence for the protein itself
   (b) a mutation in the gene for one of the specific activating enzymes
   (c) a change in the transfer RNA sequence
8. How is a regulatory protein like an old-fashioned spring clothespin? What other parallels can you suggest that might evoke the way proteins change shape and the work they do?
9. Most mutations are harmless and a very few are even beneficial to organisms. How is this possible if the usual result of a mutation is to change the amino acid sequence of a protein and affect its correct folding? (Hint: Remember that there are 20 amino acids and 64 possible combinations of triplets using the four DNA bases.)
10. Protein synthesis is a complicated and dynamic process with numerous steps, many of which require ATP for the energy to move molecules or to form new bonds. Can you think of a reason why so many steps are required?

References and Great Reading

Allen, N. S. 1974. Endoplasmic filaments generate the motive force for rotational streaming in *Nitella. Journal of Cell Biology* 63: 270–287.

Bronowski, J. 1990. *Science and Human Values.* New York: HarperCollins.

Gunning, B. E. S. and M. W. Steer, 1996. *Plant Cell Biology: Structure and Function.* Sudbury, MA: Jones and Bartlett Publishers.

Hartl, D. L. and E. W. Jones, 1999. *Essential Genetics,* 2E. Sudbury, MA: Jones and Bartlett Publishers.

Hoagland, M. 1990. *Toward the Habit of Truth: A Life in Science.* New York: W. W. Norton.

Strickberger. M. 2000. *Evolution,* 3E. Sudbury, MA: Jones and Bartlett Publishers.

Williamson, R.E. 1980. Actin in motile and other processes in plant cells. *Canadian Journal of Botany* 58: 766–772.

for more questions
and links to
web resources,
go to
WEB
Connection
www.jbpub.com/connections

CHAPTER 6

Our desired course is due north. The plane has veered to the west. We correct by turning toward the east.

Oops — we've overcorrected. We need to steer to the west again.

Arriving at the desired destination is a matter of many such corrections.

FEEDBACK

Signaling, Sensing, and Reacting

TO AN OBSERVER ON THE GROUND, AN AIRPLANE APPEARS TO FLY IN A BEELINE toward its destination. But things look very different from inside the cockpit. Buffeted by winds or shifts in air pressure, the plane regularly drifts off course. When this happens, the pilot makes a correction by steering the plane in the opposite direction. If the pilot over-corrects, then he or she must correct the correction, and so on. The plane actually flies in a zigzag.

Feedback is a central feature of life processes. All organisms share this ability to sense how they're doing and to make changes in "mid-flight" when necessary. The process of feedback governs how we grow, respond to stress and challenge, and regulate factors such as body temperature, blood pressure, and cholesterol level. This apparent purposefulness, largely unconscious, operates at every level — from the interaction of proteins in cells to the interaction of organisms in complex ecologies.

How does feedback work? The process requires two elements: first, some kind of device to measure the difference between the current state of affairs and some preset "desired" state (like our pilot's compass and the flight plan); and second, some kind of responsive machinery that can reduce that difference (like the steering mechanism). The bigger the difference, the harder that machinery must work. This is *negative feedback* — it acts to dampen instability in dynamic systems. But sometimes feedback operates as an amplifier, *increasing* the difference between the status quo and the objective. This is called *positive feedback* — and can lead to runaway and breakdown. It can also lead to creation and change, as we shall see.

Pilot, compass, and steering mechanism all represent parts of a self-corrective circuit — a feedback control system (sometimes called a cybernetic system).

6.1 Assembly Lines

The Airplane Factory

We can get an idea of how a living cell regulates itself by using an old-time airplane factory as a model. Skilled workers on multiple production lines assemble the essential components of each airplane from basic materials. Other workers stoke the fires under the steam boiler to generate the energy that powers the machinery. The bigwigs in the front office keep track of budgets, markets, and supplies, and they relay design and construction information. Floor supervisors, receiving instructions from the front office, control and coordinate how efficiently workers assemble the parts. From a distance, the production lines seem to hum along smoothly. Up close, things are a little more chaotic. Workers overshoot their production goals or miscalculate the number of parts needed and machinery breaks down. But the workers quickly correct for their mistakes, and the conveyor belts roll on.

The simplest living cell, though far more complex, is like the airplane factory, both in its organization and in its self-correcting behavior. Its workers are enzymes, teamed up in assembly lines. Some of these enzymes, the floor supervisors, possess the remarkable ability to evaluate the performance of the system and make the necessary adjustments. The cell's ultimate product is, of course, itself. It works to make more of its own components, to maintain them, to use them to further the whole organism's needs, and then to reproduce itself.

Both factories and cells are organized around a few basic rules:

1. Keep things moving in an even flow.
2. Don't allow components or products to pile up.
3. Be flexible and ready to respond to new demands.
4. Supervise every level of production.
5. Repair and replace machinery regularly.

Circular Information

The workers' assembly rate of airplane tails is satisfactory; the supervisor is content.

The tails are not being used as fast as they're made and they begin to accumulate; the supervisor takes note.

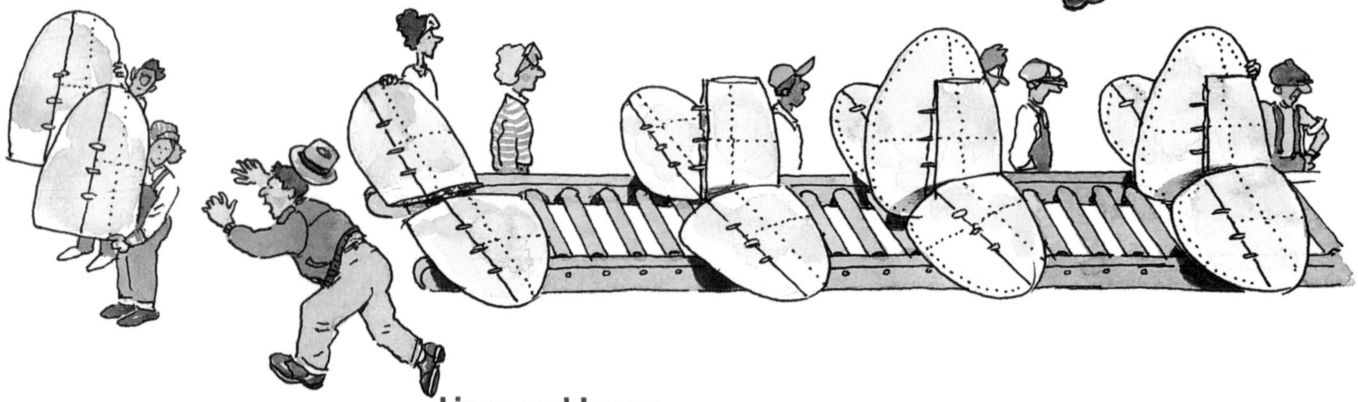

The supervisor calls for a slowdown. The inventory of finished tails decreases as workers slow production. Later, as the supply of tails falls, the supervisor signals the workers to speed up their rate of assembly.

Lines and Loops

An assembly line moves in one direction only — from input of raw materials to output of product, with the supervisor acting as the governor, or controlling agent. If too much product begins to accumulate, the supervisor slows down the input of raw materials. Conversely, if there are too many raw materials, the supervisor speeds up production.

To appreciate how feedback works, it helps to imagine the information (the signals that say "too much" or "not enough") as flowing in a loop. Bending the production line into a circle and stationing the supervisor at a strategic point overseeing both input and output gives him or her greater control. This arrangement is impractical for many factories, but it works beautifully inside cells, in molecular assembly lines like the one shown at the right.

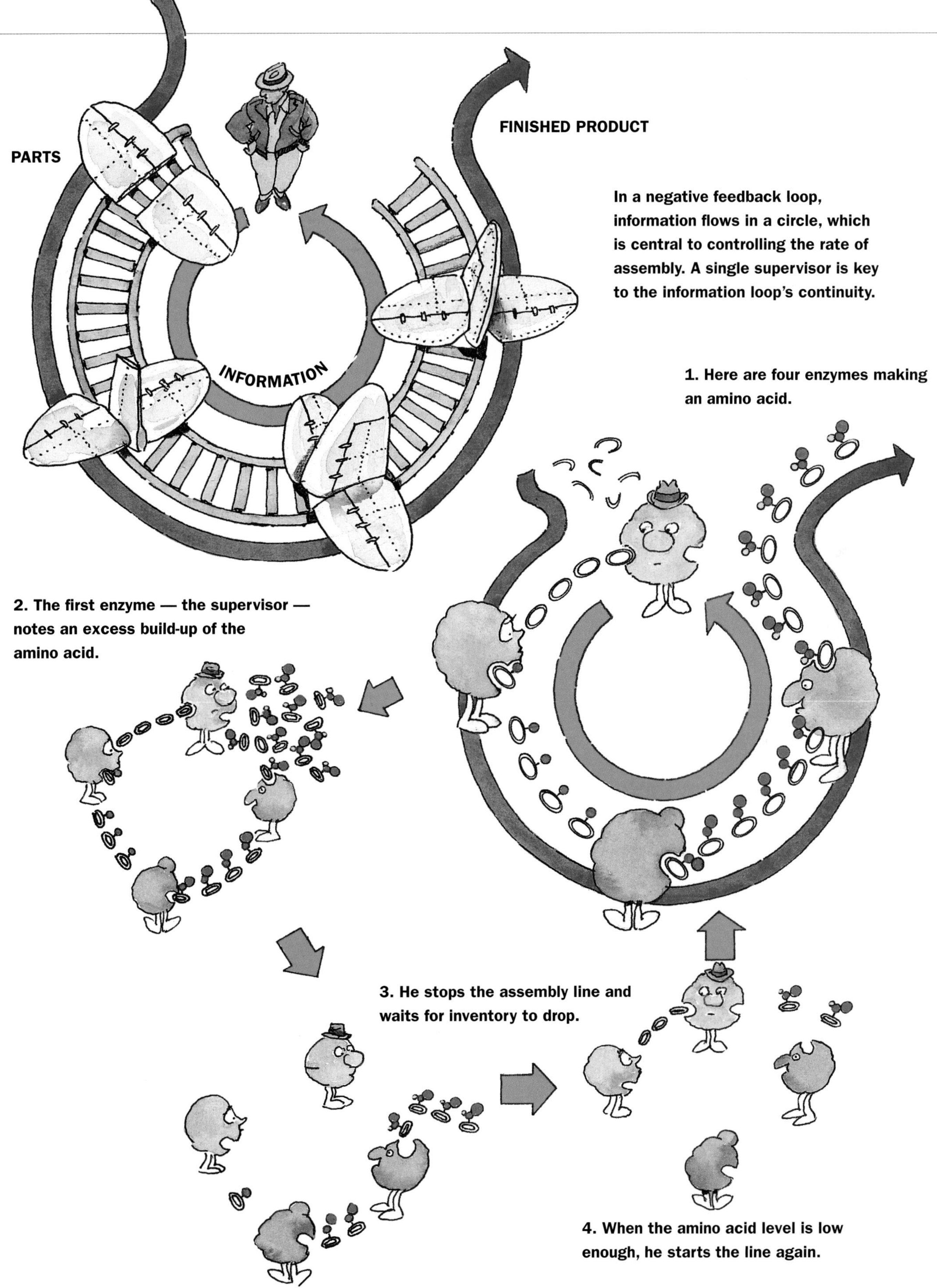
PARTS
FINISHED PRODUCT
INFORMATION
In a negative feedback loop, information flows in a circle, which is central to controlling the rate of assembly. A single supervisor is key to the information loop's continuity.
1. Here are four enzymes making an amino acid.
2. The first enzyme — the supervisor — notes an excess build-up of the amino acid.
3. He stops the assembly line and waits for inventory to drop.
4. When the amino acid level is low enough, he starts the line again.

6.3 Allostery — The Key to Feedback Control

Allostery — The Basic Idea

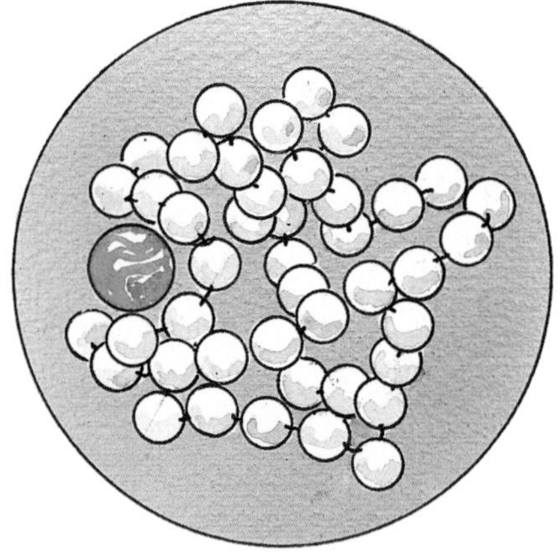

Imagine a tightly balled string of beads surrounding a marble.

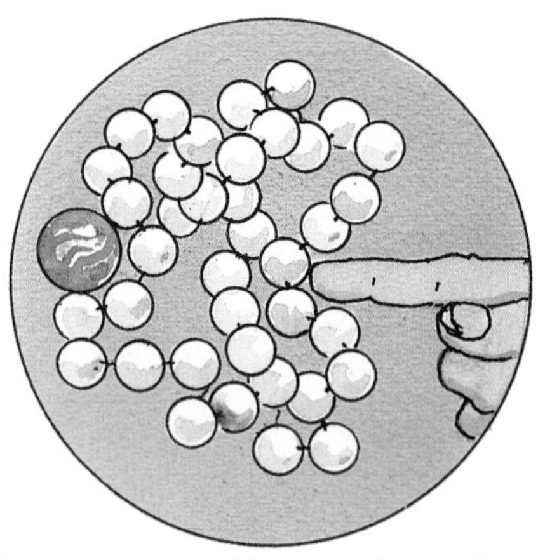

Now imagine pushing your finger into the beads on the side opposite the marble, causing the beads to shift.

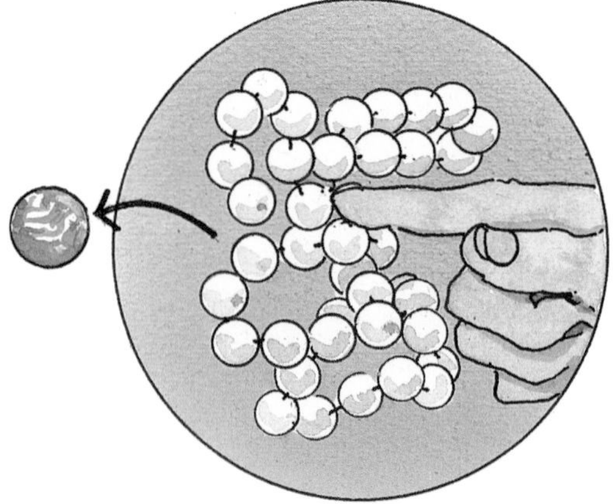

As your finger moves into its niche, the marble pops out.

Enzyme Governors

We can now better appreciate why we call enzymes "smart." Their unique chemical character gives them the ability not only to do their usual work of rearranging or disassembling other molecules but also to process information. Certain supervisory, or regulatory, enzymes do this by readily and reversibly *changing shape* in response to a signal. In addition to the working site on their surface where other molecules "dock" to get processed, regulatory enzymes have a *second* site specifically designed to hold a small signal molecule. Nestled in this special niche, the signal molecule acts like a finger on an on/off switch: It causes the enzyme to modify its shape so that its working site stops functioning. Allostery (literally "other shape"), the name given to this almost ridiculously simple behavior of switching on and off, underlies most of the unimaginably complex regulatory processes of life.

Regulatory enzymes are switched off — that is, they stop working — by literal contact with chemical information; they turn back on when that information (the chemical signal) is removed. Some regulators operate in exactly the opposite way. Like most protein behavior, these reactions are highly specific: For the most part, one and only one enzyme acts as a chemical switch for one and only one signal. But once a regulator's working site is turned off, it can shut down an entire production line. So a regulator can control a larger loop much as a governor controls the operation of a steam engine.

The behavior of allosteric enzymes offers a glimpse into the way living organisms attain their impressive complexity. As enzymes — the cell's workers — transport, build, and break down small molecules, allosteric enzymes — the regulators — control and coordinate these processes. And, as we shall see, higher-level regulators control lower-level regulators in a hierarchy of intercommunicating feedback loops.

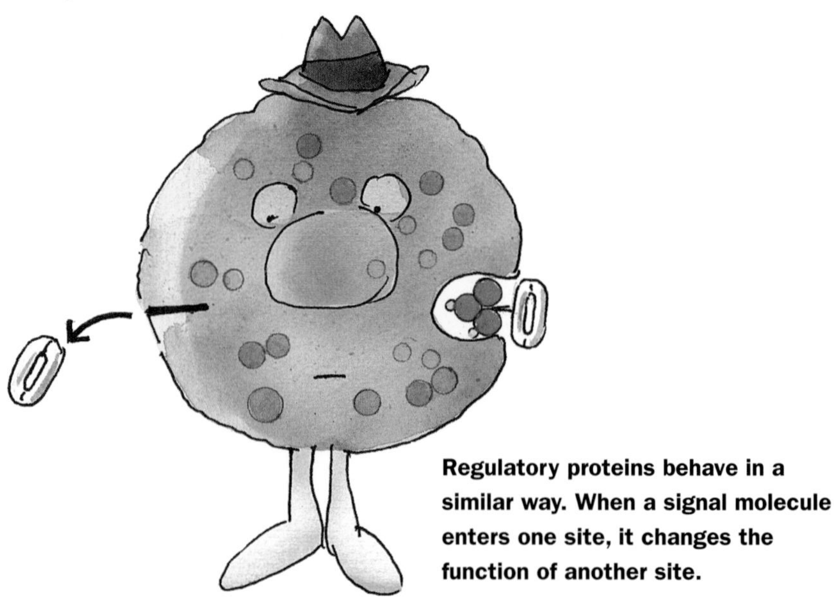

Regulatory proteins behave in a similar way. When a signal molecule enters one site, it changes the function of another site.

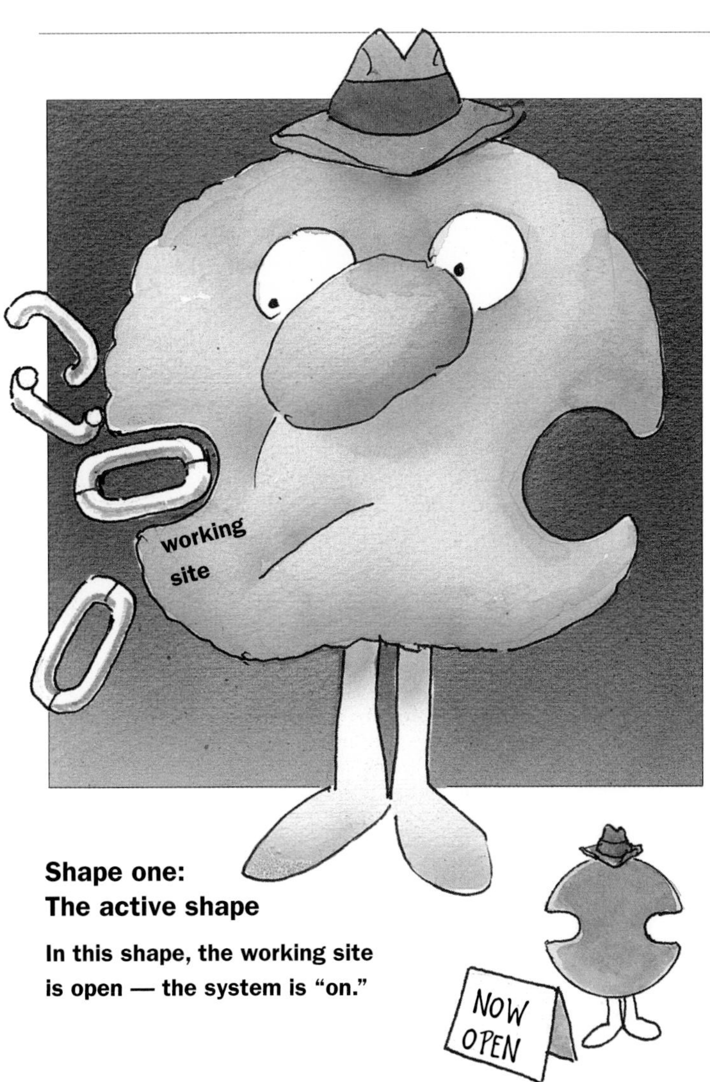

Shape one:
The active shape

In this shape, the working site is open — the system is "on."

Shape two:
The regulatory shape

In this shape, the working site is closed — the system is "off."

What exactly is the on/off signal in an enzyme assembly line? It is the final product molecule itself. This molecule is the equivalent of a message sent by the supervisor back to the first worker in the assembly line. When it fits into place on the enzyme, it says, "Enough already. Stop making us."

Site-filling by signal molecules is statistical. If lots of product molecules (signals) are around, a given set of identical sites are more likely to be filled.

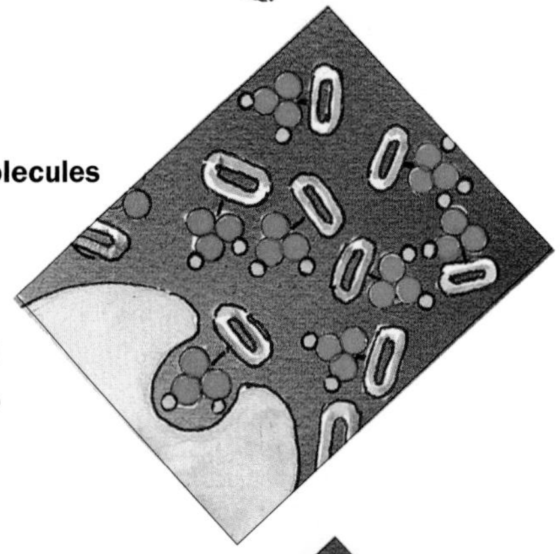

As the concentration of product molecules decreases, they fill fewer sites, leaving many of them empty.

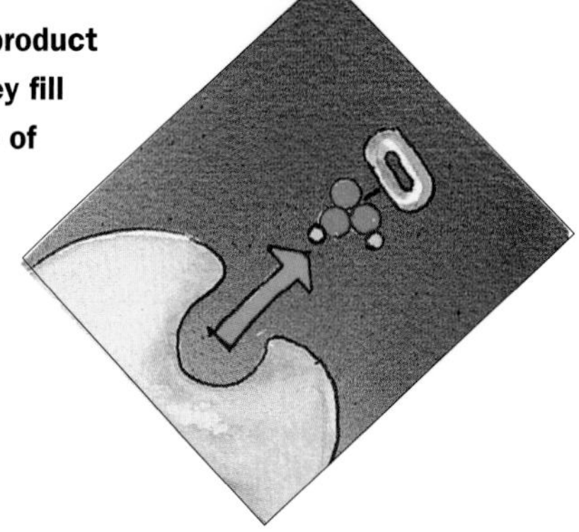

6.4 Why Allostery?

For a simple, non-allosteric enzyme, as the number of molecules it recognizes and binds to (substrate molecules) in the surroundings increases, its ability to bind and process the substrate rises rapidly to a maximum (see graph below, left).

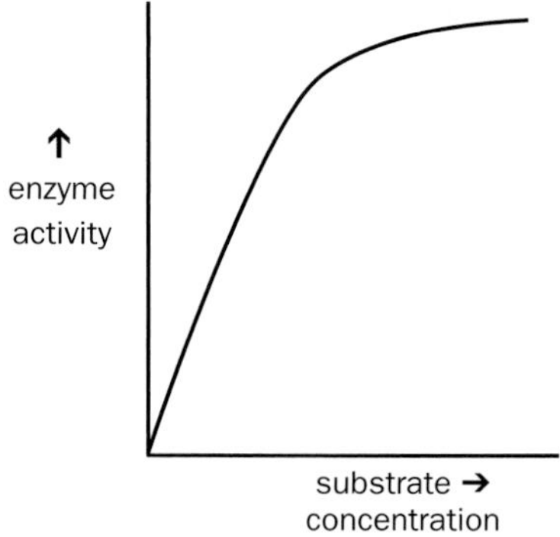

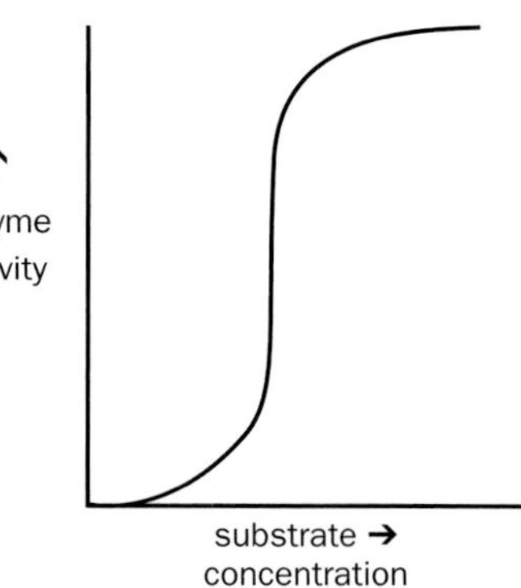

In contrast, an allosteric enzyme binds and processes substrate molecules slowly — or not at all — at low substrate concentrations, but its effectiveness rises sharply as the concentration increases (see graph above, right). This behavior is the key to the enzyme's regulatory role. At low concentrations of substrate, the enzyme is essentially non-functional — it's a switch turned off. As more and more substrate becomes available, requiring processing, the enzyme responds by steadily increasing its processing capacity — the switch turns on.

What allows this kind of enzyme to shift gears from low to high efficiency in response to increasing substrate? At low substrate concentrations, its shape — the relationship of its protein subunits to each other — is such that the substrate binding sites are relatively inaccessible. As the substrate concentration increases, binding at one site induces a change in shape that makes the other sites more available. The more substrate there is to process, the more efficient the processing.

The second essential feature of this type of regulatory protein, as we've already seen, is a susceptibility to inhibition or activation by other molecules — so-called signal, or effector, molecules. These molecules bind at special sites on the enzyme's protein subunits, changing their relationship to each other and, in consequence, affecting the enzyme's functioning. Again, the enzyme behaves like switch. The signal molecule, at low concentrations, has no effect on enzyme activity. As signal concentration rises, the enzyme's activity is turned on, or off.

Multiple proteins

Allosteric enzymes are actually more complex than those described on the preceding pages. They are not single proteins, but complexes of several — usually four to six — protein subunits, each with its own working site.

An allosteric enzyme is not just a single protein like this...

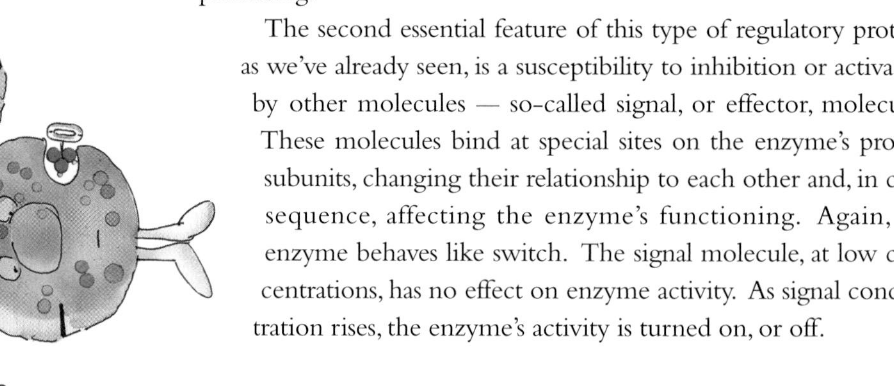

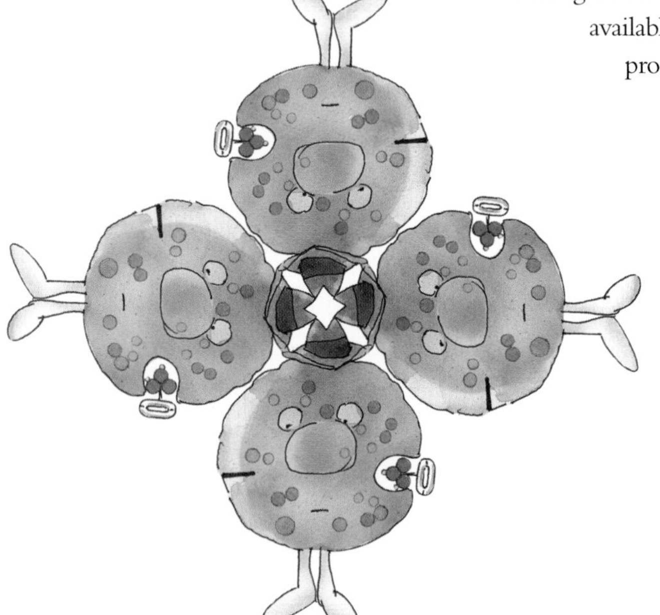

...it's a complex of proteins "stuck together," operating as a unified whole.

How Hemoglobin Works

A nice example of allostery is provided by hemoglobin, the protein inside red blood cells that carries oxygen from the lungs to body tissues and carbon dioxide from the tissues back to the lungs. Hemoglobin has four subunits, bound together by various weak bonds.

Changing relationships

Hemoglobin consists of four protein subunits of two different types, distinguished here in dark purple and light purple. The heme-iron complexes are shown as red plates at the centers of the protein subunits. The structure of oxygenated hemoglobin differs from that of deoxygenated hemoglobin only in that the protein subunits have shifted slightly relative to each other.

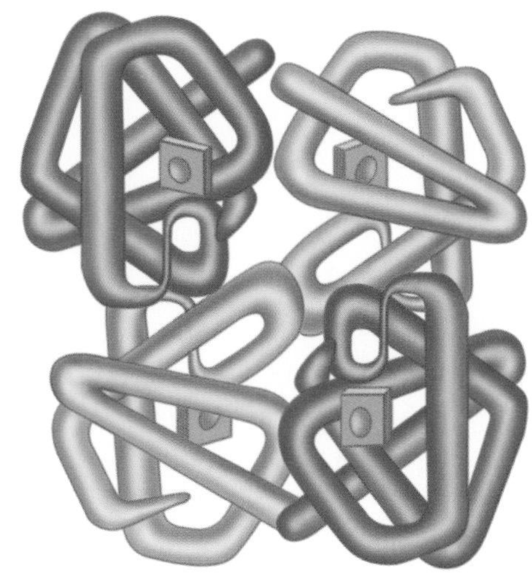

Each subunit (see the illustration above) contains a heme-iron complex, a molecule of heme with an iron atom at its center. When oxygen concentration is low, it is relatively rare for an oxygen molecule to bind to a heme-iron complex. When an oxygen does bind to one of the heme-iron complexes, this induces a change in shape in the hemoglobin molecule — a small shift in the relative positioning of its four protein subunits — which increases the affinity of the other three heme-iron complexes for oxygen. Thus, like an allosteric enzyme, hemoglobin binds with its "substrate" (oxygen) more efficiently when that substance is present at higher concentrations. Conversely, as oxygen concentration decreases, the affinity of hemoglobin for oxygen falls off, resulting in oxygen release.

The physiological implications of this behavior of hemoglobin are obvious. As blood passes through the lungs, where oxygen concentration is high, hemoglobin readily takes up oxygen and conveys it to the rest of the body. In the tissues, where oxygen is being consumed and is therefore present in relatively low concentration, hemoglobin gives up its oxygen (see page 61). Hemoglobin's relationship to oxygen follows the same S-shaped curve as an enzyme's relationship to its substrate: it's hungry for oxygen when oxygen is around, shuns it when it's scarce.

Hemoglobin also resembles an allosteric enzyme in that its capacity to bind and carry oxygen is controlled by other molecules — the effectors. One of these is carbon dioxide, which is produced constantly in the tissues as a waste product of respiration. As hemoglobin gives up oxygen, its affinity for carbon dioxide increases: It binds with carbon dioxide molecules and carries them back to the lungs. Deoxygenated hemoglobin also binds with hydrogen ions (acid), making the blood more alkaline. Alkaline blood more readily absorbs carbon dioxide, which is then transported to the lungs and breathed out into the atmosphere. All this finely tuned regulation is the result of tiny changes in the relative positioning of the four protein subunits of hemoglobin.

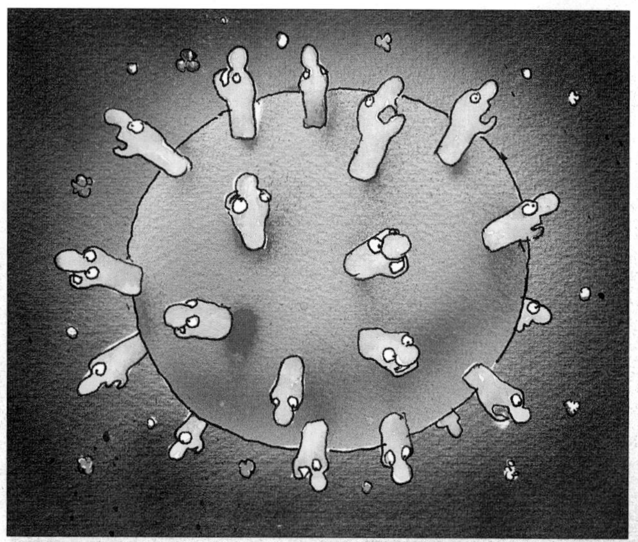

Allosteric receptors are embedded in cell membranes.

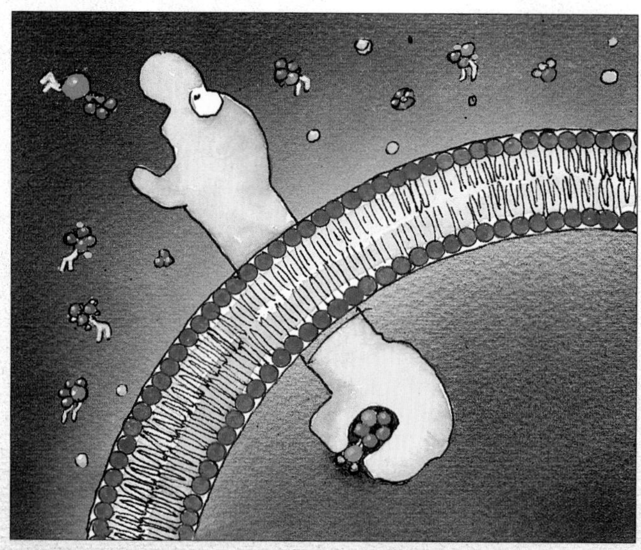

Each type of receptor matches a particular signal molecule.

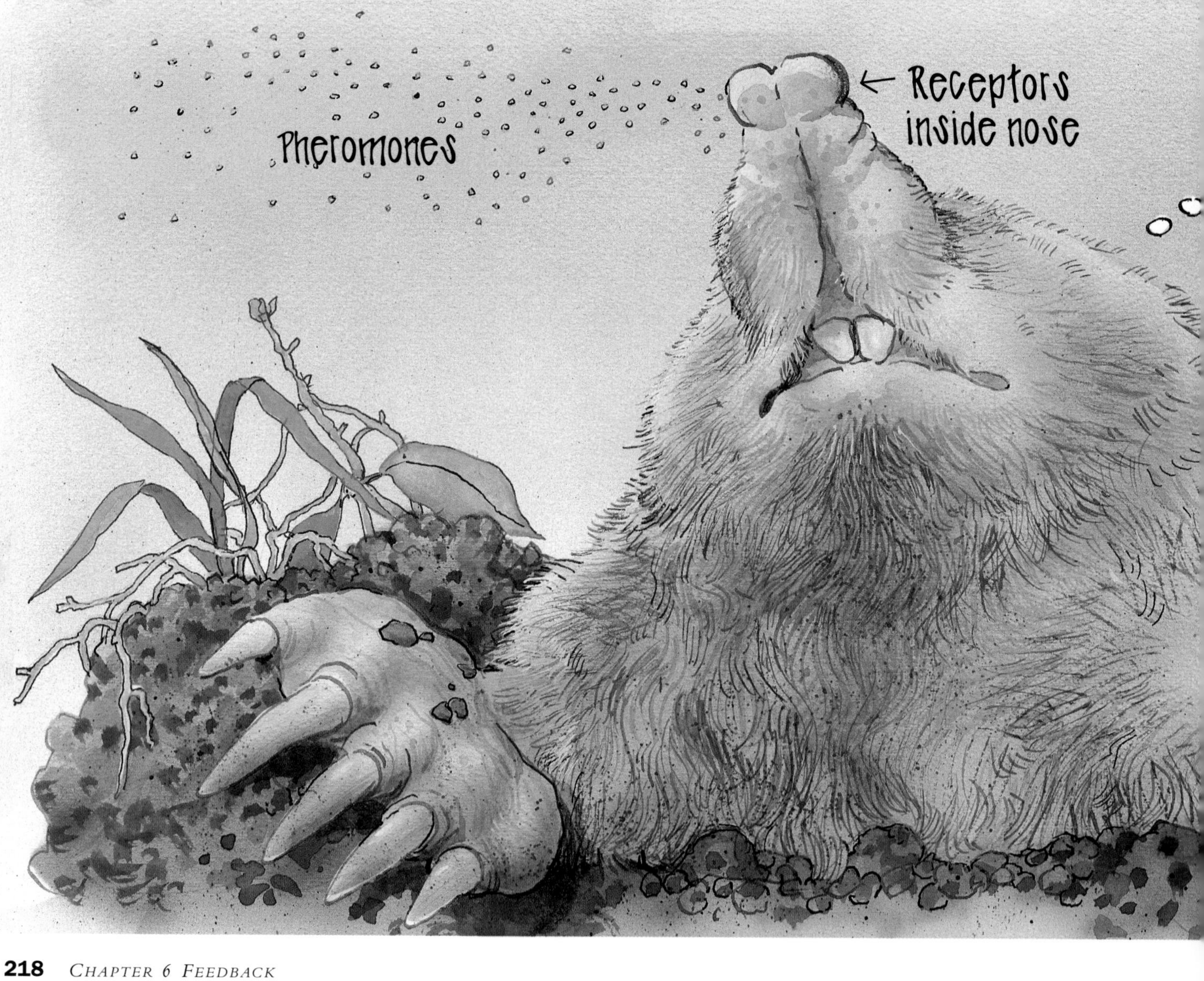

Pheromones

Receptors inside nose

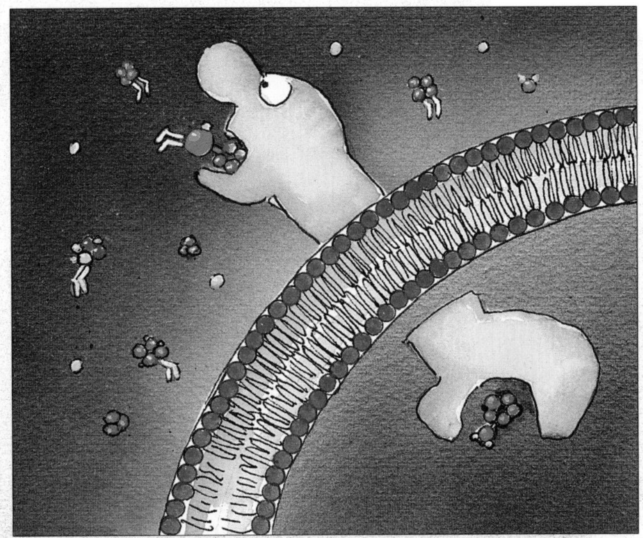

When a molecule fits...

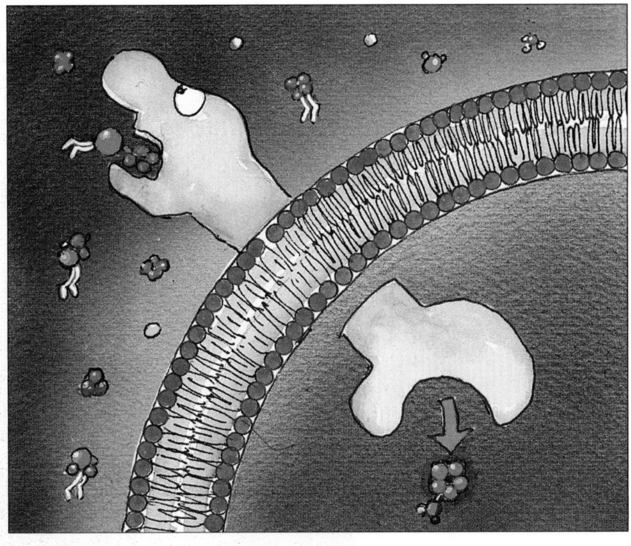

...the receptor changes shape, releasing an internal signal. This triggers cellular changes (see Chapter 7).

6.5 Allostery and Molecular Communication

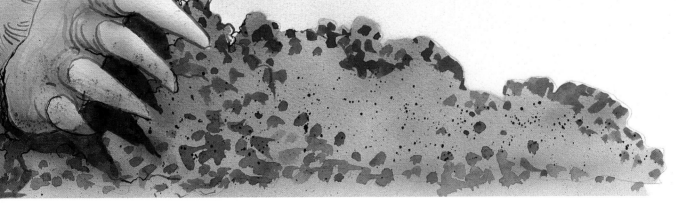

Ahhh...Spring.

Proteins Are Life's Universal Go-Betweens

Allosteric mechanisms reveal a basic characteristic of life. Though usually molecules interact only when they have some chemical affinity for one another, life's allosteric proteins can bring together molecules that have no direct chemical relationship. With such a protein acting as go-between, *any* small molecule can, in theory, act as a signal to influence *any* chemical process. A simple hormone molecule made in the brain — or thyroid, or ovary — can travel through the bloodstream to turn on chemical activities in cells all over the body. Pheromone molecules released into the air by a female mole can reach receptor proteins in the nose of a male mole and trigger a chain of events leading to mating. Allostery has enabled life to master the art not only of molecular control, but also of molecular communication. The vast evolutionary multiplication of chemical relationships brought about by receptor and regulatory proteins has created a network of interconnections within cells, between cells, and among the organs and tissues made up of those cells — a web of life.

6.6 Higher-Level Control

A Local-Level Loop

A single assembly line is controlled by a regulatory enzyme, which is one of the members of the loop.

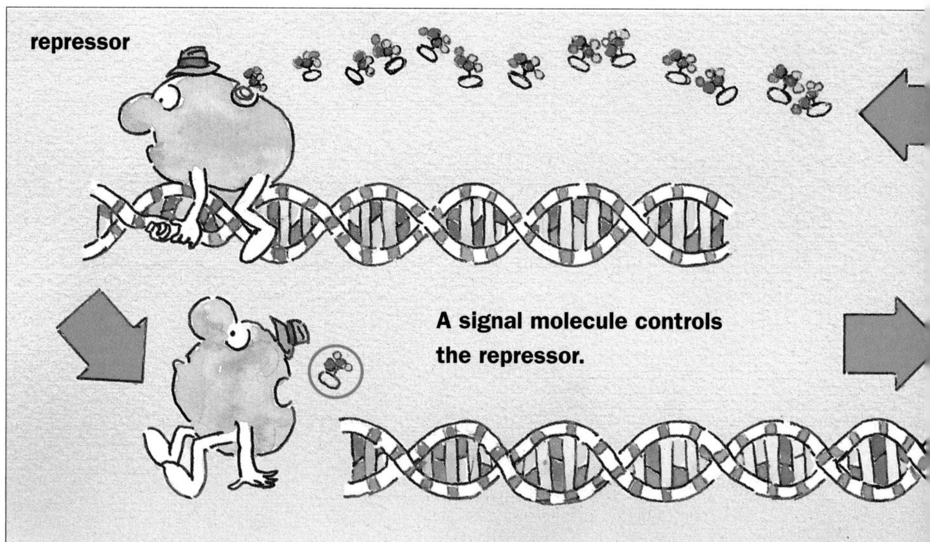
repressor
A signal molecule controls the repressor.

Local Control

Here's an imaginary rivet-making machine. Its output is controlled by the accumulation of rivets.

When the machine makes too many rivets, they turn it off. This is local control.

Controlling the Machinery That Makes the Machinery

We've seen how feedback control operates locally on the activity of each individual assembly line: The final product inhibits the activity of the first enzyme. This process is quick, sensitive, and reversible.

A higher-level process of feedback regulation controls the manufacture of the machinery that makes the product. This type of feedback control, which acts directly on the genes, is slower but far more consequential. It entails shutting down the whole process of making the enzymes involved in the assembly of a particular product — like laying off all workers on a particular assembly line.

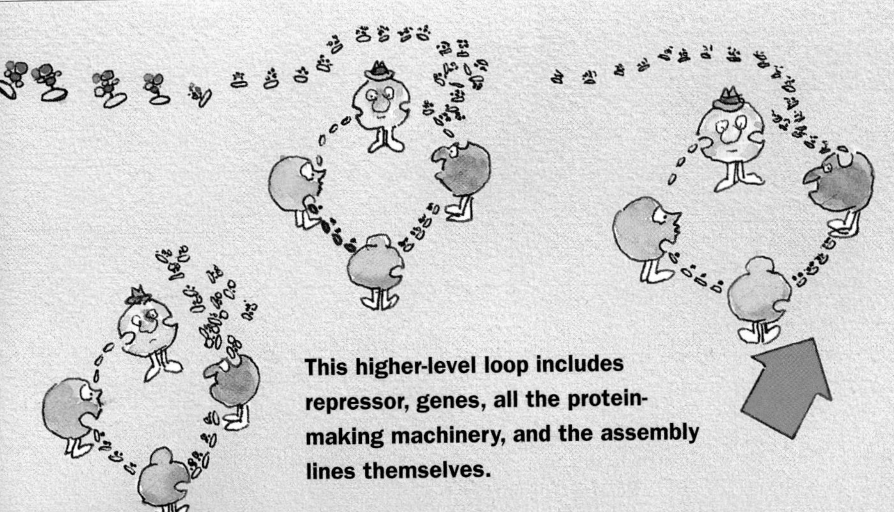

This higher-level loop includes repressor, genes, all the protein-making machinery, and the assembly lines themselves.

A Higher-Level Loop

An allosteric protein known as a repressor, which has one site that responds to a signal molecule and another site that binds to DNA, controls the synthesis of the assembly line proteins themselves. When the repressor is active, it binds to DNA and prevents the synthesis of messenger RNA. (In some cases, the signal molecule makes the repressor active; in other cases, it makes it inactive.)

This system includes the big machine, the smaller rivet makers, and the workers and operators.

Higher-Level Control

Here's a rivet-machine maker. Like the machines it manufactures, it stops or starts depending on the number of rivets that accumulate. Too many rivets block the rivet-machine maker. This is higher-level control.

Here again, the product acts as the finger on the on/off switch; but in this case the protein that is switched on or off is a gene regulator protein that controls the production of one or more enzymes. By sitting on certain genes, the gene regulator (called a repressor) blocks the production of several enzymes (including the regulatory enzyme) that produce the product. It does so by stopping transcription of messenger RNAs.

This second method of feedback control has the same purpose as the first: to avoid overproduction. But its effect is more far-reaching. It conserves materials and energy that the cell would otherwise use to make unneeded proteins, that is, those that make a currently overstocked product. This process is like controlling the conductor of the orchestra, instead of individual musicians.

6.7 When You Can't Get It, Make It Yourself

"Out of Tryptophan"

Imagine you're a bacterium, alone in a vast liquid wilderness, trying to grow big enough to divide in half and become two of yourself. You're a lean, mean machine, like a stripped-down racing car — no extras for convenience, comfort, or luxury, but superbly adapted to a single purpose. And you achieve it with only about four thousand different kinds of proteins — as compared to the fifty thousand or more in a human cell. But your most notable achievement, the finest gift you've bequeathed to all higher-level organisms, is your capacity to orchestrate your own genes, using protein governors as switches. By expressing some genes while switching off others, you can adjust to an ever-changing world — a big evolutionary step for such a little guy.

Take your impressive ability to make all the amino acids and nucleotides you need using mostly sugar — which requires hundreds of separate enzymes to do (and which, by the way, no human can do). Normally, you scavenge amino acids and nucleotides from decaying organisms you find lying around. But in lean times, when the only thing available is sugar, you must improvise (i.e., make the enzymes that allow you to produce amino acids and nucleotides from sugar) or perish.

Here, for instance, is how you make the enzymes that make just one amino acid — tryptophan. The protein governor (here called a repressor) acts as a switch for turning a gene on or off.

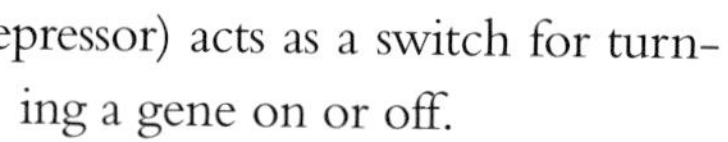

Inside the Bacterium:

The genes are blocked when tryptophan is plentiful. ▶

There are five genes involved in making the five proteins needed for the tryptophan assembly line. When tryptophan is plentiful, an allosteric repressor protein binds to the DNA.

RNA polymerase, the enzyme that transcribes messenger RNA from DNA, can't bind to the DNA and transcribe it because the repressor is sitting on the binding site.

The genes become unblocked when tryptophan becomes scarce. ▶

When tryptophan molecules are scarce — e.g., when they're being vigorously consumed for protein synthesis — they no longer fill the repressor's regulatory site. The repressor loses its grip and falls off the DNA.

RNA polymerase is now able to transcribe messenger RNAs from the five genes. The messengers proceed to ribosomes where they are translated into five enzymes. These enzymes immediately get to work making tryptophan from sugar. ▶

If tryptophan is not used rapidly enough, it will once more build up in the cell and reactivate the repressor, thus returning everything to the beginning of the control loop.

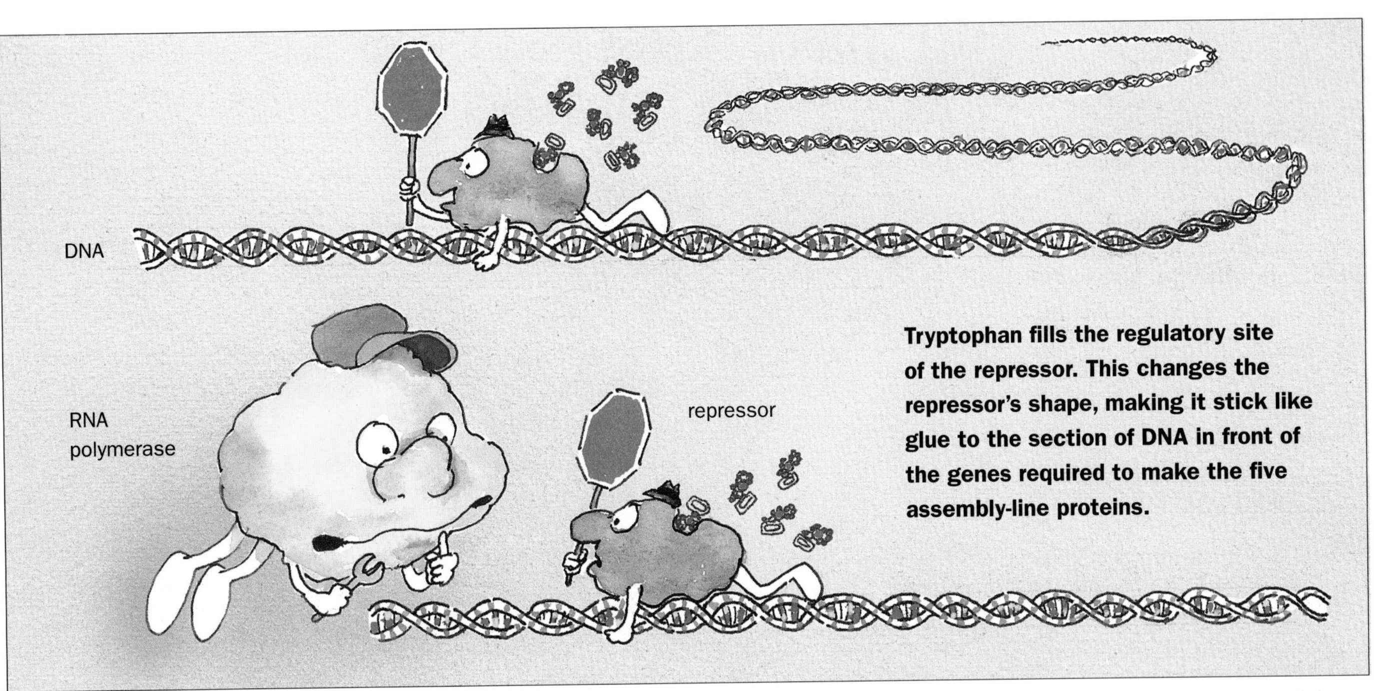

DNA
RNA polymerase
repressor
Tryptophan fills the regulatory site of the repressor. This changes the repressor's shape, making it stick like glue to the section of DNA in front of the genes required to make the five assembly-line proteins.

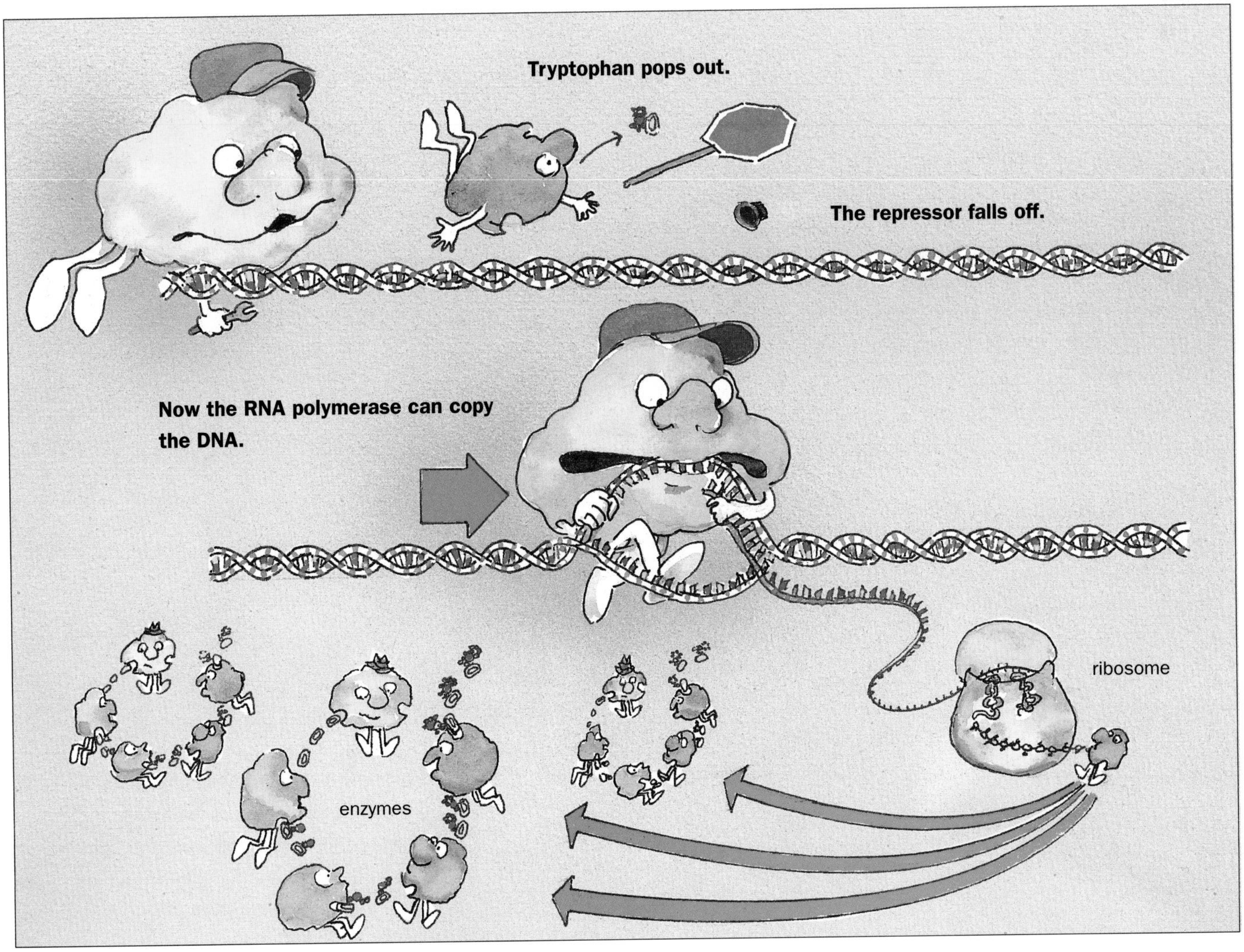

Tryptophan pops out.

The repressor falls off.

Now the RNA polymerase can copy the DNA.

ribosome

enzymes

6.8 Chemotaxis: How Chemical Signaling Creates Purposeful Movement

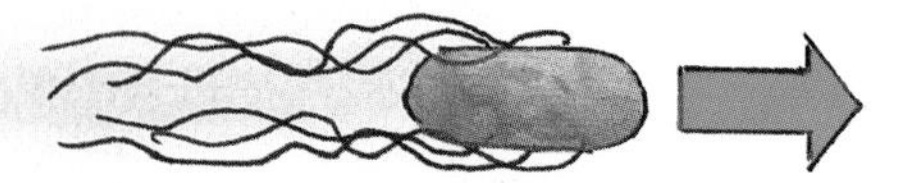

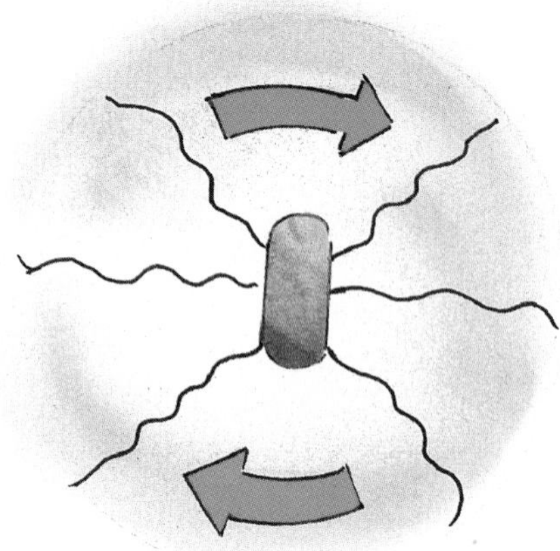

A bacterium's movement involves alternations of two kinds of action: "running" (top) and "tumbling" (bottom).

Running and Tumbling

The way bacteria find food in their environment — using chemotaxis (literally, "movement induced by chemicals") — is one of life's oldest forms of response to chemical signaling. A bacterium swims with the help of several flagella — long whip-like tails made of protein that are rotated by spinning disks, or "motors," on the bacterium's "skin." Counterclockwise rotation makes the flagella stream smoothly together to act like an outboard motor, propelling the bacterium straight ahead. Clockwise rotation makes the flagella flail about, causing the bacterium to tumble aimlessly. Normally, the direction of rotation reverses every few seconds so there's no consistent motion in any one direction. The bacterium runs a while, then tumbles a while. The result is a "random walk" (see below).

Each tumble changes the direction of the run. If it's not receiving any signals, the bacterium alternates runs and tumbles in (approximately) one-second intervals. This produces a random "walk."

Food acts as a signal that makes the bacterium run more and tumble less as it moves toward higher concentrations of food molecules.

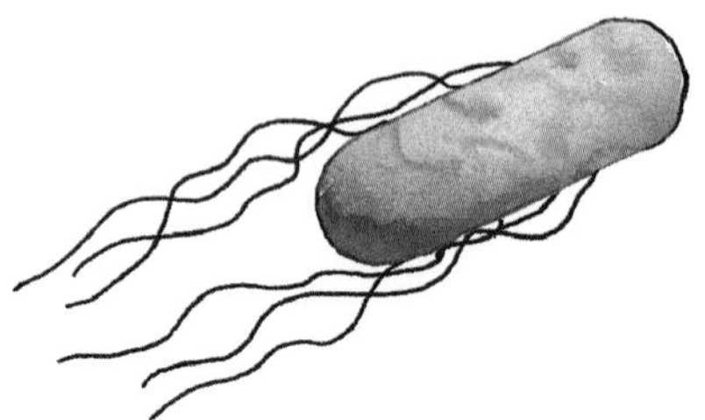

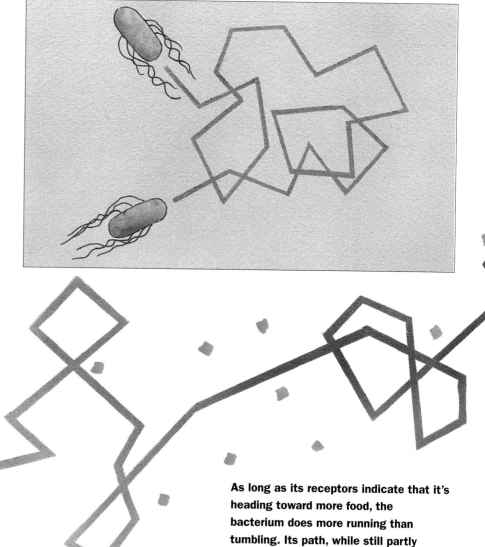

As long as its receptors indicate that it's heading toward more food, the bacterium does more running than tumbling. Its path, while still partly random, becomes more directed.

However, when the bacterium runs into food, it suddenly becomes more "purposeful." As food molecules fit into protein receptors on the outer surface of the bacterium, allosteric changes inside the bacterium signal the flagella motors to rotate counterclockwise more frequently. The result is more running, less tumbling, always in the direction of *more* signal — more food. This preponderance of "directed" running over aimless tumbling continues as long as the number of food molecules hitting the bacterium's receptors keeps *increasing*. When the rate of increase levels off, the bacterium starts tumbling again, so it stays roughly in the spot where the concentration of food is the greatest. We can recognize in this tiny creature's remarkable ability to respond to chemical signals in its environment the rudimentary beginnings of purposeful behavior. The bacterium senses differences and then acts using its own internal energy to respond to that information.

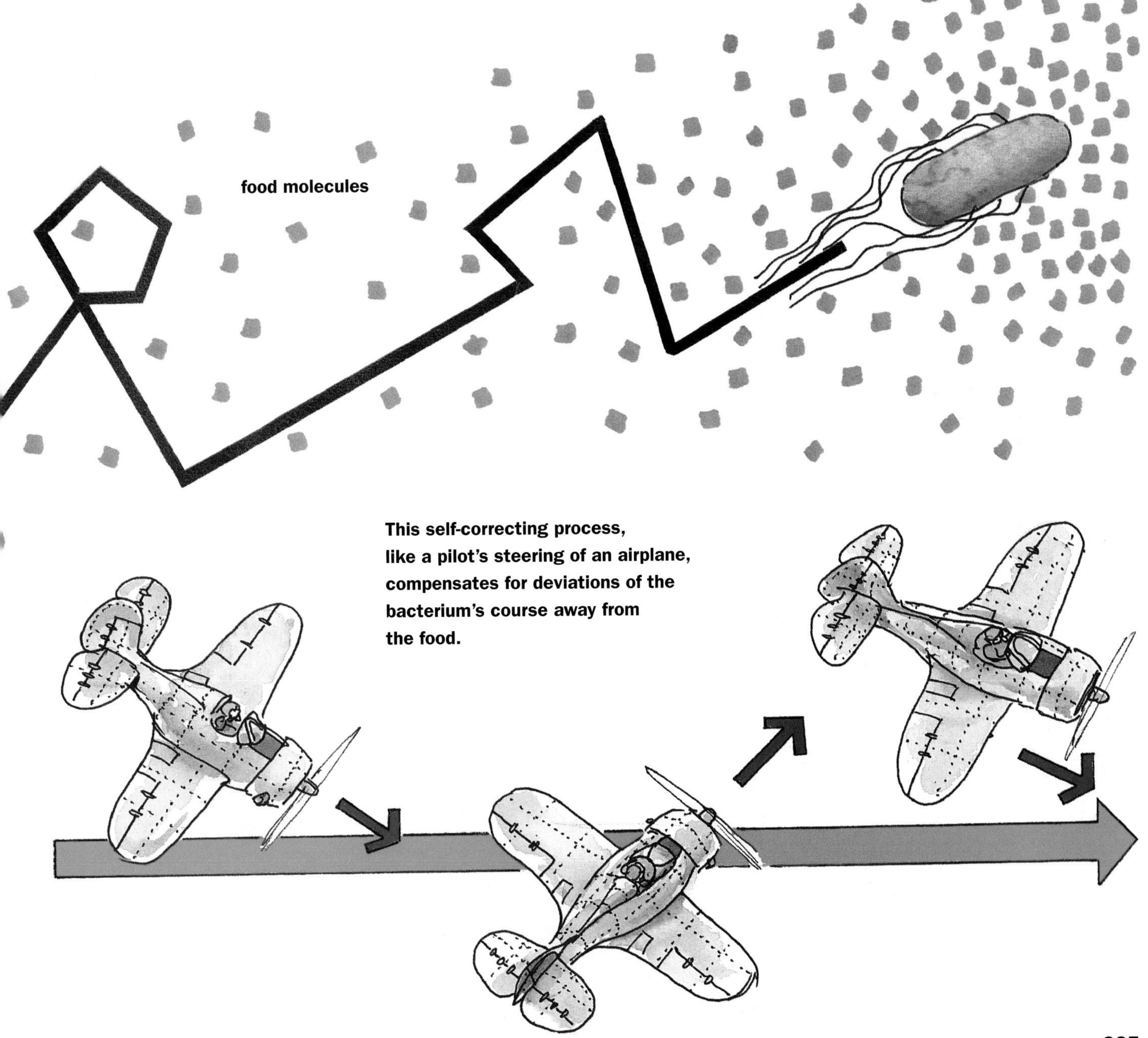

This self-correcting process, like a pilot's steering of an airplane, compensates for deviations of the bacterium's course away from the food.

6.9 Introducing the Adventures of a Slime Mold—Myxamoebic Communication

Here we begin a short exploration of the fascinating life story of the cellular slime mold, *Dictyostelium discoideum* (not to be confused with the plasmodial slime mold you met in Chapter 2). This organism, which starts out as a disconnected group of separate individual cells which, in response to a shortage of food, becomes a coordinated cellular community (see page 250), provides an elegant example of the power of molecular communication.

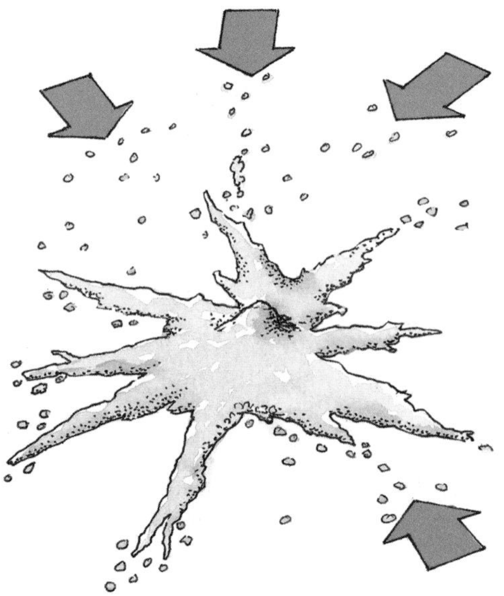

Aggregating Cells

In 1946, before much was known about such communication, graduate student John Tyler Bonner did a series of imaginative and now-classic experiments to find out just what it was that caused individual slime mold cells, called myxamoebae, to come together (to aggregate) to form a cellular community, called a slug.

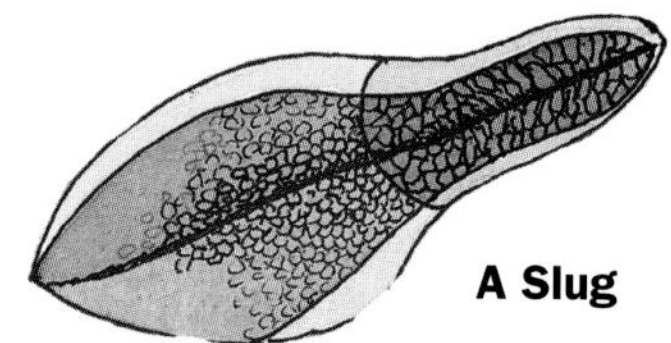

A Slug

Bonner, J. T. 1947. Evidence for the formation of cell aggregates by chemotaxis in the development of the slime mold *Dictyostelium discoideum.* *The Journal of Experimental Zoology* 106 (1).

A large number of myxamoebae were grown and washed free of nutrients. The "starving" cells began to aggregate, and the pattern of aggregation showed that the cells were attracted to a central spot from as far away as 53 cell diameters (800 microns). Several hypotheses for the mechanism of this attraction were proposed and tested:

1. An electric field might be the cause of aggregation. (Testing this hypothesis showed that individual cells had no reaction to an electric field.)
2. A magnetic field might cause the aggregation. (There was no response from individual cells to a magnetic field.)
3. The orientation of the substratum (the surface where the amoebae were aggregating) might somehow orient the individual cells. (There was no discernible effect when the substrate was reoriented.)
4. There might be structural connections between the cells. (None was found.)
5. The aggregating cells might leave a deposit on the substrate that could orient other amoebae. (No such deposit was found.)

With all these hypotheses eliminated, the possibility remained that aggregation could be the result of a diffusing chemical signal. That this

DOING *Science*

was the correct hypothesis was established by several tests: It could be shown, for example, that individual slime mold cells were attracted from one side of a thin piece of glass to the other (around the edge). Another test showed that there was attraction across a gap that the cells couldn't cross, but chemicals could. The cells responded to such gaps by "reaching out into the gaps with a sort of hopeless pseudopodial waving." The most conclusive demonstration that a chemical signal caused aggregation was an experiment that showed when a slowly moving stream of water washed over aggregating cells, further aggregation occurred only downstream from the center of aggregation: myxamoebae there moved directly upstream; myxamoebae upstream of the center moved about randomly.

These experiments showed not only that some sort of diffusible chemical is involved in slime mold aggregation, but that its concentration can be very dilute. Bonner called the chemical "acrasin," after the enchantress Acrasia (described in Edmund Spenser's *Faëry Queen* as a lovely and charming woman who lived in a "Bower of Bliss" and was adorned with everything in nature that could delight the senses).

Since these experimental results were published, the cell-attracting chemical acrasin has been identified as cyclic AMP (cAMP), a close relative of ATP whose function is to activate enzymes that initiate movement and other changes in cells. cAMP acts as a messenger for many different types of cells in other organisms, including those of humans.

Acrasin causes a positive feedback response in slime mold cells. Positive feedback loops are somewhat rare in biological systems because a regulatory "trigger" that increases the output response can lead to a "runaway" process. Under certain circumstances, when it is necessary to generate a significant response in a relatively short time, a positive feedback response makes sense for cell regulation, despite the potential for disaster.

Putting on the Brakes

Any good experiment gives rise to further questions that provoke new hypotheses to be tested. If a chemical signal causes *D. discoideum* myxamoebae to aggregate, what "brake" stops the aggregation at a point that yields a community of optimal reproductive size?

In asking themselves just this question, Debra Brock and Richard Gomer made an analogy that led them to a testable hypothesis: Individual myxamoebae are like individual people, each wearing a moderate amount of cologne. When individuals are far apart, the smell of the cologne is attractive, but as more and more individuals crowd together, the smell intensifies and grows stronger — until finally everyone says, "Stop, I can't stand the smell; it's too much!" And, in fact, Brock and Gomer found a gene in *D. discoideum* that makes a protein (which they named "countin") that acts in this way, as a stop signal.

Normal cells —
Countin protein is present

These aggregation streams break up as the aggregation reaches a certain size.

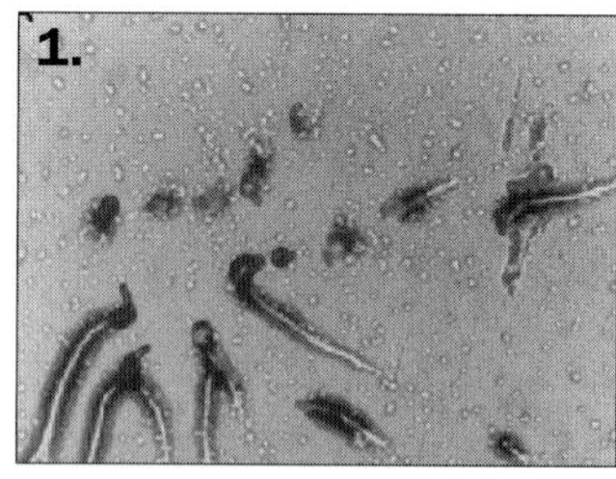

A side view of an aggregation showing the formation of a normal-sized fruiting body.

A forest of normal-sized fruiting bodies.

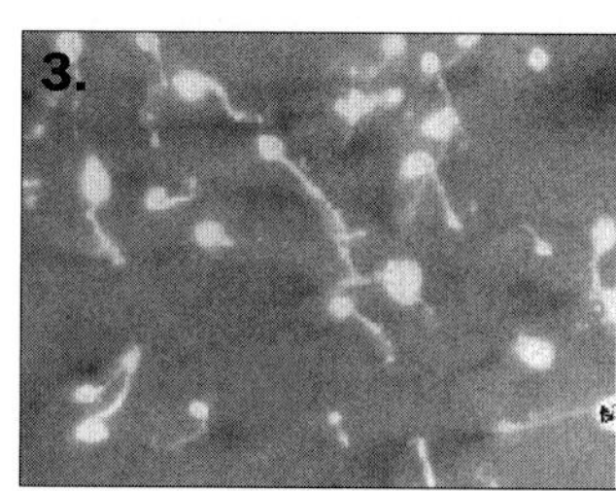

Brock, D. A., and Gomer, R. H. 1999. A cell counting factor regulating structure size in *Dictyostelium. Genes and Development,* 13(15): 1960–1969.

When there is a high density of starving *D. discoideum* cells in an area of soil from 1 to 10 mm in diameter, the cells begin to secrete a cell-density sensing factor called CMF. As the concentration of CMF within the small area increases, it reaches a point where the cells aggregate in response to pulses of cAMP (Bonner's acrasin), which acts as a chemoattractant. The aggregated cells form a migrating slug that develops a fruiting body, a mass of spore cells, at the top of a stalk. The stalk holds the spores as far above the forest floor as possible, for optimal dispersal.

It is important to the reproductive success of the slime mold that the stalk be as tall as possible, but equally important that it be strong enough not to collapse. The hypothesis underlying this experiment is that one of the factors that might lead to optimal stalk length and strength could be a chemical signal that is simultaneously secreted and sensed by individual cells. The local concentration of this chemical would rise as the number of cells diffusing it into their environment increased.

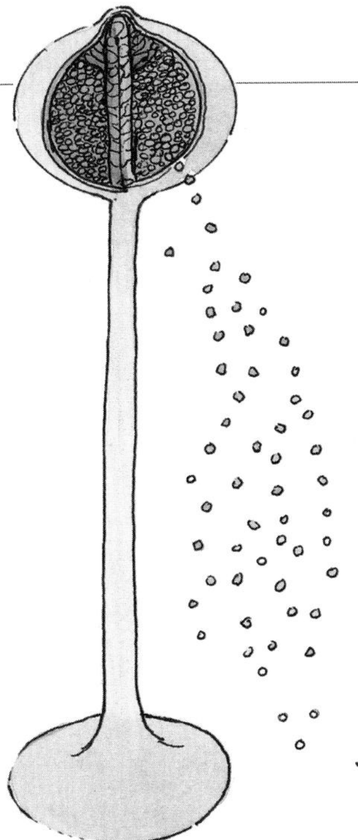

A fruiting body

DOING *Science*

By comparing the proteins of optimally aggregating cells to those of mutant cells that form excessively large aggregates with collapsing stalks, a protein was isolated (named "countin") that appeared to be involved in breaking up aggregation streams when an aggregation reached an optimal size.

The photographs show that aggregation sizes and stalk lengths are much greater when countin is absent than when it is present.

Mutant cells — Countin protein is absent

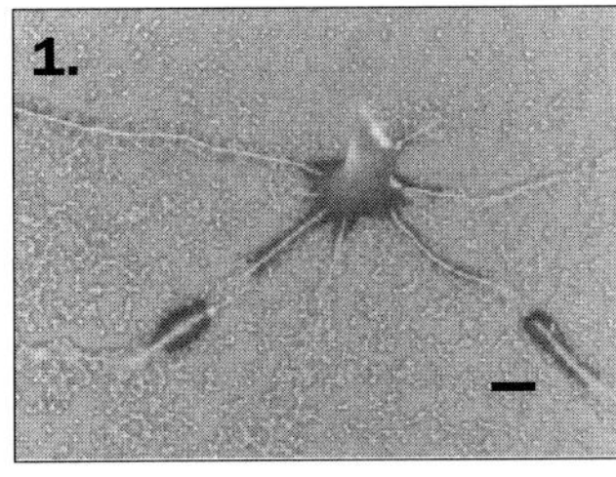

Here, unbroken aggregation streams form a larger aggregation.

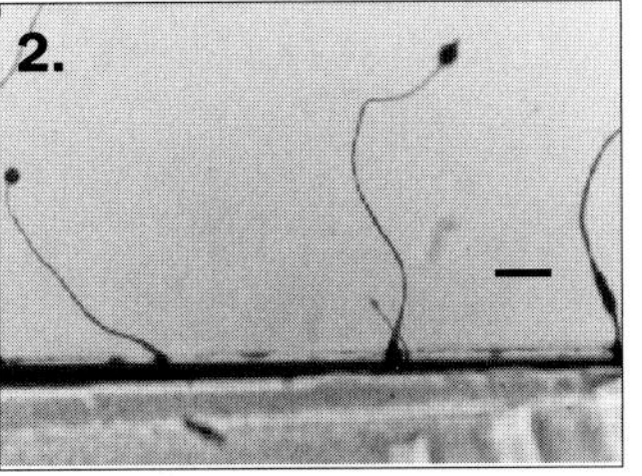

Here, fruiting bodies are much taller than those of normal-sized aggregations.

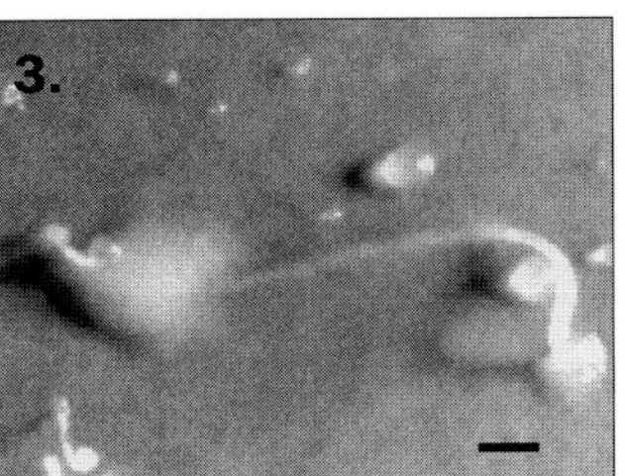

An overly large fruiting body collapses.

Bar = 500 μm or microns

1. Lock-step sequences:

At the simplest level, the spider follows a programmed sequence.

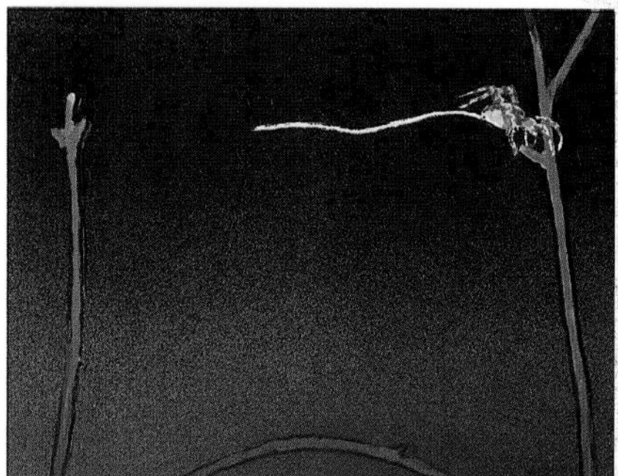

Release a strand into the breeze.

When it catches, cinch it up and attach the near end.

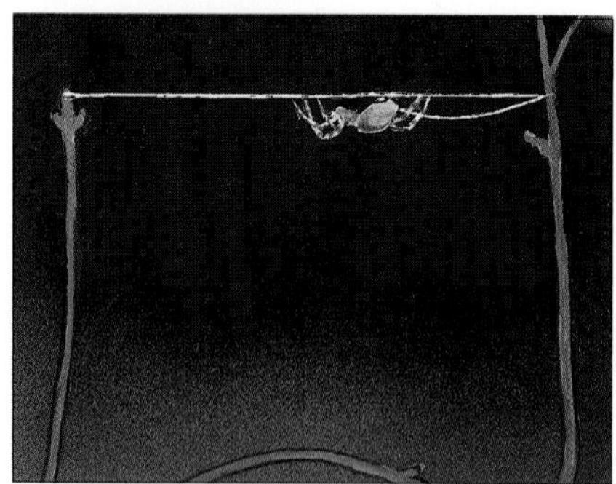

Walk across, paying out a second, looser strand. Attach at both ends.

Slide to the center of that strand and drop a third line, forming a Y.

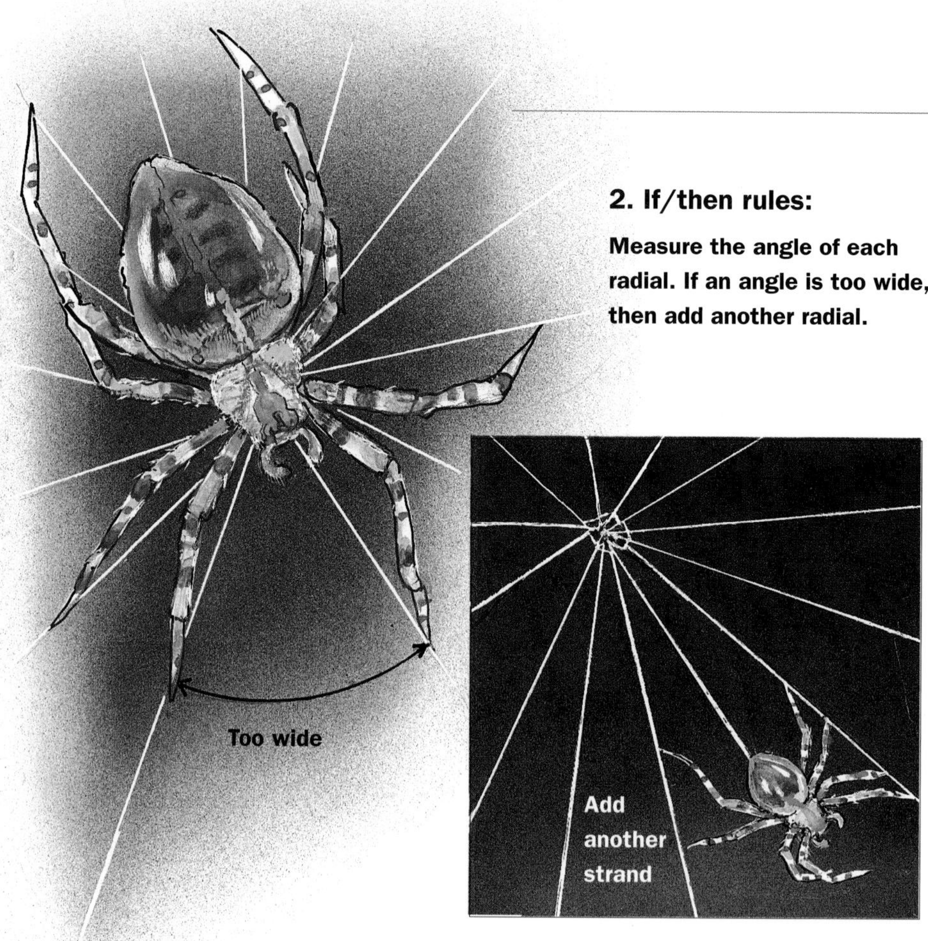

2. If/then rules:

Measure the angle of each radial. If an angle is too wide, then add another radial.

 6.10 Feedback in Neural Circuits

Making a Future

Adjusting to the environment by going into the gene bank to make new proteins takes time. Complex organisms, particularly animals, need a quicker feedback system, both to respond to current conditions and to anticipate the future. The development of the nerve cell made this possible.

A spider spinning her web is attending to her future. If she chooses a good spot, if she makes her framework strong with uniform tension on the supports, if she spaces her orb lines evenly and makes them sufficiently sticky, chances are she'll eat well in the days ahead. The web-building spider is following a set of programmed rules embedded in her nervous system. More rigid rules dictate her specific behaviors; more flexible rules govern her overall strategies. Let's imagine you're a spider following four types of rules that we'll call (1) lock-step sequences, (2) if/then rules, (3) trial and error, and (4) simulation.

Here's a lock-step sequence: Cast a thread into the breeze and when it catches something, pull it tight and fasten the end nearest to you. Walk across this new span, paying out a loose second strand. Go back to the center of this second strand and drop a third strand straight down, forming a "Y," etc. Lock-step sequences dictate an unvarying chain of specific procedures, like a soufflé recipe that must be followed to the letter for success. Here there's little room for feedback. The end point of one sequence is the starting point for the next.

At the next level, where feedback enters, are if/then rules: First, check each line for tension. If a line feels slack, then cinch it up. From the web's center, measure each angle between radial lines. If an angle is too great (meaning there's too much space between radials), then add another radial, etc. If/then rules, like lock-step rules, govern specific actions, but they permit modification based on feedback from sensory input.

A third type of rule doesn't specify an operation but says, more generally, "Repeat what works; stop what doesn't." Observers have noticed that if a spider's web is destroyed more than a couple of times while she's working on it, she'll abandon the job and start a new one. This is trial and error — or, better, trial and feedback.

The final type of rule is simulation: Rather than carrying out an elaborate process to see if it works, make a mental model of it and imagine the consequences. Such modeling requires more sophisticated nervous circuitry. If spiders can do it at all, it is in a very rudimentary way. They do, in fact, make a loose "sketch" of the final spiral pattern of a web; then, retracing their steps and using their legs to measure, they make a more precise version — and eat up the original "sketch."

All animals, and maybe even other organisms, work with a mixture of rigid and flexible rules. The more complex the animal, the more flexible the rules, which leads increasingly to the ability to modify behavior based on feedback from experience — a process more commonly called learning.

4. Simulation:

Make a preliminary "sketch" of the spiral using temporary webbing. Then, retracing your steps, make a permanent spiral, measuring carefully as you go. (Eat the original "sketch.")

3. Trial and error:

If the web sways too much in the wind, try adding weights to it. If this doesn't work, abandon the site.

Oh, What a Web We Weave . . .

The programmed industriousness of an orb-weaving spider, *Plesiometa argyra*, which lives in the forests of Costa Rica, is taken advantage of by a certain parasitic ichneumon wasp. Parasitic wasps have the nasty habit of laying their eggs in the bodies of other insects, where the wasp larvae mature and use their host as a food supply. The ichneumon wasp attacks the orb weaver, temporarily paralyzing it and laying an egg on its abdomen.

After the egg hatches, the growing wasp larva subsists on fluids drawn from small punctures in the spider's abdomen. This goes on for several weeks, with no apparent change in the spider's daily routine; it spins its web and catches flying insects for food. Then comes the final stage. The night before the mature wasp larva kills its host, it induces the spider to build — instead of its usual orb — a platform suspended by strong "cables" of spider silk, which resists wind and rain. After the spider has finished, the larva kills and eats it, and then spins its own cocoon on the relatively safe platform.

You've just read (page 230) that a web-building spider follows a preprogrammed set of rules. How, then, does the ichneumon wasp get the orb weaver to do its bidding instead? Researcher William Eberhard, who first noticed this phenomenon, speculated that the wasp injects into the spider some chemical that reprograms its web-building routine.

The orb-weaving spider's usual web-building program has a five-part subroutine of lock-step sequences. Once the injected chemical signal takes effect, the spider simply repeats the first two parts of this sequence over and over to construct the sturdy platform for the wasp. Eberhard experimented by removing the wasp larva from the spider's abdomen on the final evening; he discovered that the spider built a platform web that night and the next — but then reverted to spinning its usual orb.

Such specific manipulation of host behavior by a parasite has biologists amazed and bemused. What kind of "drug" is involved, and exactly how does it affect the nervous system? New hypotheses are being proposed, with results of experimentation to follow.

a. Normal web

b. Reprogrammed web

The wasp pupa hangs from this rudimentary web.

Berdoy, M., Webster, J. P., and MacDonald, D. W. 2000. Fatal attraction in rats infected with *Toxoplasma gondii*. *Proceedings of the Royal Society of London Series B*, 267(1452):1591–1594.

The protist *Toxoplasma gondii* infects rats (and other mammals, including humans), but needs a feline host to complete its life cycle. Researchers monitored the behavior of healthy and infected rats to test the hypothesis that the protozoan manipulates its host so as to increase the rat's risk of predation by a cat. Healthy rats avoided places that signaled the presence of cats. Rats infected with *Toxoplasma,* however, were observed to have an altered perception of risk and, in some cases, were even attracted to cat odors. The behavioral effect is very selective; other behavioral and general health factors were intact in infected rats.

This research may help epidemiologists in their efforts to control toxoplasmosis, the infection caused by this protist. Toxoplasmosis is one of the most common infections found in humans. An estimated 22% of the U.S. population is infected with *Toxoplasma,* and the proportion is 85% in France. Most people show no symptoms and are generally advised not to worry about it. There are two important exceptions: Toxoplasmosis can cause eye and brain damage in people with severely weakened immune systems, and infants infected before birth can develop mental retardation or other serious mental and physical problems.

And researchers suggest another potential implication of toxoplasmosis: Neurological bases of cognitive processes and of anxiety may be compromised in humans and other mammals. Certainly, the *Toxoplasma*-infected rats showed such effects. If this suggestion turns out to be the case, we may become very worried indeed.

Under Attack

Human white blood cells are under attack by the crescent-shaped *Toxoplasma gondii.*

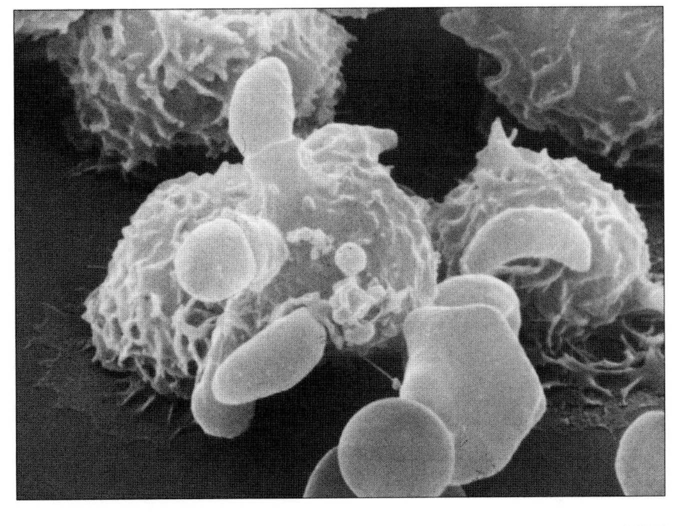

6.11 Cascading

Positive Feedback

Up to now, the feedback we've primarily discussed is negative feedback. *Positive* feedback occurs in a loop when a signal stimulates the production of more, not less, of a product. The more there is of the product, the more of it that is produced. Imagine an opera in which a certain note in the diva's aria acts as a cue to bring additional cast members onstage. As these new singers add their voices, still more performers come onstage. Thus, more brings on more.

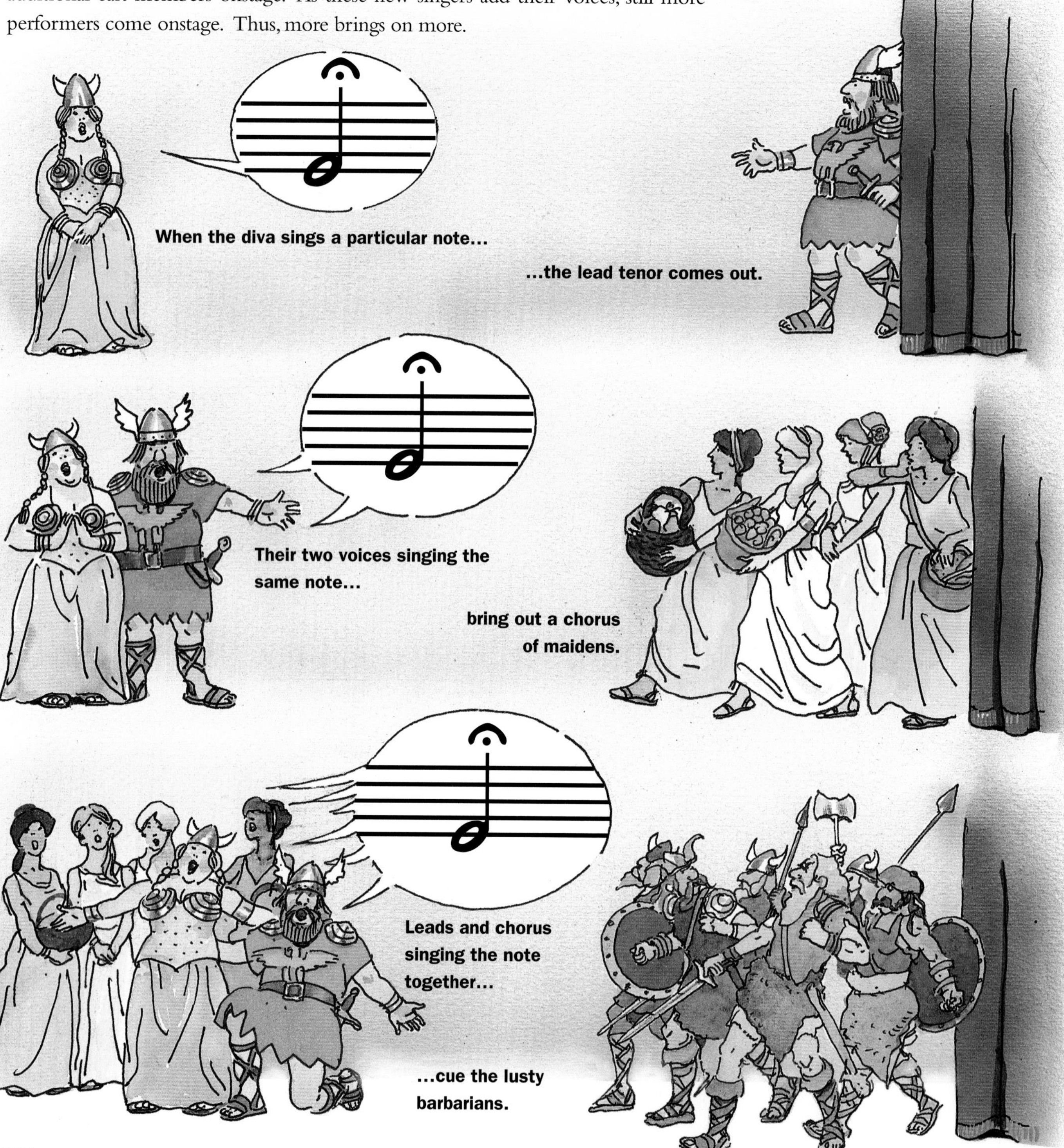

In biology, the process of positive feedback is called cascading. Events trigger other events in ever-growing amplification, as when a hormone signal triggers childbirth. This amplication process can lead to a dangerous situation of "runaway," as in the case of addiction or the unrestrained cell growth of cancer. But cascading can also be a source of creativity, breaking a system out of the straitjacket of the usual and into something new. Like interest compounding itself, learning leads to more learning; success breeds more success. In Chapter 7, *Community,* we examine a vital cascade — the growth of an embryo from a single cell. And in Chapter 8, *Evolution,* we'll look at the grand cascade of evolution.

Each successive addition of voices summons an even larger contingent. Soon the stage is filled with singers.

6.12 Ecology Loops

Self-correcting Systems

From a cybernetic perspective, an ecosystem is one huge feedback loop — a set of interconnected parts that act on one another so that a change in any part of the loop affects the other parts. In fresh water, for instance, some fish eat algae and excrete organic waste; bacteria eat the waste and excrete inorganic materials; algae "eat" the inorganic molecules — each population thrives and multiplies in an interdependent cycle. Such balanced ecosystems have flexibility: Troublesome imbalances are corrected at some point within the loop.

A rise in nutrients, as when fertilizer runoff enters the water, may encourage an imbalance — an overgrowth of algae, for example. If the algae grow too dense, sunlight will fail to penetrate to their lower layers, and these layers will die. The resulting increase in organic waste will lead to explosive growth of the bacterial population, depleting the water of oxygen. Normally, the fish will restore the balance by increasing their numbers in response to the excess food — the algae.

We've seen how cyclical systems inside cells are controlled by allosteric proteins that act as supervisors, or governors. In an ecological system, the "governor" is usually the largest organism — the one with the slowest metabolic activity. The fresh-water ecosystem can't correct itself any faster than the fish are able to respond to the increased growth of algae. So even though ecosystems self-correct, they can be overwhelmed by sudden and extreme changes. Too much organic waste — in the form of sewage, for example — can totally deplete the oxygen in the fresh water and cause a collapse of the entire system.

This is a simple model. In most cases, ecosystems operate not as single loops but as networks of interconnecting loops in which both positive and negative feedback play a role. If we could look deep inside those circuits, within each member, we would witness a never-ending making and breaking of molecules — the basic processes of life — presided over by myriad microscopic, bustling protein governors.

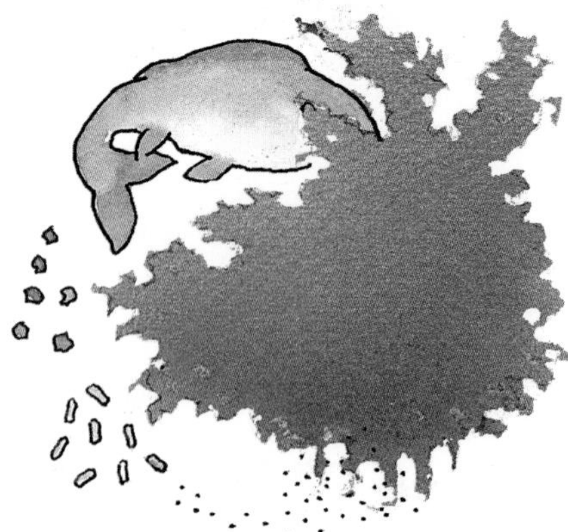

Water temperature rises, encouraging the overgrowth of algae. The bottom layer dies, increasing organic waste.

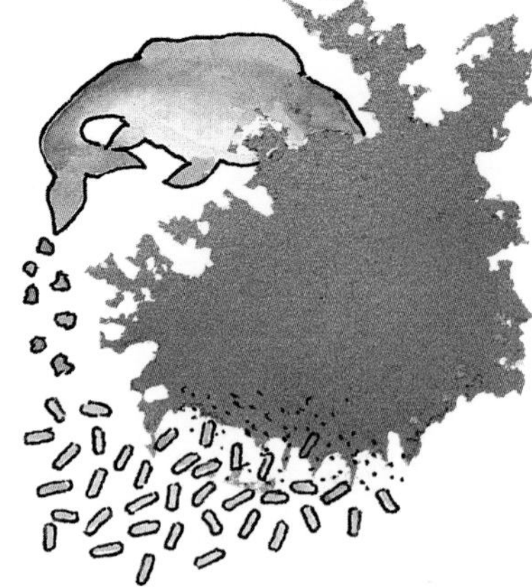

Bacteria multiply, depleting the water's oxygen content.

If fish multiply before too much oxygen is used, they reduce algae overgrowth and restore balance.

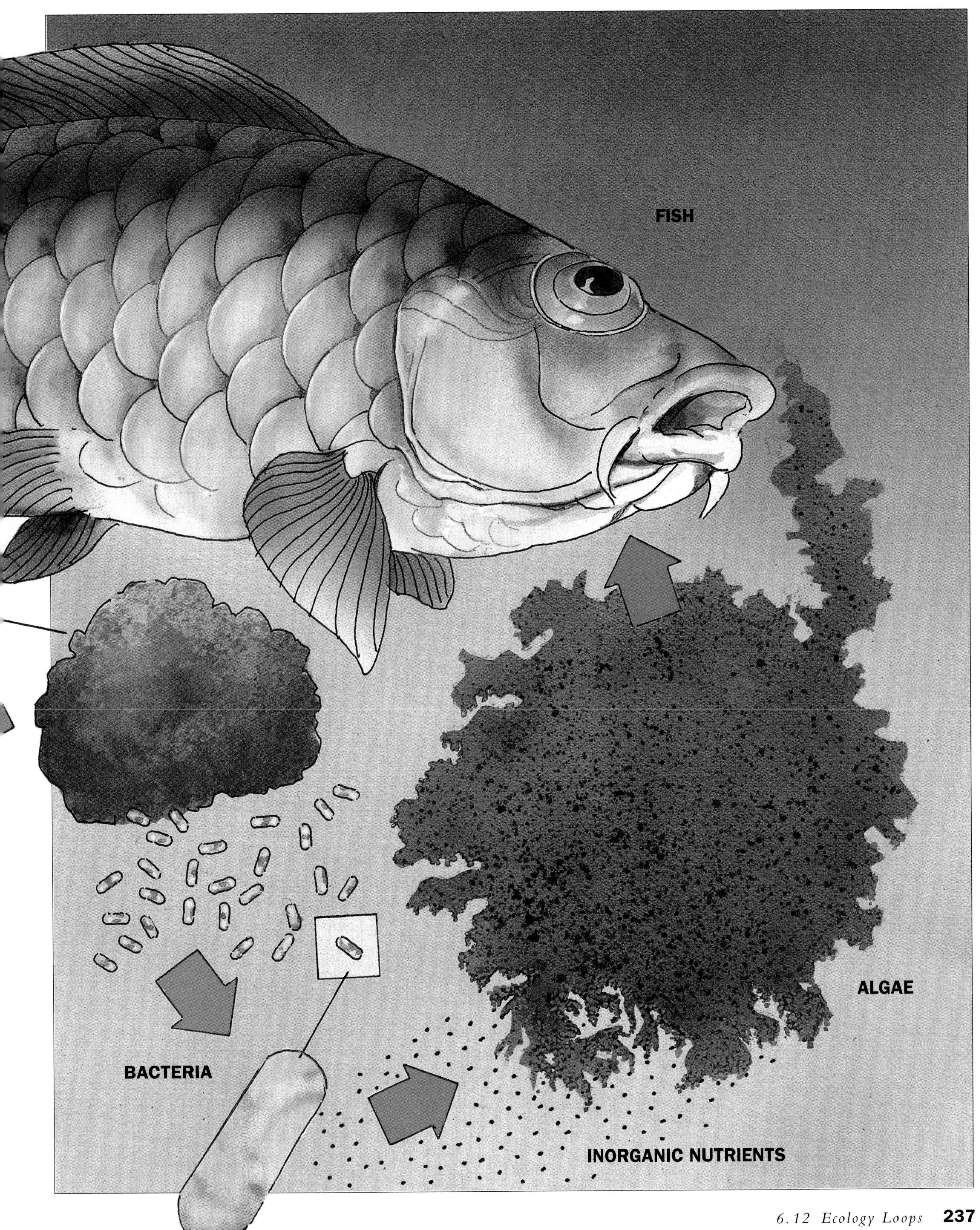
FISH
ALGAE
BACTERIA
INORGANIC NUTRIENTS

6.13 Death on the Wing

A feedback loop works efficiently when populations of the different organisms that make up the ecosystem are kept in check by predatory, parasitic, and competitive interactions. No one species overwhelms another. For example, species of fire ants native to the Southeastern United States are not prolific enough to be considered pests, because they are kept in check by forces in their "home" ecosystem. But a species that invades or is introduced into such an ecosystem from outside can wreak havoc. Freed of the natural checks and balances that exist in its native habitat, an introduced species can experience very rapid population growth. This happened after a species of South American fire ant was accidentally imported into the United States, probably through the port of Mobile, Alabama in the 1920s. The population of this introduced species grew rapidly, and the ants spread to other Southern states, becoming a dangerous pest. These fire ants have a sting so potent it can kill small mammals. They pose serious danger to humans due to their tendency to nest in electrical systems, where they cause short circuits and fires.

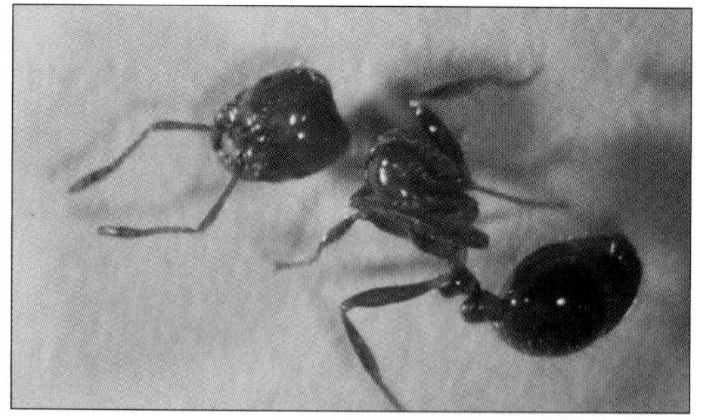

Fire ant, top, and its foe, the phorid fly, below.

Entomologists (scientists who study insects) at the University of Texas in Austin have found a species of fly (family Phoridae) that may help control the invading fire ants. There are many species of phorid flies, but each attacks only one species of fire ant. The fly species under study attacks only the imported ants. A female fly of this species attacks a worker ant by landing briefly on its neck and inserting a needle-like egg-laying device (an ovipositor). All this takes a fraction of a second and leaves the worker ant partly paralyzed and briefly disoriented. The larva that hatches from the deposited egg begins devouring the contents of the worker's head and neck. It continues feeding until the worker's head literally falls off. The larva then develops into an adult fly, following the same kind of cycle you saw for *Drosophila* on page 150.

Entomologists in Texas are exploring ways to use these phorid flies to control the dangerous imported fire ants. But, surprisingly, it's not the flies' fatal attacks on worker ants that the researchers think will be effective. Worker ants are relatively easy for a colony to replace, and it's not likely that the attacks on them could permanently reduce the ant population. The promise lies in tilting the playing field in favor of competing native species of ants. Female phorids cause severe disarray among members of a colony as they fly above the ants that are moving along trails as they forage for food or swarming out of a disturbed mound. When workers sense the presence of phorids, the colony becomes a mob scene rather than a smoothly running community. A colony under attack cannot provision or protect itself as effectively as an undisturbed colony.

Just like everything in an ecosystem, the imported fire ants must compete for space and resources with other organisms, particularly other species of ants. A disorganized colony is a less effective competitor. If the normal functioning of the fire ants' colony is disrupted, other organisms, among them species of less pestiferous ants, may gain a competitive advantage. The overall population of the imported fire ants may then grow more slowly, or even shrink. Native fire ant species may reclaim lost territory. Entomologists don't think the imported ants will be eliminated, but they hope to gain some control of the population through this method of biological control.

www.jbpub.com/connections

Saving the Landscape

One of the most tragic examples of our unthinking bludgeoning of the landscape is to be seen in the sagebrush lands of the West, where a vast campaign is on to destroy the sage and to substitute grasslands. If ever an enterprise needed to be illuminated with a sense of the history and meaning of the landscape, it is this. For here the natural landscape is eloquent of the interplay of forces that have created it. . . .

As the landscape evolved, there must have been a long period of trial and error in which plants attempted the colonization of this high and windswept land. One after another must have failed. At last one group of plants evolved which combined all of the qualities needed to survive. The sage — low-growing and shrubby — could hold its place on the mountain slopes and on the plains, and within its small gray leaves it could hold moisture enough to defy the thieving winds. . . .

Along with the plants, animal life, too, was evolving in harmony with the searching requirements of the land. In time there were two as perfectly adjusted to their habitat as the sage. One was a mammal, the fleet and graceful pronghorn antelope. The other was a bird, the sage grouse. . . .

The antelope . . . have adjusted their lives to the sage. . . . The sage provides them with the food that tides them over the winter. Where all other plants have shed their leaves, the sage remains ever green — the gray-green leaves — bitter, aromatic, rich in proteins, fats and needed minerals — clinging to the stems of the dense and shrubby plants. Though the snows pile up, the tops of the sage remain exposed, or can be reached by the sharp, pawing hoofs of the antelopes. Then grouse feed on them, too, finding them on bare and windswept ledges or following the antelope to feed where they have scratched away the snow.

Rachel Carson, *Silent Spring,* 1962

Thomas Moran
Cliffs of the Upper Colorado River, Wyoming Territory, 1882

Thomas Moran's paintings of the astonishing splendor of the American West may have been one of the factors that influenced Congress to create the National Park System.

Question.

Gypsy moth caterpillars defoliate the hardwood forests of the Northeastern United States every few years, causing extensive damage and killing many trees. For several years after a peak infestation, the number of moths decreases; it then remains low for a while before climbing again — a continuous cycle. What type of system does this represent? What might be causing the initial decrease in the moth population?

Answer...

This is a self-correcting system based on cyclic predator/prey interactions. The moth population increases until the available food sources (certain species of trees) are depleted. Concurrently, as the gypsy moths increase, they provide a growing food source for various birds and insects that feed on moths. With less food and increased predation, the gypsy moth population will decrease for a while. Once the population bottoms out, predators will look elsewhere for a meal and the forests will recover — until conditions can support another population increase. This type of cyclic self-correction is critical in maintaining balance among organisms in communities (see Chapter 7).

Some of the Things You Learned About in Chapter 6

Questions About the Ideas in Chapter 6

1. Why, in biological processes, are apparently straight lines really zig-zags?
2. Compare positive and negative feedback. Provide an example of each.
3. What do we mean when we say, in connection with control systems, that information flows in a circle?
4. Show parallelism among an airplane flight, a steam engine, chemotaxis, and an owl catching its prey.
5. How does a protein's ability to change shape allow it to control a cellular assembly line?
6. A molecular assembly line can be controlled by a protein governor. What controls the existence of the assembly line?

References and Great Reading

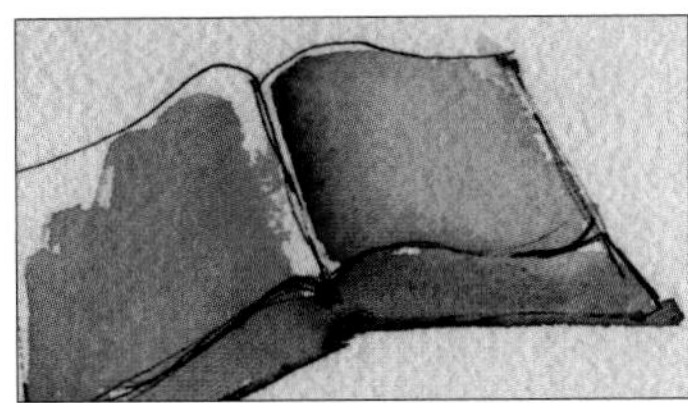

Alcamo, I. E. 2001. *Fundamentals of Microbiology,* 6E. Sudbury, MA: Jones and Bartlett Publishers.

Berdoy, M., J. P. Webster, D. W. MacDonald, 2000. Fatal attraction in rats infected with *Toxoplasma gondii. Proceedings of the Royal Society of London Series B,* 267(1452): 1591–1594.

Bonner, J. T. 1947. Evidence for the formation of cell aggregates by chemotaxis in the development of the slime mold *Dictyostelium discoideum. The Journal of Experimental Zoology,* 106(1).

Brock, D. A., and R. H. Gomer, 1999. A cell counting factor regulating structure size in *Dictyostelium. Genes and Development,* 13(15): 1960–1969.

Carson, R. 1962. *Silent Spring.* Cambridge: Houghton Mifflin Company.

Commoner, B. 1971. *The Closing Circle.* New York: Bantam Books.

Eberhard, W. G. 2000. Spider manipulation by a wasp larva. *Nature,* 406 (July 20): 255.

Gilbert, L. E. 1996. Prospects of controlling fire ants with parasitoid flies: The perspective from research based at Brackenridge Field Laboratory. Presented March 23, Proceedings of Second Conference on Quail Management.

Judson, H. F. 1996. *The Eighth Day of Creation, Makers of the Revolution in Biology (expanded edition).* New York: Cold Spring Harbor Laboratory Press.

Lovelock, J. 1990. *The Ages of Gaia.* New York: W. W. Norton.

Monod, J. 1971. *Chance and Necessity.* New York: Alfred A Knopf.

Needham, C. et al. 2000. *Intimate Strangers, Unseen Life on Earth.* Washington: ASM Press.

Weiner, N. 1967. *The Human Use of Human Beings.* New York: Avon Books.

Wade, N. 2000. Wasp invades a spider and puts it to work. *The New York Times* (July 25): 141.

For more questions and links to web resources, go to
WEB
Connection
www.jbpub.com/connections

CHAPTER 7

COMMUNITY

E Pluribus Unum

Consider a face: Its basic features change very little in the course of a year. Yet in that time, most of its original cells and all of the molecules of which those cells are made will be replaced by new ones. The fabric changes, but not the pattern.

AMONG THE MOST ASTONISHING REVELATIONS IN BIOLOGY WAS THE discovery that all visible living creatures are themselves made up of living "creatures" called cells. Cells are not merely inert structural units or building blocks, but individual beings with lives of their own — living, reproducing, and dying just as we do. We are, in a sense, hives of cells. We move, eat, and speak thanks to the coordinated effort of specialized groups of individuals within our cellular community.

Cells are small. *Caenorhabditis elegans*, a worm so small it can barely be seen, is made of precisely 969 of them. You are made of an astronomical 5 trillion cells, give or take a few. Obviously, an aggregate that large requires an extraordinary degree of communication and cooperation. Cells must continuously "talk" to each other. They use electrical and chemical signals to control every action you take.

In a community, cells organize themselves into specific patterns, maintaining their three-dimensional relationships with great precision. Two separated heart muscle cells pulsating at different rates will, if placed side by side, fall into synchronous rhythm. Cells from different tissues will, if mixed in a blender, re-sort themselves into their original tissues in a short time.

Throughout evolution, the more cells tended to stick together, the more information they shared. Slowly, the rudimentary systems of primitive single cells became connected and elaborated on, ushering in higher-level behavior such as seeing, feeling, and thinking.

> It is the sweeping simplicity of [Nature's] means that overwhelms with a sense of awe. This is what makes nature beautiful . . . the simplicity of the materials which make so many patterns, the unity under the surface chaos. Unity is the scientist's definition of beauty, and it makes Nature beautiful to him all his life.
>
> Jacob Bronowski

7.1 Emergence

The Whole and the Parts

Life is more than the sum of its parts. If you throw a number of highly predictable individuals together, they will almost certainly interact and organize in as yet completely unforeseeable and complex ways.

Take the information molecule, DNA. We've seen that DNA is a long chain molecule composed of four kinds of nucleotides (see Chapter 4, *Information*). There's nothing in the basic makeup or chemistry of those nucleotides that gives a hint of DNA's remarkable role in life. Only when they get strung together in the specific sequences of DNA do we perceive a wholly new quality — information. The true meaning of DNA lies not in its parts, but in the *organization* of those parts.

This truth holds for a whole host of phenomena, living and nonliving. For example, an individual water molecule by itself has no wetness. Wetness results when billions of water molecules slide and tumble over each other, forming and breaking lattices. Individual atoms have no color. Color arises when atoms are organized in molecules, each absorbing certain light waves and transmitting others. A single brain cell (neuron) contains no "thought." Thought emerges as millions of neurons shuttle electrochemical impulses through ordered networks.

So the whole, rather than being simply the sum of its parts, is more like the *product* of its parts — a multiplication of the interactions among all the parts. In a true community, whether a plant, a human, or a city, individuals somehow manage to transcend themselves and become part of something much greater — even though each one still myopically works for its own ends and by its own simple local rules.

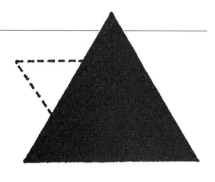

Emergent Patterns: When Simple Units Follow Simple Rules

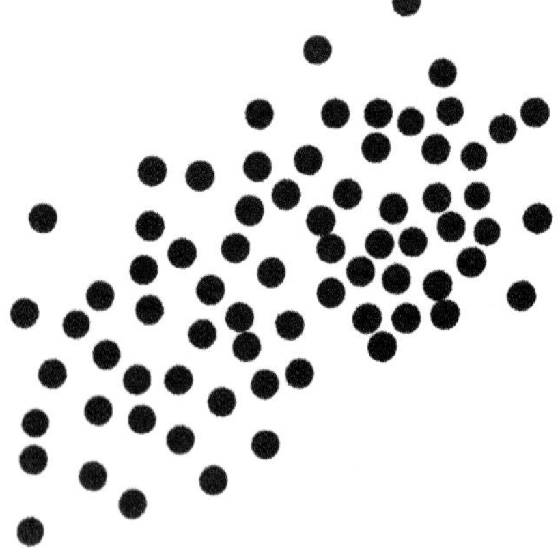

Imagine that each dot in this random assortment follows only two rules:

1. Keep exactly one dot-width from any neighbor.

2. Stay as close as possible to the center.

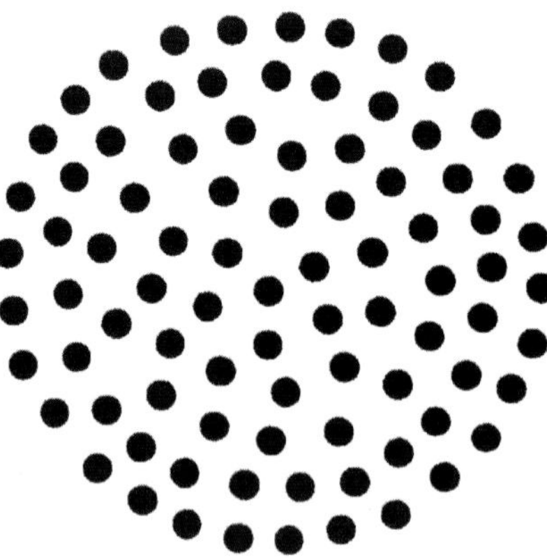

Emergent pattern: A disk

Rule: Add successively smaller triangles to the middle of each side of a triangle.

Emergent pattern: An increasingly intricate "snowflake"

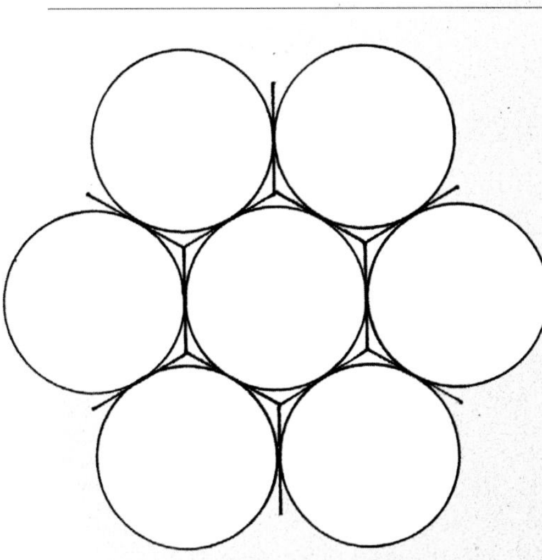

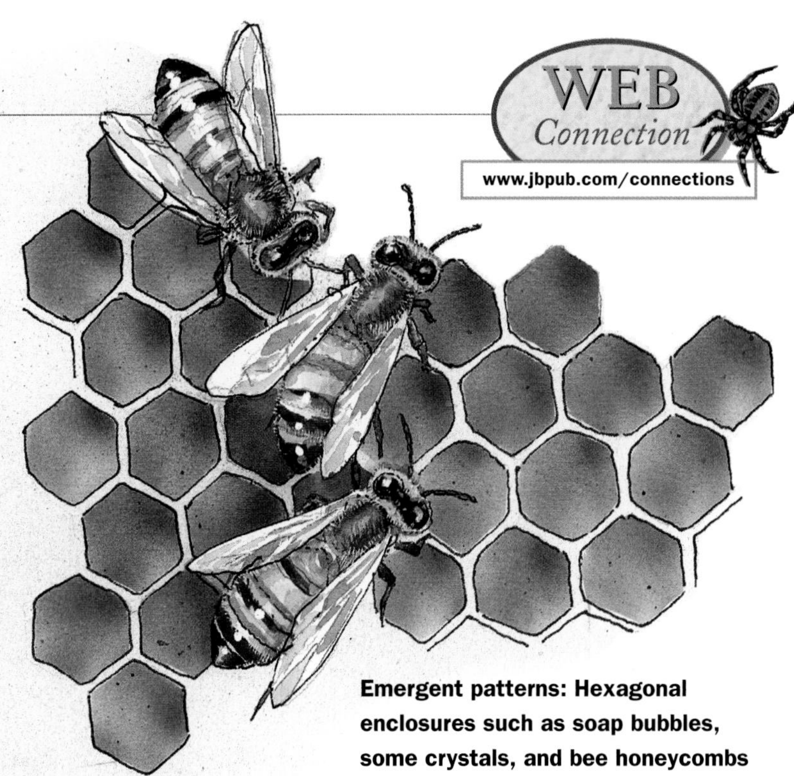

Rule: Assemble a group of spheres or cylinders so that they share walls, and use the least amount of material.

Emergent patterns: Hexagonal enclosures such as soap bubbles, some crystals, and bee honeycombs

Rule: Allow one of two connected, initially parallel surfaces to grow faster than the other.

Emergent patterns: Rams' horns, plant tendrils, and chambered nautilus

Love Songs and Other Patterns

Layer upon layer of patterns emerge as we examine the structure and behavior of living things. Patterns in the sequences of nucleotides in DNA lead to specialized patterns of interaction among proteins that, in turn, lead to patterns of interactions among cells in communities, as we see in ram horns and fern fronds. And many of life's most intriguing patterns are those we observe in animal behavior.

A female fruitfly won't express serious interest in a male until he croons the proper tune. (Different species of fruitflies have distinctive song patterns, which probably help females recognize males of their own species.) The males produce their "love song" by vibrating a wing. The pulses of sound have a clear pattern, as you can see in the illustration below. Geneticists have found a gene associated with this sound pattern. If there is a mutation in the fruitfly "song" gene, the "music" changes.

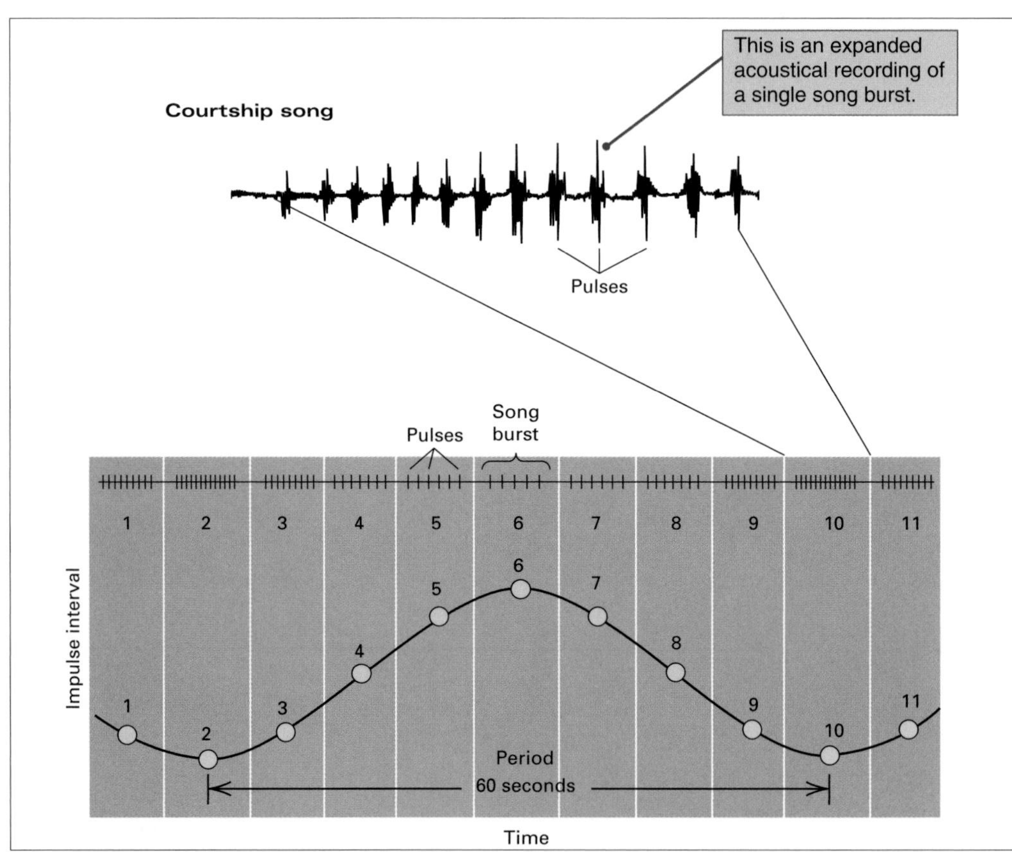

Analyzing a love song

The courtship song consists of a series of song bursts in which the frequency of pulses in each burst rises and fall.

Even more complex innate patterns of behavior show up in honeybee communities. A worker bee that has discovered a food source returns to the hive and performs a "waggle dance" for other workers. The basic pattern of this dance is always the same, and it means: "I've found a food source, and I'm going to tell you where it is." The bee begins the dance by crawling along the surface of a honeycomb in a straight line while buzzing its wings and shaking (wagging) its abdomen. At the end of the straight run, it circles back and repeats the dance along the same straight line. In the hive, each honeycomb stands vertically. The angle between an imaginary vertical line on the surface of the comb and the straight-line part of the "waggle" dance matches the angle between the direction to the food source and the Sun. When a newly informed bee leaves the hive, it will fly at that angle relative to the Sun. The length of time the straight run takes indicates the distance to the food source.

These songs and dances are innate behaviors. Fruitflies and bees don't have to learn how to perform them or how to interpret them. The patterns of nerve connections that generate them are as genetically programmed as is the fly's or bee's anatomy.

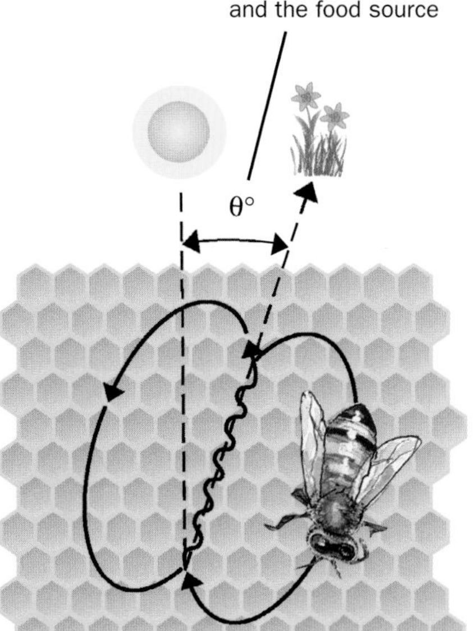

The waggle dance.

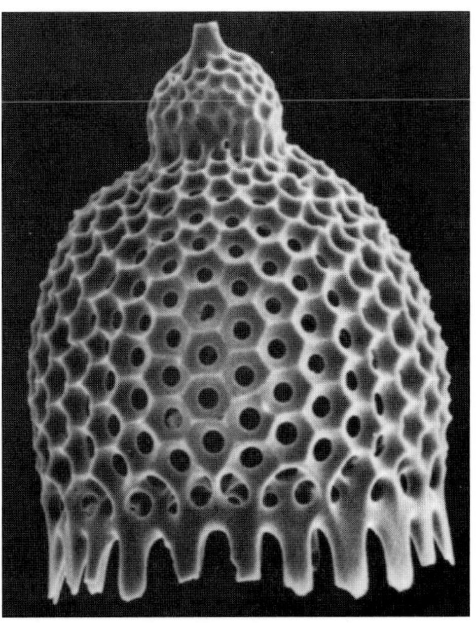

Structural symmetry

Humans had to use a blueprint derived from words, equations, drawings, and individual and communal behavior to create the Eiffel Tower. The equally complex pattern of the minuscule radiolarian skeleton (magnified 695 times) came into being driven by a DNA recipe that evolved over millions of years of chance variation and environmental selection.

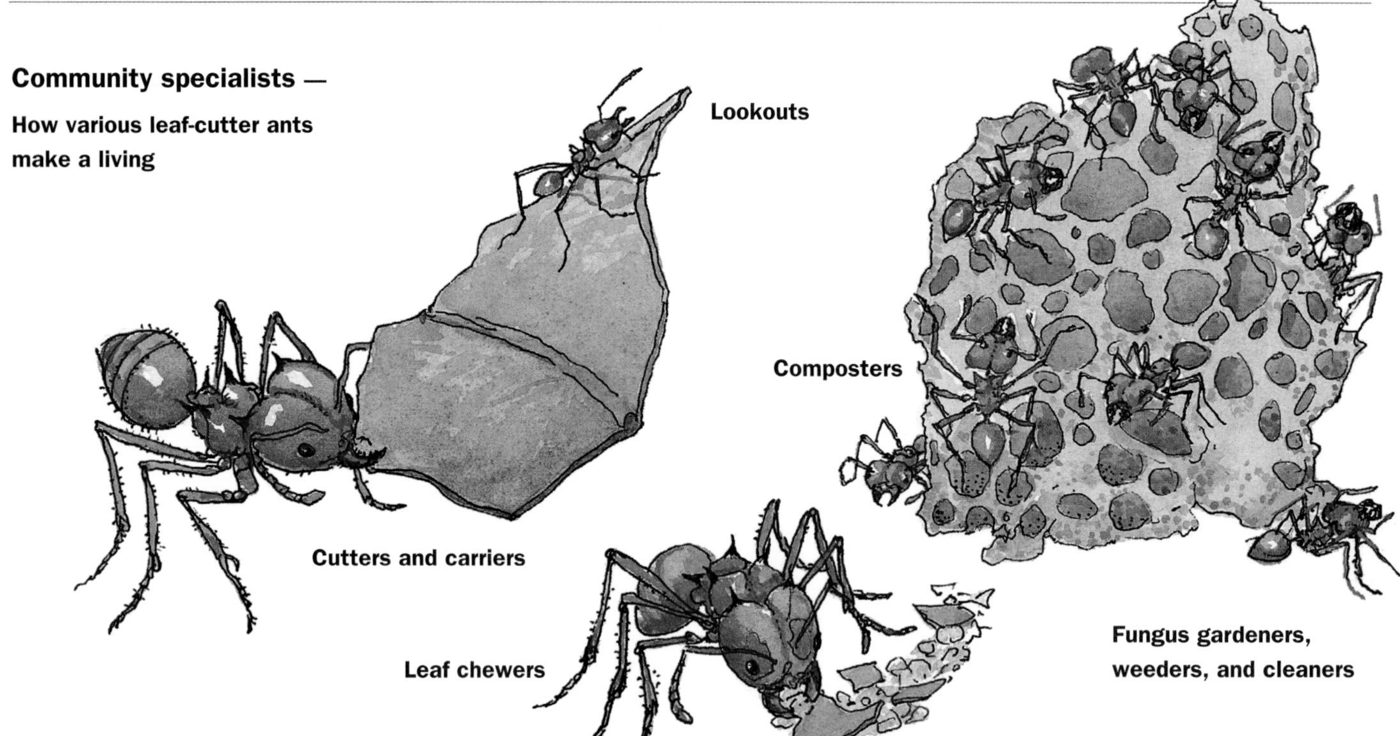

7.2 The Emergence of Higher-Level Behavior

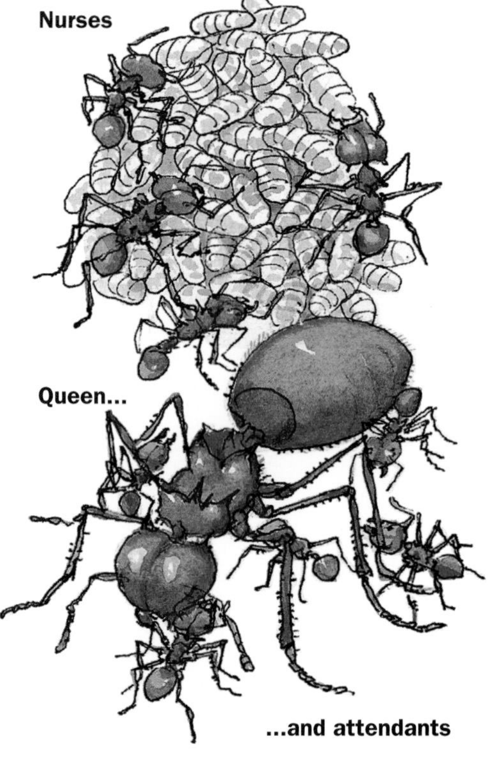

Community specialization has given social insects an enormous evolutionary advantage. While they represent only 2 percent of the world's insect species, they comprise more than half of the total mass of insect life on Earth.

Superorganisms?

Social insects — such as bees, termites, and ants — provide one of the best examples of the way in which sophisticated behavior emerges from the interactions of simpler entities. Just as an organism "knows" more than any of its individual cells, an ant colony knows more than any single ant.

Ants, though nearly blind, display a remarkable sensitivity to chemical signals. They use chemical substances to send simple messages, such as "follow my trail," "I'm a colony member," "on guard!," "help!," "I'm over here," etc. Biologist E. O. Wilson, a noted expert in ant behavior, suggests that an individual ant can probably send and receive some 15 different messages.

Imagine several scout ants out foraging for food. One stumbles upon some honey, and drawing upon her repertoire, she deposits a chemical that says "follow my trail" as she heads for home. The other scouts, having searched for food in vain, leave no returning trail. Sisters in the colony immediately pick up the trail left by the successful ant and go directly to the honey — each one reinforcing the trail on her return. Very quickly, a long column of ants makes its way directly to the food. (This is an example of positive feedback, see page 234.) They appear to follow each other, but in fact each one follows its nose (or, more accurately, its antennae), bumping into and stumbling over those returning with food. Notice that in this scenario, a random search quickly becomes an organized effort, even though each ant simply follows its own rules.

Clearly, the sharing of information brings the ant colony to a level of complexity (some even call it intelligence) not found in the individual ant. This is why some biologists refer to colonies of ants, bees, and termites as "superorganisms."

How to Build Without a Plan:

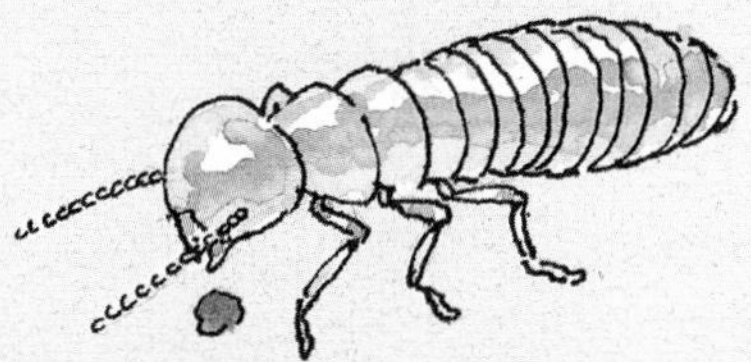

1. Individual termites randomly deposit dollops of mud and processed wood mixed with saliva that contains a chemical signal: "spit here."

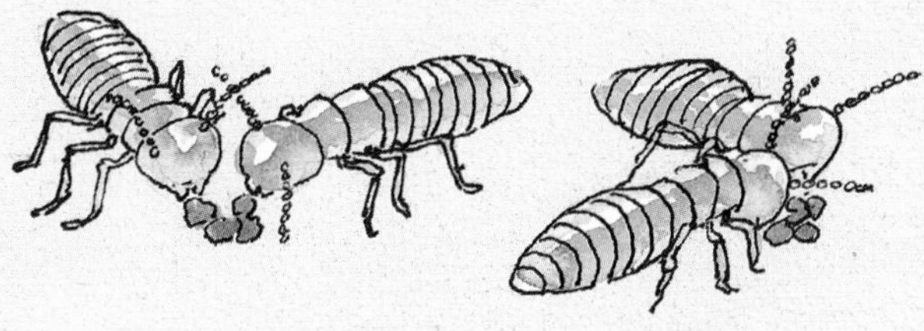

2. Other termites add their deposits to the original ones.

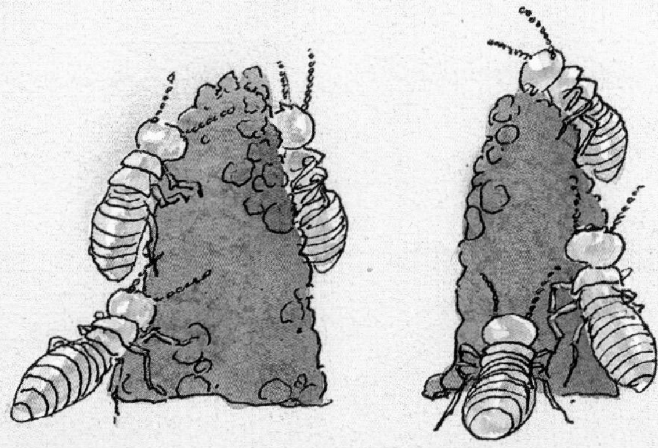

3. Soon the little piles grow into columns. When the columns reach a certain height, the termites, still guided by a chemical signal, shift to a second message...

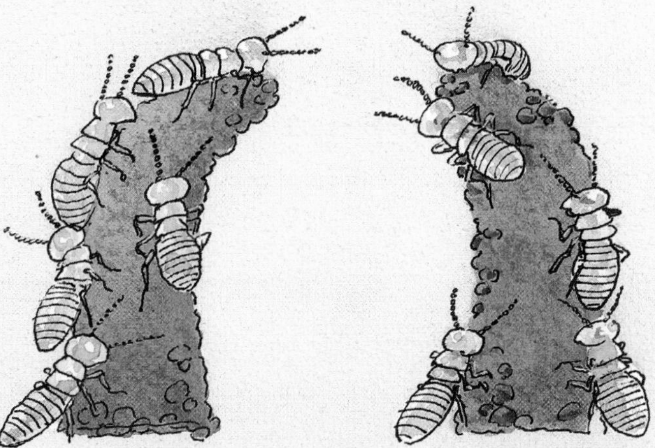

4. ..."spit on the side nearest a neighboring column." Responding to this signal, the termites add new material to the columns so as to make them "bend" toward each other.

5. By following such local rules, termites can construct an elaborate layered network of arches and tunnels — their skyscraper nest — without the need for specific blueprints.

The Continuing Adventures of a Slime Mold: Individual to Aggregate to Individual

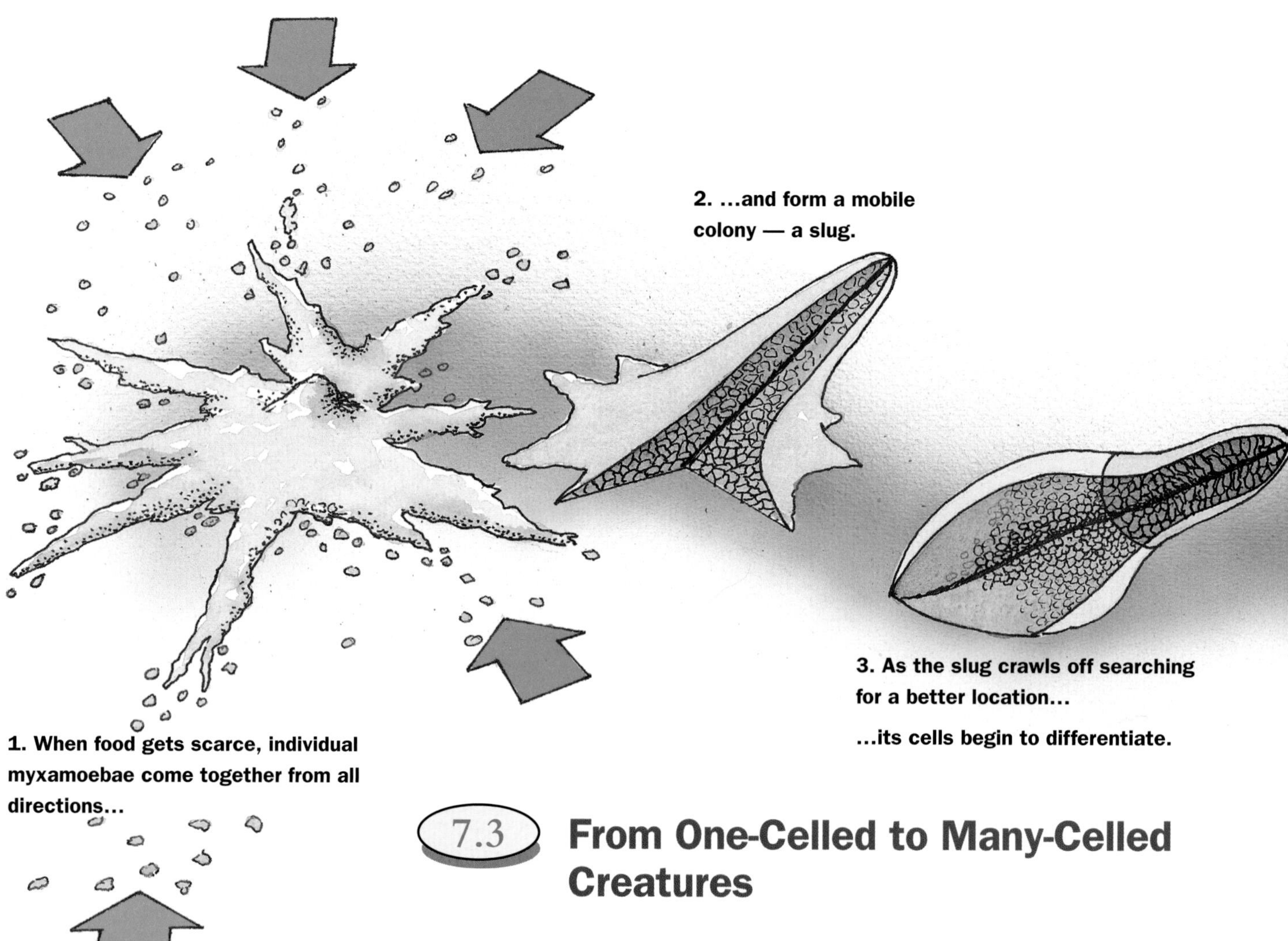

2. ...and form a mobile colony — a slug.
3. As the slug crawls off searching for a better location...
...its cells begin to differentiate.
1. When food gets scarce, individual myxamoebae come together from all directions...

7.3 From One-Celled to Many-Celled Creatures

Two-Faced and Slimy

According to the late author-philosopher Arthur Koestler, we all possess, in effect, two faces. With one face, we look inward and see ourselves as individuals. We express this face in our autonomy and independence. With our other face, we look outward and see that we are members of a larger community. We express this face in our communication and interaction. Having two faces is a matter not of choice, but of biology. Every living organism is both a whole unto itself and a part of something larger.

Few creatures exhibit this duality more dramatically than the strange, lowly slime mold to which you were introduced in the last chapter. As free-living single cells — myxamoebae — residing on the forest floor, slime molds thrive on a diet of bacteria and yeast. But, when food becomes scarce, something unusual happens. A single amoeba, apparently self-appointed, begins to emit a chemical signal. Nearby neighbors, irresistibly drawn to the signal, "ooze" over and attach themselves to the signaler. Each new member of the cluster amplifies the signal by releasing its own signal (a good example of positive feedback). More myxamoebae arrive, eventually forming a colony of up to 20,000 cells. Then a startling transformation occurs: The

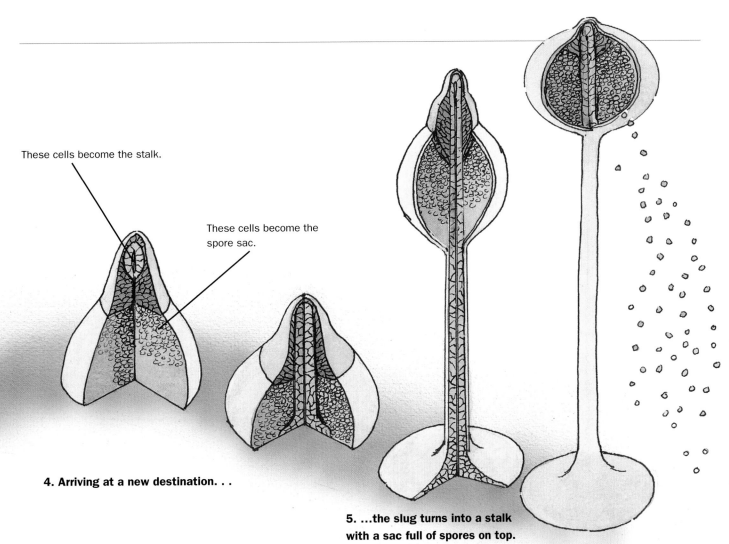

These cells become the stalk.

These cells become the spore sac.

4. Arriving at a new destination. . .

5. ...the slug turns into a stalk with a sac full of spores on top.

6. Soon the dry spores are released, and each grows into a new, independent amoeba.

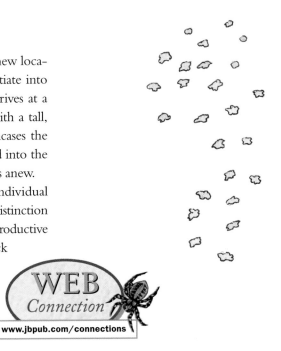

aggregate shapes itself into what is called a slug and begins to migrate to a new location, leaving a trail of slime behind it. As the slug moves, its cells differentiate into three distinct types, whose purposes become clear only when the slug arrives at a suitable new spot. One group of cells forms into a floor plate, or foot, with a tall, reedy stalk extending upward. A second group becomes a sac, which encases the third group, a cluster of spores. When these spores are eventually dispersed into the surroundings, they turn into new myxamoebae. And then the cycle begins anew.

The change from individual to community member and back to individual echoes another primal cycle: egg to organism to egg again. In blurring the distinction between a part and the whole and in anticipating life's more complex reproductive strategies, the slime mold provides some tantalizing clues about how cells stick together, communicate, and differentiate — all key factors in the development of embryos in more complex organisms. The lowly slime mold demonstrates the awesome power of community, that is: how a group can cooperate to accomplish tasks unimaginable by single individuals.

WEB Connection
www.jbpub.com/connections

7.4 Embryo Development — From One Cell to Many

A family builds a house in the middle of nowhere.

Later, using the same set of plans, the family's grown children build next door.

In time, the grandchildren add houses, using their own versions of the original blueprint.

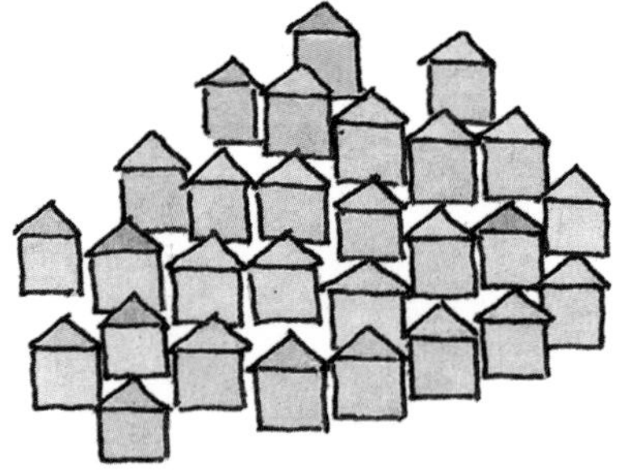

...and the cluster grows into a larger community.

By combining different parts of the building plans, some builders create new kinds of houses...

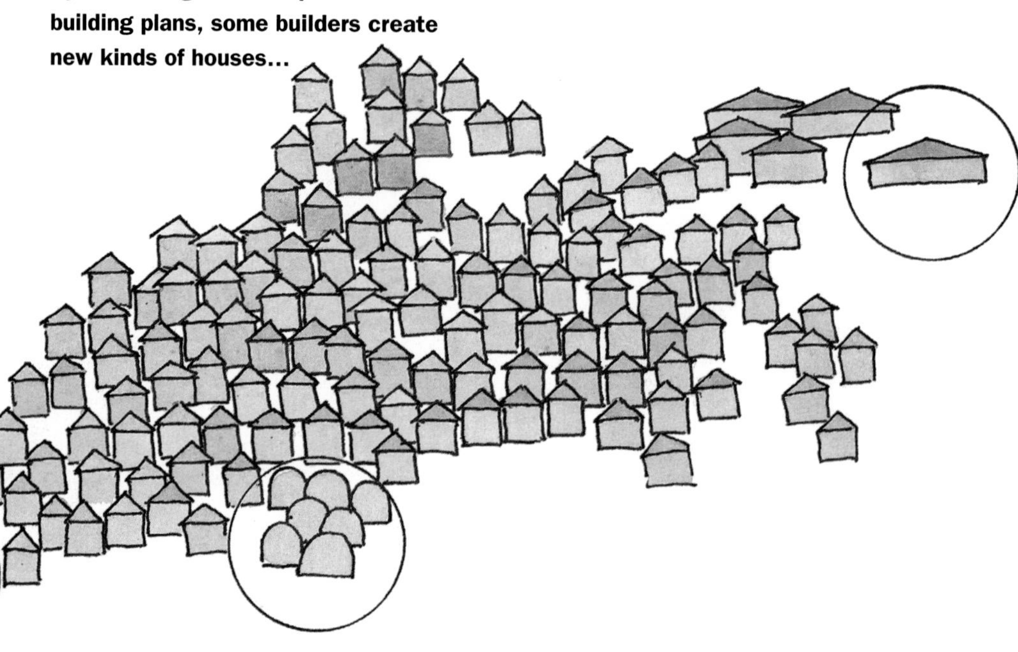

A Self-organizing Community

Few puzzles in biology have been more challenging than the question of how a single cell can grow within the space of a few days, weeks, or months, into a complex organism of millions, billions, or even trillions of cells.

We know that embryo development — life's building program — is based on information located primarily in the genes of each individual cell. We are still learning precisely how the proteins produced by these genes interact to coordinate such a complex event. It's difficult for us to envision a process in which so many things go on at once. First, cells grow and divide. Second, they begin to change, to differentiate, becoming specialists such as bone, skin, nerve, and all the other cell types in the body. Third, they migrate to various locations. And, fourth, they influence the behavior of their neighbors.

The simultaneous progression and interaction of these four activities lead quickly to extraordinary complexity. A growing cluster of houses offers a simple metaphor for this process.

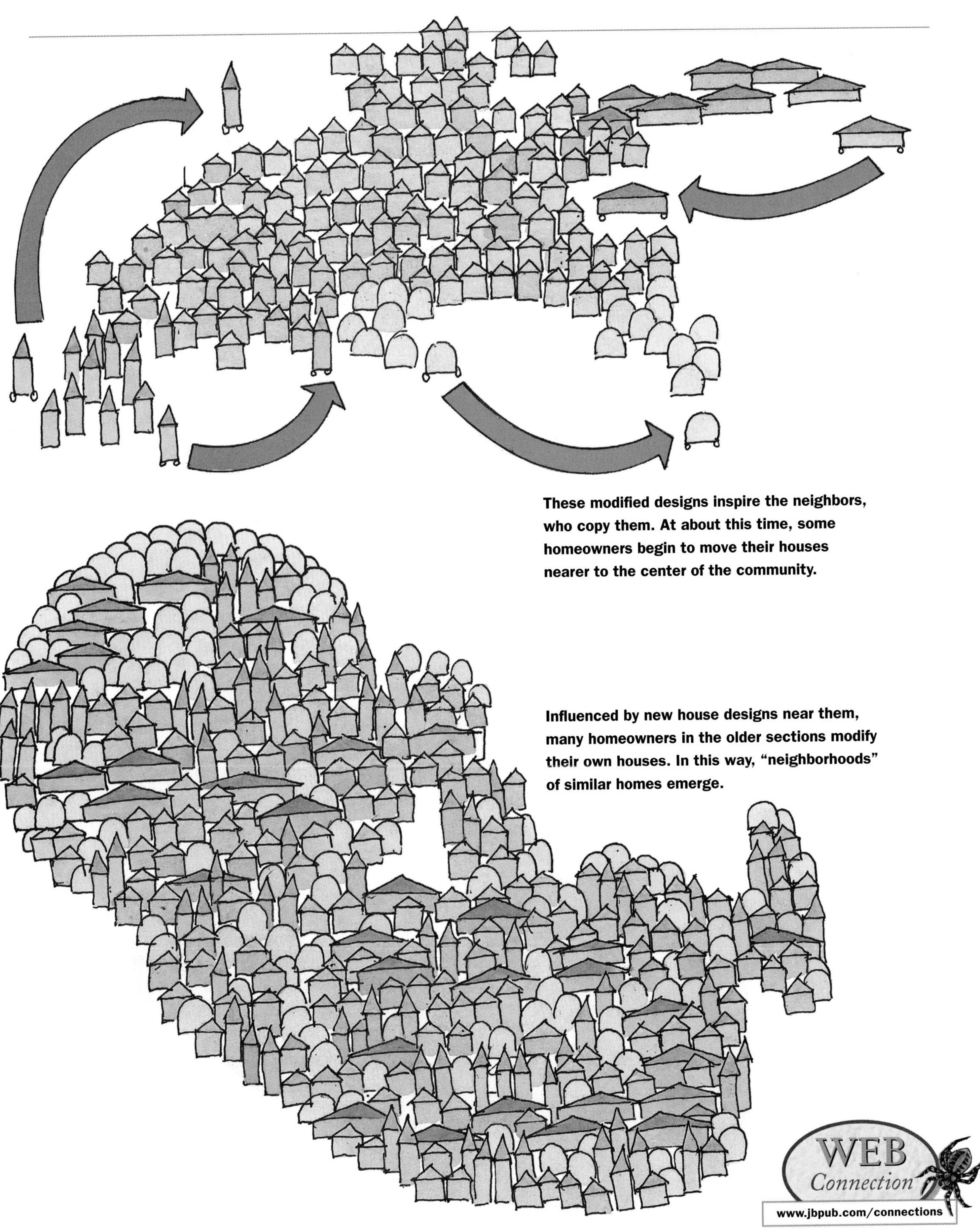

These modified designs inspire the neighbors, who copy them. At about this time, some homeowners begin to move their houses nearer to the center of the community.

Influenced by new house designs near them, many homeowners in the older sections modify their own houses. In this way, "neighborhoods" of similar homes emerge.

WEB Connection
www.jbpub.com/connections

7.5 Organizing a Body — Part I

Preserving Information

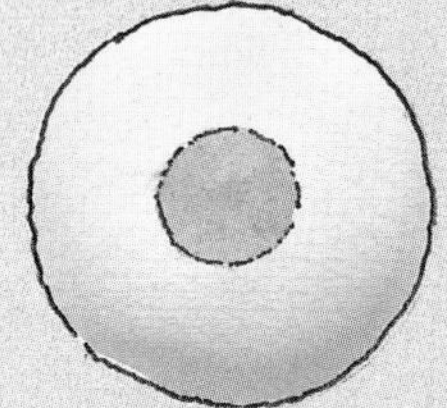

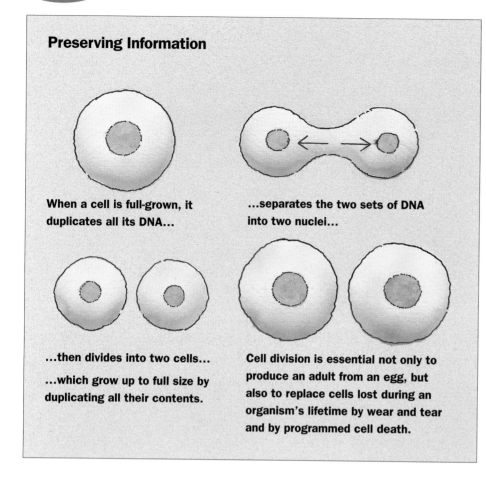

When a cell is full-grown, it duplicates all its DNA...

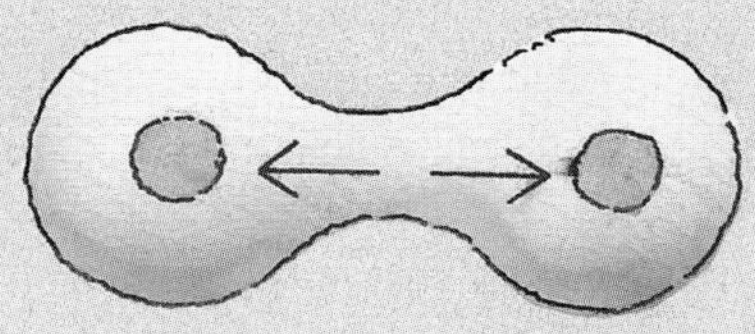

...separates the two sets of DNA into two nuclei...

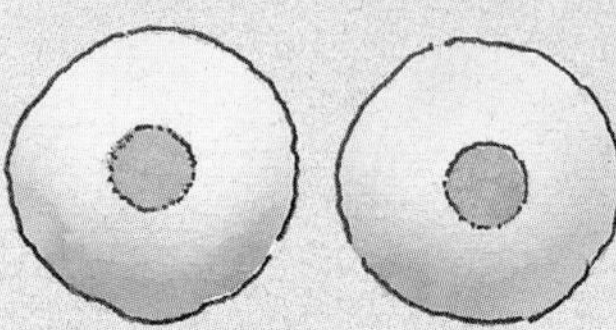

...then divides into two cells...

...which grow up to full size by duplicating all their contents.

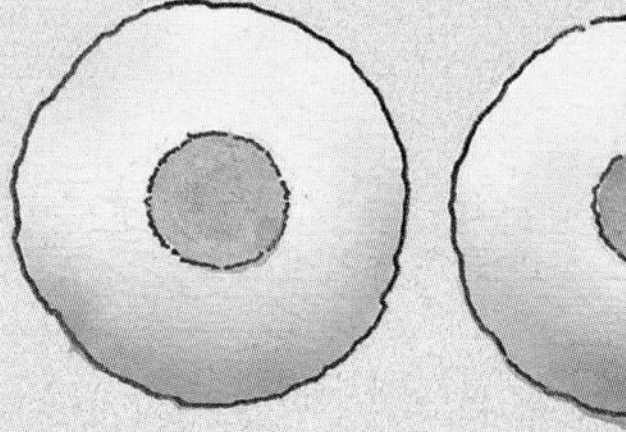

Cell division is essential not only to produce an adult from an egg, but also to replace cells lost during an organism's lifetime by wear and tear and by programmed cell death.

Forming a Hollow Ball ▶

We each began as an egg fertilized by a sperm, a single cell. This cell divided, then divided again, and then again. Doubling over and over — 2 producing 4, producing 8, etc. — quickly leads to large numbers. If all these early cells divided at the same rate, it would take only about 30 divisions to make the many billions of cells of a newborn human.

Early on, cells divide but don't seem to change. By the time there are about a hundred cells, which takes about five days in human development, the cells have formed into a hollow ball, a blastula. A cluster on one side will develop into the embryo, while the single outer layer will become the nourishing sac called the placenta.

Dividing to Multiply

Every cell of your body contains a complete set of all the information that went into building it. At first, this seems unnecessarily cumbersome. After all, wouldn't it be more sensible for a skin cell to contain only the information required for it to function as part of the skin? Why should it bother to carry information needed for brain cells or liver cells? If you were an architect planning to build an entire city, you wouldn't include in the blueprints for each building the entire set of blueprints for all the other buildings. But life does just that!

To understand *why,* it helps to know *how* cell division happens. All cells grow: They double their size by doubling everything they're made of; then they exactly double their DNA and divide in half. Two completely new cells replace each parent. Each of these new cells has received a complete genome — a complete set of genes — an exact copy of all the information in the parent. Enzymes perform this genome-doubling with great precision. Evolution has clearly found this process preferable to some mind-bogglingly complicated mechanism for divvying up the genes during cell division. Each cell, endowed with a complete library, selects those books (genes) it needs for its particular purposes, and leaves the rest of the books on the shelves.

Gene doubling and apportionment into two cells (mitosis), as discussed here, should be distinguished from what happens when eggs and sperm are made (meiosis). This is a special case in which the mother's and father's sets of genes double as usual, then mix (think back to the charm bracelet analogy in Chapter 2). Next, they divide into 4 cells, instead of two. These are sperm or eggs, carrying half of the parents' genes. (For more on gene mixing in sex cells, see page 314.)

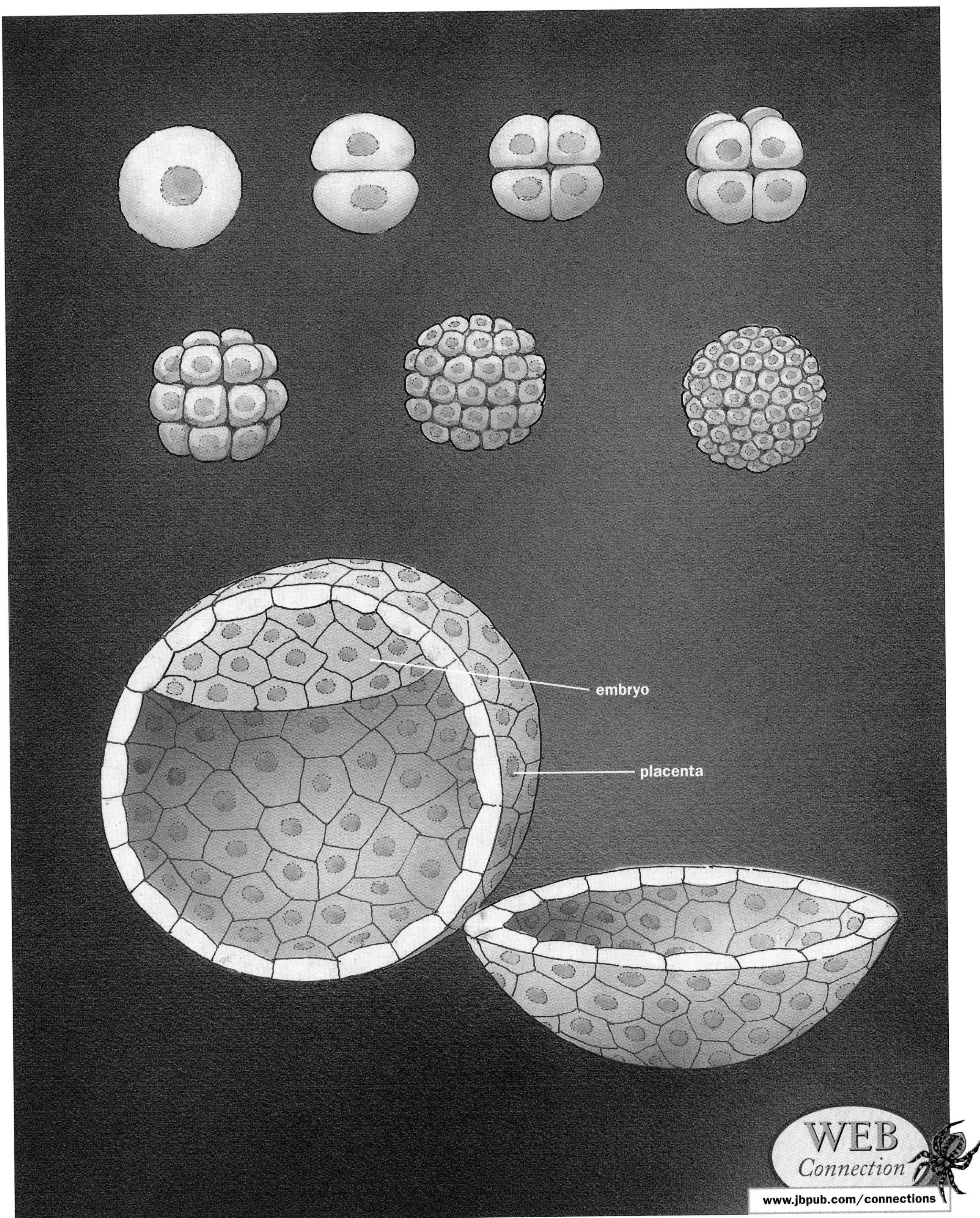
embryo
placenta
WEB
Connection
www.jbpub.com/connections

A Microscopic View of the Cell-Division Cycle — Mitosis

These confocal scanning photomicrographs allow such an up-close view that you can actually see a wheat root tip cell dividing. Here the nuclear material is blue, and the microtubules that play a role in separating the duplicated DNA into two cells are yellow.

Preface to cell division

The first sign that a cell is preparing for division: A band of microtubules forms at the cell's "equator." This band predicts where the cell wall will divide. (Magnified 2000 times.)

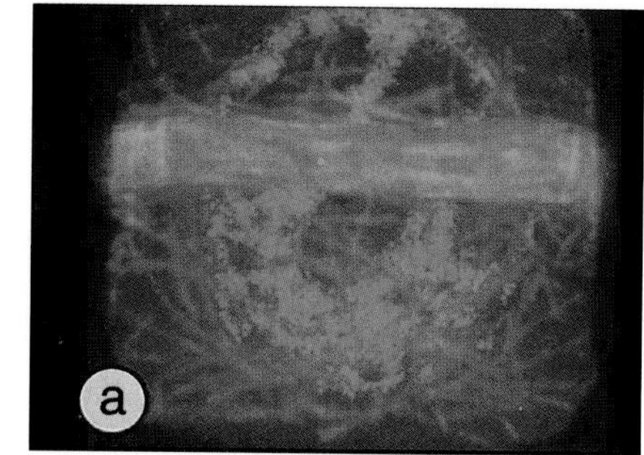

Prophase

In the first stage of mitosis, microtubules called spindle fibers start to form outside the nucleus at the north and south poles of the cell. The band of microtubules at the cell equator begins to disappear. Inside the nucleus, the DNA is duplicating itself. Then the paired stands shorten and thicken. (B magnified 1700 times.)

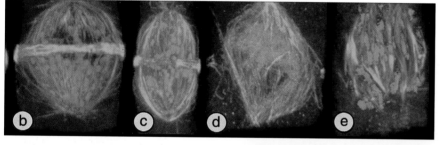

Metaphase–Anaphase

In the second stage, metaphase, the duplicated DNA, each pair joined at a centromere, lines up across the center of the cell. Next, in anaphase, the spindle fibers, now attached to the DNA pairs, begin to pull them apart. A member of each pair is pulled to an opposite pole of the cell. (F magnified 1700 times)

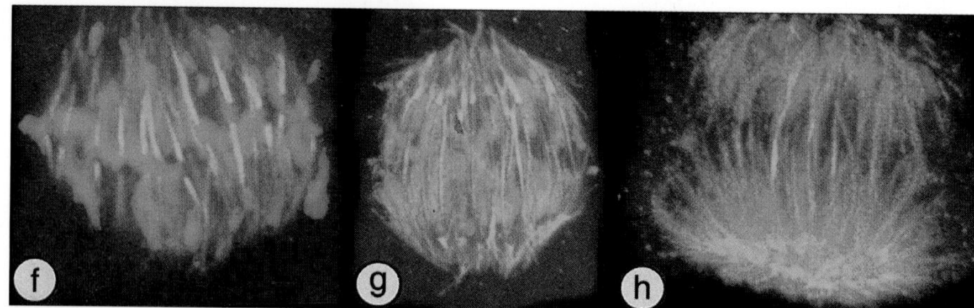

Telophase

With the chromosomes pulled away from the center, here you can see the microtubules running from the poles to the equator. More and more of them form, moving into place the materials that will form the new cell wall. The chromosomes start to uncoil. (J magnified 1300 times.)

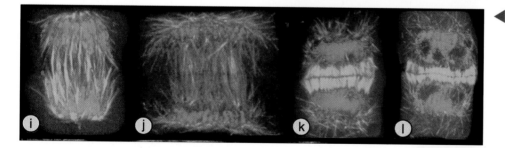

As the daughter cells separate, the microtubules retreat from the division site. They reorient themselves around each daughter nucleus and then at the new cells' surfaces.

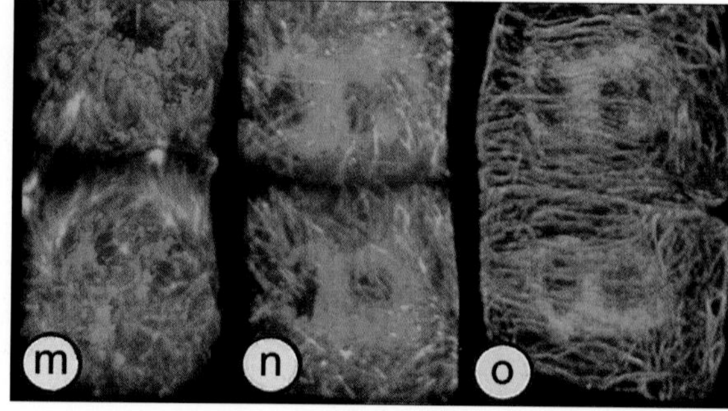

The Visible Zebrafish

This common freshwater fish, often an inhabitant of home aquariums, is especially handy in developmental research because its cell division and differentiation patterns are easily visible (these photographs show the fish egg and embryo magnified only about 60 times.)

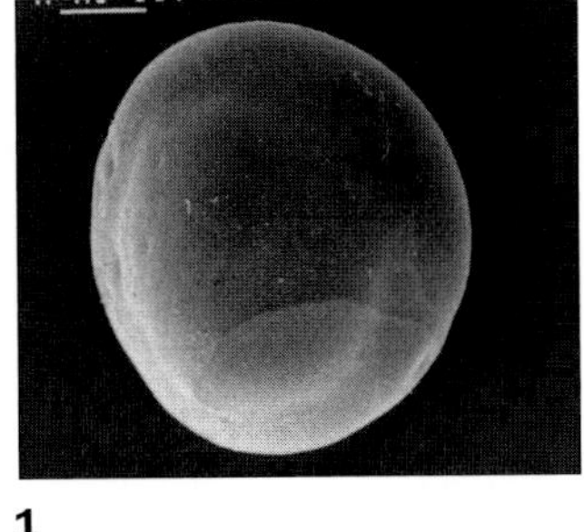

1

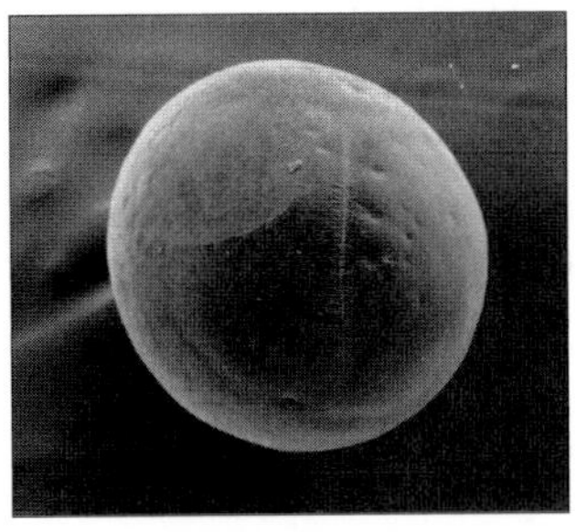

2

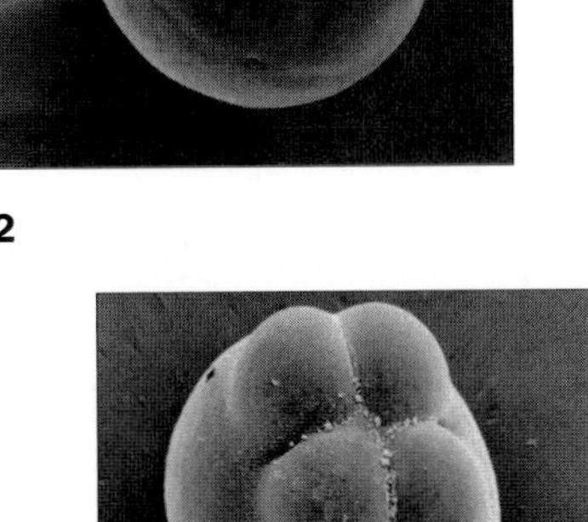

3

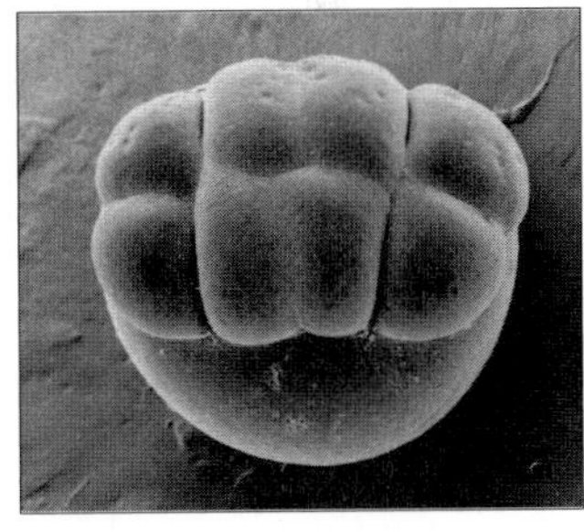

4

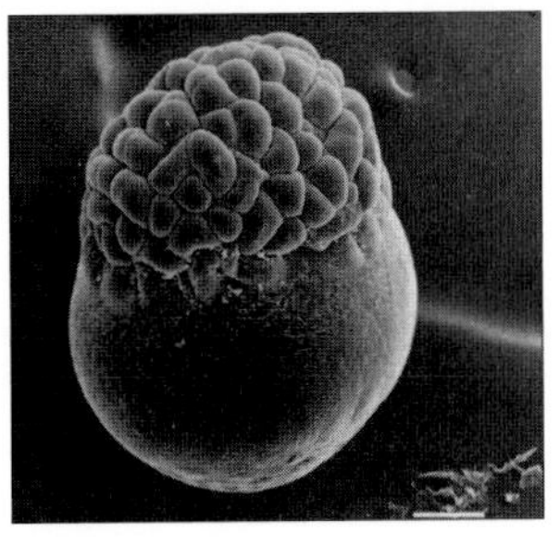

5

6

7

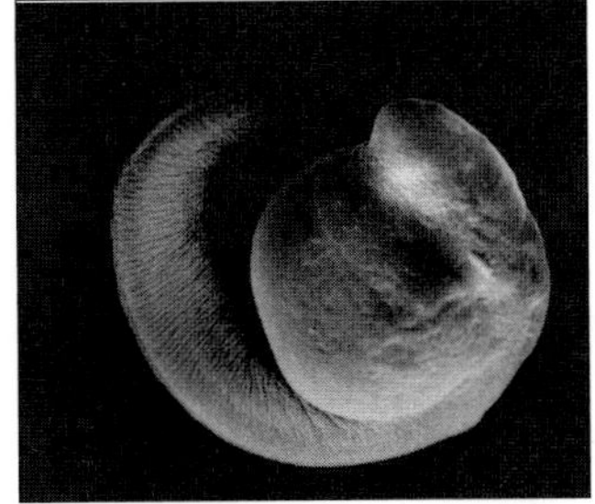

8

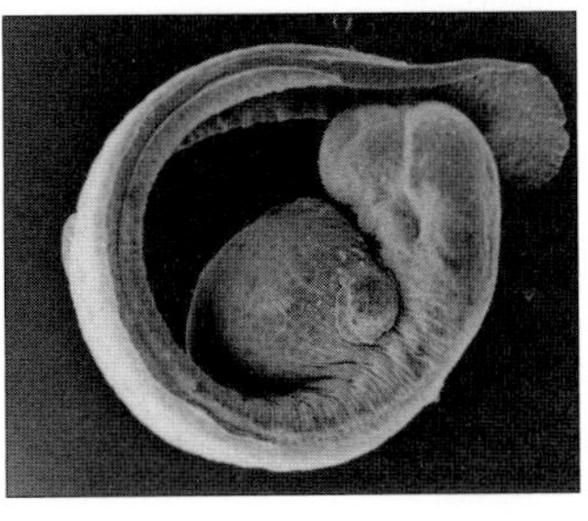

9

10

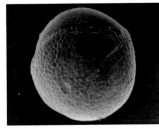

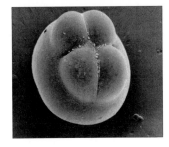

1. All the non-yolk material of the egg moves to one side of the egg and forms a bulge.
2. A depression forms down the middle of the bulge—soon two cells can be seen.
3. As the cells enlarge, a second cleavage line develops—leading to four cells.
4. Two more cleavage lines appear—and the result is eight cells (notice that the yolk is decreasing in proportion as the cells multiply).
5. Division continues (notice new cells creeping down to cover the yolk sac).
6. The yolk is now enclosed by a sphere of cells.
7. Cells begin to migrate to a ridge-like region on the sphere's surface (at right).
8. Continuing cell migration and cell differentiation form an embryo.
9. The tail as well as the eye and fin bumps are evident.
10. Four days after fertilization, the still immature fish hatches.

7.6 Cell Signaling

Cascading Inside the Cell

In Chapter 6 we asked you to imagine being a bacterium — a single cell, bounding independently through life, dividing to become two of yourself whenever conditions seem favorable. Now, try being one of the billions of cells of a multicellular organism. You still have inside you, in your DNA, the know-how to do most things a bacterium can do — and more — but now you have a role to play that is larger than yourself. You're one small part of a huge enterprise. That means you're at the beck and call of signals. You can't grow and divide except in response to commands from somewhere else. These signals may be hormone molecules, coming via the bloodstream from a distant gland, or they may be proteins produced by neighboring cells. Your surface bristles with an array of receptor proteins, each responsive to a separate signal.

When a cell-division signal binds to the end of a receptor on the cell's surface, it causes the other end, inside the cell, to change shape. This in turn switches on a rapid sequence of protein-to-protein contacts, which eventually activates the genes that turn on the machinery of cell division. (The energy for these contacts and activations is supplied by ATP.)

A major step in cell division, shown here, is the start of DNA duplication.

Such signal-induced sequences of molecular events resemble a cartoon in which a crowing rooster startles the cat who jumps off his perch, causing a steel ball to roll down a chute, striking a switch that turns on the stove, heating up the coffee.

Many such protein-to-protein relays pass signal information to genes each time cells undergo any of the multitudinous changes needed to transform themselves from tiny blob into newborn baby.

Sometimes a cell turns "bad"; it becomes a sociopath, dividing at will, destroying its neighbors, and taking off for distant parts. This cell-run-amok is a cancer cell. Inside it, one or more of the signaling proteins have been damaged. Such damage, caused by a mutation in the gene responsible for making that particular protein, can have devastating consequences for the harmonious operation of the whole cellular community.

All these relays are currently the subject of intense biological investigation. Unlocking their secrets will help us solve the mysteries of embryonic development — how life is re-created at each generation — and of cancer — how life can be destroyed by damaged relays.

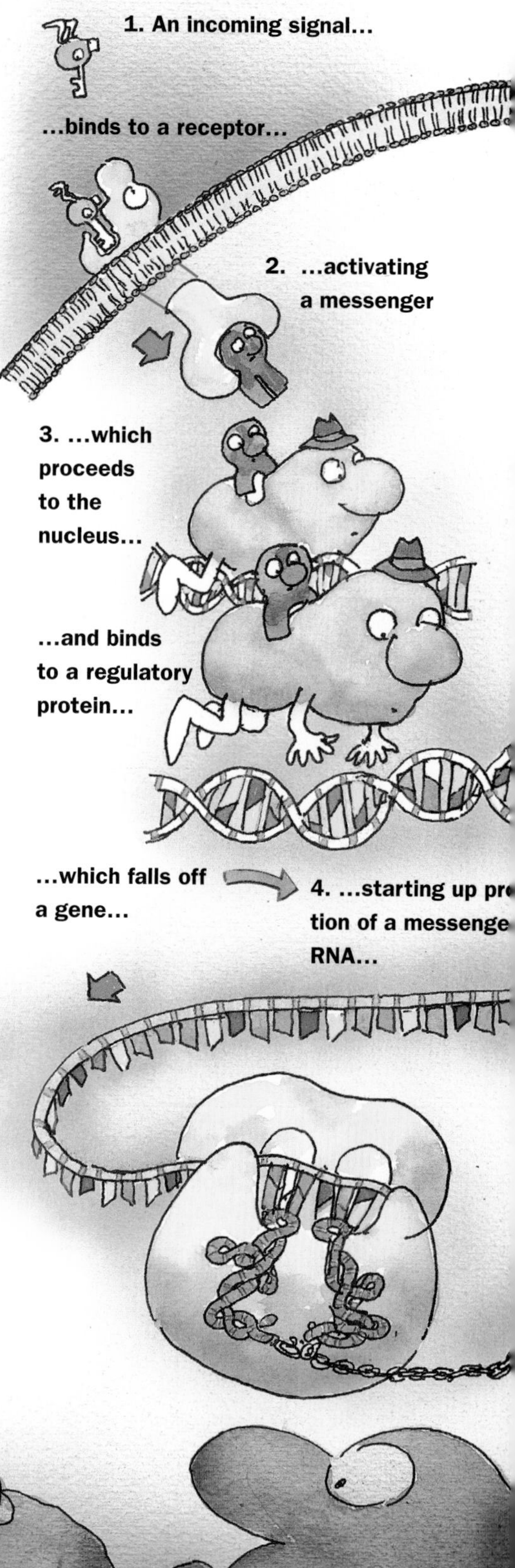

1. An incoming signal...
...binds to a receptor...
2. ...activating a messenger
3. ...which proceeds to the nucleus...
...and binds to a regulatory protein...
...which falls off a gene...
4. ...starting up
RNA...

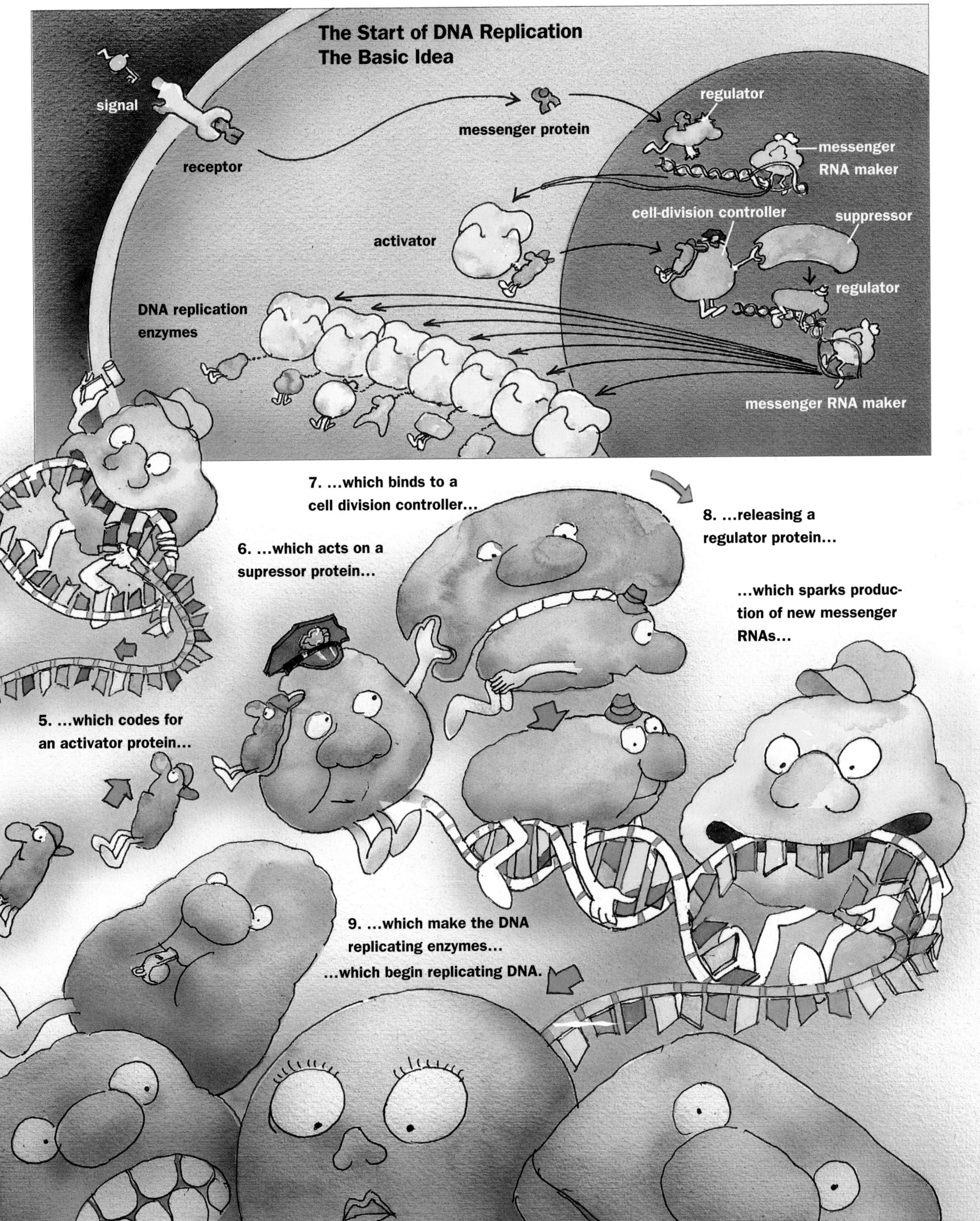
The Start of DNA Replication
The Basic Idea
signal
receptor
messenger protein
regulator
messenger
RNA maker
activator
cell-division controller
suppressor
regulator
DNA replication
enzymes
messenger RNA maker
7. ...which binds to a
cell division controller...
8. ...releasing a
regulator protein...
6. ...which acts on a
supressor protein...
...which sparks produc-
tion of new messenger
RNAs...
5. ...which codes for
an activator protein...
9. ...which make the DNA
replicating enzymes...
...which begin replicating DNA.

7.7 Organizing a Body — Part II

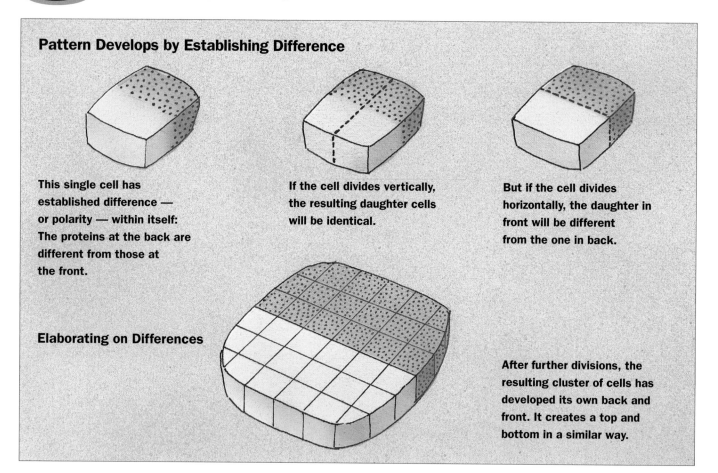

Pattern Develops by Establishing Difference

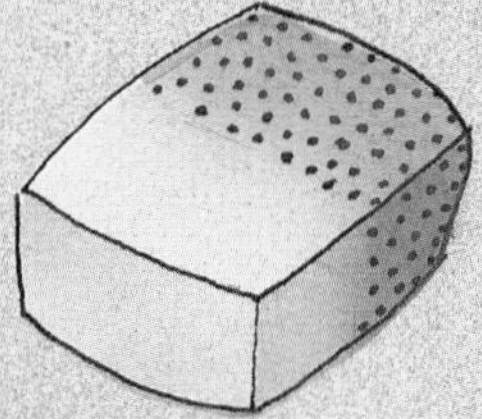

This single cell has established difference — or polarity — within itself: The proteins at the back are different from those at the front.

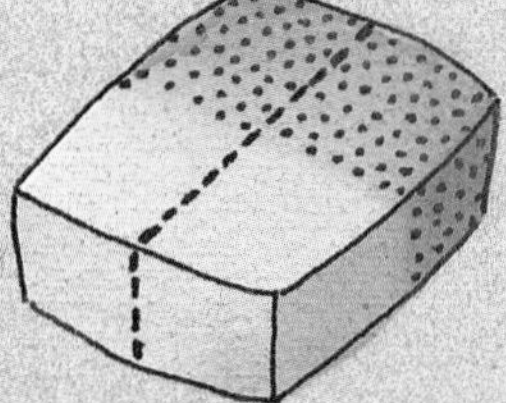

If the cell divides vertically, the resulting daughter cells will be identical.

But if the cell divides horizontally, the daughter in front will be different from the one in back.

Elaborating on Differences

After further divisions, the resulting cluster of cells has developed its own back and front. It creates a top and bottom in a similar way.

The Beginnings of Pattern

Scientists in the past, peering through their crude microscopes, swore they could see a tiny, fully-formed human figure huddled inside each human sperm cell. But this was wishful thinking. In embryo development, simplicity precedes complexity; similarity precedes difference.

The first step toward the creation of a body comes when a few cells occupying a space no bigger than the point of a pin begin to take on the general character of what we'll call "topness," and others adopt "bottomness"; still others assume "frontness" or "backness"; "outsideness" or "insideness." There's no sign yet of a head or a tail, a backbone or belly, skin or internal organs — to say nothing of things in-between.

Multiplying cells that have made the commitment to, say, topness will, generation by generation, make small changes in their character and forge new relationships with their neighbors so as eventually to become a recognizable head. Each step depends on the specific changes that have occurred earlier. This is why we say "memory" is essential to embryonic development. Cells can't arrive at the final version of *what* they will be until their forebears have determined *where* they will be.

The Embryo Begins to Take Shape

On page 255 we showed the embryo as a cluster of cells inside a hollow ball. At the top right of page 261 we show how this cluster begins to flatten out as a disk (top row) which then elongates and forms three layers: one for skin and nerve cells; one that becomes the digestive system; and one for all other cell types. (The layers are shown separated in the diagram at lower right.) We use a simplified grid to portray the growing number of the embryo's cells.

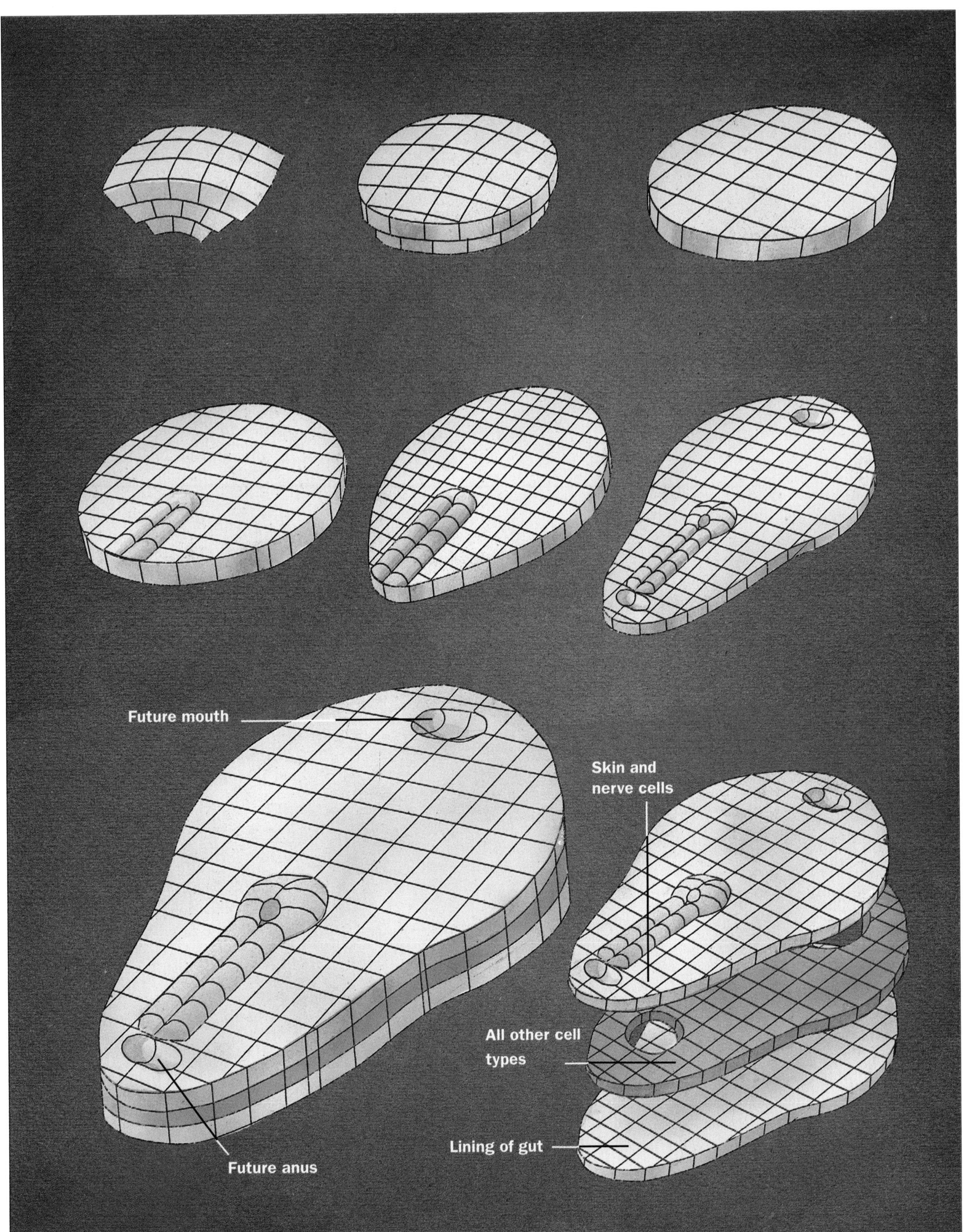
Future mouth
Skin and
nerve cells
All other cell
types
Lining of gut
Future anus

Organizing a Body — Part III

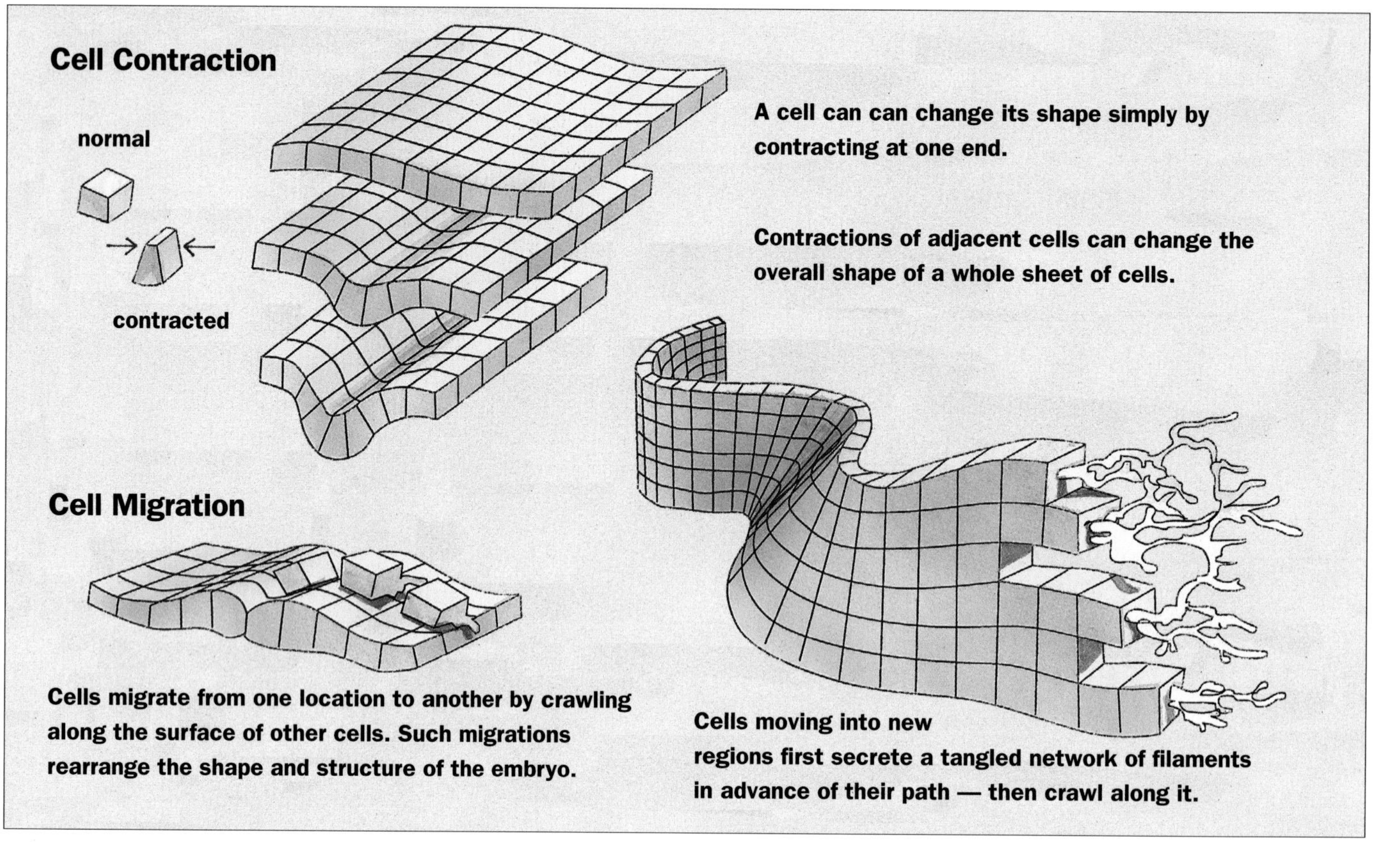

Cell Contraction

normal

contracted

A cell can can change its shape simply by contracting at one end.

Contractions of adjacent cells can change the overall shape of a whole sheet of cells.

Cell Migration

Cells migrate from one location to another by crawling along the surface of other cells. Such migrations rearrange the shape and structure of the embryo.

Cells moving into new regions first secrete a tangled network of filaments in advance of their path — then crawl along it.

Contracting and Moving

As the embryo begins to take shape, its cells become even more active. In addition to simply dividing, they also begin to contract and move.

A group of cells, contracting in unison, can change the entire shape of the embryo, as when cells on the embryo's back fold up to form the neural tube, the channel that will house the spinal cord. Other groups of cells detach themselves from their neighbors and move to new locations. Time-lapse movies of developing embryos reveal sheets of cells streaming past each other in simultaneous migration. Cells destined to form the gut move upward toward the mouth region. Two ridges on either side of the back move toward the center to form the embryonic spinal cord and brain. Amazingly, the migrating cells seem to know just where to go. It appears that they follow a chemical trail, as ants and bacteria do to reach a food source.

Pioneer cells setting out for new regions first lay down a matrix — a tangled network of filaments that give the cells something to cling to. This is a little like a growing hedge first assembling a picket fence, and then using the fence for support as it grows over it. For a look at recent research on this topic, see page 284.

Head Start ▶

The flat embryo rolls up into a tube with the gut-lining layer inside (top row) as a wide ridge begins to form around a centerline groove. The ridge then "zips up" (center row) to form the future spinal cord. The tube then curls into a "comma" as head and tail begin to emerge (bottom row).

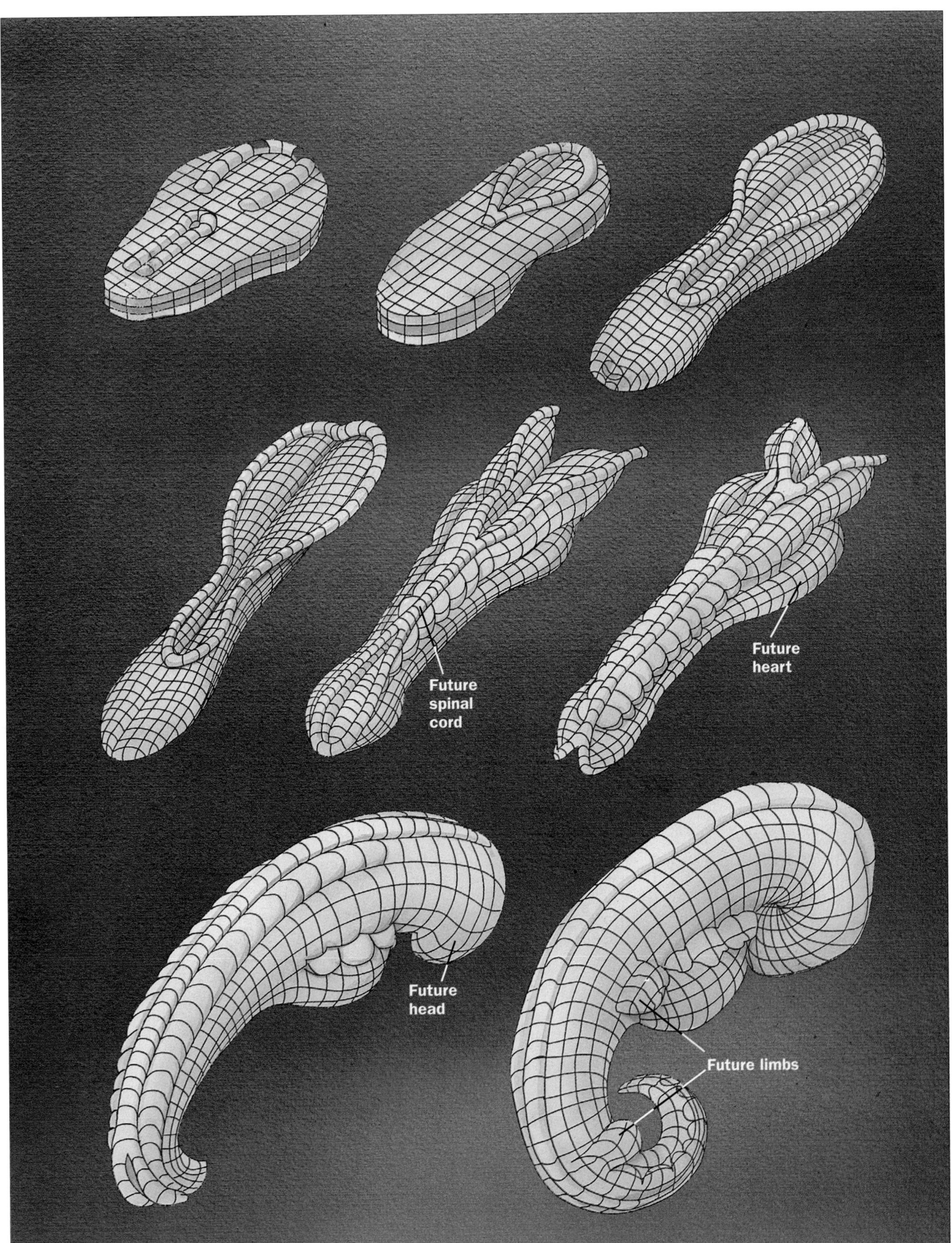
Future
spinal
cord
Future
heart
Future
head
Future limbs

7.9 Genes as Switches

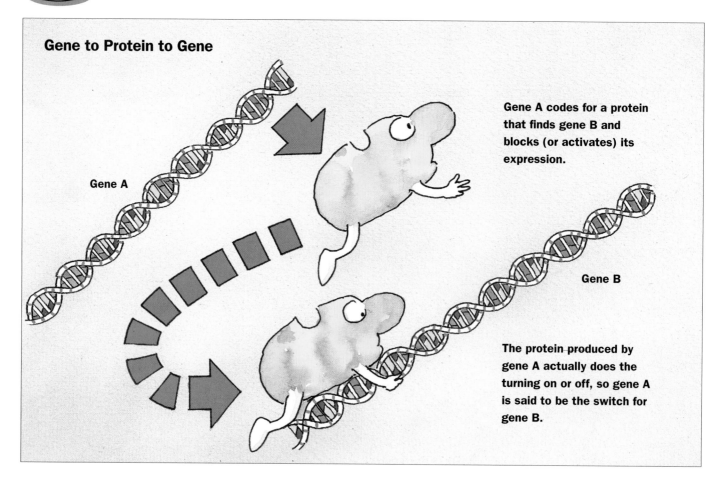

Gene to Protein to Gene

Gene A

Gene A codes for a protein that finds gene B and blocks (or activates) its expression.

Gene B

The protein produced by gene A actually does the turning on or off, so gene A is said to be the switch for gene B.

How One Gene Can Turn On Another

To understand how an embryo develops, you need to appreciate how genes act as on/off switches. In Chapter 6, *Feedback*, we saw how some genes carry instructions for making worker proteins and other genes code for regulator proteins. Regulator proteins don't make anything or hold anything together. Instead, they go into the nucleus where the DNA is, find a specific gene, and sit on it — thereby blocking it from creating a particular protein. (Some work the opposite way: By sitting on a gene, they signal it to start producing instructions to make a protein. This is what happens, for instance, in the start-up of cell division.) In short, regulator genes act as switches, turning on and turning off the genes that code for worker proteins.

The Origin of Genetic Circuits

A key evolutionary breakthrough occurred when one gene began to control another.

When you consider that each individual cell carries all the genes for the entire organism, the necessity of switches becomes clear. The organism wouldn't function properly if, for example, its muscle cells made liver cell proteins. So muscle cells must turn on the genes that make the proteins that will turn *off* liver genes. They keep those genes off for the muscle cells' entire lives. In other words, each cell type — whether muscle, liver, skin, or some other kind, has its own active network of appropriate genes. The rest of its genes — the majority, in fact — lie silent, repressed forever by stubborn regulatory proteins.

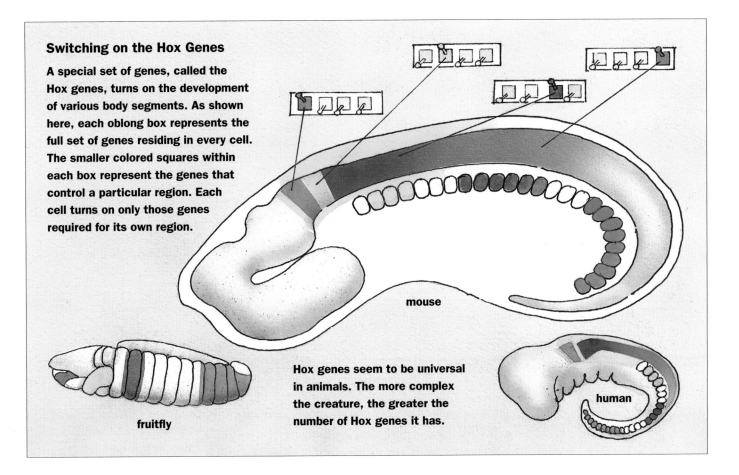

Switching on the Hox Genes

A special set of genes, called the Hox genes, turns on the development of various body segments. As shown here, each oblong box represents the full set of genes residing in every cell. The smaller colored squares within each box represent the genes that control a particular region. Each cell turns on only those genes required for its own region.

mouse

fruitfly

Hox genes seem to be universal in animals. The more complex the creature, the greater the number of Hox genes it has.

human

Switches Build on Switches

A single master switch can control a number of subordinate switches. This feature simplifies control over complex operations.

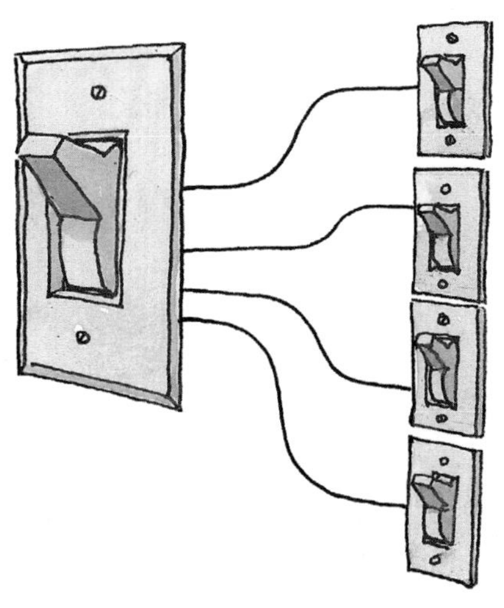

Body-shaping Switches

Perhaps the most interesting protein switches operate during an embryo's development. As the growing embryo changes shape, each cell needs to know just when to bring the appropriate proteins on line and when to shut them down. Timing is crucial.

If the regulatory genes could talk, they might sound something like this: Gene A: "OK, start making the proteins that define the front end...good, now shut those off." Gene B: "Excellent, now bring in the head-forming proteins," and so forth.

Feedback signals operate the protein switches. Each stage produces the signal molecules that set the next stage in motion. Cascades of worker proteins, governed by a much smaller number of protein switches, swarm into action in wave after wave.

Among the most intriguing of the "master" regulating genes discovered so far are those known as the Hox genes. Becoming active early in the embryo's development, they tell its cells where its head, chest, and lower body should be — and consequently, where its eyes, arms, legs, etc., will go. If you have ever doubted your kinship with the rest of the world's creatures, you should know that these body-shaping Hox genes are found in insects, worms, fish, frogs, chickens, and cows as well as in humans.

7.10 Organizing a Body — Part IV

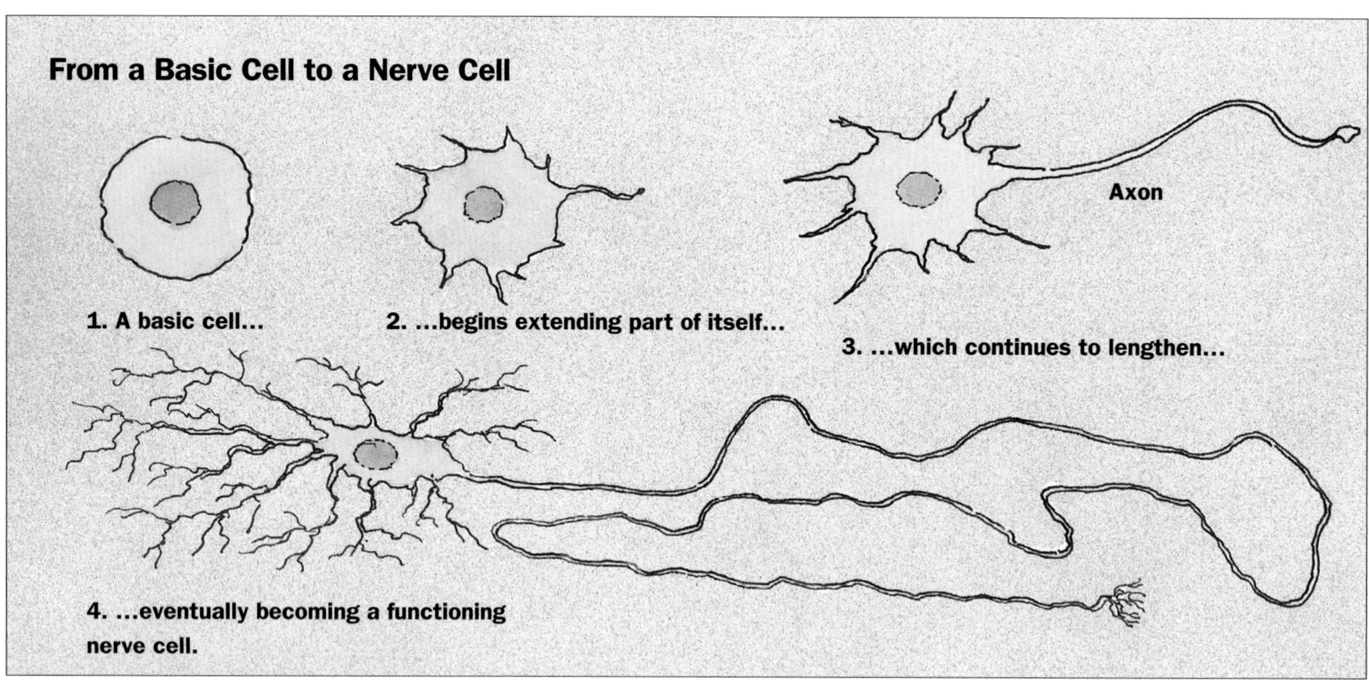
From a Basic Cell to a Nerve Cell
Axon
1. A basic cell...
2. ...begins extending part of itself...
3. ...which continues to lengthen...
4. ...eventually becoming a functioning nerve cell.

A Community of Specialists

We've seen how the embryo progresses from a hollow ball of cells to the beginnings of a recognizable body. Organization proceeds from the general to the specific. First, a blob acquires a top, bottom, front, and back. Next, it develops bands, or rows of cells that define specific regions of the body, and then bumps, which will become head, tail, and limbs. In this process, the embryo's cells polarize, contract, and migrate. They also *differentiate,* taking different paths toward specialization.

The complexity of the tasks facing even the simplest multicellular organism demands specialization. This, we might say, is the cell's trade-off — its way of contributing to the whole organism in exchange for food and shelter. Cells carrying within them the capacity to become the whole organism "decide" to become just a small working part of that organism. They do this gradually. During embryonic development, each new generation of cells becomes a little different from the generation before.

At first, the changes are too miniscule to see; then, after a few generations, they're obvious. A small group of cells in an embryo will look like all its neighbors, but as those cells multiply, they become progressively longer and thinner than nearby cells. Within them appear increasing numbers of long stringy proteins (actin and myosin) that can shorten and lengthen themselves, contracting and relaxing the whole cell. These cells are becoming *muscle.*

Not only must cells do their special jobs in the organism, they must, of course, do them in the proper locations and with the help of the right neighbors. Muscle cells move to a site where a limb will emerge, and there find neighbor cells that will become bone, nerves, and blood vessels.

Skin and Nerve

Nerve cells at first seem no different from their companion skin cells. Yet, as they and their descendants grow and divide, their destiny becomes progressively and irreversibly determined, even though their outward appearance seems unchanged. Then, suddenly, they stop dividing and change becomes obvious: They begin to push out long extensions of themselves — axons — which snake off to make connections with other nerve cells, setting up the brain's wiring circuitry. Thereafter they live as long as the organism.

In contrast, other members of the original cell population will become skin cells. Unlike their sister nerve cells that have a single "birthday," and a long life, skin cells serve a short time, die, and are continually replaced from persistent parent cells.

WEB Connection
www.jbpub.com/connections

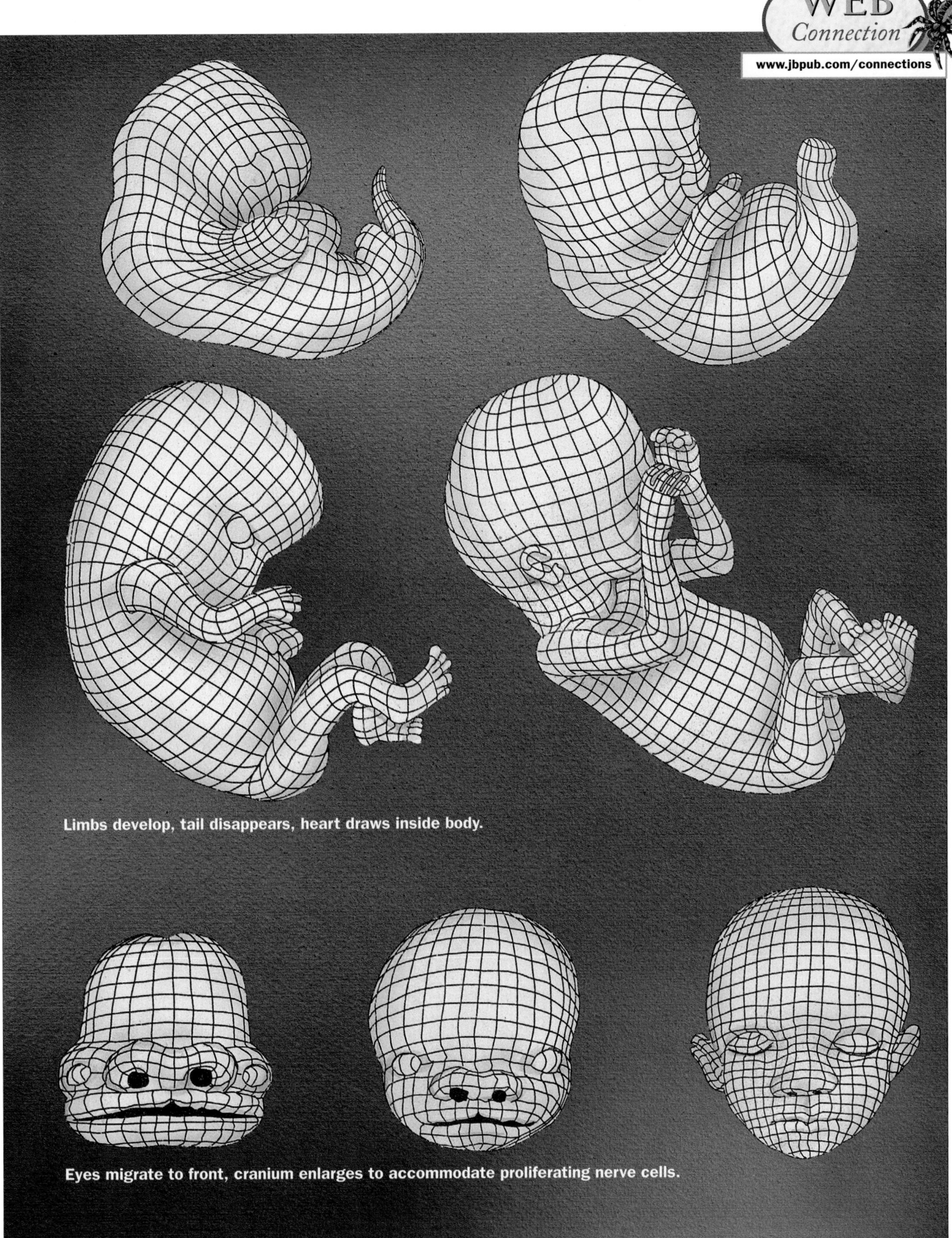
Limbs develop, tail disappears, heart draws inside body.
Eyes migrate to front, cranium enlarges to accommodate proliferating nerve cells.

Specialties and Subspecialties

Specialized cellular communities, such as our skin, need to form many subspecialist cell groupings in order to perform efficiently. First, since skin is our "outside" it has cells specialized for protection. The skin's surface cells, the epidermis, form a barrier against an external environment teeming with potentially dangerous microorganisms — bacteria and fungi. These microorganisms can't easily penetrate healthy skin. But when skin is damaged, they can get into the body's "inside" and cause infections or illnesses. Also, the skin is practically waterproof and acts as a barrier in the opposite direction, preventing the loss of water from the inside out. (Our bodies are 75 percent water, and air is a drying agent.)

As beautifully specialized as it is to perform its protective function, skin does far more. It also has interactive functions: It's a sense organ that allows us to feel temperature, pain, and pressure. And it's a temperature-regulation device.

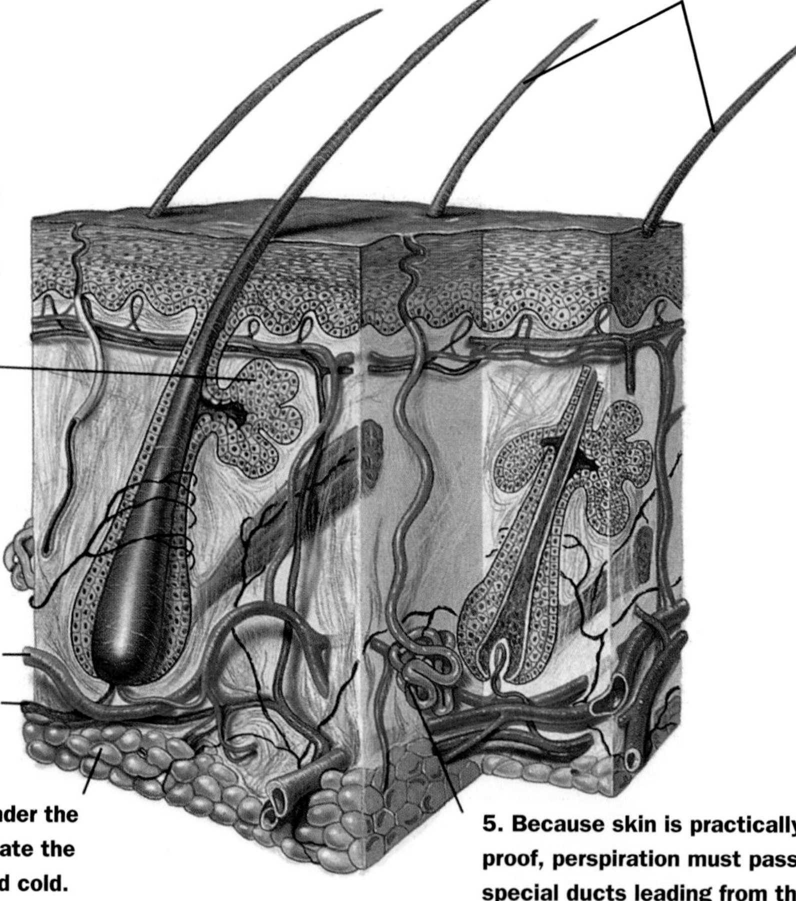

1. Hairs grow out of tube-like follicles that are attached to nerves, which send messages to another specialized cellular community known as the central nervous system. When a breeze moves the hairs, we feel it.
2. Oil glands, called sebaceous glands, produce natural oils that help keep the skin pliable and waterproof.
3. Arteries course through the skin, bringing blood containing oxygen and nutrients to all the cells. Blood in veins carries away carbon dioxide and other wastes.
Vein
Artery
4. A layer of fat under the skin helps to insulate the body from heat and cold.
5. Because skin is practically waterproof, perspiration must pass through special ducts leading from the sweat glands to the surface. When this moisture evaporates, it cools the skin, helping to regulate body temperature.

Every organism needs a protective, interactive barrier between its inside and its outside, and each organism's DNA provides a unique recipe adapted to its needs and its environment. The resulting variety of protective and functional coverings (integuments) is considerable.

REPRESENTATIVE ANIMAL	INTEGUMENT
Sponge	Layer of epidermal cells

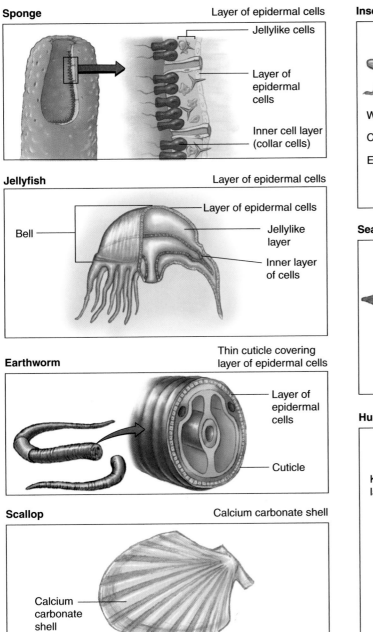

Jellylike cells
Layer of epidermal cells
Inner cell layer (collar cells)

REPRESENTATIVE ANIMAL	INTEGUMENT
Jellyfish	Layer of epidermal cells

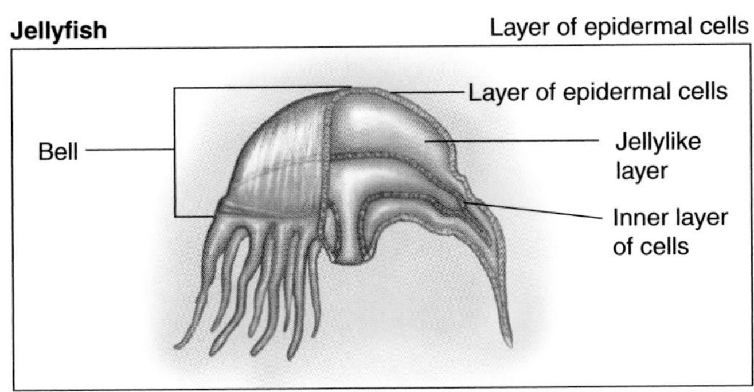

REPRESENTATIVE ANIMAL	INTEGUMENT
Earthworm	Thin cuticle covering layer of epidermal cells

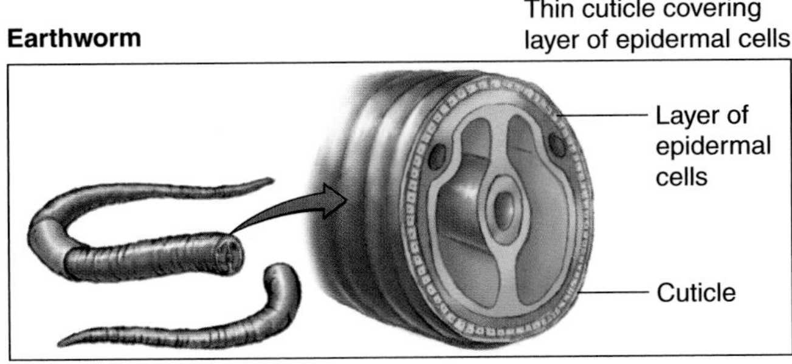

REPRESENTATIVE ANIMAL	INTEGUMENT
Scallop	Calcium carbonate shell

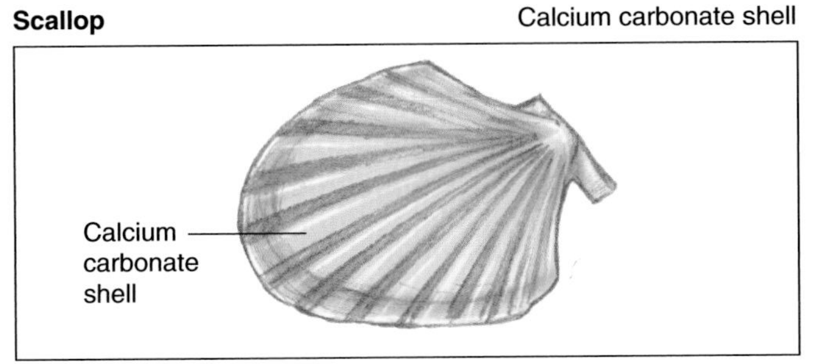

REPRESENTATIVE ANIMAL	INTEGUMENT
Insect	Chitin cuticle with a waxy surface and underlying epidermis

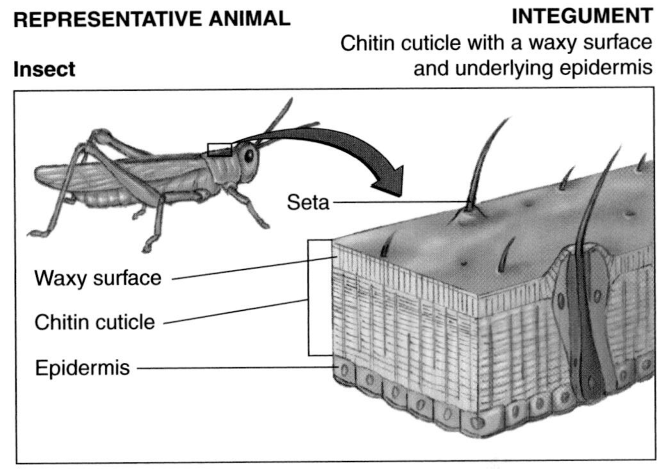

REPRESENTATIVE ANIMAL	INTEGUMENT
Sea star	Thin, prickly epidermis

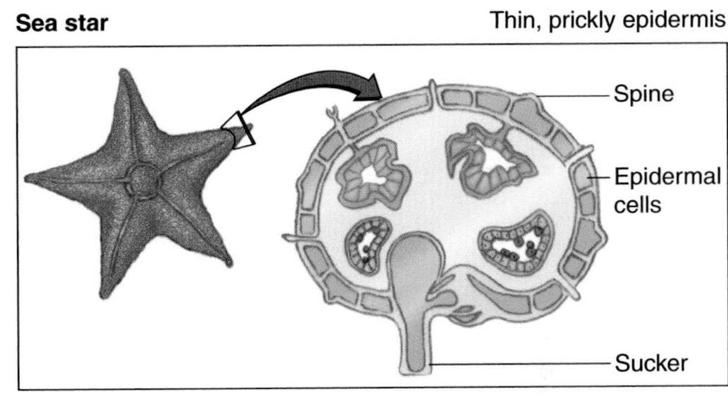

REPRESENTATIVE ANIMAL	INTEGUMENT
Human	Keratinized skin

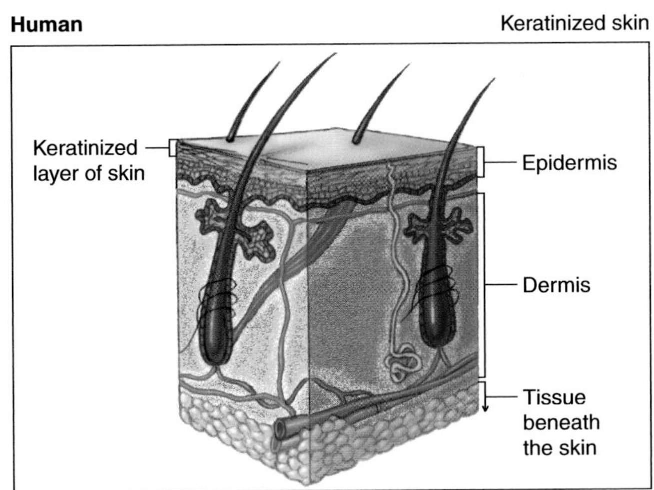

7.11 How Cells Differentiate

The Lineage Plan

Cells change from generation to generation because new ingredients are added to each successive generation and then are apportioned unequally when the cell divides.

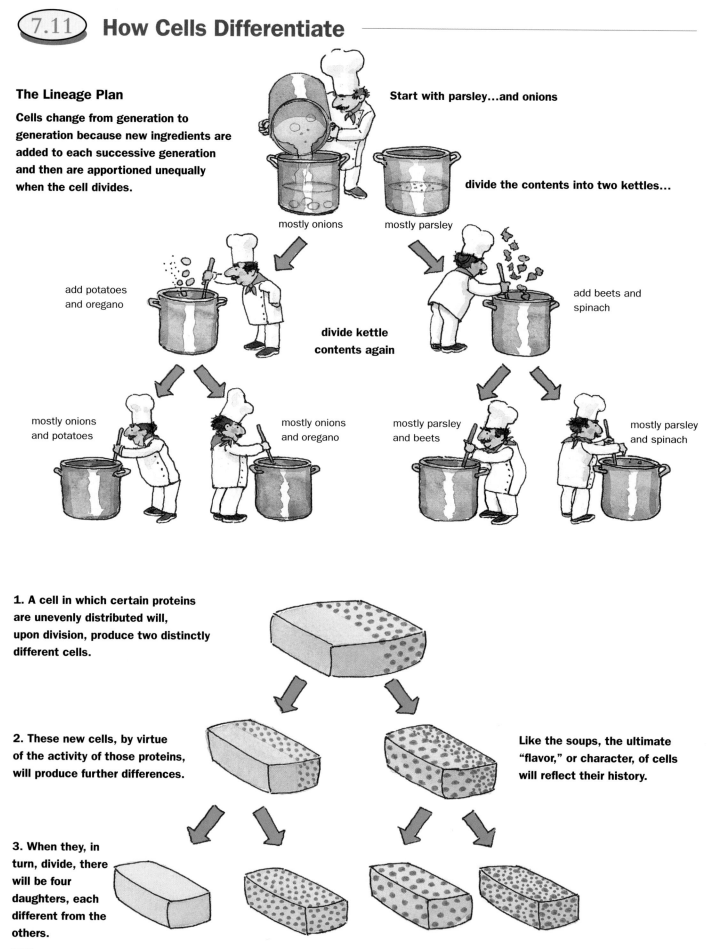

Start with parsley...and onions

divide the contents into two kettles...

1. A cell in which certain proteins are unevenly distributed will, upon division, produce two distinctly different cells.

2. These new cells, by virtue of the activity of those proteins, will produce further differences.

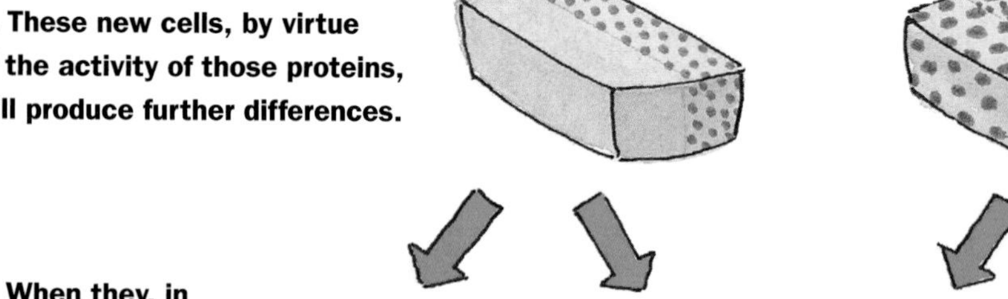

Like the soups, the ultimate "flavor," or character, of cells will reflect their history.

3. When they, in turn, divide, there will be four daughters, each different from the others.

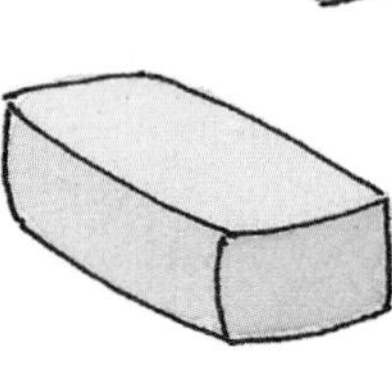

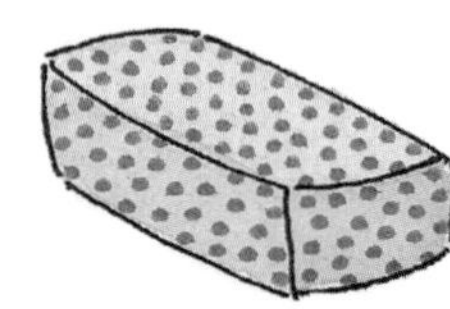

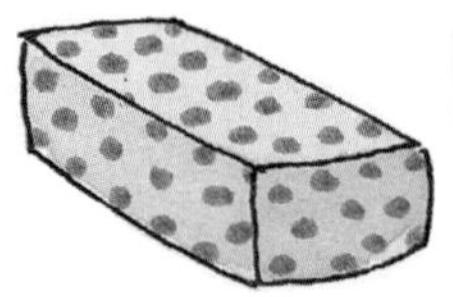

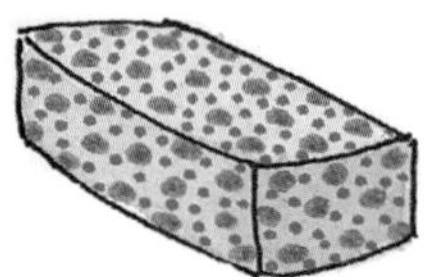

The Lineage Plan and the Contact Plan

If every cell in your body comes from a single fertilized egg, how do you end up with so many *kinds* of cells (humans have about 250 kinds)? Apparently, this happens in either of two ways: A cell can progressively change itself over generations (the lineage plan), or it can be induced to change by a neighboring cell (the contact plan).

The lineage idea is illustrated on the left with a chef and soup model. The chef starts with a simple broth containing one heavy ingredient (onions), which sinks to the bottom of the pot, and one light ingredient (parsley), which floats. When he pours this soup, without stirring it, into separate kettles, the contents of the two new kettles will be different: each will have a different proportion of onions and parsley.

Next, the chef adds more light and heavy ingredients to each of the two kettles. Then he subdivides the contents of the kettles as before, producing four completely different soups. You can see how each "generation" of soup is a little different from that before it, yet there's always a little of the original broth in every kettle. Thus the final state of a soup — or a cell — is determined by its history.

At the lower left we show how this principle works with cells. Instead of vegetables, different proteins are being produced in only half of the cell, so when it divides the proteins are partitioned unequally. This leads to different cell types.

The contact plan, whereby a cell signals its immediate neighbor to change, is simpler. It works by a process called *induction:* One cell sends a message to its next-door neighbor, instructing it to make a particular protein (or proteins). The neighbor responds by making that protein, but only in the part of it nearest the signaler. As shown below, when this neighbor cell divides, the two daughter cells will be different because they will have different amounts of the protein. Repetition of this process over many generations results in an assortment of new and different cells.

The Contact Plan

Cells signal their immediate neighbors to change.

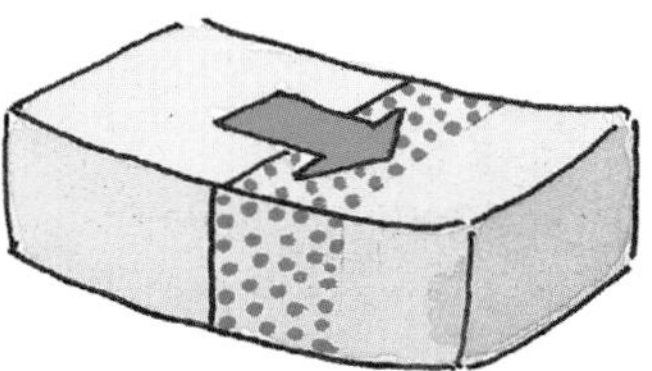

1. One cell induces its neighbor to make proteins in the region nearest it. This makes the neighbor asymmetrical.

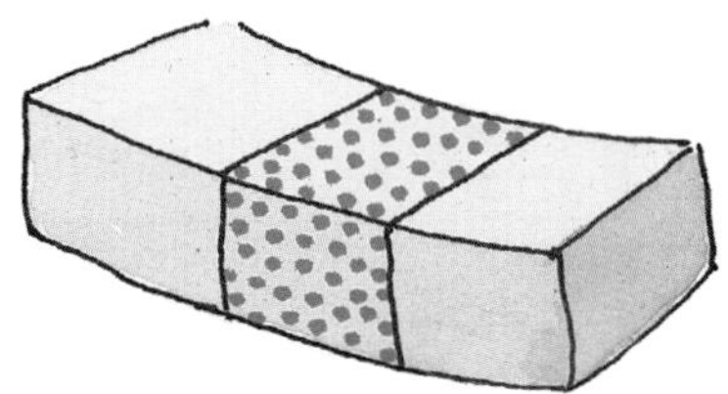

2. The neighbor divides, creating two different daughter cells.

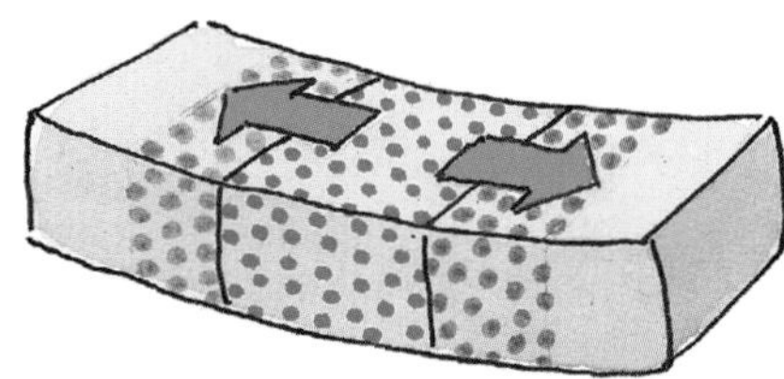

3. The altered daughter cell — the cell in the middle — similarly induces changes in its neighbors.

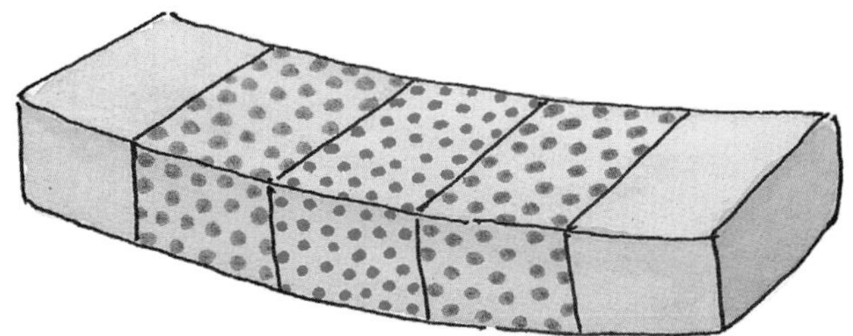

4. When these divide, they, in turn, induce their neighbors to create different daughter cells — and so on.

Question.

Why is the position of a cell in a tissue important in differentiation?

Answer...

Where a cell "sits," or ends up after migration determines which other cells it can contact, both to send or receive information. These signals will trigger the production of various components necessary in a differentiated cell or tissue.

ODE TO GROWTH

Like an awl-tip breaking ice
The green shoot cleaves the gray spring air.
The young boy finds his school-pants cuffs
Too high above his shoes when fall returns.
The penciled marks on the bathroom doorframe climb.
The cells replicate,
Somatotropin
Comes bubbling down the bloodstream, a busybody
With instructions for the fingernails,
another set for the epiderm,
a third for the budding mammae.
All hot from the hypothalamus
And admitting of no editing,
Lest dwarves result, or cretins, or neoplasms.
In spineless crustaceans
The machinery of molting is controlled
By phasing signals from nervous ganglia
Located, often, in the eyestalks, where these exist.
In plants
A family of auxins
Shuttling up and down
Inhibit or encourage cell elongation
As eventual shapeliness demands,
And veto lateral budding while apical growth proceeds,
And even determine abcission—
The falling of leaves.
For death and surrender
Are part of growth's package....

John Updike, *Facing Nature Poems,* 1986

Growth: Plants vs. Animals

A fertilized human egg cell differentiates in the stable, protected environment of the mother's uterus. Under normal circumstances, outside influences on this developing organism are exerted only through temperature and chemicals in the mother's blood. Cell differentiation in plants, on the other hand, is subject to a number of constantly changing external influences. Plant cells differentiate in response to signals generated by gravity, the position of the Sun, the length of the day, the temperature, and the amount of moisture, to name a few. In order to succeed and multiply, plants must be able to sense these signals and develop — in fact, shape themselves — in response to them. The various parts of a plant must communicate with one another so that cell growth and differentiation can proceed in response to environmental change. All this flexibility is programmed in the plant's genes as is, of course, its habit of breeding true: an acorn does, after all, become an oak tree.

Question.

Plants have far fewer signaling hormones than animals do, and the hormones move through the plant tissues at a much slower rate. What characteristics of plants and animals might explain these differences?

Answer...

Both plants and animals have evolved signaling systems that are advantageous to their specific survival and reproductive needs. Everything in animals must respond more rapidly than in plants. Animals move around, while plants are stationary. An animal needs to be much more rapidly perceptive and responsive than a plant does in order to find food and a mate.

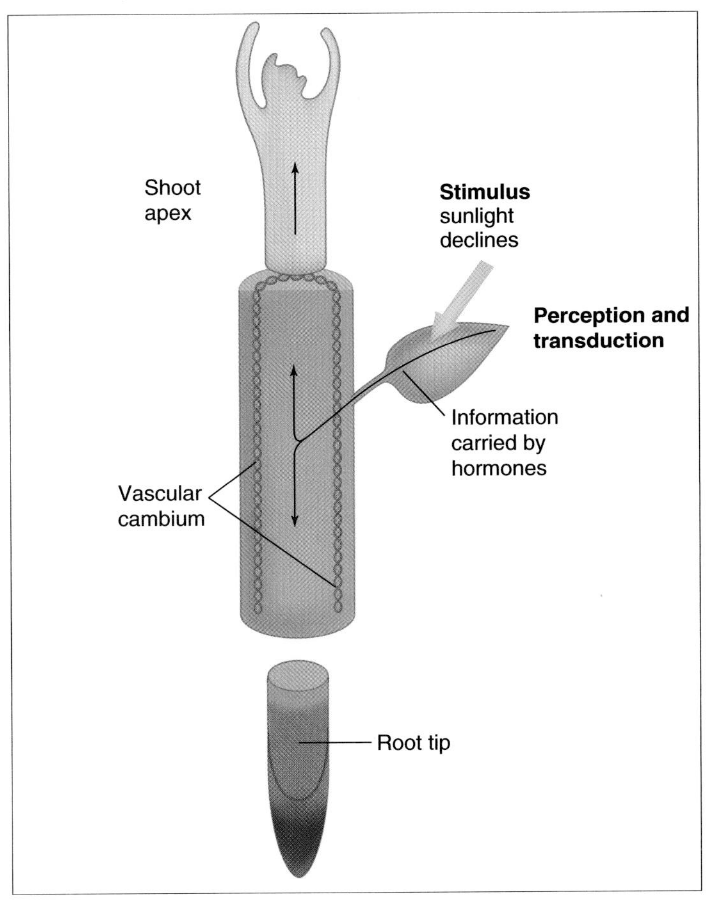

The leaf is in charge

Here a plant's leaf acts as the orchestrator of changes in the rate and kind of cell differentiation. Sensing a decline in the amount of sunlight as autumn approaches, the leaf generates a hormonal message that is picked up by receptors in the cell membrane of other plant cells. The hormones—auxins—move slowly (only about 11 mm per hour) and cause the growing shoot tip, which had been producing leaves, to produce bud scales instead. The vascular cambium (which ordinarily produces the plant's structural parts) stops making new cells. Water- and mineral-conducting cells become dormant, and root growth slows, but does not stop completely.

Ending dormancy

These two groups of seeds show how environmental conditions work to end dormancy and signal seeds to germinate. The seeds start out containing a chemical that inhibits cell growth. This chemical is washed out by rain or destroyed by cold winter temperatures. Once this happens, light signals the start of cell division. The seeds respond only to light in the red part of the solar spectrum—and respond more rapidly as Spring approaches and the Sun gets higher in the sky. Thus, they will germinate only under favorable conditions: when Winter is ending and not when they are too deeply buried in soil.

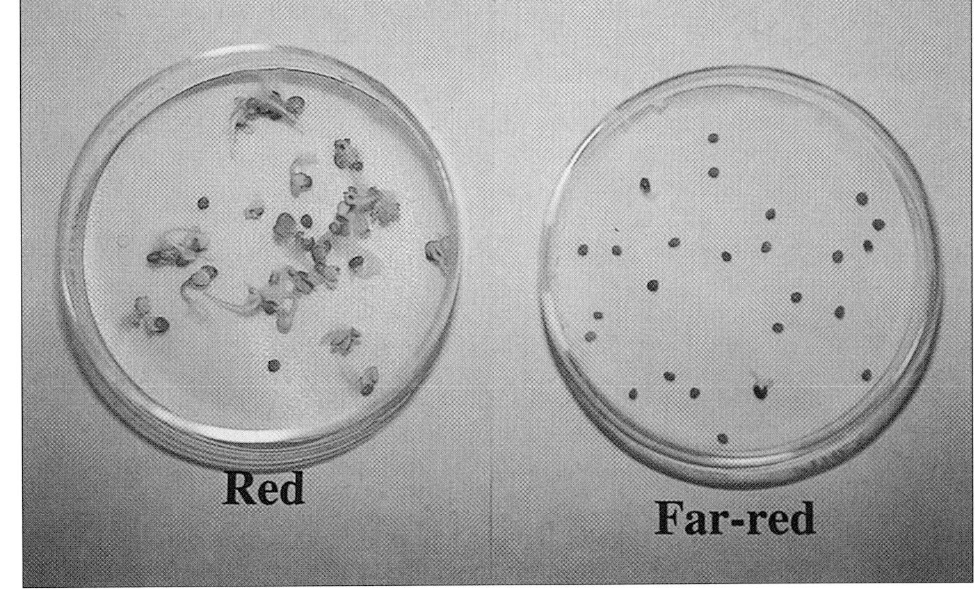

7.12 Positional Signals

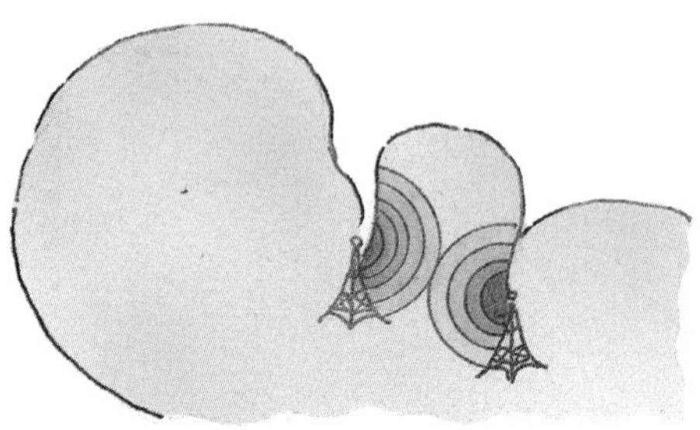

Morphogen gradients develop a limb bud from the trunk of an embryo.

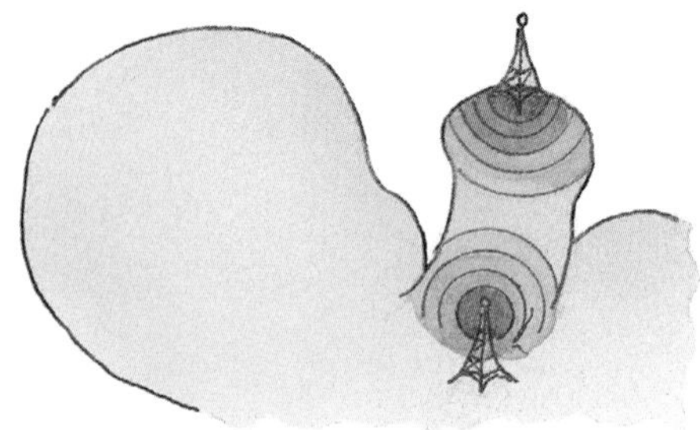

Signal gradients, which originate from key points, command the cells to form the top, bottom, and near and far ends of the limb.

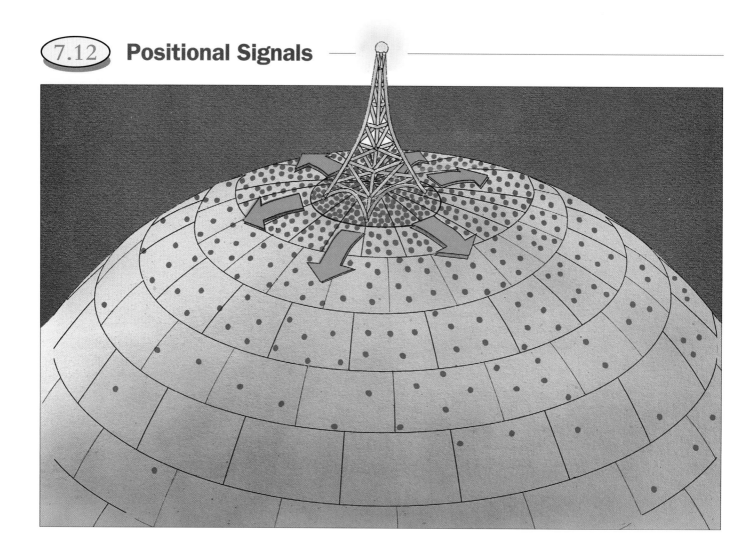

When What You Do Depends on Where You Are

In studying how primordial buds of tissue in embryos are sculpted into recognizable body parts, scientists have come to appreciate a special class of molecules called morphogens (the word means "makers of shape"). Morphogens, which are usually proteins, do not act locally in cell-to-cell contacts; instead, they affect all cells over an area of perhaps a square millimeter or two. As their concentration varies, so does their effect on cells.

Imagine a radio tower broadcasting from its position in a single cell. The message received by neighboring cells depends on their distance from the transmitter. Developing cells anywhere within range of the broadcast will read the signal differently, depending on its strength. Nearby cells, since they receive a stronger signal, act in one way; cells further away, since they receive a weaker signal, respond differently, and so on. Beyond the tower's range there is no response.

The resulting morphogen gradients display an impressive versatility. A single family of genes makes morphogens that direct the development of limbs, sex organs, and the brain.

7.13 Cell Death

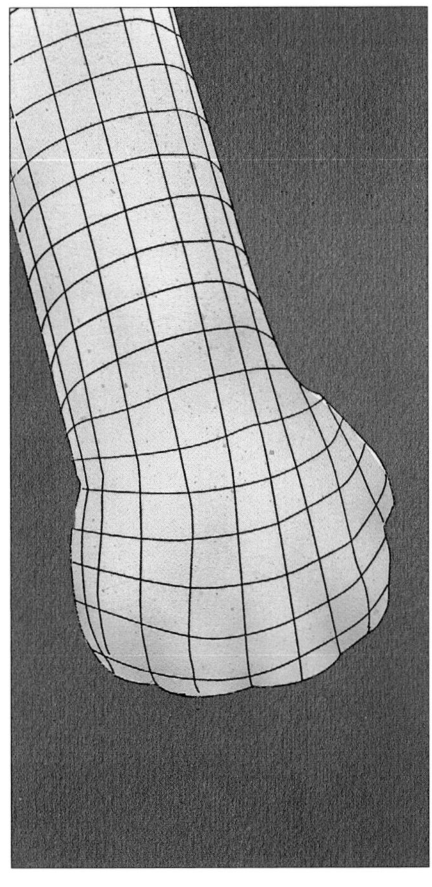

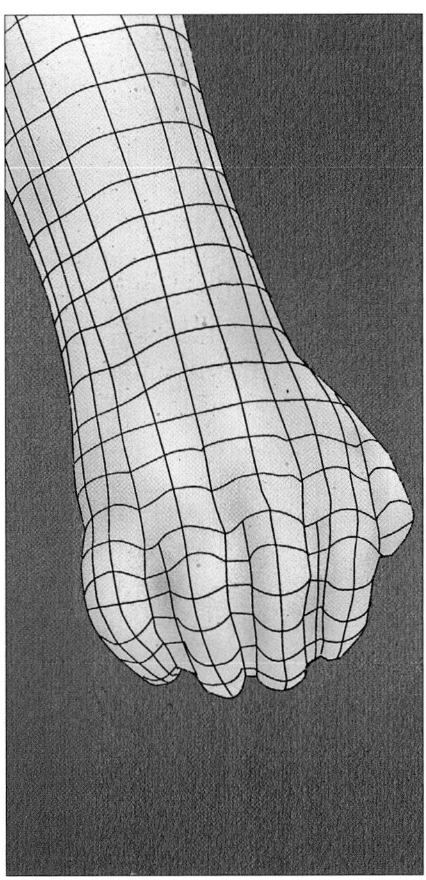

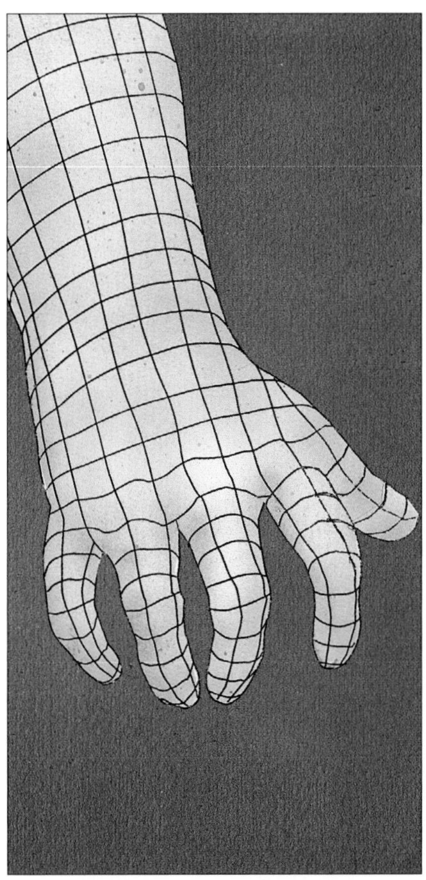

Dying to Contribute

We think of death as the end of life. But cell death plays an essential role in the creation of living bodies — and self-destruction is part of the program.

While the brain is being built, neurons are produced in profusion — far more of them than the brain will ever use. This overproduction of brain cells, called "blooming," is followed by a massive "pruning" in adolescence. Those brain cells that have made only weak connections to other neurons — or no connections at all — simply die. Some parts of the nervous system lose 85 percent of their neurons in this way! But don't worry; the many billions that are left have been organized and connected by your early experience and are more than ample to see you through adult life.

Your hands also owe their final form to the programmed death of cells. Cells are signaled to create not five fingers but four gaps — the spaces between the fingers. Cells residing in these gaps take their own lives, freeing up the fingers in the embryo.

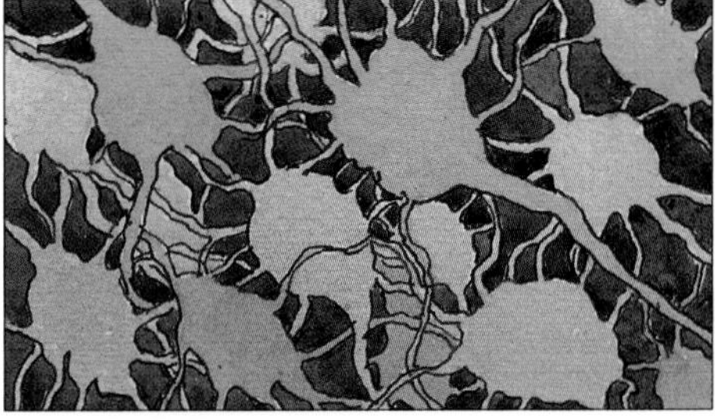

BLOOMING: The brain overproduces neurons until adolescence.

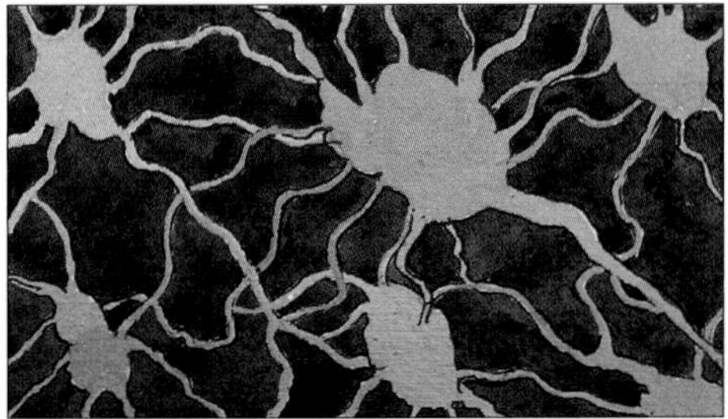

PRUNING: Unconnected neurons self-destruct.

Cell Death and Aging

What makes a body age? Why don't our cells just keep regenerating themselves forever? One of several possible answers seems to lie, like so much else, in our chromosomes. Chromosomes have at their ends special repeat sequences of DNA, called telomeres. During embryogenesis, cells divide many times. Embryonic cells are able to maintain intact telomeres, but adult cells lose this ability. Once that happens, the telomeres shorten with each cell division. You might think of the telomere repeats as tickets. Each cell division costs one ticket. At some point, the tickets are used up — the telomeres are too short to permit further cell division, and the cell ages and dies.

Embryonic cells produce an enzyme, telomerase, that maintains the protective chromosome ends. But telomerase production is repressed (the switch is turned off) in adult cells. Interestingly, and unfortunately, cancer cells have the ability to produce telomerase, which is probably what allows them to keep growing indefinitely. Current cancer research is targeting ways to block this ability.

Cloning

Dolly is undoubtedly the world's most famous sheep. She has the distinction of being the first mammal cloned from an adult cell, and the first newborn with six-year-old telomeres. (The few earlier successful animal clones originated from undifferentiated embryonic cells.)

A clone is a genetically identical copy of a single individual. Clones are produced in nature by asexual reproduction. (Unicellular organisms reproduce asexually, as do some multi- cellular organisms — like aphids and starfish. Many plants reproduce both sexually and asexually.) Recently, researchers have been working on the cloning of more complex organisms, such as mice, cows, monkeys, and some endangered wild species.

Cloning involves the transfer of DNA from the nucleus of a donor cell (in Dolly's case, an udder cell from a six-year-old pregnant ewe) into a recipient cell — an unfertilized egg that has had its chromosomes removed by micropipette suction. The donor cell (with its DNA-containing nucleus) is stimulated with pulses of electricity to fuse with the chromosome-less egg cell.

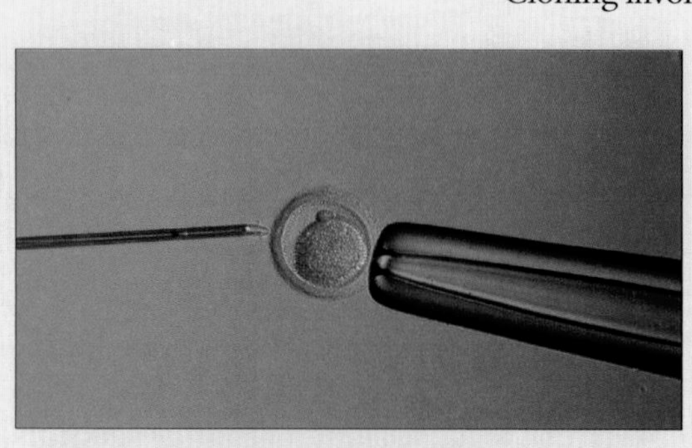

A micropipette, right, holds an egg, from which the DNA has been removed. The microneedle on the left injects new DNA. The egg with its new DNA will be implanted in the oviduct of a surrogate mother.

Under appropriate conditions, embryogenesis begins, and the developing embryo is then implanted in a host mother (see the illustration at right). Those conditions are clearly not easy to attain: Ian Wilmut and colleagues at the Roslin Institute in Scotland fused 277 donor cells with eggs to produce just one viable animal.

For an adult donor cell, one of the required conditions is nuclear reprogramming. That is, the cell must be induced to *de*differentiate — to free all its DNA from the restraints of differentiation. You recall that genes, probably thousands of them, switch on and off at all stages of development. The donor cell's nucleus has to be reprogrammed to begin

again at the beginning. Roslin Institute researchers achieved this by starving the cultured donor cells of almost all nutrients until they had ceased normal growth and cell division and had entered a quiescent state, in which few or no genes were thought to have been switched on. This "switching off" may have initiated nuclear reprogramming.

Dolly's offspring—a single lamb (Bonnie) born in 1998, and triplets born in 1999—were conceived naturally. Bonnie's telomeres are normal. All the offspring, of course, got half their DNA from their father.

Since Dolly was cloned, successes with a number of other mammals have been reported. In 2000, investigators at Advanced Cell Technology in Worcester, Massachusetts, reported that they had successfully cloned six calves from adult cow cells. The calves' telomeres, rather than being shortened, are actually longer than those of uncloned calves of the same age. The researchers don't know why. It may be due to some critical unknown difference between cows and sheep, or to the use of a different cloning technique from that used to produce Dolly.

Dolly and Bonnie

Dolly was cloned from what was essentially a six-year-old cell. After her existence was revealed in 1997, scientists were very eager to learn whether her cells were "young-looking" or "old-looking." In fact, Dolly's cells have the shortened telomeres typical of cells of an older sheep. It remains to be seen how that will affect her life expectancy. Will she age prematurely? Or will she live out the normal sheep life span of about thirteen years?

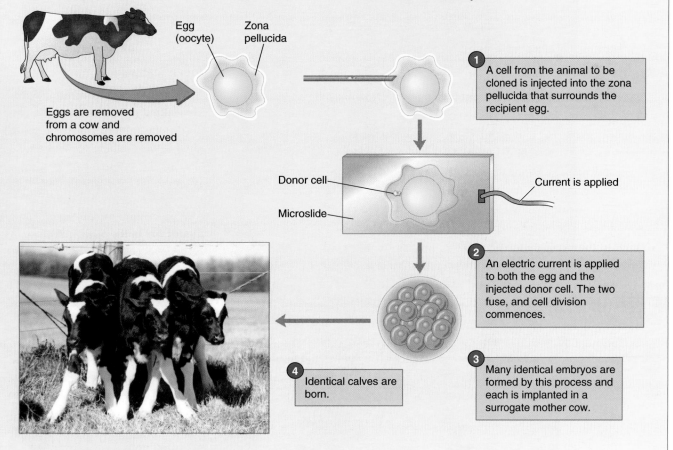

Egg (oocyte)

Zona pellucida

Eggs are removed from a cow and chromosomes are removed

1. A cell from the animal to be cloned is injected into the zona pellucida that surrounds the recipient egg.

Donor cell

Microslide

Current is applied

2. An electric current is applied to both the egg and the injected donor cell. The two fuse, and cell division commences.

3. Many identical embryos are formed by this process and each is implanted in a surrogate mother cow.

4. Identical calves are born.

7.14 Organizing a Body — Part V

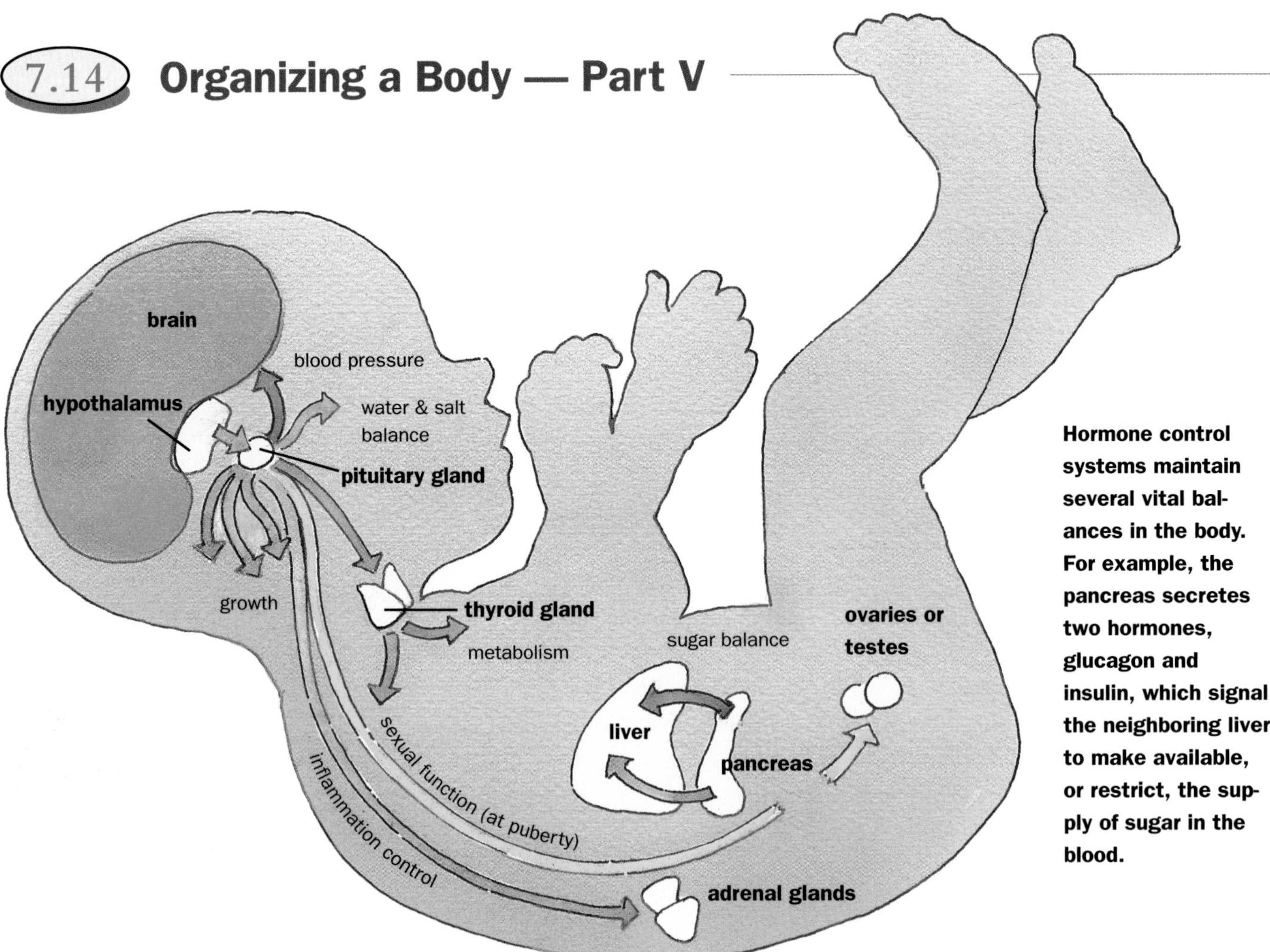

Hormone control systems maintain several vital balances in the body. For example, the pancreas secretes two hormones, glucagon and insulin, which signal the neighboring liver to make available, or restrict, the supply of sugar in the blood.

A Chain of Command

An embryo's development is shepherded by an increasing hierarchy of commands. In the early stages, cells are guided by local interactions; chemical signals pass between only a few cells at a time. Different regions of the body develop semi-independently. Gradually, central channels of communication develop. As cells become more specialized, they become dependent on other cells to do things for them. Cells packed close together in tissues can no longer snatch their food and building materials from their environment the way independent bacteria do. Tissue cells need an elaborate network of blood vessels to bring things to them. They act increasingly in obedience to distant signals traveling in the bloodstream or on nerve pathways.

For example, the thyroid gland makes a hormone that enters the blood and is picked up by cell receptors on target cells, causing their metabolism to speed up. Nerve cells develop long extensions of themselves that reach muscle cells and stimulate them to contract or relax. The brain gradually takes over as master controller of the nerves and glands, first regulating automatic functions such as heartbeat and blood pressure, then becoming sensitive to sensory signals such as sounds and feelings.

The birth of a newborn infant completes a profound emergence: A quarter of a trillion individual cells, following their local rules, have become one unique being.

More Signals

Hormones, proteins secreted by specialized groups of cells (glands) located in different parts of the body, carry instructions to all the cells through the blood. A master gland, the pituitary, receives signals from the brain and, in response, sends out its own hormones that switch on hormone production in the other glands. As the blood levels of these hormones rise, they signal the pituitary to reduce its hormone output, thereby lowering, by negative feedback, their own level.

In later steps, the embryo yokes together the body's chemical signaling system — endocrine glands — and electrical signaling system — nerves and brain. As these links are formed, the embryo increasingly responds as a whole being, rather than as separate groups of cells.

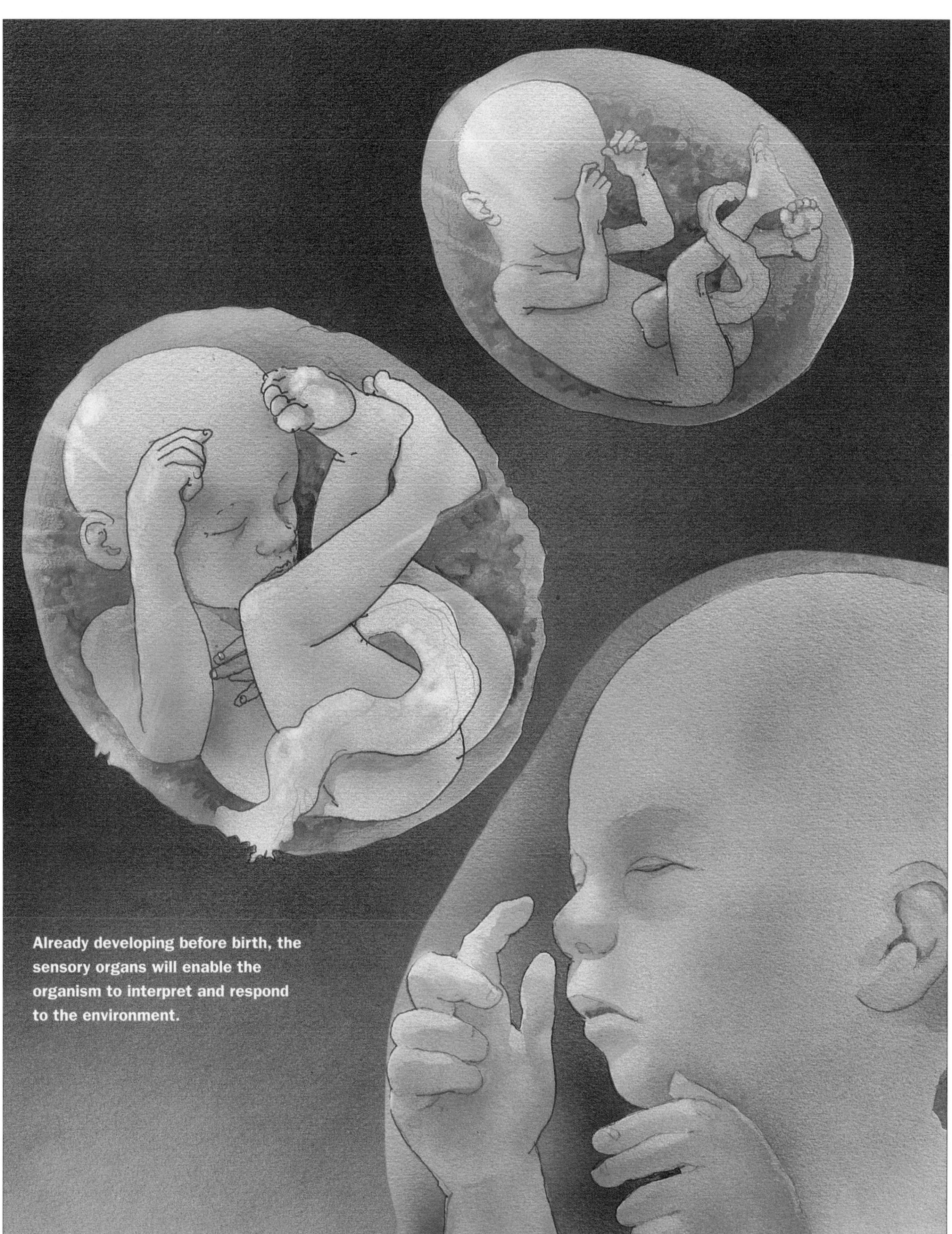

Already developing before birth, the sensory organs will enable the organism to interpret and respond to the environment.

Labor Management

The birth process is another example of efficiently programmed nerve and hormonal communication and feedback. Two types of messages course back and forth between the fetus and mother and between the mother's uterus and cervix and her brain: One type consists of nerve impulses that arise from the pressure the fetus exerts on the mother's cervix. The other type of message is delivered by hormones produced by both the fetus and the mother. These messages target the uterus — the muscular organ that provides, through its contractions, the force that propels the baby into the world.

Changing hormone levels in the fetus and mother initiate the positive-feedback loop of labor. These changes signal cells in the mother's placenta to begin manufacturing hormone-like substances called prostaglandins, which cause the smooth muscle in her uterus to begin contracting. The contractions push the fetus toward the birth canal. This, in turn, increases pressure on the mother's cervix. The pressure sends nerve impulses to the mother's brain. These signals tell the hypothalamus to instruct the posterior pituitary gland to begin releasing another hormone, oxytocin. Oxytocin causes even stronger uterine contractions. In this positive feedback loop, the increased contractions produce more pressure on the mother's cervix, sending more nerve signals to the hypothalamus, causing more oxytocin to be released. The positive feedback leads to ever increasing contractions. Oxytocin and prostaglandins work together to intensify contractions that eventually result in birth. Once the

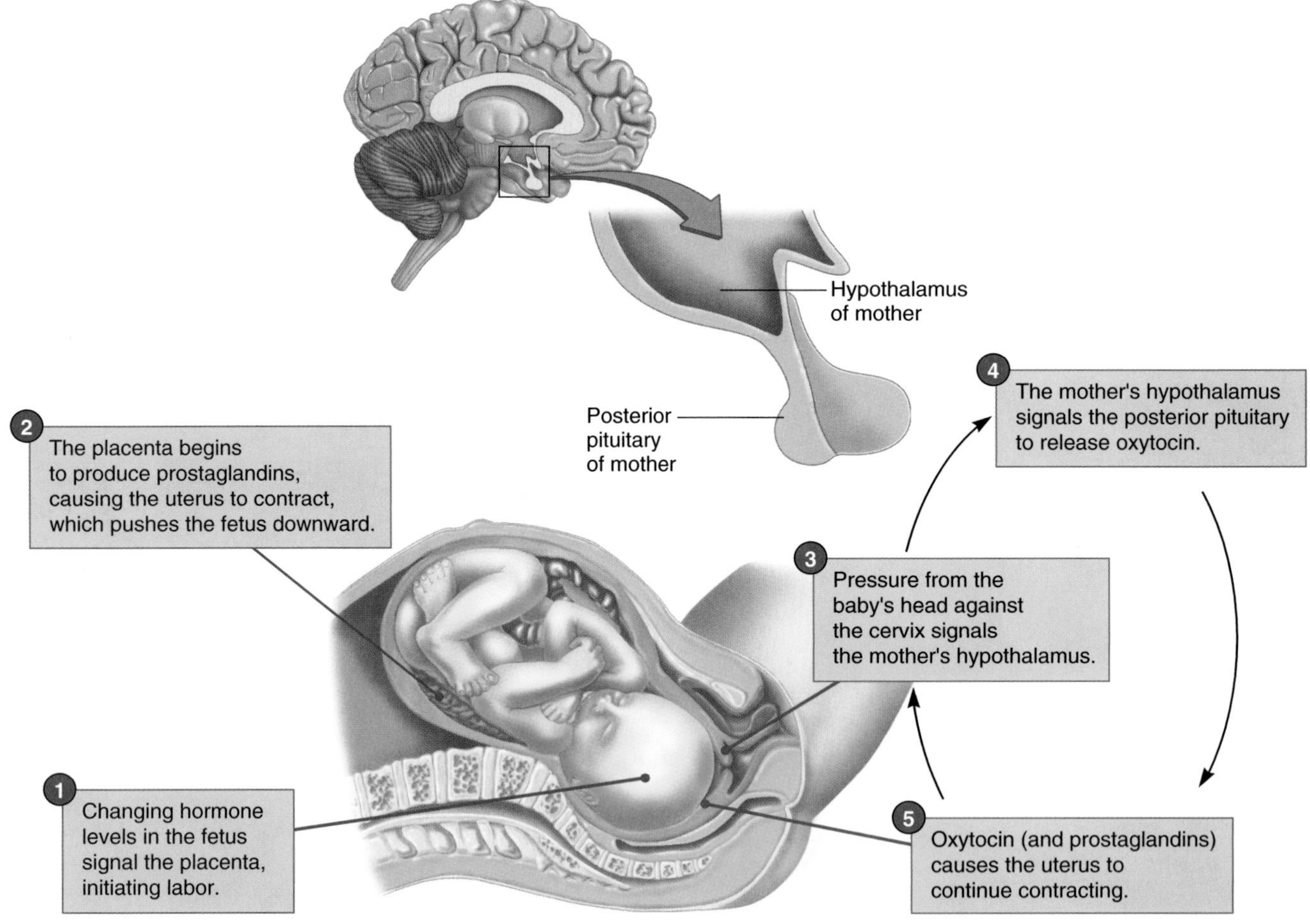

Hypothalamus of mother
Posterior pituitary of mother
1 Changing hormone levels in the fetus signal the placenta, initiating labor.
2 The placenta begins to produce prostaglandins, causing the uterus to contract, which pushes the fetus downward.
3 Pressure from the baby's head against the cervix signals the mother's hypothalamus.
4 The mother's hypothalamus signals the posterior pituitary to release oxytocin.
5 Oxytocin (and prostaglandins) causes the uterus to continue contracting.

baby is born, the pressure on the cervix stops, and the levels of the hormones that have been stimulating contractions begin to drop. But the contractions continue for a time, until the placenta, or afterbirth, is expelled.

Compared to the newborn child, the placenta seems like a by-product, or even a waste product, of birth. But the placenta has, for about nine months, been the site of a life-giving connection between mother and child that constitutes the most intimate physical relationship they will ever share. Here, the blood of mother and child come into such close proximity that the two systems are virtually one. All the nutrients and oxygen the fetus needs are transferred from the mother's blood through the placenta. All the wastes the fetus must dispose of are transferred to the mother's blood in the placenta. This relationship abruptly ends at birth, and the placenta is expelled.

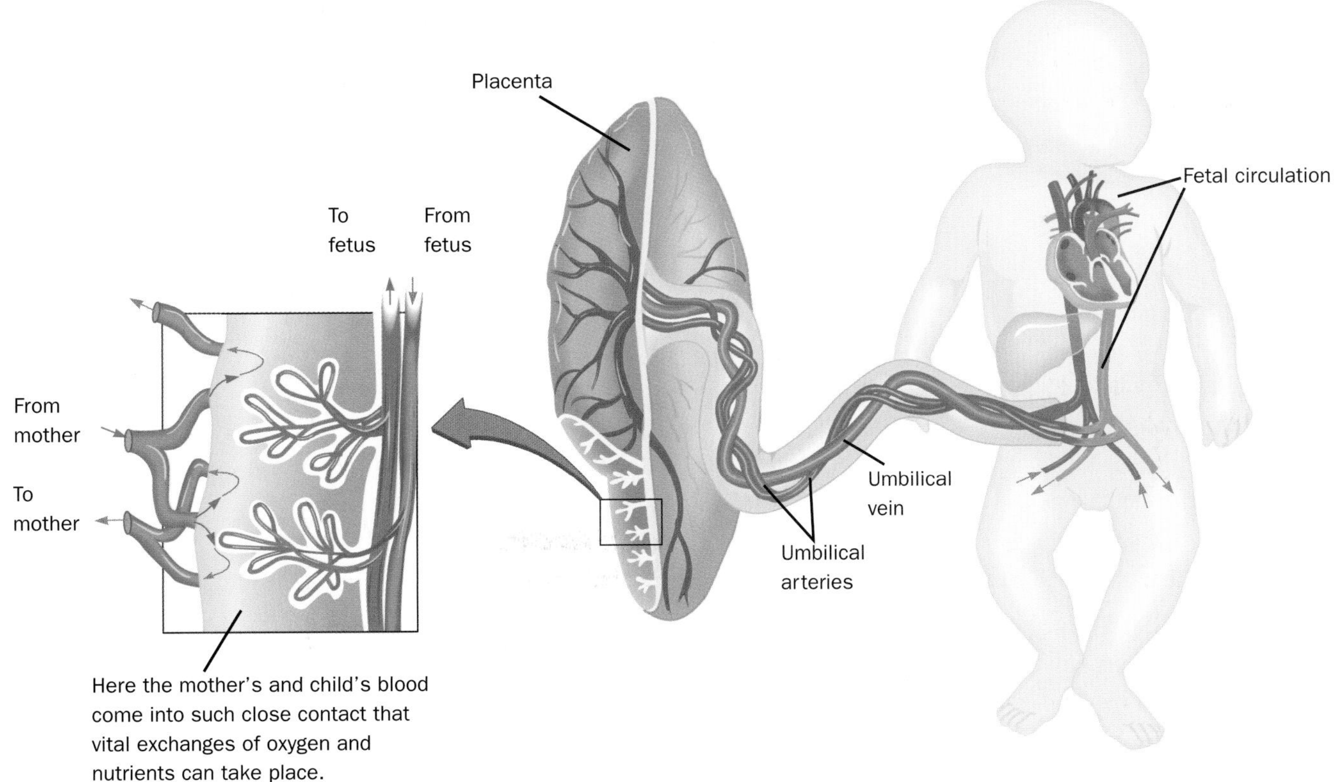
Placenta
Fetal circulation
To fetus
From fetus
From mother
To mother
Umbilical vein
Umbilical arteries
Here the mother's and child's blood come into such close contact that vital exchanges of oxygen and nutrients can take place.

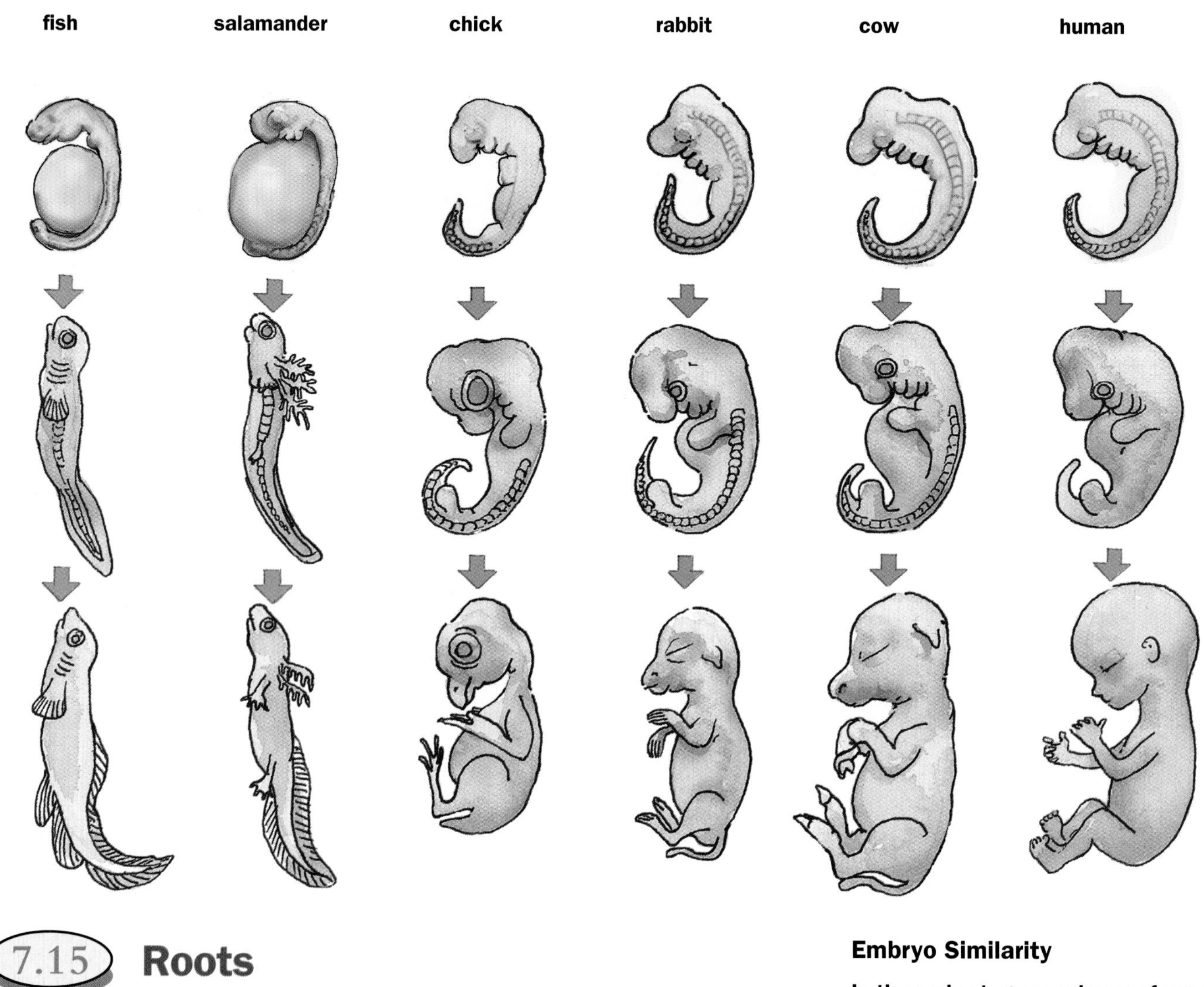

Embryo Similarity

In the early stages, embryos of most vertebrates are remarkably similar.

7.15 Roots

Another Look at Nature's Unity

Years ago, scientists noticed with surprise that widely different organisms may appear remarkably similar during certain early stages of their development. Experts in embryology couldn't distinguish, for instance, between the early embryos of a bird and a human. Why should embryos so closely resemble each other?

This question has led to an appreciation of evolution as a master tinkerer. A tinkerer does not start from scratch each time, but instead makes use of suitable old bits and pieces to build new things. Once a program works well for building a fish, keep it and use it later as a foundation for building a person. It's easier to get rid of a tail and gills than it is to devise a whole new construction scheme.

Over and over in our study of genes and proteins, we discover the effects of tinkering during evolution. The Hox genes of fruitflies are very similar in nucleotide sequence to the genes that perform the same body-shaping function in many other animals, although each species has its own unique sequence. Clearly, these varied genes evolved from a single gene in a remote common ancestor.

At the right are some other examples of tinkering — structures that originally served one purpose and were later co-opted for a completely different purpose.

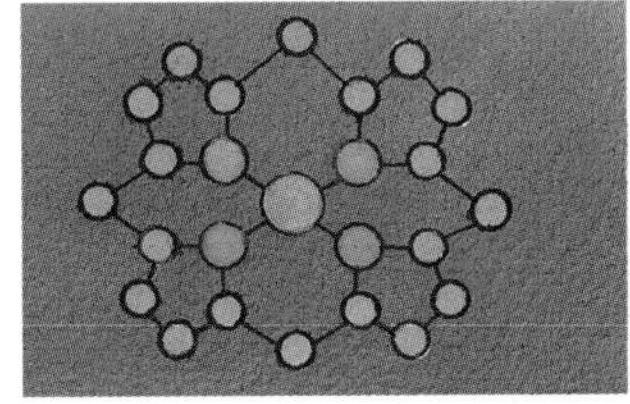
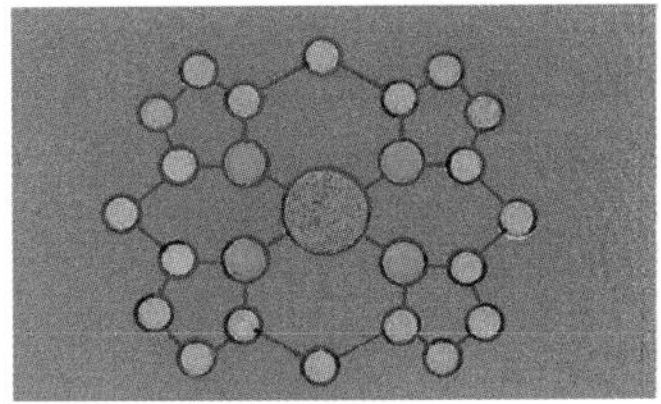

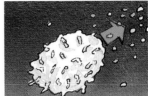

From Photosynthesis to Oxygen Transport

The molecule that captures sunlight in plants is remarkably similar to the molecule in hemoglobin that carries oxygen in the animal's bloodstream.

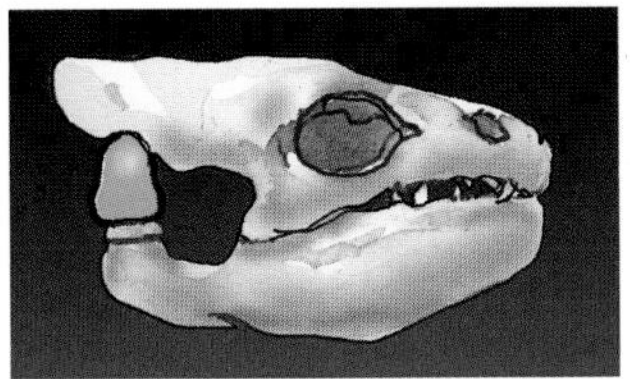
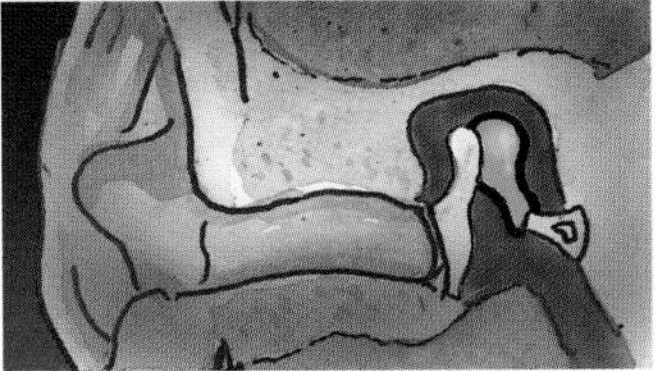

From Chewing to Hearing

A bone that was once part of a reptile's jaw has evolved into a device in the ear for transmitting sound waves.

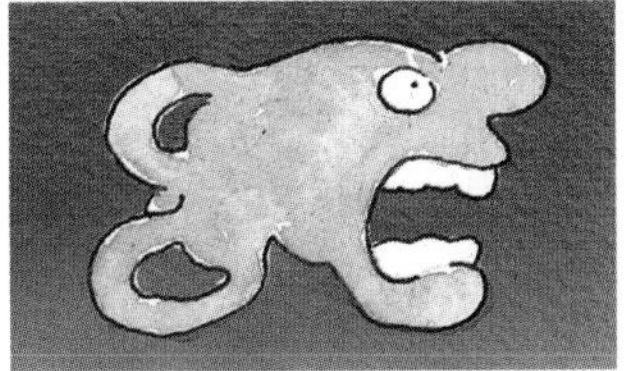
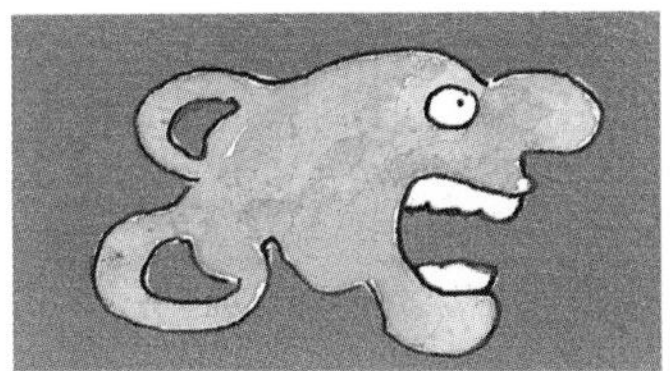

From Digestion to Blood Clotting

The final folded shape of certain digestive enzymes is nearly identical to that of proteins involved in blood clotting.

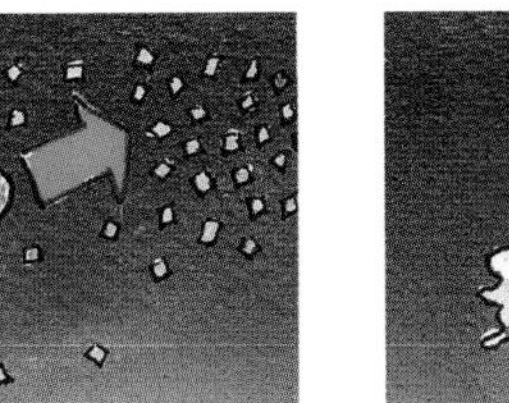

From Foraging to Defending

Like bacteria drawn to food signals, our white blood cells are attracted by chemotaxis to sites of inflammation.

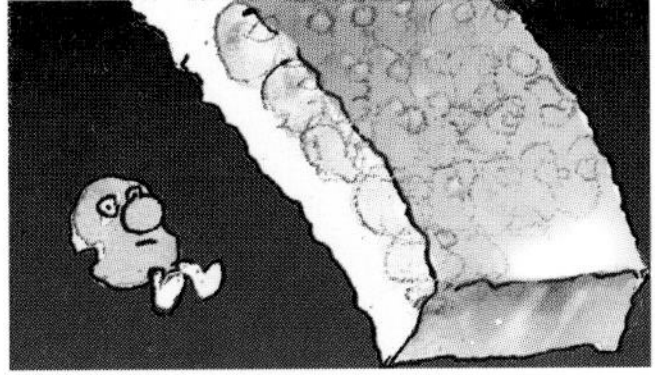
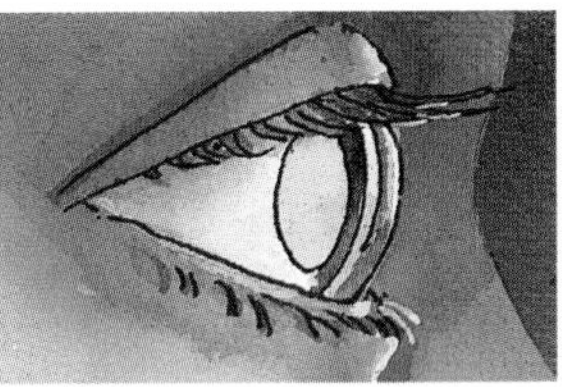

From Catalyzing to Seeing

An enzyme involved in energy production can, in large numbers, give transparency to a cell. From this, the lens of our eye evolved.

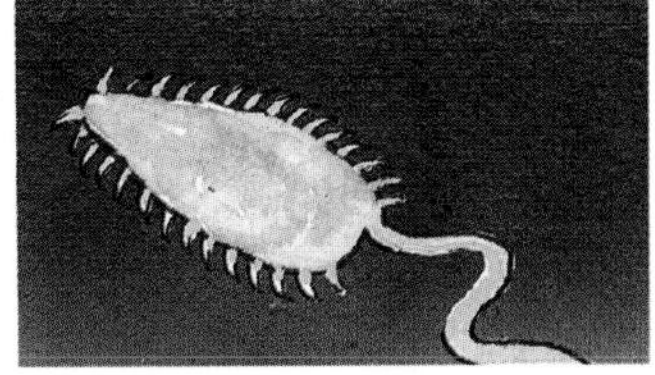
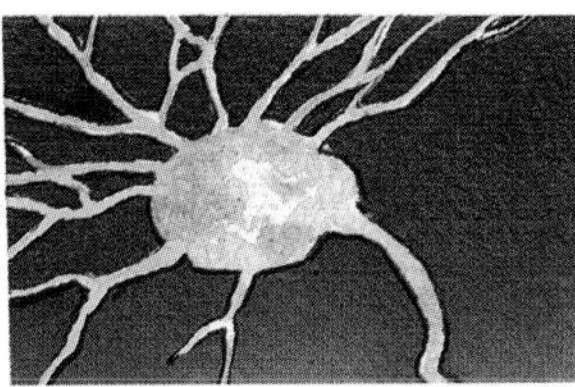

From Locomotion to Housekeeping

The cilia that propel many single cell creatures also serve to protect our lungs by sweeping up dirt particles.

Doing *Science*

McKeown, C., J. Austin, and V. Praitis, 1998.
sma-1 encodes a β_H-spectrin homolog required for
Caenorhabditis elegans morphogenesis.
Development 125: 2087–2098.

A picket fence of filaments provides a lattice on which cells can grow, but it also plays another important role in development: defining the shapes of individual cells, which can, in turn, affect the shape of an entire organ or animal. In some elongated animals, individual cells lengthen at a particular time during development. Judith Austin of the University of Chicago and her colleagues studied this process, called morphogenesis, in the worm *Caenorhabditis elegans.*

The whole lengthening process takes a couple of hours, during which the worm more than triples its length without adding any new cells. Instead, the cells it already has change their shape. During this period, the filaments attached to the inner surfaces of the worm's cells are reorganized into parallel bundles, and the cells become elongated, squeezed from a cuboidal shape to a longer, flatter, rectangular shape. In a mutant worm whose reorganizer protein is defective (the *sma-1* mutant), the filaments reorganize at a much slower rate. This causes the adult worm to end up shorter and fatter than a normal worm, as you can see in the photos at right.

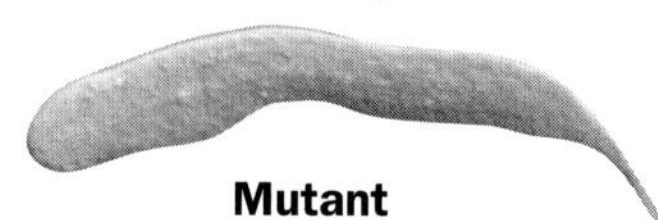

Some of the Things You Learned About in Chapter 7

Questions About the Ideas in Chapter 7

1. Our bodies are made up of about 250 different kinds of cells. Are there 250 different sets of DNA in the human body?
2. Which is "smarter," an ant or an ant colony? Why?

3. Pattern develops by establishing difference. What does this mean?

4. How can one gene be a switch for another?

5. How does one cell communicate with another?

6. You had more brain cells when you were born than you have now. Why might that fact be advantageous?

7. As cells in an embryo grow and become different, they change in two fundamental ways: by lineage and by contact. How do these two ways differ?

8. Why do the embryos of widely different animals look so similar?

9. Cellular self-destruction is an important process in the differentiation and development of a multicellular organism. Isn't it wasteful of materials and energy to produce so many cells only to lose them later?

10. Why is the position of a cell in a tissue important in differentiation?

11. How is a cancer cell in a person's body like an animal introduced into a new environment where there are no predators?

References and Great Reading

Bonner, J. T. 1955. *Cells and Society*. Princeton: Princeton University Press.

Dawkins, R. 1987. *The Blind Watchmaker.* New York: W. W. Norton.

Gunning, B. E. S., and M. W. Steer, 1996. *Plant Cell Biology: Structure and Function.* Sudbury, MA: Jones and Bartlett Publishers.

Hölldobler, B. and E.O. Wilson, 1990. *The Ants.* Cambridge: Harvard University Press.

Mauseth, J. D. 1999. *Botany: An Introduction to Plant Biology.* Sudbury, MA: Jones and Bartlett Publishers.

McKeown, C., V. Praetis, and J. Austin, 1998. *sma-1* encodes a β_H-spectrin homolog required for *Caenorhabditis elegans* morphogenesis. *Development* 125: 2087–2098.

Rukeyser, M. 1968. *The Speed of Darkness.* New York: Macmillan.

Shakespeare, W. 1968. Sonnet 148 from *The Complete Works,* Ed. G. B. Harrison. New York: Harcourt, Brace, and World, Inc.

Shih, G., and R. Kessel, 1982. *Living Images: Biological Microstructures Revealed by Scanning Electron Microscopy.* Boston: Science Books International.

Updike, J. 1986. "Ode to Growth" from *Facing Nature: Poems.* New York: Alfred A. Knopf.

Winston, M.L. 1987. *The Biology of the Honey Bee.* Cambridge: Harvard University Press.

Wright, R. 1988. *Three Scientists and Their Gods.* New York: Times Books.

For more questions and links to web resources, go to
WEB
Connection
www.jbpub.com/connections

CHAPTER 8

EVOLUTION

The Pattern for Creation

FROM AN EVOLUTIONARY VIEWPOINT, LIFE IS A RIVER OF INFORMATION. INFORMATION arises, branches into countless tributaries, and pools in endless combinations. It flows, generation to generation, through the bodies of living creatures, shaping and organizing them along the way. The success of each living thing determines the future of the information it carries. The information is sorted and sifted, with the most useful preserved as it moves downstream. This flow is evolution.

Evolution's principal mechanism, natural selection, is not one process but two: chance and selection. Operating in tandem, chance creates random changes in the information (gene) pool of a population, while selection non-randomly keeps much of what "works" (that is, what contributes to survival and the production of offspring) and eliminates much of what doesn't. Nature generates changes in information; changes in information alter life forms; life forms interact with their environment; the environment selects the changes most likely to help the life form survive. Thus, successful changes persevere and are improved upon, which explains why the creatures around us seem so remarkably adapted to their environments. They, and we, are the success stories — at least so far. For, although we are momentary winners, we inherit no guarantees about the future. Of all life forms that ever existed, over 99 percent of species have died out, and we too are only a work in progress, on the way to change or extinction.

Chance and selection are fundamental to any creative act. Chance generates novelty. Selection chooses those innovations that fit existing conditions. Operating together, chance and selection can produce results so remarkably well-suited to the environment that they give every appearance of having been designed in advance. But while evolution tends toward greater complexity, its mechanism is one that can create without a plan.

8.1 Evolution of a New World View

An Ancient Earth

James Hutton (1726–1797) pioneered the science of geology. He hypothesized that the Earth is very much older than the 6000 years allotted by Christian dogma. Moreover, it is regularly subject to *slow,* not just catastrophic, erosion and sedimentation, as well as to periodic earthquakes and volcanic upheaval — changes similar to those we see happening *now*.

Life Has Evolved from Simplicity to Complexity

Jean-Baptiste Lamarck (1744–1829) theorized that living creatures possessed a built-in drive to become increasingly complex — their efforts culminating in humans.

Familiar-looking Fossils

Fossils often resemble present-day creatures. New life forms don't arise from thin air; there must be connection, progression, transformation.

Selective Breeding

Animal and plant breeders have demonstrated that life forms are not stable and unchanging. By selecting breeding stock carefully, they can produce change readily.

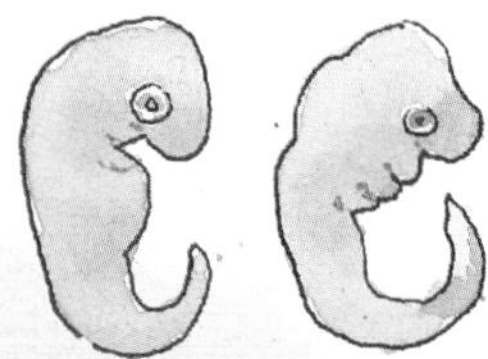

Look-alike Embryos

Embryos of fish, amphibians, reptiles, birds, and mammals — as noted in *Community* — are very similar in their early stages, suggesting that these organisms follow similar developmental patterns and share a common ancestor.

Charles Darwin, 1809–1882

Common Body Plan

Modern life forms share a common body plan. If an organism possesses a rudimentary body part, such as tiny, useless wings on an insect, it suggests that the organism's ancestors had a more useful version of that part.

Similarity of Geographically Isolated Creatures

Creatures living on different continents show related characteristics, suggesting that species migrated long ago and then developed in distinctly different ways.

The Struggle for Existence

Thomas Malthus (1766–1834) suggested that we humans produce more offspring than our food supply can sustain, which raised the possibility that competition for available resources causes creatures to adapt and change.

Before Darwin

Throughout most of our history, humans have viewed the Earth as a static creation of God (or gods), unchanging except for a few worldwide catastrophes like the Biblical flood. Surely the complexity and beauty and fitness of living things meant that they were divinely designed and produced. For centuries, the prevailing world view, first articulated by Aristotle, was that everything had a fixed place in a natural hierarchy — from the most complex Celestial Beings down to the simplest and lowliest of creatures. Fossils were thought to be the remains of earlier creatures produced and then wiped out by God. They had no connection, it seemed, with each other or with living creatures.

The early 1800s was a time of dramatic change: the rise of capitalism, secularism, science, skepticism, and the start of the Industrial Revolution. Our belief in a finite, Earth-centered universe eroded as evidence mounted that our planet is a minor player in a universe that reaches far into the unknown. Scientists began to question the assumption that supernatural causes governed natural events.

In rapid succession, a number of discoveries and realizations challenged the older ideas of permanence and the special creation of every natural phenomenon.

Darwin's Insights

Many able scientists had noted these developments but Charles Darwin (1809-1882) put them together into a coherent theory. Darwin wrote out his basic theory in 1844, but delayed publication out of caution because he saw it as too revolutionary. Alfred Russel Wallace (1823-1913) came up with the same idea independently in the 1850s and wrote Darwin about it. He and Darwin jointly published their theory in 1858. We may summarize Darwin's ideas as follows:

- Life had a *common beginning;* new forms of life branch off from earlier forms.
- There is *random variation* among individuals in populations and differences continue to arise by chance.
- The pressure of a constantly changing environment in which individuals must compete for survival results in *selection of favorable traits.* Traits that fit well with the environment survive and get passed on to offspring, while traits that do not fit perish.
- While each adaptation is small, *cumulative selection* of favorable traits leads over time to increasingly different forms of life and eventually to new species.

These, together, are evolution.

Further Evidence Supporting Natural Selection

Strengthening the Theory

Darwin's theory has turned out to be one of humankind's most brilliant leaps of the imagination. While Darwin amassed a prodigious body of evidence to support his theory of natural selection, it wasn't until the twentieth century that scientists made the discoveries that revealed the hidden mechanisms of the process. We summarize here some of the landmark research.

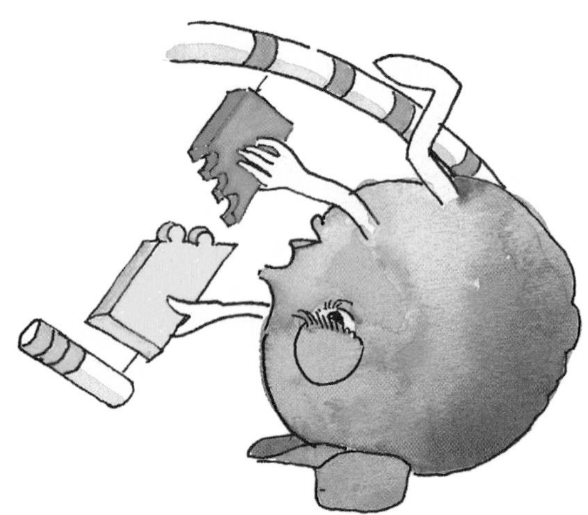

Chemical Mechanisms

Study of the nucleotides in DNA and the amino acids in proteins has shown that life's potential to create diversity is far greater than the appearance of organisms would suggest. Furthermore, scientists continue to discover more about how genes get altered, moved around, duplicated, and passed from one organism to another. In spite of all its diversity, molecular studies show that a remarkable degree of unity underlies all of life — powerful support for Darwin's view that all life shares a common origin. **(See page 320.)**

Chemical Relatedness

Scientists can determine how closely related two species are by comparing their anatomy and examining fossil remains. This method has been corroborated by comparing sequences of amino acids in the proteins and nucleotides in the DNA of different species. The more similar the sequences, the more closely related the species. **(See page 345.)**

Genetics

While convinced that the process of selection created diversity, Darwin had no idea *how* life forms actually changed. The physical basis of evolution became clear with increasing knowledge of genetics — the study of the nature of inheritance, sexual recombination, and mutation.

Natural Selection Observed

Recent studies of finches on a single island show that natural selection can happen very rapidly. In a large population, birds have various beak sizes. When major climatic changes affected the kinds of seeds available as food, birds with better adapted beaks out-reproduced their peers in a single generation. Scientists have observed similar adaptive changes in moth, fruitfly, and bacterial populations. **(See page 342.)**

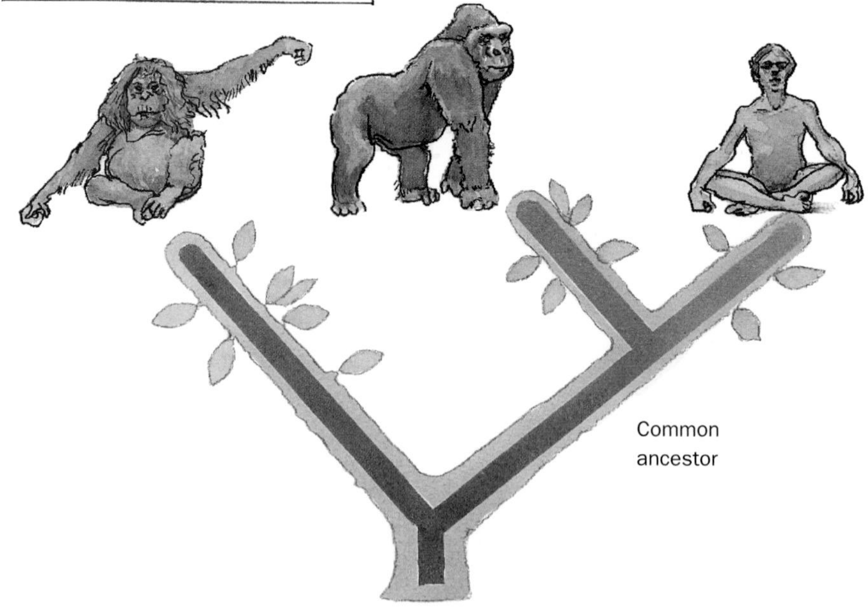

Population Genetics

Geneticists view populations of organisms of the same species as "pools" of genes. In the 1930s, scientists began to apply statistical methods to measure numbers of genes in certain populations and how they change over generations. They learned that species of organisms conserve in their gene pools a great capacity for diversity, which makes them enormously adaptable.

Geographic Separation

Naturalists have observed that if a small pool of genes becomes separated from a larger parent pool, for example, if a small flock of birds of one species migrates to an island, the genes in the small pool change relatively rapidly over the generations and eventually become the pool of a new species. **(See page 330.)**

8.3 The Origins of Life

Chains in a Chemical Soup

Before life began, nucleotides were plentiful on Earth. Some of them began linking up into chains of RNA, which were able to act as both template (a model for copying) and enzyme (a catalyst to facilitate copying).

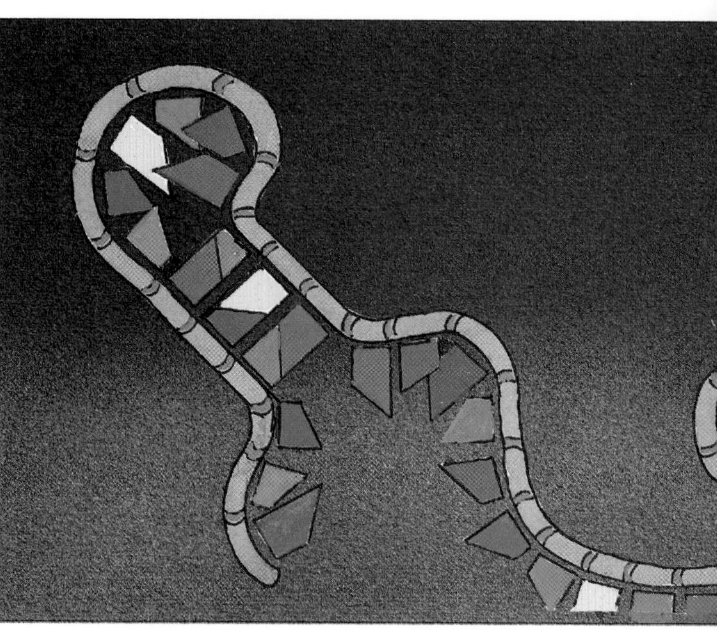

Template Meets Enzyme

The RNA chains assumed various shapes depending on the order of nucleotides. Occasionally, two similar chains met up, with one acting as enzyme and using the other as a template.

A Replication Production Line

Linking nucleotides together along the template's length, the enzyme created a complementary copy of the template (and thus of itself). Subsequent copying of the copy led to a chain identical to the first one — i.e., a replicated enzyme. Over time, millions of copies of the original enzyme would be churned out.

Mistakes Create Diversity

These "replicators" inevitably made copying mistakes which, when copied, resulted in differing RNA chains. Some of these variants were improvements on the original; others weren't. Those better able to compete for nucleotides multiplied more rapidly and became the dominant chemical species. Efficient copying had established itself as a means of information propagation and exchange.

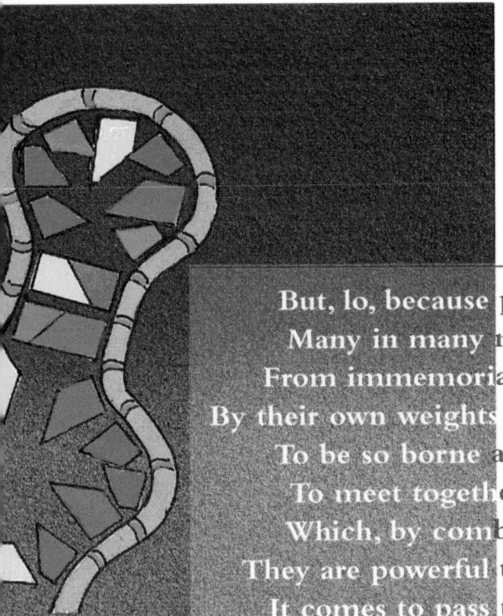

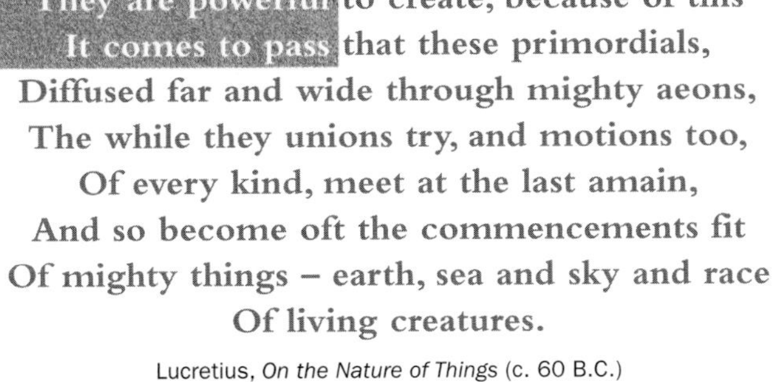

But, lo, because primordials of things,
Many in many modes, astir by blows
From immemorial aeons, in motion too
By their own weights have evermore been wont
To be so borne along and in all modes
To meet together and to try all sorts
Which, by combining one with other
They are powerful to create; because of this
It comes to pass that these primordials,
Diffused far and wide through mighty aeons,
The while they unions try, and motions too,
Of every kind, meet at the last amain,
And so become oft the commencements fit
Of mighty things – earth, sea and sky and race
Of living creatures.

Lucretius, *On the Nature of Things* (c. 60 B.C.)

Self-Replicating Chains

Evolution, broadly speaking, is the self-organizing process, not only of life, but of the universe itself. The ordering of matter into elementary particles and then into planets and stars was a necessary prelude to life on Earth. Life arose out of the conditions set by those earlier events. Of course, there exists no fossil record before the emergence of cells (microfossils of Archaebacteria), but we can make reasonable guesses about how life began.

Our story begins on the steaming, turbulent surface of early Earth, some 4 billion years ago, in locations like the hot springs that exist today, where we still find Archaea bacteria. These organisms are known to be very ancient, and they thrive in temperatures near the boiling point of water. Nucleotides and amino acids were probably plentiful before life appeared. Not only can these essential building blocks for DNA, RNA, and protein be made surprisingly easily and thus may have been assembled spontaneously right here on Earth, but they have been found in space dust and meteorites that are likely to have showered down on Earth over long stretches of time.

Long chains of phosphate molecules, called Poly P, found today in volcanic condensates and oceanic steam vents, may have provided the triphosphate ends of the early nucleotides, thereby giving these molecules the energy they needed to bond to other nucleotides. Once the first nucleotide chains — probably RNA — had been formed, some of them developed a remarkable ability: They could copy themselves. They were not alive in any sense, but simply floated about in the "prebiotic" soup, mindlessly self-replicating.

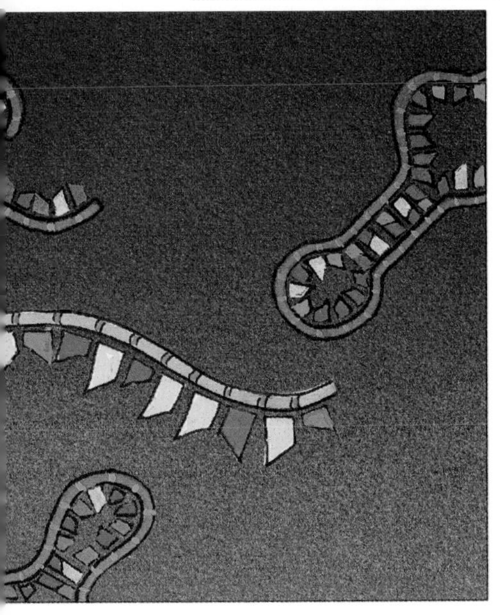

A self-replicating molecule needs at least two special properties: (1) It must be a *template* — a sequence of units (nucleotides) along which a complementary sequence of similar units can be ordered. (2) It must be an *enzyme,* able to pull free nucleotides from the surroundings and bond them together along the template. We now know that RNA, and only RNA, can perform *both* of these functions. Thus, the earliest self-copying system may have been a mix of similar RNA chains, some acting as enzymes, others as templates, able to perpetuate themselves endlessly.

How could such a rudimentary self-copying system *evolve* into something that could link amino acids together to make proteins and finally surround itself with a membrane to become a living cell? Quite simply by making mistakes. Occasional inevitable errors in copying, "nature's typos," produced a *variety* of RNA molecules, some better at copying than others. The efficient copiers prospered because they could interact with amino acids and begin to order them so as to produce more effective protein catalysts, transfer RNAs, ribosomes, and other cell parts.

WEB Connection
www.jbpub.com/connections

Evolutionary Time Line

4.5 Billion Years Ago | **4 Billion Years Ago**

8.4 A Brief History of Life

Accumulating Information — from Soup to Brains

During most of life's nearly four billion years on Earth, tiny unicellular and multicellular creatures living in water were hard at work setting the stage for the big, showy creatures that appeared only in the last half billion years of the drama. Frogs, dinosaurs, trees, birds, mammals, and all the rest arose as elaborations of developmental scenarios worked out by players too small to be seen.

A Condensing Cloud of Gas

Gravity compresses the particles in hot gases, forming our planet.

The Earth Cools

As the outermost crust cools, heat and gases escape through cracks and volcanoes.

Cell Division

Under pressure from its accumulating contents, a single compartment divides into two.

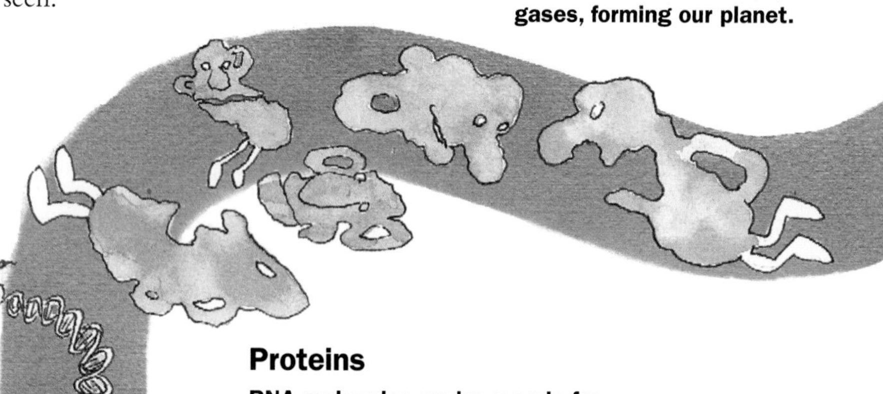

Proteins

RNA molecules evolve a code for amino acid sequences and begin to assemble crude proteins.

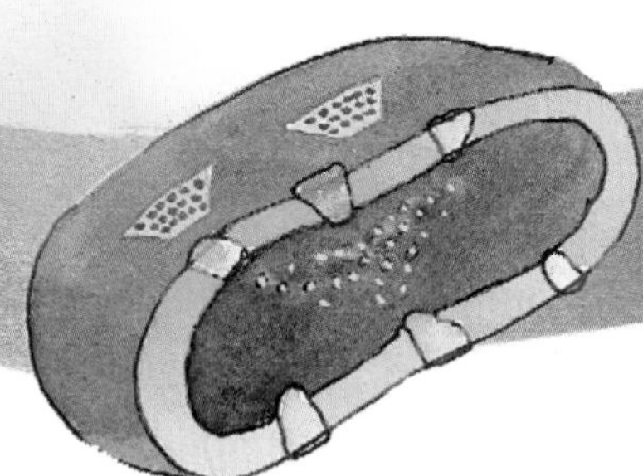

DNA

DNA takes over as the information carrier. RNA becomes the functional link between DNA and amino acids.

Fermentation

Sugar-converting enzymes make limited amounts of ATP, which supplies energy for cells' activities.

Photosynthesis

Some microorganisms "learn" to convert sunlight to sugar, thus tapping an inexhaustible energy source to make food.

3 Billion Years Ago

(Colors on this time line correspond to the colors on the path of evolution below.)

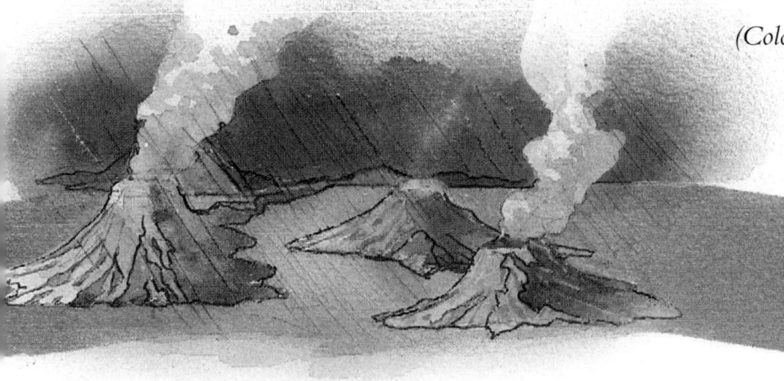

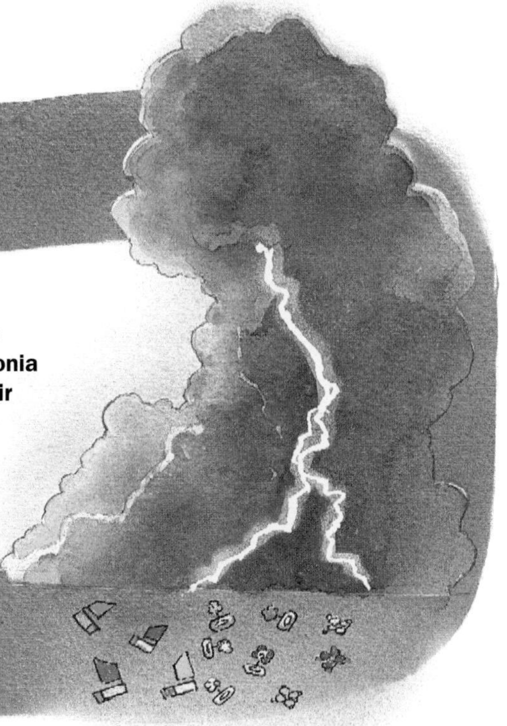

Water and Clay Deposits

Rain and steam create oceans and ponds. Evaporation produces a rich, soupy breeding ground.

Atmosphere

Hydrogen, nitrogen, carbon dioxide, and possibly ammonia and methane hang in the air and dissolve in the water.

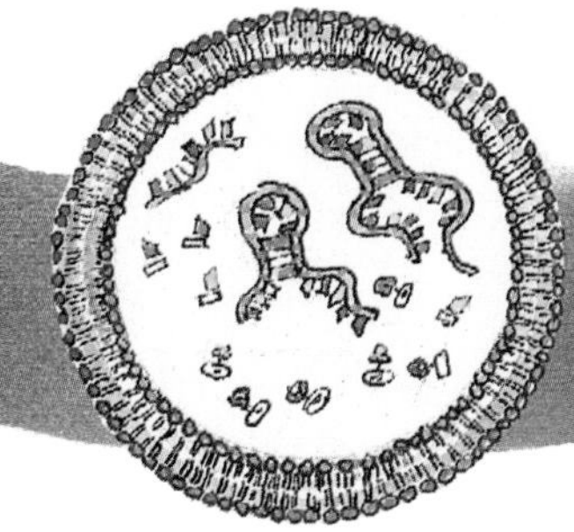

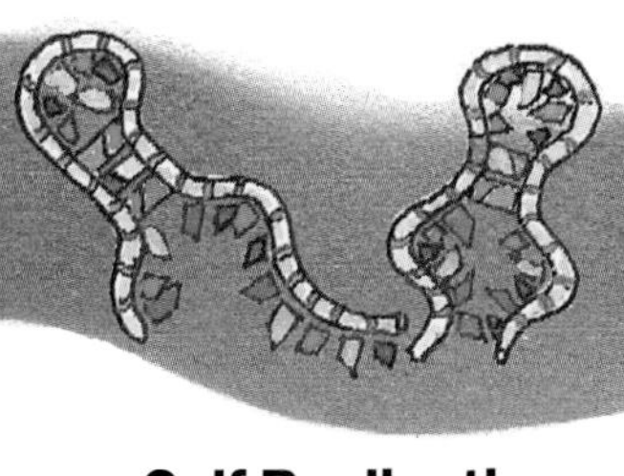

Compartmentalization

Fat molecules spontaneously assemble into bubbles or compartments, sometimes trapping RNA molecules inside.

Self-Replication

Nucleotides begin forming RNA chains. One chain can copy another.

Life's Simple Molecules

Amino acids and nucleotides arrive on space debris or, perhaps, are formed on Earth, aided by lightning and ultraviolet light.

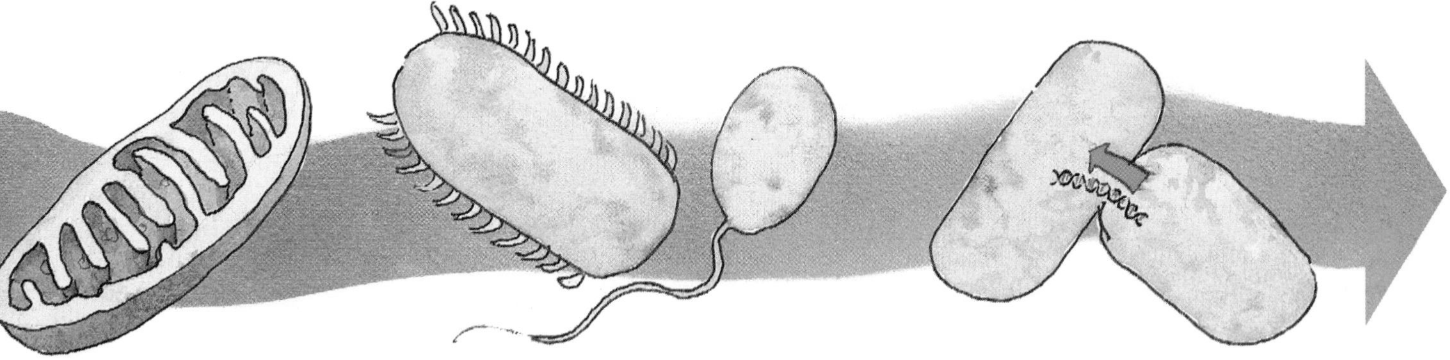

Oxygen Breathing

A few microorganisms "learn" to use the waste oxygen of photosynthesis to make large amounts of ATP.

Locomotion

Cells develop hair-like cilia and whip-like flagella, allowing them to move around in search of food.

Primitive Sex

One cell injects bits of its DNA into another. New gene combinations proliferate.

Evolutionary Time Line

2 Billion Years Ago

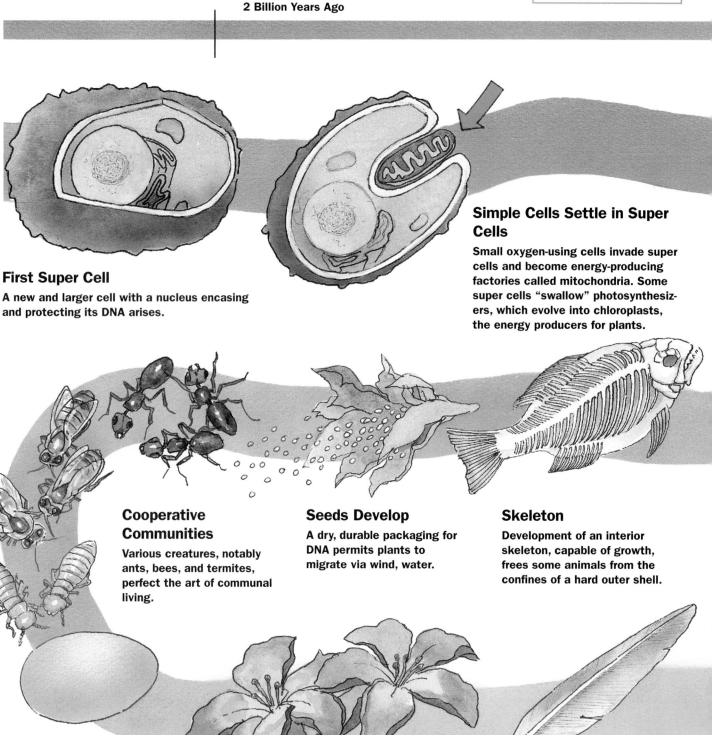

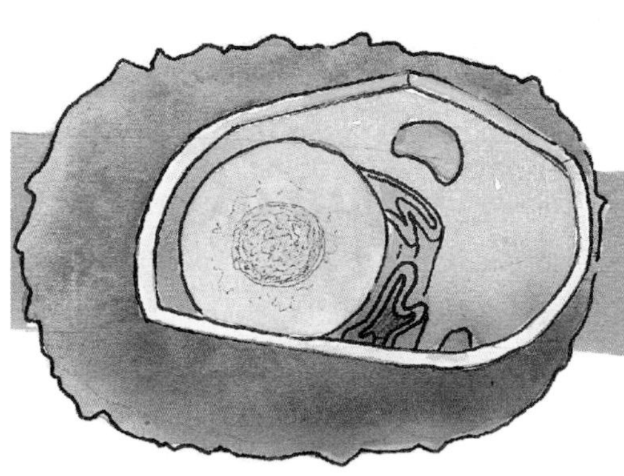

Simple Cells Settle in Super Cells

Small oxygen-using cells invade super cells and become energy-producing factories called mitochondria. Some super cells "swallow" photosynthesizers, which evolve into chloroplasts, the energy producers for plants.

First Super Cell

A new and larger cell with a nucleus encasing and protecting its DNA arises.

Cooperative Communities

Various creatures, notably ants, bees, and termites, perfect the art of communal living.

Seeds Develop

A dry, durable packaging for DNA permits plants to migrate via wind, water.

Skeleton

Development of an interior skeleton, capable of growth, frees some animals from the confines of a hard outer shell.

Waterproof Egg

A waterproof container enclosed embryos in their own portable "sea," enabling animals to migrate permanently to land.

Flowers

Flowering plants develop in symbiosis with animals, exchanging nectar for pollen dispersal.

Feathers

Some creatures develop a complex, lightweight modification of a scale, providing warmth and, ultimately, the gift of flight.

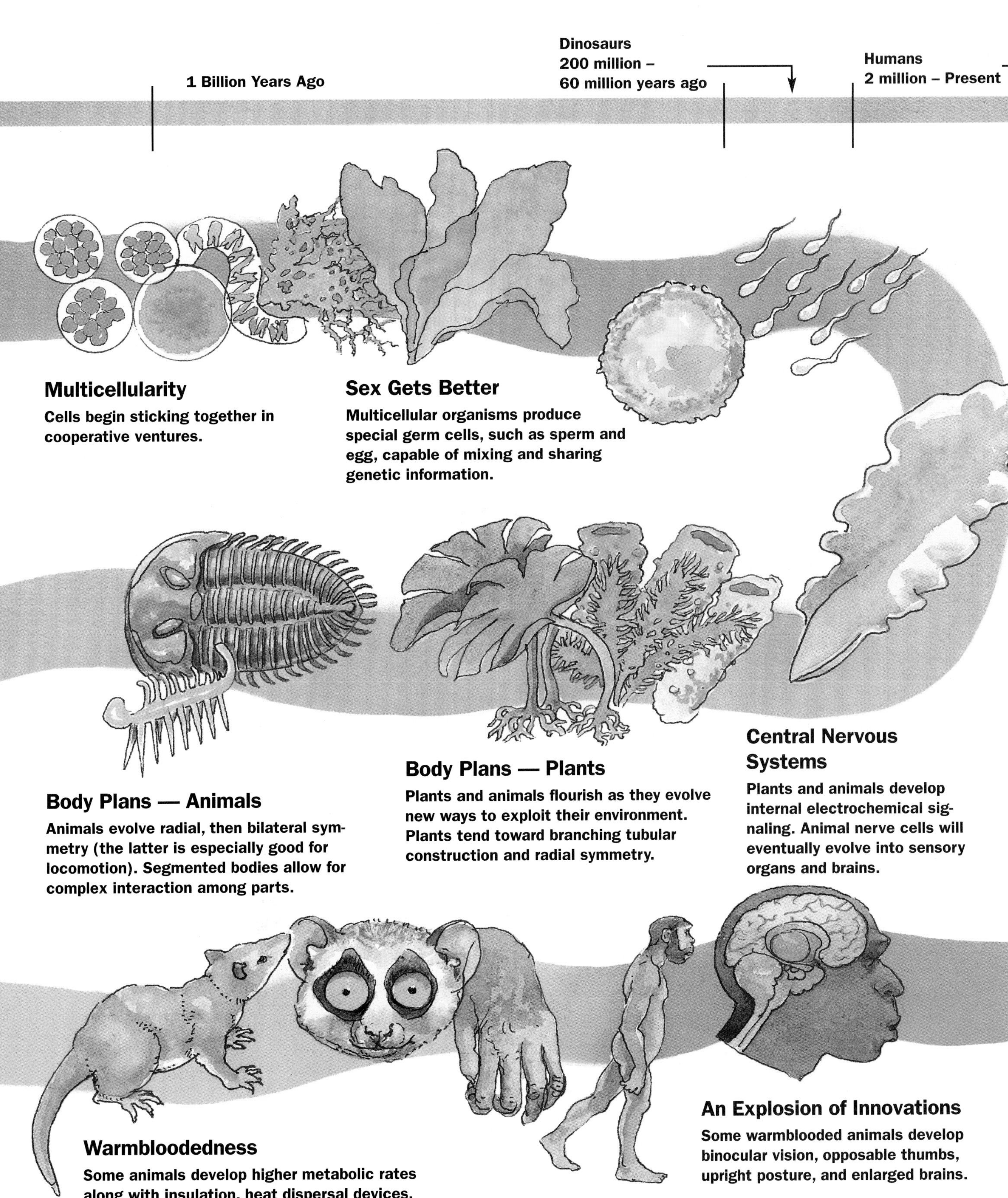
Dinosaurs
200 million –
60 million years ago
Humans
2 million – Present
1 Billion Years Ago
Multicellularity
Cells begin sticking together in cooperative ventures.
Sex Gets Better
Multicellular organisms produce special germ cells, such as sperm and egg, capable of mixing and sharing genetic information.
Body Plans — Animals
Animals evolve radial, then bilateral symmetry (the latter is especially good for locomotion). Segmented bodies allow for complex interaction among parts.
Body Plans — Plants
Plants and animals flourish as they evolve new ways to exploit their environment. Plants tend toward branching tubular construction and radial symmetry.
Central Nervous Systems
Plants and animals develop internal electrochemical signaling. Animal nerve cells will eventually evolve into sensory organs and brains.
Warmbloodedness
Some animals develop higher metabolic rates along with insulation, heat dispersal devices, and internal temperature control.
An Explosion of Innovations
Some warmblooded animals develop binocular vision, opposable thumbs, upright posture, and enlarged brains.

An Elephant-Sized Mouse

Life on Earth has existed for nearly 4 billion years — such a stupendous stretch of time that it's hard to comprehend its implication for evolutionary change. Here's one example that may help.

Imagine a population of mice that are, for whatever reason, increasing in weight by 0.1 percent each generation. In 12,000 generations, the mice will be as big as elephants. If we call a generation about 5 years — a figure between the actual spans for a mouse and for an elephant — this 100,000-fold increase in body weight will take 60,000 years. This is a very brief period on the evolutionary time scale: if life's almost 4 billion years are seen as comparable to the human life span of 80 years, 60,000 years is about the same as five hours.

8.5 Small Changes Add Up to Big Differences

Combined Innovations

Evolution proceeds by gradual tinkering. Complex organisms had, at one time, simpler predecessors. Small improvements then accumulated in such a way as to produce big changes over time. Car design offers a useful analogy. For example, the first headlights were dim, removable, oil-fired lanterns; today's headlights are brilliant, battery-powered, fixed flood beams. As in nature, these changes occurred in small increments, punctuated sometimes by big leaps — as when manufacturers moved the lights from the car's side to the front bumper. At the same time, novelties such as rumble seats and running boards became obsolete and disappeared. Customers' preferences acted as the driving force of selection.

Significant changes over time often involve the cobbling together of the fruits of several independent and unrelated developments. Production of modern headlights required the invention of the battery, the generator, and plastic "glass," just as development of an eye needed photorecepter cells, an optic nerve, and a transparent lens and cornea.

It's important, when comparing car design to organism "design," to remember that *evolution proceeds without a foreseen purpose or direction*. Random changes, cumulative selection (i.e., innovations that build on top of prior innovations), and long stretches of time are what allow evolution to work.

The Evolution of Headlights

Detachable lanterns hang alongside the driver's seat.

Lanterns move down front to better show the road ahead.

Electrified headlamps are powered by the car.

Headlights are mounted on fenders.

Headlights are embedded in fenders.

Headlights become an integral part of the front end.

Big Changes Over Time: The Results of Combined Innovations

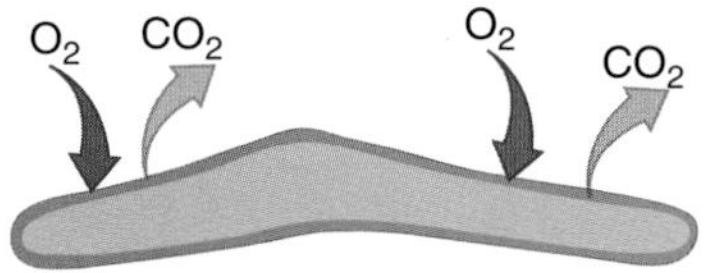

unicellular gas exchange

In multicellular organisms like earthworms, which are huge compared to, say, a bacterium, exchanging oxygen and carbon dioxide became an evolutionary challenge. Most cells in an earthworm need pickup-and-delivery service between the organism's surroundings and its inner reaches. Many large organisms, such as earthworms, frogs, fish, and mammals, have evolved some means of doing this based on the same general plan. They have pumps — hearts — connected to a system of closed vessels that circulate blood to all their cells and through a gas-exchange device that has a lot of surface area in contact with the external environment (see below).

In earthworms, capillary beds in the skin pick up oxygen from the soil and off-load carbon dioxide. Other capillary beds, not shown, exchange gases with body cells.

Fishes' gills are dense mats of capillaries that pick up dissolved oxygen and release carbon dioxide.

Mammals and birds have an additional loop in their circulatory systems, consisting of vessels dedicated to carrying oxygen-depleted blood from the heart to the lungs and returning oxygen-rich blood back to the heart, which pumps it to body tissues.

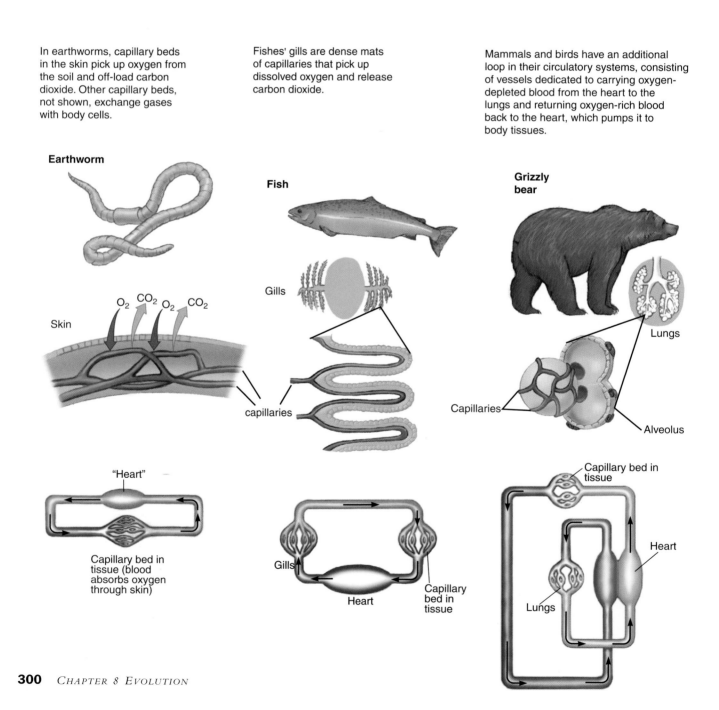

Earthworm

Fish

Grizzly bear

Skin

O_2 CO_2 O_2 CO_2

Gills

capillaries

Lungs

Capillaries

Alveolus

"Heart"

Capillary bed in tissue (blood absorbs oxygen through skin)

Gills

Heart

Capillary bed in tissue

Capillary bed in tissue

Heart

Lungs

The capillary beds in the figure on the previous page are networks of tiny vessels where "delivery to" becomes "pickup from." Through these capillaries, blood delivers oxygen and food (from the separate digestive system) and picks up carbon dioxide. Each of these circulatory systems evolved over time as random changes in each kind of organism's genes produced tiny innovations that were successful in that creature's environment.

As changes accumulated, elaborations arose in the hearts and in the gas-exchange devices. Specialized gas-exchange structures called gills and lungs evolved. These provided much more surface area for obtaining oxygen and releasing carbon dioxide. If you had to breathe through your skin, as earthworms do, you would have a "breathing surface" only a few square feet in area. But you possess a set of lungs, a highly evolved pair of gas-exchange structures. If your lungs' breathing surface were spread out flat, it would cover an area about the size of a tennis court!

The more extensive surface area allows animals with lungs to exchange large amounts of oxygen and carbon dioxide within a relatively small volume of space, a chest cavity. This permits the consumption of a huge amount of energy in a relatively small space. That ability, in turn, allows animals with lungs to be big: It takes a lot of energy to "run" a big animal like a whale, and it takes a lot of energy for a lion to run down a gazelle. When you think about it, you realize that all of the very large animals that have ever lived on this planet have either lungs or gills.

The active center of circulatory systems is a heart. Earthworms have five simple pumps composed of muscular arteries (basically just tubes of muscle tissue) that compress and send blood coursing throughout their bodies. Oxygen gets into earthworms and carbon dioxide gets out through their skin's surface. Fish have two-chambered hearts. Frogs have three-chambered hearts, and so do reptiles, though the latter have more elaborate "dividers." Birds and mammals have four-chambered hearts that allow for the added loop of vessels to and from their lungs.

Without these more sophisticated pumps and more extensive surface areas for exchanging gases, animals as big as dinosaurs, whales, elephants, ostriches, and humans probably couldn't have evolved.

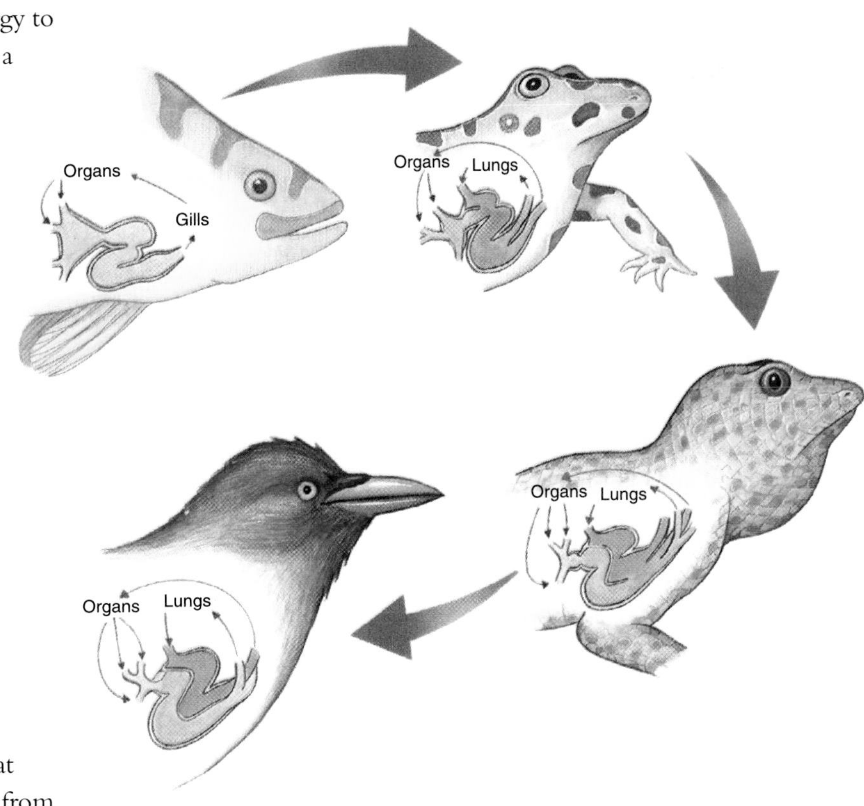

Increasing complexity

The path of evolution from fish to amphibian to reptiles to birds shows hearts of increasing complexity.

8.7 Writing Poetry Evolution's Way

Monkeys and Word Processors

Could a roomful of monkeys randomly pecking at their typewriters eventually write a Shakespearean sonnet? This question has been used to challenge the idea that life arose by chance. The odds that a hundred monkeys typing away for a million years could accidentally produce such a work of art are vanishingly small. But if we impose some rules of evolution on this random process, as evolution does on random genetic variation, we can see how nature increases the chance of success — indeed, how it makes success inevitable. First, let's stipulate that the monkeys will type out not Shakespeare's actual sonnets but original sonnets of comparable complexity. That is, we won't demand a specific outcome — just a general pattern. Let's have the monkeys work with word processors programmed to keep successful results and throw out everything else. This rule is the evolutionary principle of selection. Let's arrange for the work to proceed in progressive stages of increasing complexity — another characteristic of evolution. By combining random typing with cumulative "capturing" of successful results, monkeys can write beautiful poetry!

Cumulative Selection

1.
First Team: Making Words

Whenever a monkey accidentally types a sequence of letters that the computer recognizes as a valid word, it is saved. "Roses" is acceptable. "Rosgbz" is not. Saved words accumulate over time.

2.
Second Team: Making Sentences

Words generated by team 1 are coded and put into the computers of team 2. When the monkeys strike the keys, the words are strung together in random sequences. The computer saves only those sequences with subjects and predicates — i.e., sentences. "Roses are red" is acceptable. "Roses salad bleakly" is not.

3.
Third Team: Making Sonnets

Sentences generated by team 2 are coded into the computers of team 3. These monkeys randomly order the sentences. Only fourteen-line sequences conforming to the sonnet form are saved.

4.
Fourth Team: Publishing Sonnet Collections

Monkeys in team 4 randomly collect team 3's sonnets into groups, which are printed in bound books. Most sonnets would be nonsense, but a few would be coherent. A tiny fraction of a large enough sample might even be beautiful.

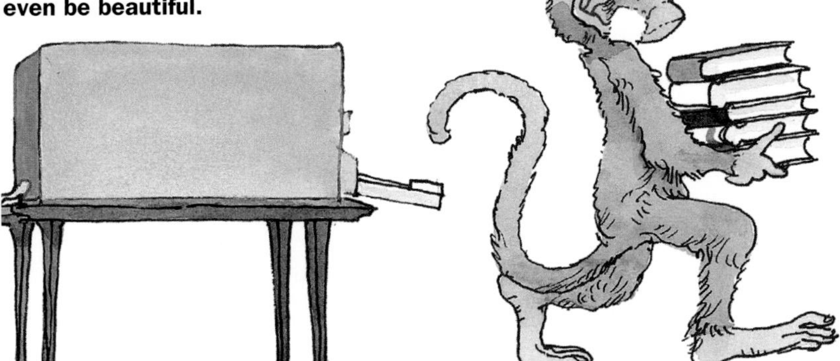

5.
Books Are Offered To The Public

Only those books that sell out are reprinted. Thus, the worst poetry is "selected out." The best is kept. Eventually, given enough time, a good collection of sonnets will emerge.

8.8 Even a Small Advantage Tends to Survive and Multiply

The "Ingenuity" of Chance

When we admire a bird in flight and are told that this ingenious being's ancestor was an earthbound dinosaur, it's pretty hard to swallow. Yet, as demonstrated by our sonnet-writing monkeys, small changes that arise fortuitously can be "saved" and even amplified at each generation by the addition of further advantages. Given enough time, such changes can produce something that's never been seen before.

A possible scenario for a bird's beginnings is pictured on the opposite page. In any population of individuals, one born with a slight advantage is more likely to grow up and have offspring that will be like it. Evolution's rules are such that even the slightest advantage will take hold, spread, and eventually dominate in a reproducing population.

WEB Connection
www.jbpub.com/connections

As some reptiles moved toward warmbloodedness, their scales evolved into feathers, providing insulation.

Imagine a baby born with lighter bones than its nestmates — a "freak" produced by a chance variation in bone cells.

When the nest is invaded by a predator, the variant offspring's lighter bones and feather-like scales give it a little extra lift, allowing it to jump away from death's jaws. The double advantage will be passed on to its offspring.

8.9 Multiple Changes

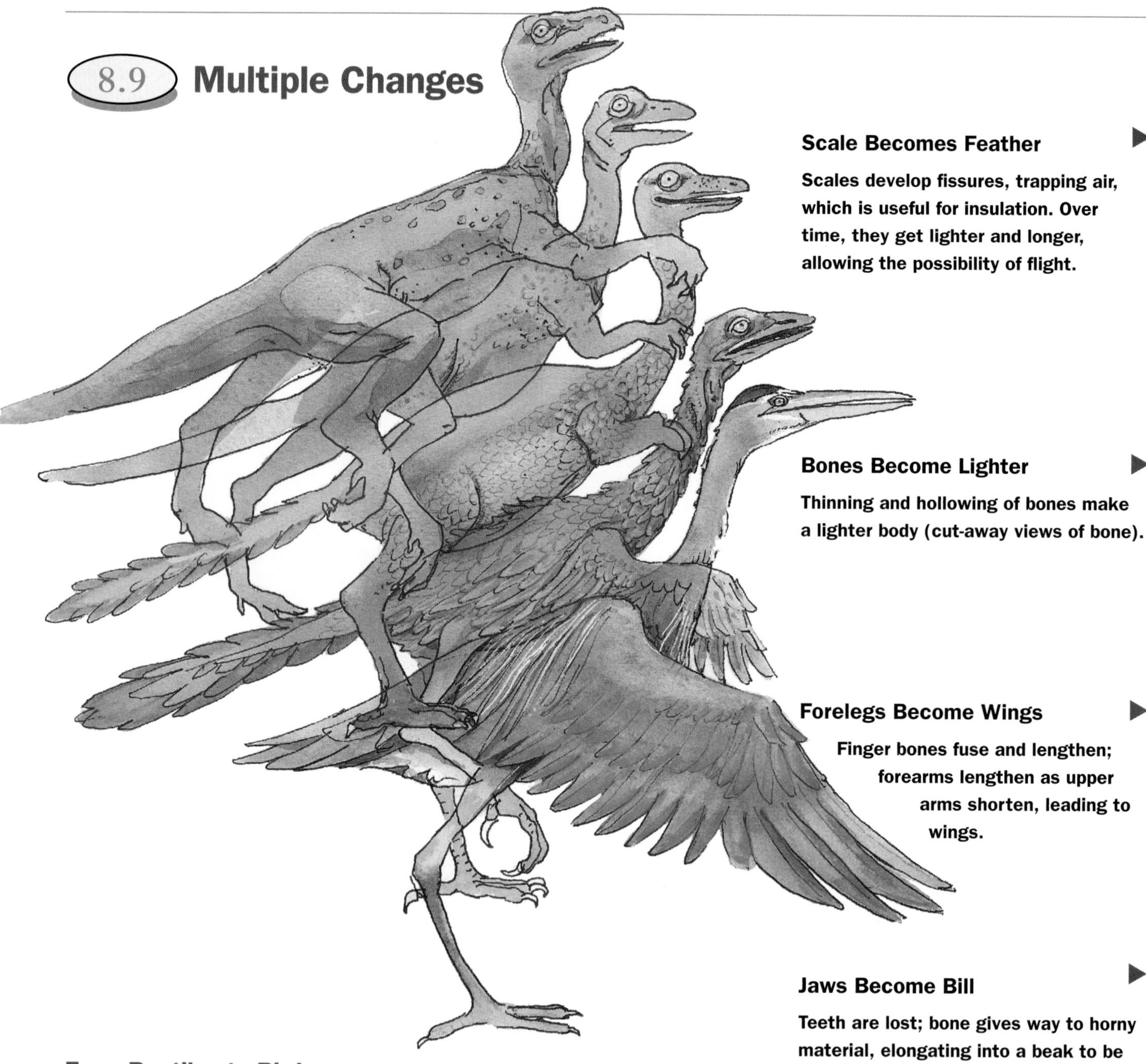

Scale Becomes Feather
Scales develop fissures, trapping air, which is useful for insulation. Over time, they get lighter and longer, allowing the possibility of flight.
Bones Become Lighter
Thinning and hollowing of bones make a lighter body (cut-away views of bone).
Forelegs Become Wings
Finger bones fuse and lengthen; forearms lengthen as upper arms shorten, leading to wings.
Jaws Become Bill
Teeth are lost; bone gives way to horny material, elongating into a beak to be used for grasping, preening, and probing.
A Toe Turns Backward
The first of four toes swings backward, making it useful initially as a weapon and later for perching and grasping.

From Reptiles to Birds

It takes a lot more than feathers and light bones to turn a reptile into a bird. Birds, especially migratory ones, need internal navigational equipment, responsive to celestial patterns and the Earth's magnetic field. Their eyesight would have to become keener than a reptile's to spot food on the ground. The greater and more sustained energy demands of flight would require major adjustments in its body's ability to maintain a steady temperature and to pump blood. Its reptilian forelegs would have to become aerodynamically efficient as they evolved into wings, and its breastbone would have to become keeled to provide leverage for wing muscles.

While these modifications would arise independently, with each small step conferring some advantage (or, at least, doing no harm), they would also reinforce one another in contributing, generation after generation, to the proto-bird's long-term survival.

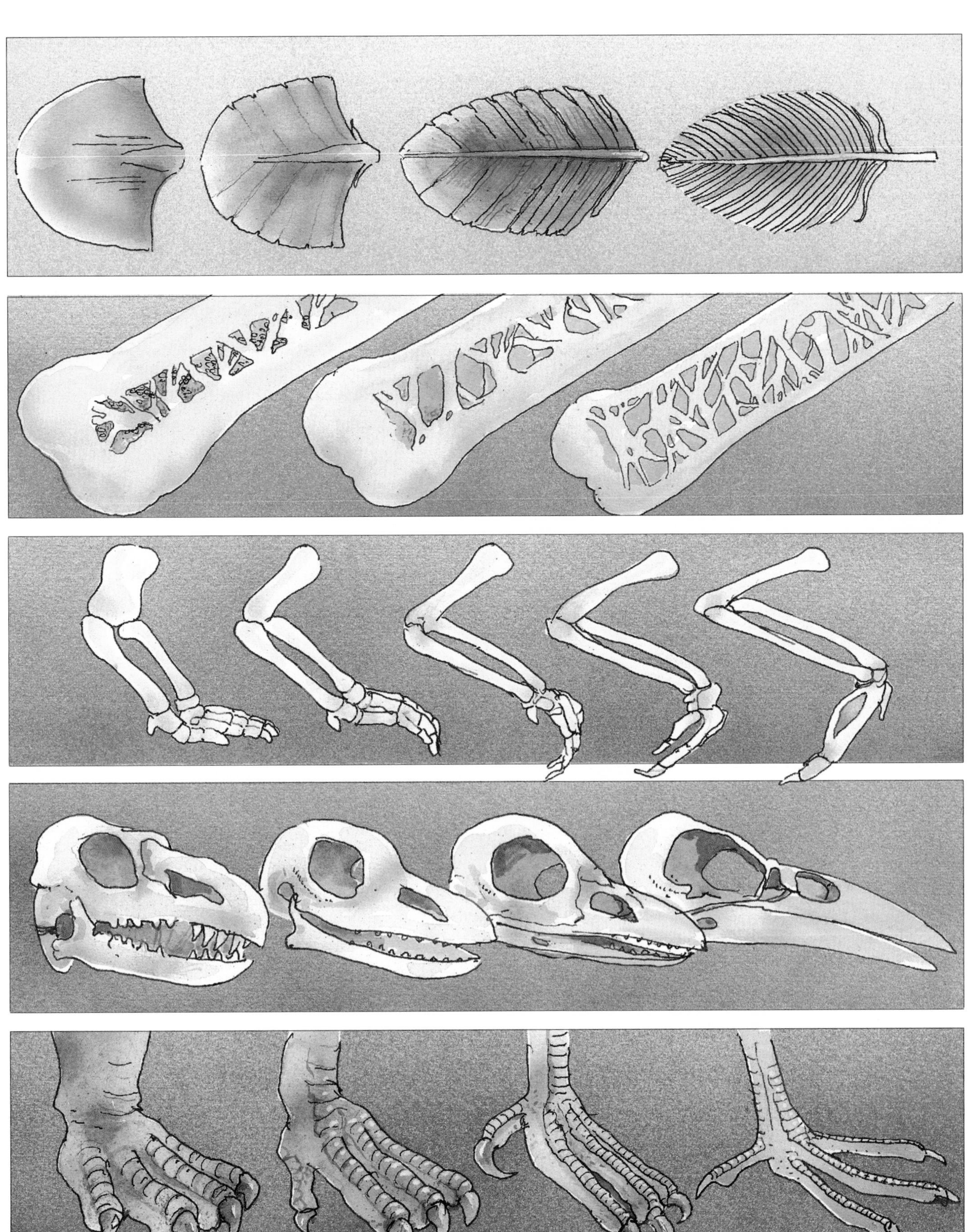

Finding Evidence of Relatedness

The first physical evidence of an evolutionary link between reptiles and birds, a fossil Archaeopteryx, was discovered in 1862. The 150-million-year-old fossil bird, about the size of a crow, had features of both birds and reptiles.

Along with birdlike wings, feathers and skeletal structures, Archaeopteryx had reptilian teeth and a long jointed tail. Charles Darwin cited this discovery in later editions of his *On the Origin of Species by Means of Natural Selection* as evidence of the evolutionary relatedness of birds and reptiles.

Archaeopteryx

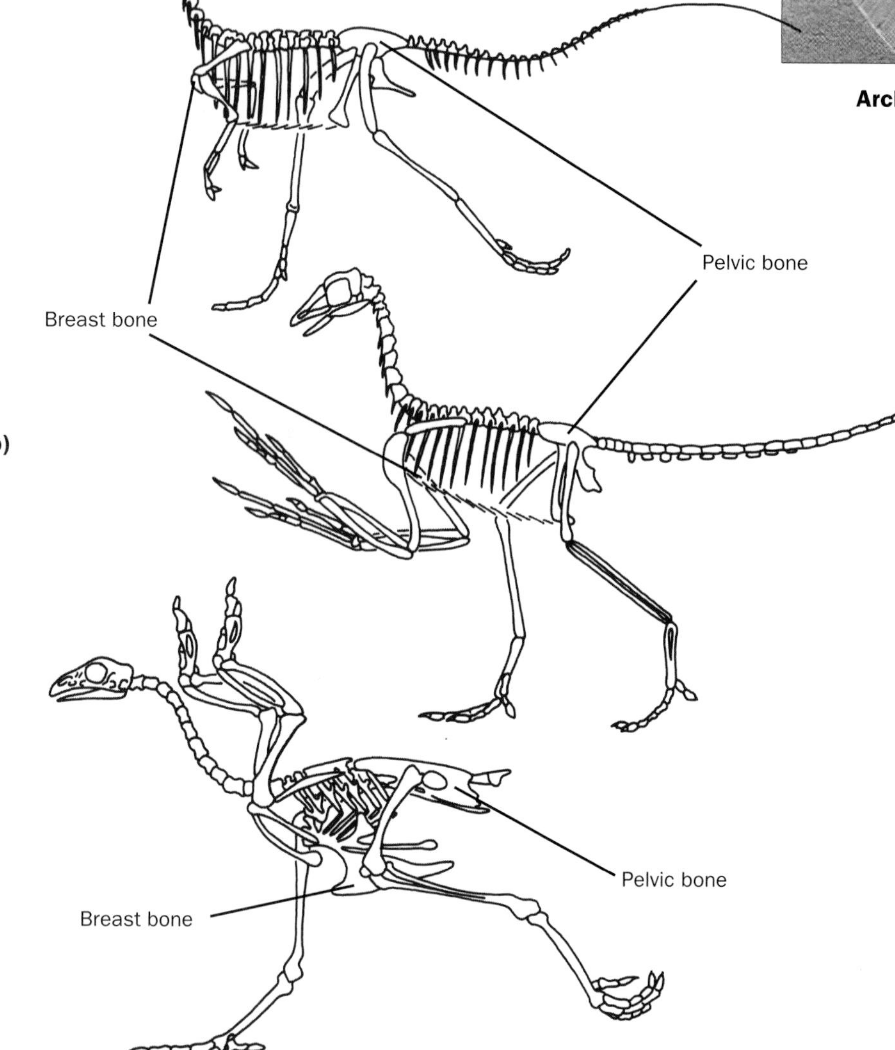

Compare the bones

These skeletons of (a) *Compsognathus*, a dinosaur, (b) *Archaeopteryx* and (c) *Gallus,* a modern chicken, are aligned to compare the bony features of these animals, so widely separated in time. You can see that, while there are many similarities among these skeletons, it's clear that birds have changed noticeably from the fossil ancestor to the modern form. Especially note the difference in the pelvic bones of (b) and (c) and the very well-developed breastbone of the modern chicken (for the attachment of flight muscles) and the absence of a jointed tail.

Since Darwin's time, more and more fossil remains have been discovered, enough to establish a strong evolutionary relatedness between birds and dinosaurs. There is still much debate, though, about whether birds are directly descended from dinosaurs, or simply share a reptilian ancestor with them.

Microscopic Evidence of Dinosaur Descent

In trying to establish the ancestral lineage of modern birds, many different aspects of dinosaur, reptile, and bird skeletons have been studied, on many different scales. Very recently, microscopic studies have been made of the arrangements of the tiny vessels (called canaliculi) that carry nutrients from one bone cell to another. The comparisons show strong similarity between canilicular orientations in the bones of modern birds and those of the Theropod order of dinosaurs.

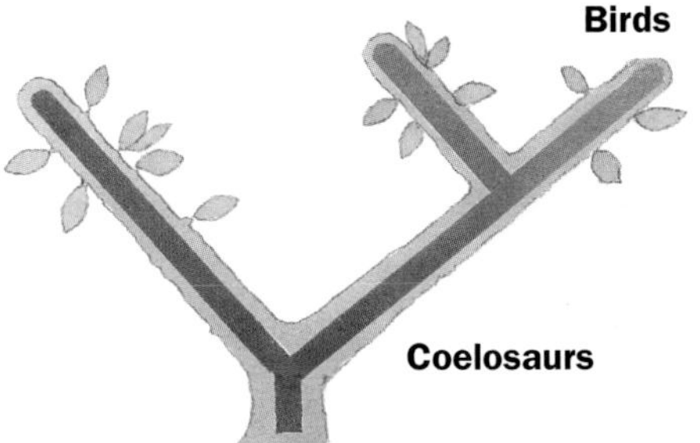

Doing Science

John M. Rensberger & Wahito Watabe
Fine structure of bone in dinosaurs, birds and mammals.
Nature, Vol 406 10 August 2000, 619-622.

Cross-sections of bone were taken from living species of birds and mammals, and ground down to about one hundredth of an inch thick. The patterns of the bone cells themselves were examined. A striking difference was found in the way bone canaliculi were laid out in the bones of birds and mammals: mammal canaliculi were laid out in an organized pattern of parallel lines radiating out from a central canal, while in birds the layout was random — as in a dried-up mud puddle.

Drs. Rensberger and Watabe then hypothesized that, if birds are descended from some order of dinosaurs, dinosaur canalicular structure should resemble that of birds. They examined the canalicular structure of bone in 17 species of birds, all of which showed a characteristic random pattern. Several fossil Theropod dinosaur bone fragments also showed this pattern, as did three specimens from different Ornithomimid dinosaurs. Specimens from other branches of the dinosaur family tree, on the other hand, showed parallel canaliculi.

Patterns of canaliculi in bone sections

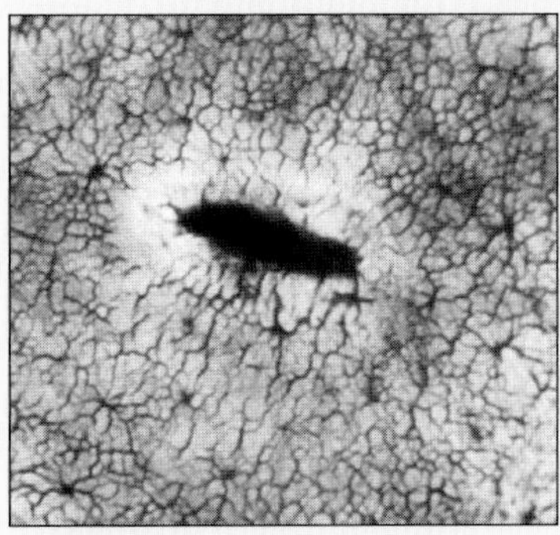
Bald eagle

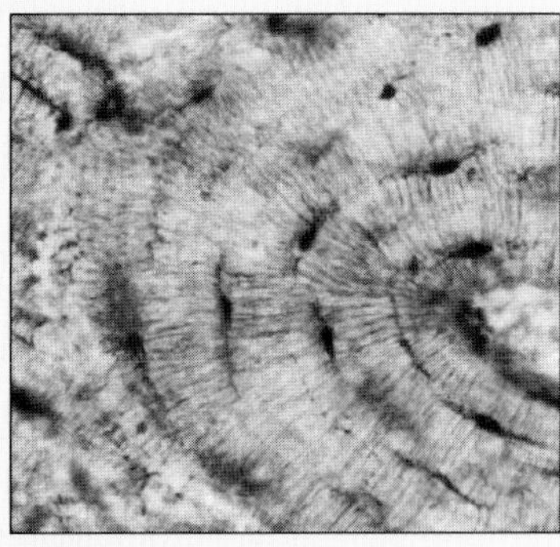
Mammal

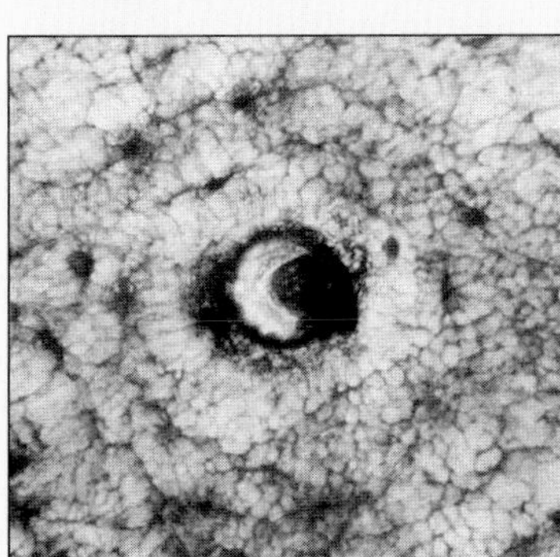
Ornithomimid dinosaur

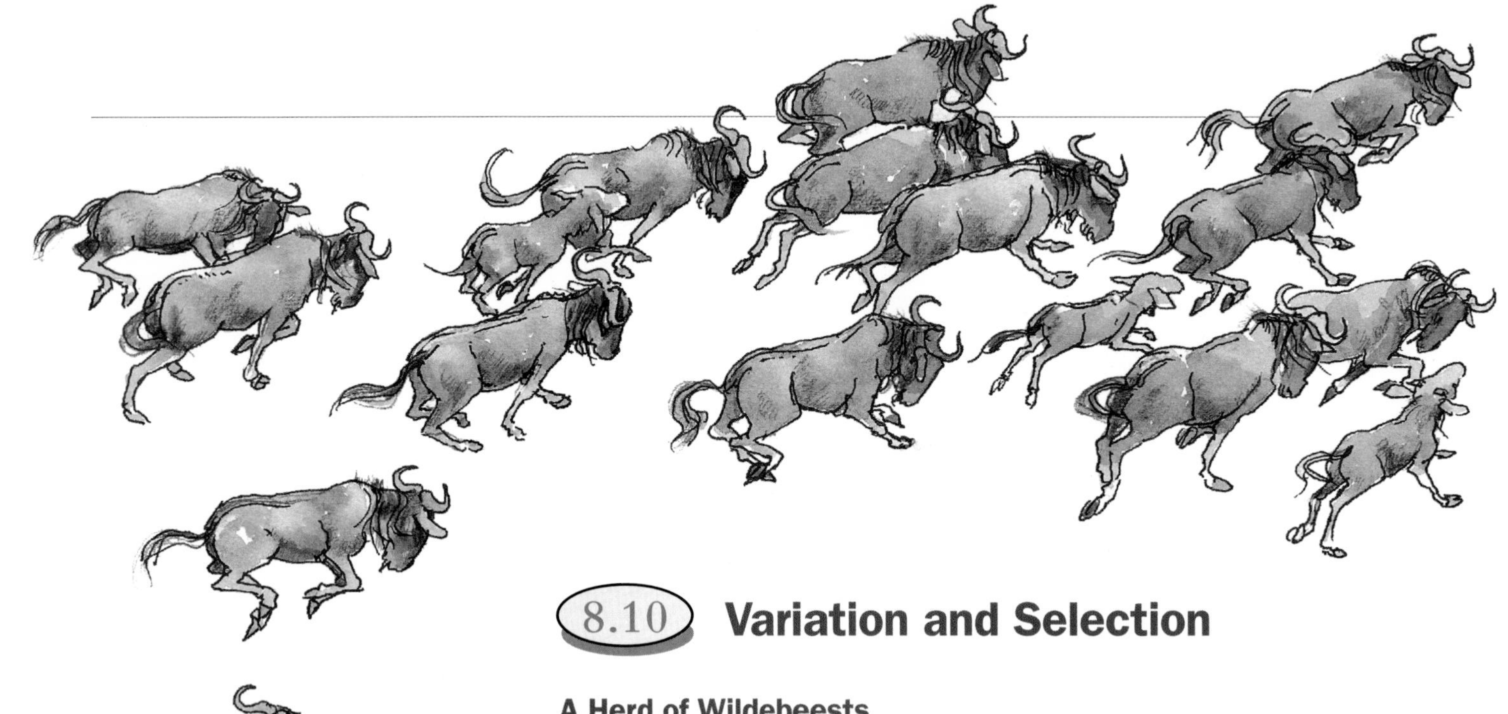

8.10 Variation and Selection

A Herd of Wildebeests

Each year in Africa's Serengeti, great herds of wildebeest begin a 600-mile migration. During the journey, some of the animals are eaten by predators, some drown crossing rivers, some die of injury or disease. Some of the deaths are due to back luck; but, overall, the faster, stronger, and more alert survive the trip, while the less well-endowed are weeded out along the way.

A Herd of DNA

Now let's shift focus and consider the wildebeests not as a herd of animals but as a vast pool of information. The information exists as separate sets of genes called genomes — each of which resides in an individual animal. While these information sets are similar — each one of them, after all, spells out "wildebeest" — *each one is also unique.* There could be no evolution without such individual differences among genomes. The information in each wildebeest's genome provides the tools — the capabilities of that individual animal — for negotiating the trip. Along the way, some genomes are destroyed. The "best" information sets are those that survive the journey. The genes that are removed by death are, therefore, not just any genes. They are, on average, those that make wildebeests less likely to survive. Individuals die, but the population benefits by an average improvement in its gene pool.

Information Sets

Every cell of every wildebeest contains two sets of genes (simplified here as two short DNA helixes) — one from its mother, one from its father. Every set differs in small ways from every other set, as indicated by the color differences.

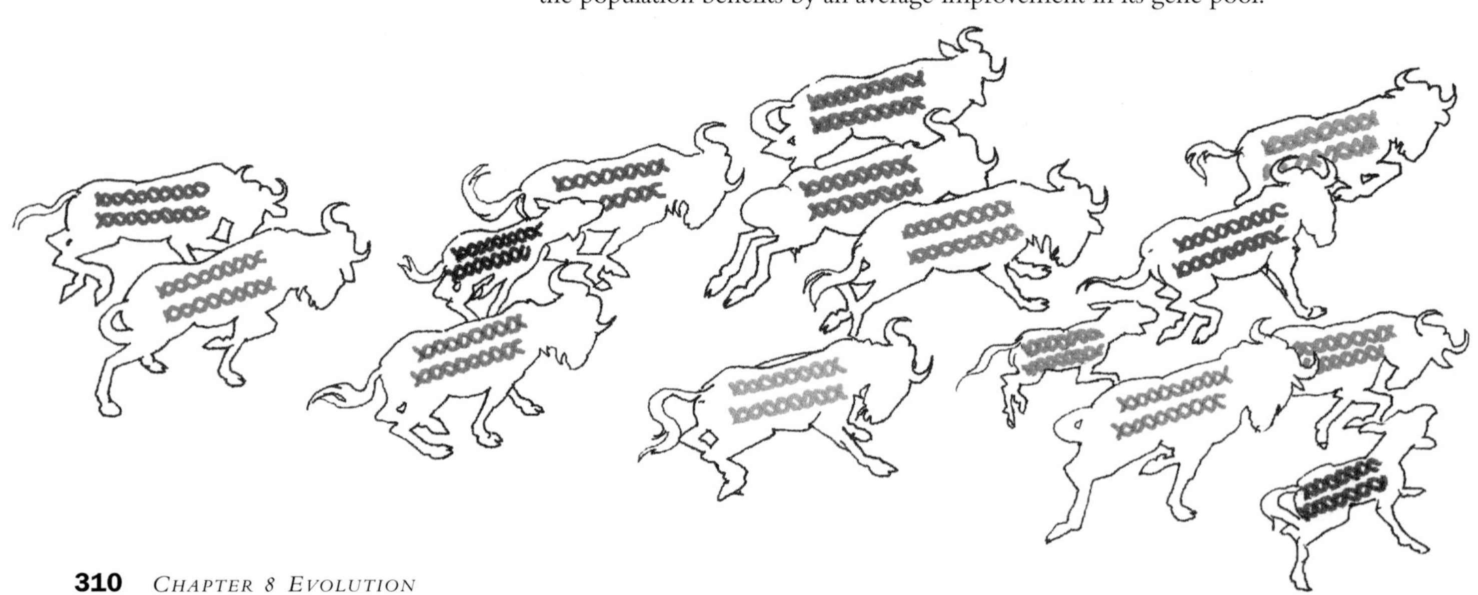

Death on the Journey

Some animals are eaten...

Some drown...

Some succumb to disease.

A Loss of Information

Predation, drowning, and disease remove some information sets from the pool.

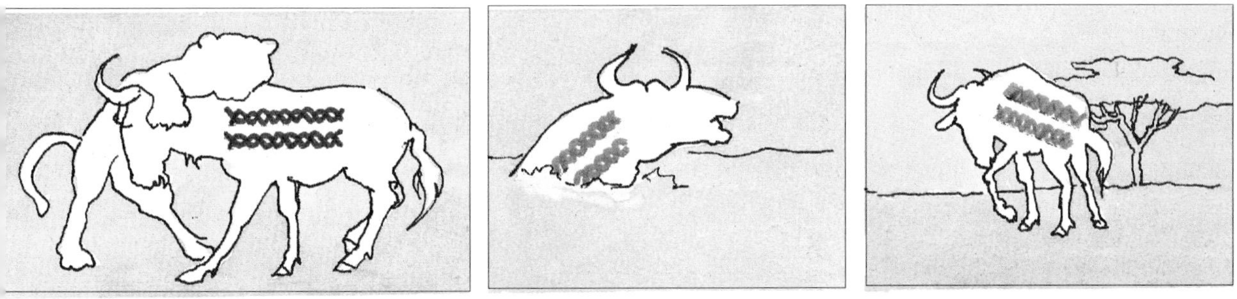

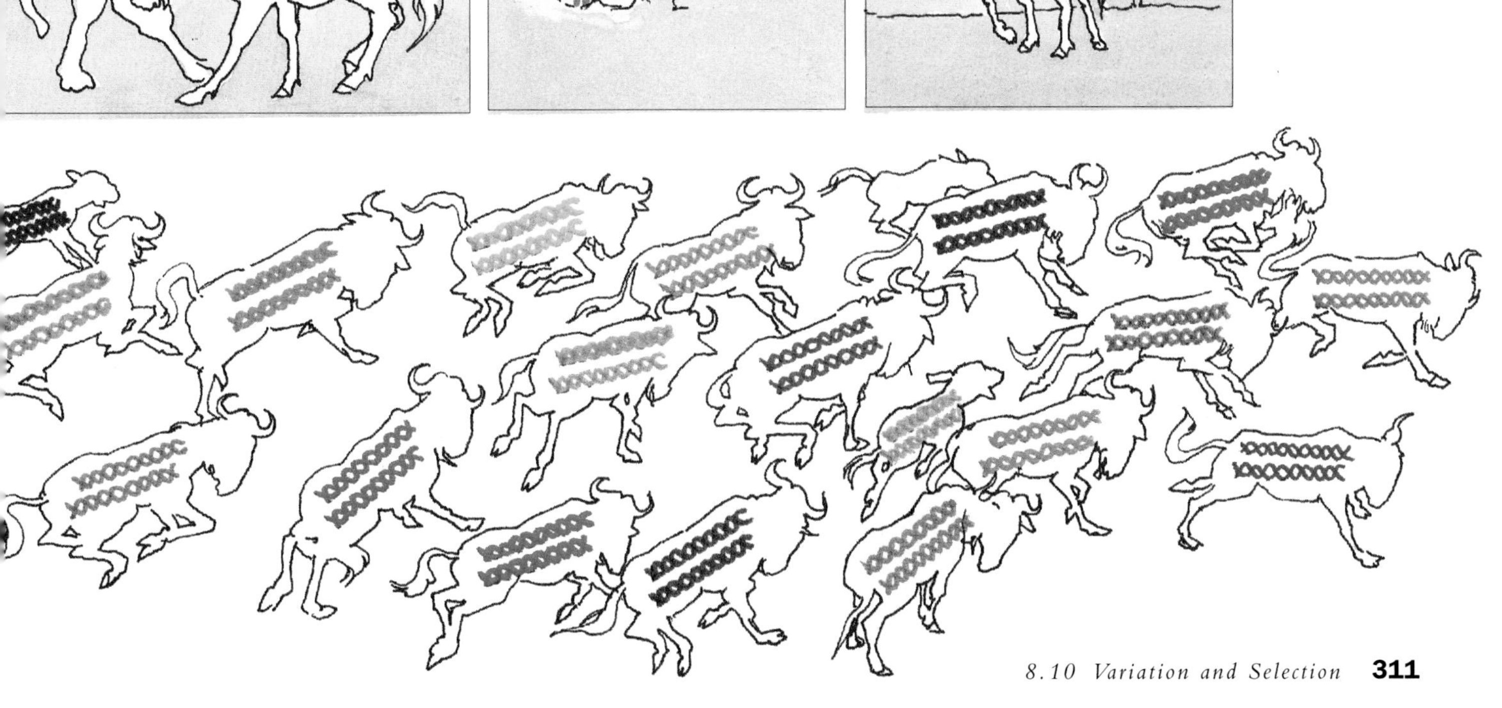

Mixing the Survivors' Genes

During migration, the wildebeests mate. A female chooses from among her suitors the male that most impresses her by displays of prowess indicating his dominance over other males — a final selective test among the herd's information sets. Mating is a mingling of sets of genes.

Even before mating, in the ovaries of each female and in the testes of each male, the mother-father pairs of gene sets are first thoroughly mixed and then randomly packaged into eggs and sperms (see page 314). After fertilization, the eggs develop into calves. Every cell of a calf contains a new genome, and all these genes are arranged in new combinations, ripe with the potential for novel characteristics. This is evolution's way of ensuring that any improvements that arise in individuals are distributed as widely as possible to later generations.

Each parent contributes half of its genes to the offspring. Although the parents' individual genes have been successfully "road-tested," the unique combination of genes in their offspring is not yet tried and true.

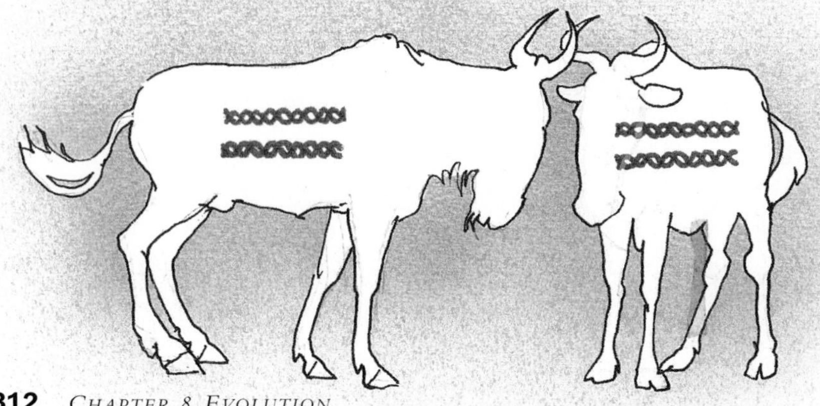

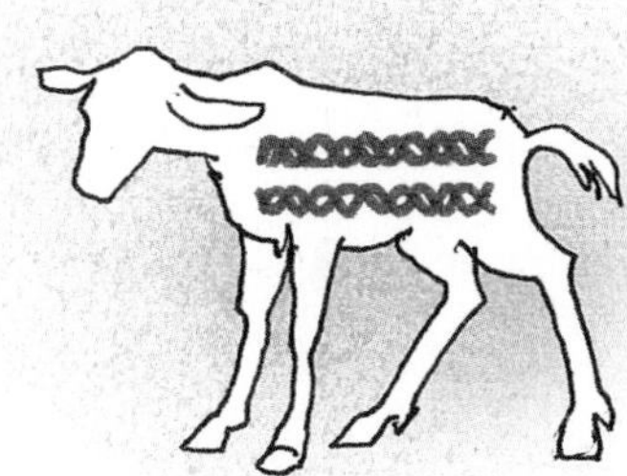

Male and Female — Two Separate Strategies

The making of new life requires an equal contribution of genes from both a male and a female. But typically the female animal makes a substantially greater overall contribution to the ultimate success of the endeavor. While a sperm cell is not much more than a set of genes, an egg supplies, in addition to genes, the food, the energy-generating machinery, and the protein-making capability needed to launch the new life.

Furthermore, in many species, the female's body provides the environment for development of the fetus, and, among mammals, the female often continues to nourish the offspring after it has left her body. Owing to her larger reproductive investment, the female is choosier about her mates. Conversely, the lower investment of the male makes him less discriminating and more available. Thus, animals' basic sexual pattern: aggressive males propose; discerning females choose.

Sex

The Mechanics of Gene Mixing

Sex is widely understood as the means of mixing the genes of a male and a female. But few people realize that the actual mixing occurs prior to mating in the body of each animal (or plant). Mating is simply the means of bringing together already randomly mixed gene combinations. Here's a simplified picture of how it works.

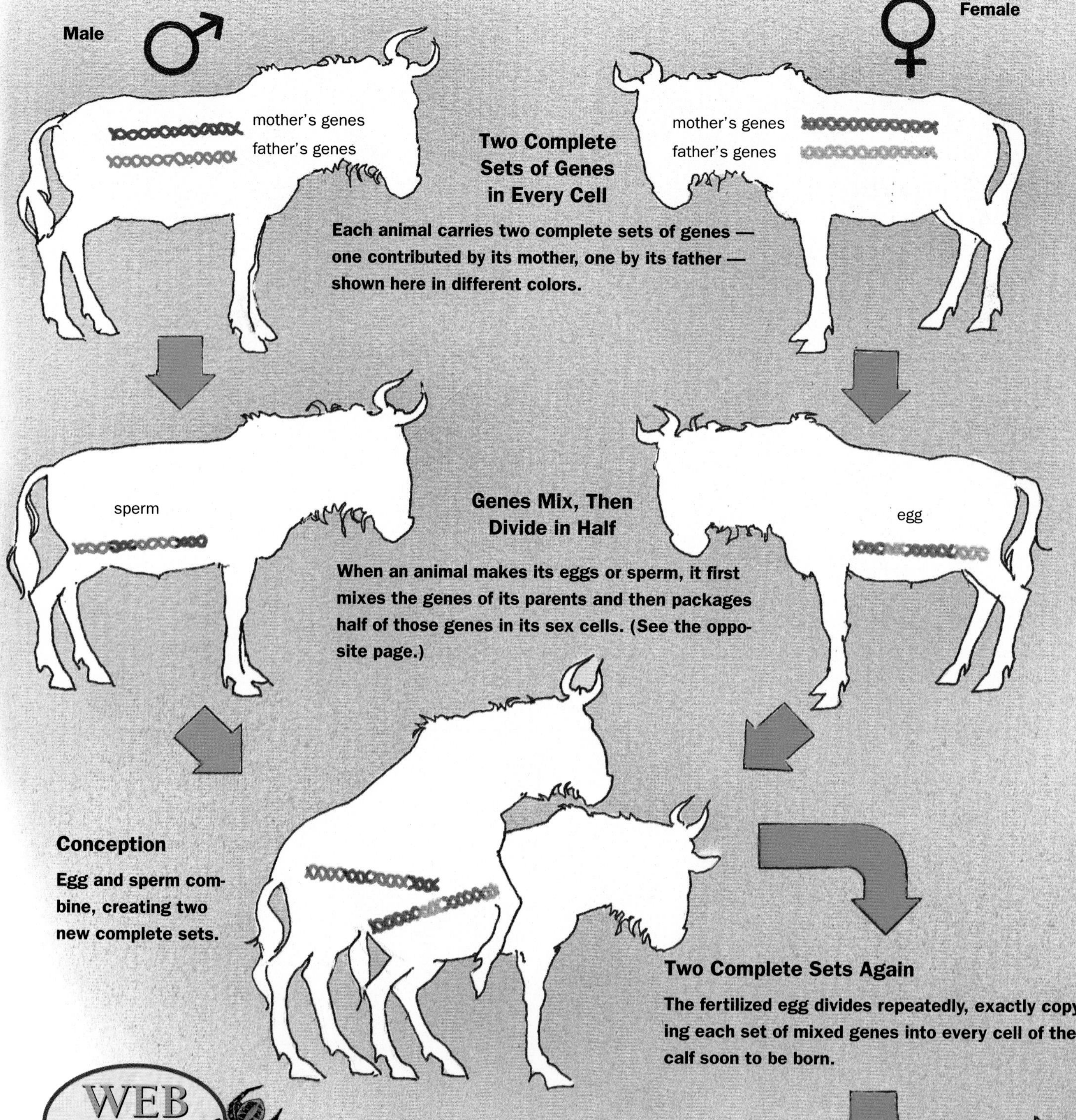

Male

mother's genes

father's genes

Female

mother's genes

father's genes

Two Complete Sets of Genes in Every Cell

Each animal carries two complete sets of genes — one contributed by its mother, one by its father — shown here in different colors.

sperm

egg

Genes Mix, Then Divide in Half

When an animal makes its eggs or sperm, it first mixes the genes of its parents and then packages half of those genes in its sex cells. (See the opposite page.)

Conception

Egg and sperm combine, creating two new complete sets.

Two Complete Sets Again

The fertilized egg divides repeatedly, exactly copying each set of mixed genes into every cell of the calf soon to be born.

WEB Connection

www.jbpub.com/connections

8.12 Why Do It?

Sexual Reproduction Creates New Gene Combinations

Since all your cells carry all the information to create a duplicate of you, any cell of your body could, theoretically, develop into a perfect copy of you. Many plants form offshoots that detach and become new plants; single cells taken from any part of a plant can grow into a whole new plant. Animal cells can be induced to do a similar thing in the laboratory. For instance, frog skin cells inserted into a frog egg from which all DNA has been removed will develop into a new frog. Thus, the skin cell's DNA contains all the information to make a complete frog. Furthermore, many kinds of multicellular organisms simply "bud" identical offspring from their bodies.

So why then do most organisms reproduce using sperm and eggs? Why do we need sex? It seems wasteful for half a population to produce eggs and half to produce sperm to fertilize those eggs: It means that only half the population can actually produce offspring. Why don't we simply "bud" children directly from our bodies? It seems much more tidy and efficient; and wouldn't it allow us to rapidly out-reproduce sexually reproducing organisms?

Any organism that produces exact copies of itself — whether a single cell dividing or a multicellular organism budding — can't readily adapt to changes in its environment; the only changes the gene pool of its species will undergo are those caused by mutations. Given this limitation, the species is likely to evolve relatively slowly. Because a sexually reproducing organism has two copies of every gene in its cells, from the point of evolution, it has a *spare*. (This spare gene can undergo change by mutation and may become useful to a later generation of the organism.)

Furthermore, there is clearly a tremendous evolutionary value in mixing genetic information, in making all sorts of new combinations of already "proven" genes. Some combinations of genes are bound to be winners and lead to new ways of adapting to the changing environment. For example, parasites have a harder time causing trouble if the species they're trying to infect is constantly changing by mixing genes. Sex takes advantage of the favorable attributes of two individuals to create newness.

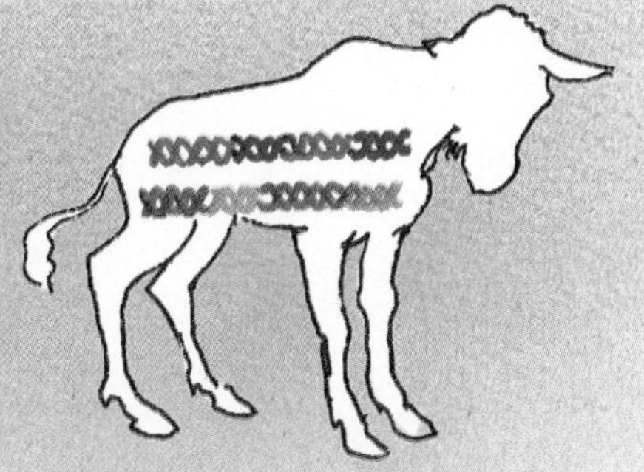

Making Eggs and Sperm

A chromosome is a spool of DNA containing thousands of genes.

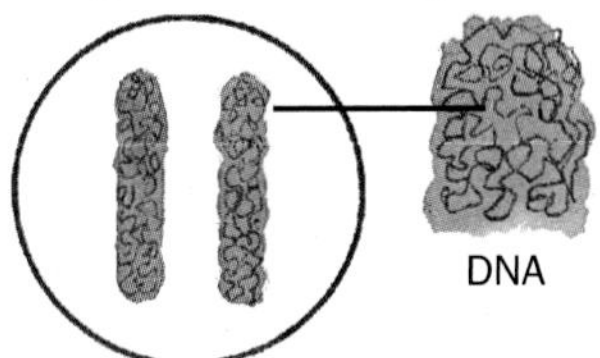

In special cells in the ovary and testis, chromosome pairs double...

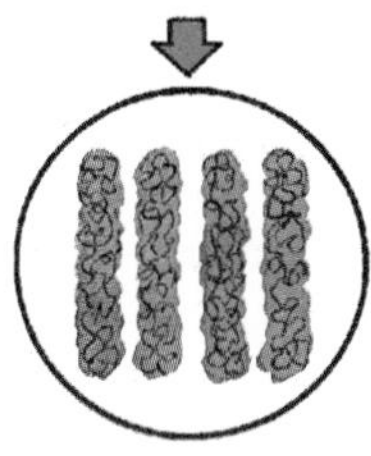

...embrace...

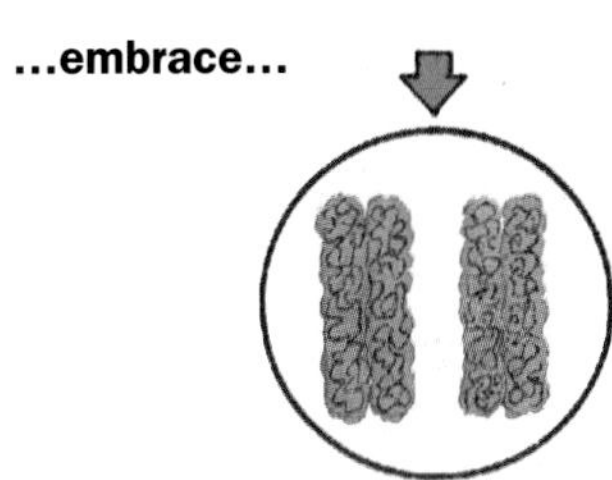

...transfer segments...

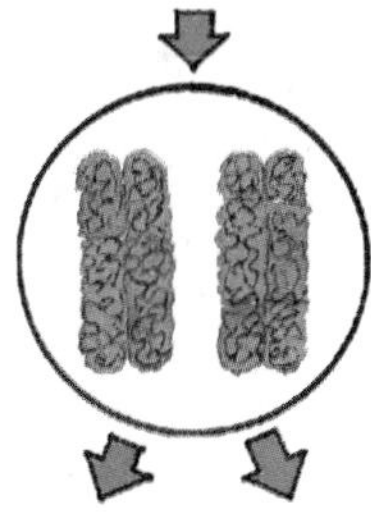

...separate into two cells...

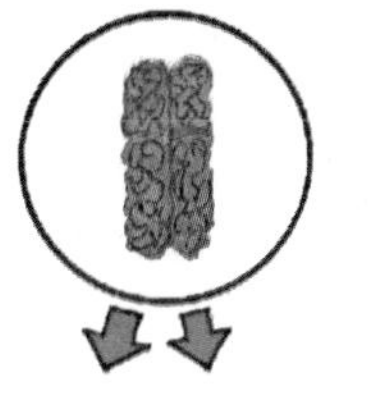

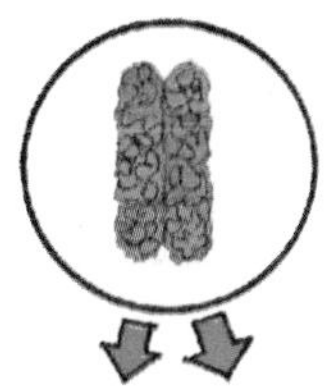

...then into four eggs or sperm...

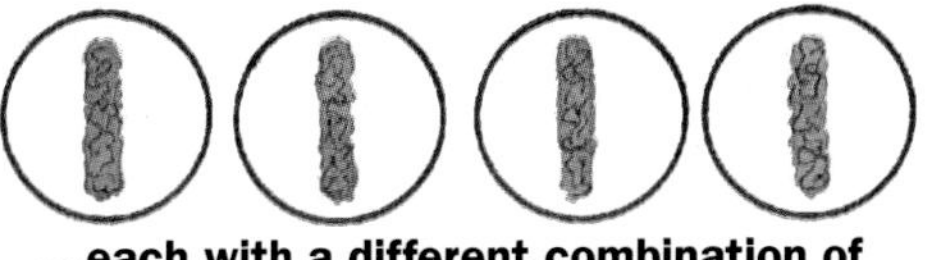

...each with a different combination of mother's and father's genes.

How Fireflies Find a Mate: The Lure of Light

Firefly behavior includes an interesting example of a strategy for finding a mate of the right species. The selective test females apply to screen prospective mates concerns the timing of the males' light production and the accompanying flight pattern. The seemingly random blinks of light you see moving about in a backyard on a summer evening are meaningful to fireflies; they are also the product of natural selection operating on millions of generations of these insects.

A male firefly

Fireflies, often called lightning bugs, belong to a family of beetles called *Lampyridae.* Active at night, they would be invisible to each other were it not for the little bioluminescent lamps in their abdomens (see page 134 for information on how light is produced).

There are more than 2000 species of fireflies, and a given locale is often home to more than one species. Therefore, successful firefly mating depends on being able to recognize and find a partner of the same species in an energy-efficient way. Each species of firefly has evolved its own particular way to tell who's who.

In certain species, only male lightning bugs can fly. The grublike and wingless females live in the vegetation close to the ground. Males in search of a mate emit light flashes of specific duration, intensity, and timing. Also, while flashing, they fly in a characteristic pattern: One species may fly in a comma-shaped loop during a flash; another may emit three quick flashes while flying straight upward. Each species' unique flash-and-flight pattern has been tightly choreographed by millions of years of mating success and genetic recombinations. This behavior evolved along with receptors so sensitive to light and motion that females of some species can distinguish between male flashes that differ in length by as little as four-hundredths of a second.

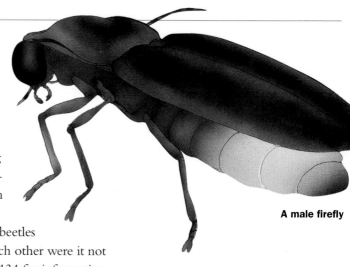

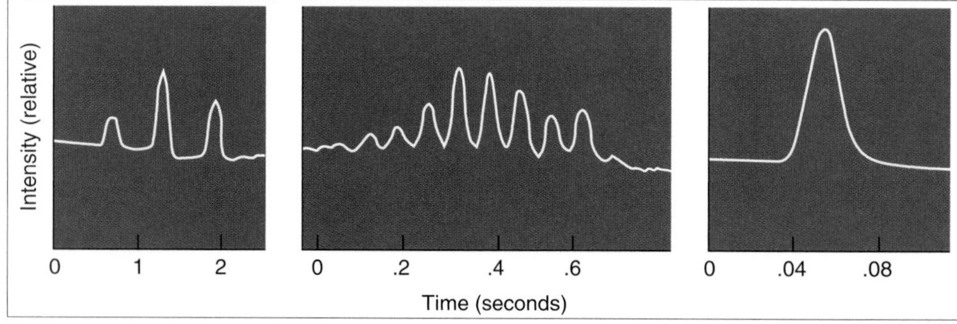

Flash patterns

These patterns show the variations in timing and intensity of flashes among three different firefly species. Note that timing can differ by a factor of ten, while intensity is less variable.

A female firefly thus easily recognizes the pattern used by a male of the same species and flashes back. The length of time it takes her to respond, and the duration of her response, alert the male that she is a potential mate, and he flies to her.

In southeast Asia, the males of some firefly species gather in crowds in trees and flash in unison every two seconds. This coordinated, amplified signal has the power to attract the attention of females, and other males, from great distances. The huge turnout increases chances for successful matings.

On a less romantic note, the females of some species of fireflies use their signals in another behavioral strategy having to do with survival: They attract males of different species as prey. In one species, the females respond to the characteristic flash-and-flight pattern produced by males of their own species. After mating, they begin to mimic the responses of females of other species. When a male of one of those other species comes to call, he is pounced upon and devoured; for him, a really unfortunate blind date ends in becoming fodder for the female's continued survival and her production of offspring.

A bacterium, *Photorhabdus luminescens,* is the only terrestrial microbe known to luminesce. As in fireflies, the enzyme luciferase triggers the light-producing reaction in these bacteria. They live in the gut of a microscopically small worm that preys on caterpillars and some other insects. The worm-invaded prey insects glow brightly as the bacteria multiply within them.

Question.

What survival advantage might the ability to luminesce bring to these bacteria?

Answer...

The light created by the bacteria in the prey insects could attract additional prey for their worm host. It could also attract passing birds, which would pick up the insect and its resident bacteria and deposit them in an entirely new territory, thus extending the bacteria's territory and access to prey.

THE CONJUGATION OF THE PARAMECIUM

This has nothing
to do with
propagating

The species
is continued
as so many are
(among the smaller creatures)
by fission

A paramecium

(and this species
is very small
next in order to
the amoeba, the beginning one)

The paramecium
achieves, then,
immortality
by dividing

But when
the paramecium
desires renewal
strength another joy
this is what
the paramecium does:

The paramecium
lies down beside
another
paramecium

Slowly inexplicably
the exchange
takes place
in which
some bits
of the nucleus of each
are exchanged

for some bits
of the nucleus
of the other

This is called
the conjugation of the paramecium.

An amoeba

Muriel Rukeyser, *The Speed of Darkness*, 1968

Variations on the Gene-Mixing Theme: The Reproductive Habits of Unicellular Creatures

Unicellular organisms, such as bacteria and paramecia, don't have to exchange genetic information in order to reproduce; they can reproduce asexually. By and large, they produce complete new organisms the way multicellular organisms produce new cells: by doubling their DNA, growing, and then dividing. This process, illustrated below, is called fission.

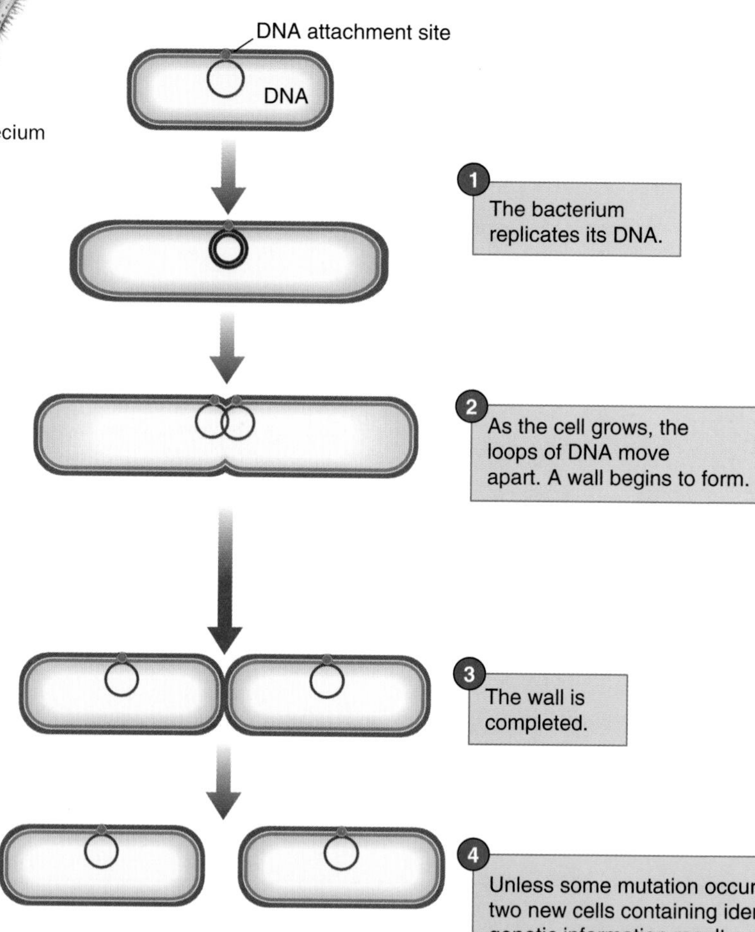

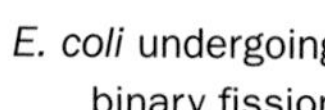

DNA attachment site

1 The bacterium replicates its DNA.
2 As the cell grows, the loops of DNA move apart. A wall begins to form.
3 The wall is completed.
4 Unless some mutation occurs, two new cells containing identical genetic information result.

E. coli undergoing binary fission

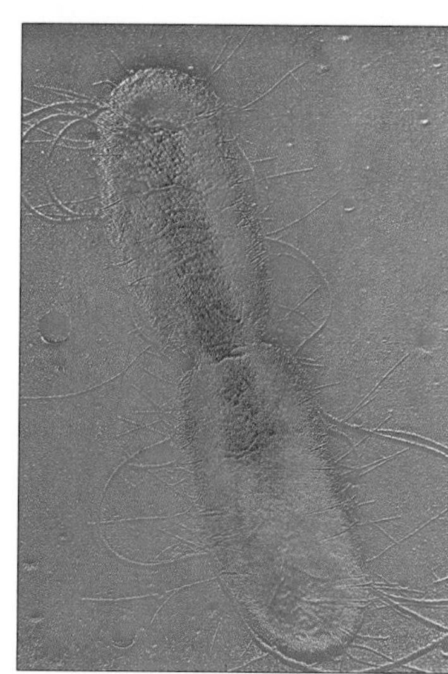

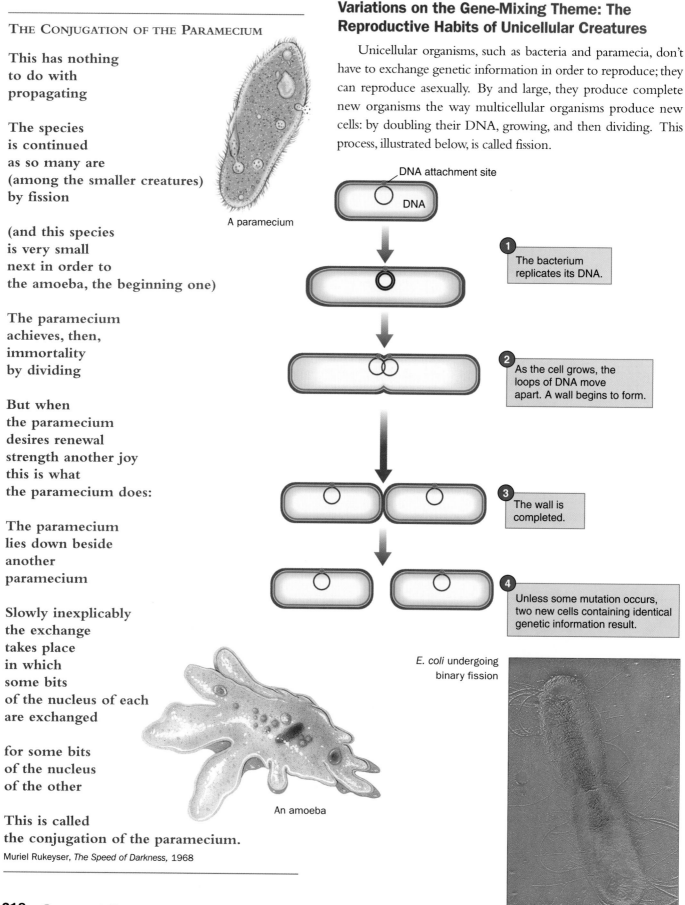

This kind of reproduction leads to large populations consisting of genetically identical organisms, clones. Since such a population lacks genetic diversity, a single major change in the environment could wipe it out entirely. While there's always the chance there will be a mutant survivor, these unicellular organisms have evolved their own means of protecting themselves from extinction by exchanging genetic information. They can, for instance, pick up and incorporate into their own genomes DNA released when other cells around them die and break up. This is called transformation. (This process can be observed in the laboratory and it provided the first convincing evidence that DNA was, in fact, the genetic material). In addition, bacteria can join together, by tubes — pili — and transmit DNA from one to another — a process called conjugation.

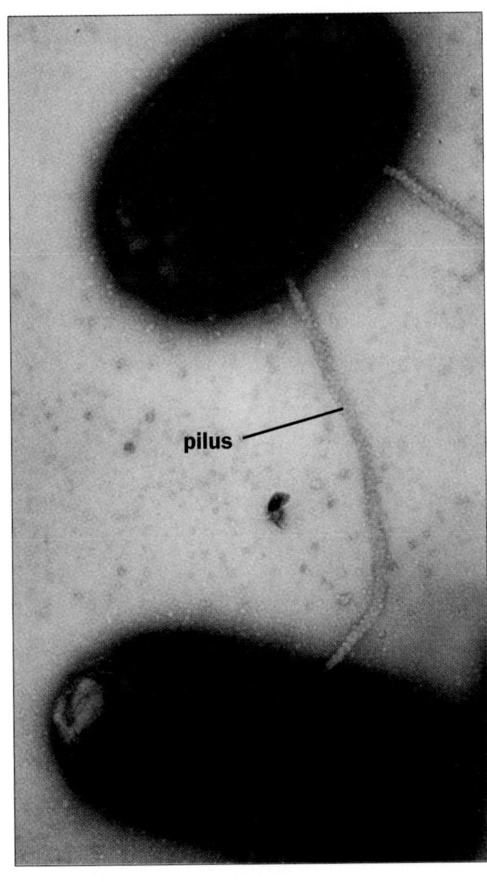

Conjugation

The purple structures in this electron micrograph are *E.coli* bacteria exchanging short segments of DNA via a pilus (the green threadlike structure).

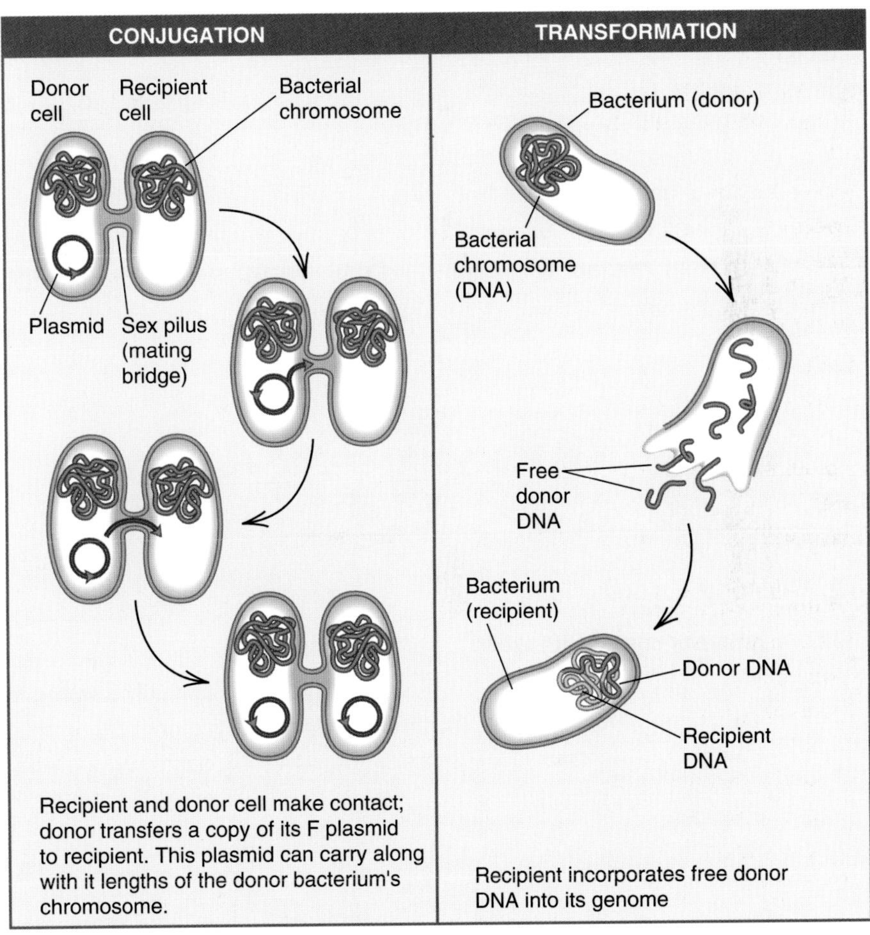

Transformation and conjugation are two mechanisms that result in increased diversity of bacterial gene pools.

Question.

How can genetic diversity result from DNA exchange between two identical cells?

Answer...

If the DNA segment from one cell is incorporated into the other cell's DNA in the middle of a gene, it produces a new nucleotide sequence. The result is a mutant gene and a different protein—genetic diversity has been introduced.

8.13 Mutations

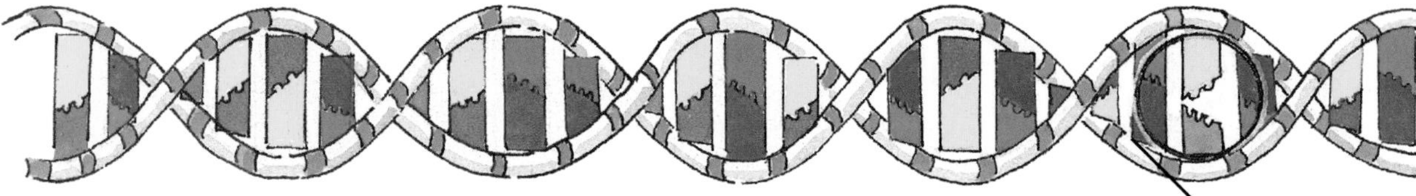

How Chance Events Introduce Novelty

Mutations are chance changes in the nucleotide sequence of DNA. When DNA is being doubled during cell division, mistakes — like typographical errors — are occasionally made: The wrong nucleotide is inserted into the growing strand. That error will be copied into all future generations of DNA.

Any change in DNA is automatically copied into messenger RNA, so the protein made from such a messenger may have one differing amino acid. Depending on what part of the protein is altered, this change may have no effect on the protein's function, damage it, or, rarely, improve it.

Most mutations do not improve the capabilities of an organism that has already been perfected by evolution over millions of years — any more than random letter substitutions improve a poem. But, once in a while, a mutation does confer an advantage, which will then be passed on to offspring. It is these rare novel improvements in protein function that account for evolutionary innovation. Randomness introduces newness.

GENETIC TYPOS

A mutation is a mistake that changes the information content of a gene — just as a small typographic change can alter the meaning of a sentence:

A stitch in time saves < nine. / none.

He who laughs < last / least > laughs best.

1. Copying Errors

DNA is generally copied with great accuracy...

...but occasionally a wrong nucleotide gets inserted.

2. Damage to DNA

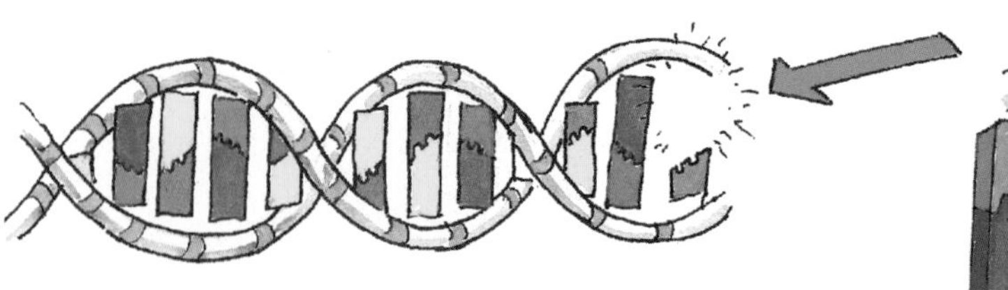

Radiation (ultraviolet, X-rays, etc.) or toxic chemicals can occasionally damage a nucleotide...

...breaking it so it's "unreadable."

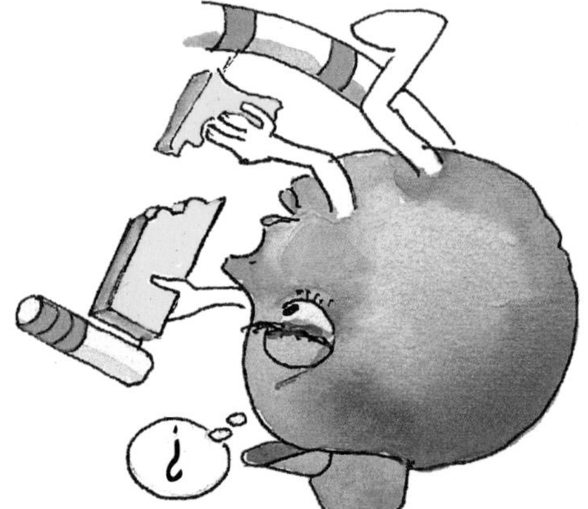

During repair, the wrong nucleotide may get inserted.

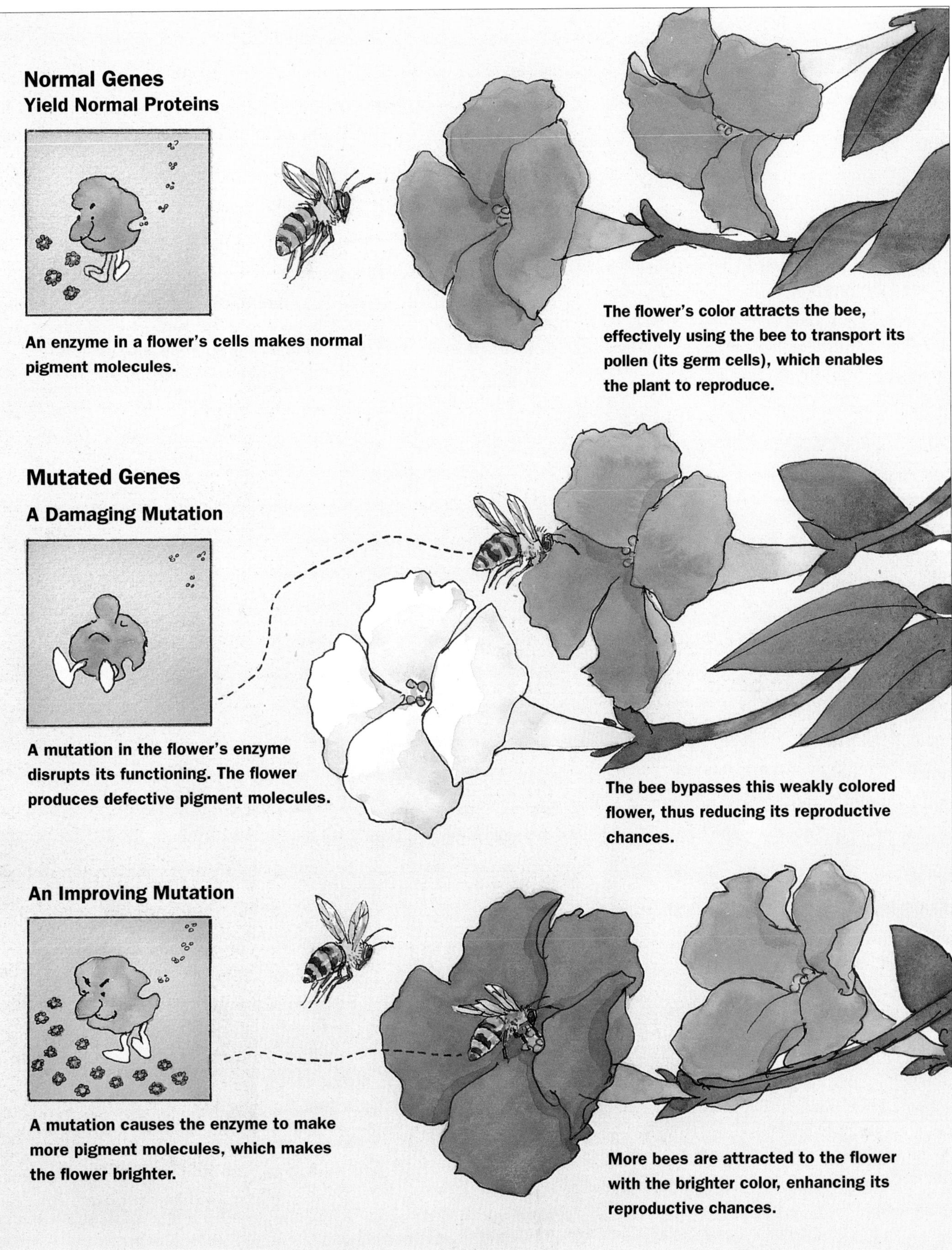
Normal Genes
Yield Normal Proteins
An enzyme in a flower's cells makes normal pigment molecules.
The flower's color attracts the bee, effectively using the bee to transport its pollen (its germ cells), which enables the plant to reproduce.
Mutated Genes
A Damaging Mutation
A mutation in the flower's enzyme disrupts its functioning. The flower produces defective pigment molecules.
The bee bypasses this weakly colored flower, thus reducing its reproductive chances.
An Improving Mutation
A mutation causes the enzyme to make more pigment molecules, which makes the flower brighter.
More bees are attracted to the flower with the brighter color, enhancing its reproductive chances.

8.14 Evolutionary Breakthroughs

Creating New Patterns

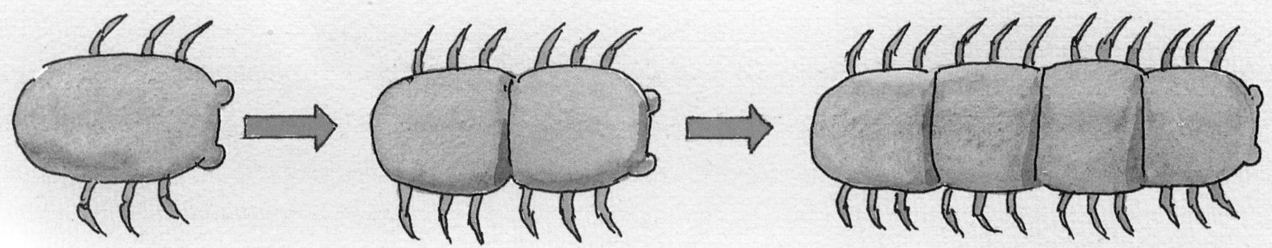

Added Segments

A mutation in a gene controlling the body organization of this hypothetical organism accidentally makes a duplicate body... creating a "Siamese twin" offspring. In future generations, the gene repeats its mistake, adding more body segments.

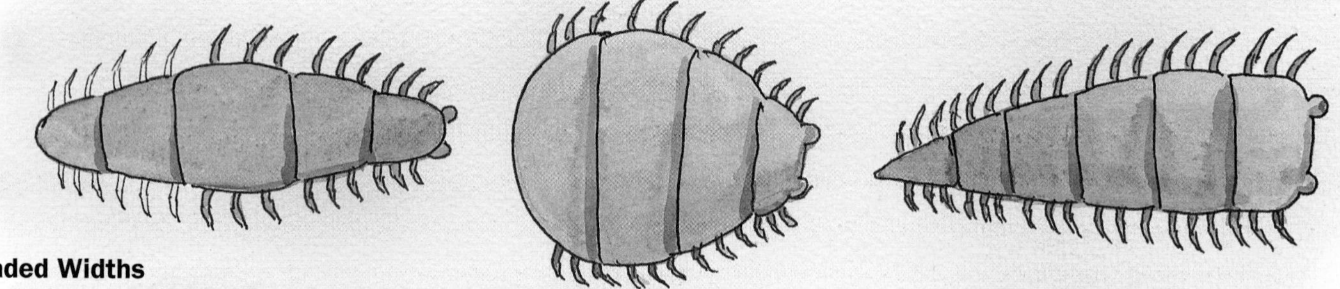

Graded Widths

Mutations in another gene create progressive gradations of segment widths, which cause the organisms to taper or bulge in various ways.

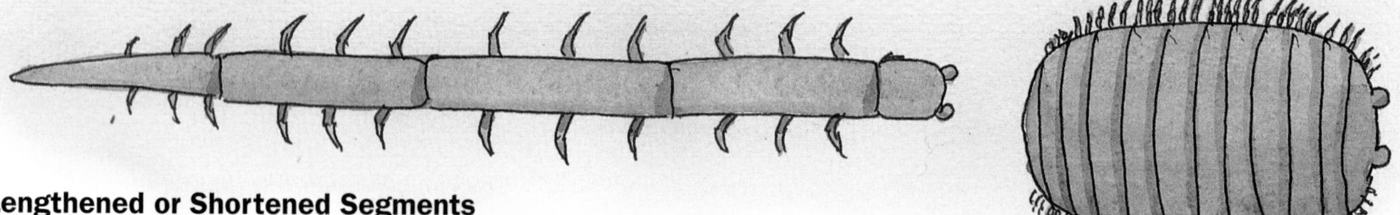

Lengthened or Shortened Segments

Mutations in still another gene lengthen or shorten the segments (see page 284).

Sometimes mutations in developmental genes give rise to a whole repertoire of body plans.

Small Mutations — Big Jumps

Ninety-nine percent of the genes of chimpanzees are identical to ours. The remaining one percent of genes somehow give us erect postures and less hair, larger skulls, and bigger brains than chimpanzees. These are almost certainly "switch" genes — ones that turn on or off other genes during embryonic development (see Chapter 7, *Community*). Small delays in timing during the formation of our skulls and brains, for instance, could contribute to our larger craniums and greater reasoning ability.

Similarly, consider the difference between giraffe necks and human necks. Both have the same number of bones (seven), but a giraffe neck vertebra can be seven inches long while a human one is less than an inch. Imagine, in the embryo of a primitive giraffe, a heritable gene "defect" in a switch gene in neck bone cells that causes it to be stuck in the "on" position, thus creating longer-than-usual vertebrae.

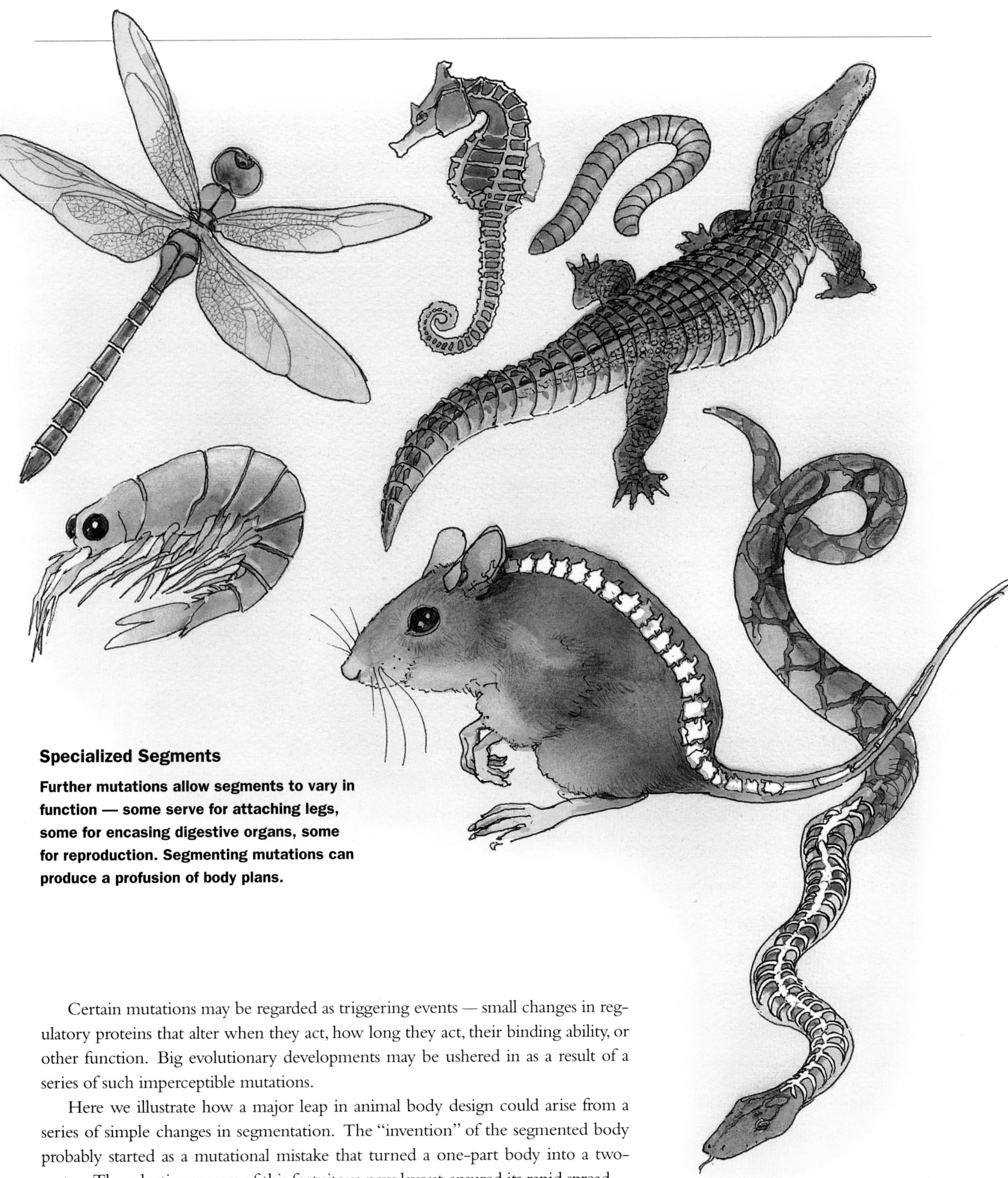

Specialized Segments

Further mutations allow segments to vary in function — some serve for attaching legs, some for encasing digestive organs, some for reproduction. Segmenting mutations can produce a profusion of body plans.

Certain mutations may be regarded as triggering events — small changes in regulatory proteins that alter when they act, how long they act, their binding ability, or other function. Big evolutionary developments may be ushered in as a result of a series of such imperceptible mutations.

Here we illustrate how a major leap in animal body design could arise from a series of simple changes in segmentation. The "invention" of the segmented body probably started as a mutational mistake that turned a one-part body into a two-parter. The selective success of this fortuitous new layout ensured its rapid spread — and made it more likely that additional segmentations, when they arose by chance, would also be successful.

8.15 Mobile Information

"Jumping Genes"

Suppose that every now and then someone took one of your books, randomly ripped out a page, and reinserted it in another chapter. Something like this happens in unusual alterations in DNA. Certain enzymes act like scissors, snipping out short lengths of DNA. Other enzymes splice these segments into new locations, similar to what happens during recombination of female and male genes to produce sex cells. (See page 315.) Such *transpositions* happen rarely, but when they do, these "jumping genes" can affect the proper functioning of nearby genes. Scientists aren't sure why these transpositions happen. Just as the torn-out and reinserted page of your book will muddle the text, most transpositions muddle the information in the genes. Occasionally, however, some lead to useful innovations.

Hmmm . . .
what to do?
Cut here,
maybe?

Occasionally a "cutter" enzyme mistakenly grabs a segment of DNA containing one or more genes...

snip
snip

...and cuts it out of its usual location.

stitch
stitch

The segment then breaks free, rolls into a circle (a "stitcher" enzyme splices the cut ends together)...

...and moves to a new location on the chromosomes.

CELL NUCLEUS

Certain circular segments of DNA that have split off from the host genome can replicate independently. These are called plasmids.

Corn and Cold Spring Harbor

Remarkably, most of what we know about the molecular mechanisms of inheritance and genetic disease in humans has come from studies in peas, fruitflies, yeast, bacteria, and corn.

Barbara McClintock, a geneticist at the Cold Spring Harbor Laboratory on Long Island, New York, was the pioneer who discovered jumping genes in the chromosomes of corn plants. Her work revealed that genes are not static but can be rearranged by events that occur naturally inside cells or by outside agents such as X-rays that cause trauma to cells. McClintock's work also led to the realization that there are two kinds of genes: ones that give instructions for specific functions (that is, code for worker proteins) and ones that turn those functions on and off (that is, code for regulator proteins). McClintock was honored with a Nobel Prize in 1982.

Plasmids

Sometimes segments of DNA are cut loose by enzymes but are not reinserted elsewhere in the genome. Instead, they roll up into circles and replicate indefinitely as separate genetic units called plasmids, functioning like tiny extra chromosomes.

Plasmids may consist of only a few thousand nucleotides, having just enough genetic information to allow them to replicate themselves independently of their host cell's chromosomes. Or they may have genes that code for proteins useful to the host. For instance, some plasmids in bacteria carry genes that code for proteins that destroy antibiotics — making the bacteria immune to these drugs. Other plasmids enable their hosts to produce poisons that kill other bacteria. Still others, as you saw on page 319, make it possible for one bacterium to inject its DNA into another — a kind of primitive sex.

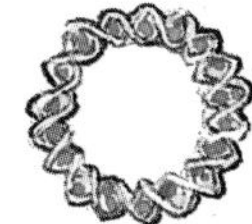

Antibiotics: Too Much of a Good Thing

Microbes are the largest mass of living matter on our planet. And most of those in contact with us and our fellow creatures are benign. They not only pose no threat to us but they often bestow benefits. We are home to millions of them, and over vast stretches of evolutionary time we have comfortably adapted to one another.

The trouble, of course, is with those few species that have developed ways of evading our defenses and causing disease. We owe much to antibiotics for protecting us from many of these virulent pathogens. And since the introduction to widespread use of the first powerfully effective antibiotic, penicillin, in the early 1940s, we have witnessed the saving of many millions of lives. The researching, production and marketing of antibiotics has since become a multibillion dollar industry, and not just for the treatment of human disease. Now the animals we raise for food are routinely dosed with antibiotics to prevent diseases from spreading in their often very crowded quarters. Antibiotics also help farm animals grow faster and larger by keeping low grade bacterial infections at bay.

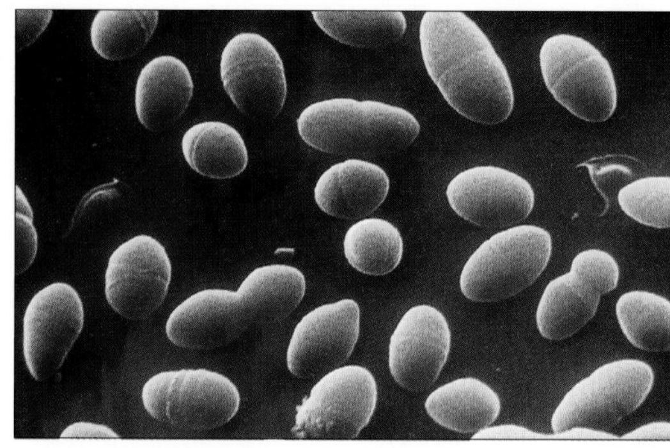

Enterococci bacteria

The downside to all this success, however, is serious. We are accelerating a dangerous experiment in evolution akin to the one performed by Salvadore Luria on page 342. For by inundating our environment and the food we eat with antibiotics, we are encouraging the natural selection of disease-causing bacteria that have gained the ability to resist the lethal effects of antibiotics through mutation.

Antibiotic resistance is not a new phenomenon. From the advent of penicillin, antibiotics have encountered the occasional bacterium resistant to it. Lately, however, the use of antibiotics has become so widespread that resistant bacteria are undergoing natural selection pressure. By our practices we are encouraging resistant forms to survive and flourish.

Why not just use different antibiotics? That was the reasoning followed for many years. It worked. But today, several strains of bacteria are resistant to every known antibiotic. How did this happen?

Since bacteria multiply very rapidly (some double as often as every half hour) and since, as naked single cells, bacteria are maximally exposed to their environment, they accumulate mutations very rapidly. Furthermore, as we have discussed earlier, bacteria can transfer genetic information (including antibiotic resistance) between cells.

Treating a person (or an animal) with antibiotics has several consequences: if the dose is less than optimal for killing off all the disease-producers, the occasional resistant mutant has a chance to

arise, survive, and multiply. Furthermore, we kill many "good" bacteria in the process — bacteria that inhabit our bodies and protect us. Over time, more bacteria become resistant to more of the antibiotics that medical researchers develop.

Perhaps you've heard of vancomycin, often called the antibiotic of last resort because it isn't used until everything else has been tried. So far, vancomycin has been largely successful, even against methicillin-resistant *Staphylococcus aureus* (MRSA), one of the most serious hospital-acquired infections. But now the plot thickens.

Enterococci are strains of bacteria that inhabit the gastrointestinal and genital tracts of humans and many animals. They don't normally cause problems, but recently the use of avoparcin, an antibiotic and growth promoter that was developed specifically for farm animals and is chemically very similar to vancomycin, has dramatically increased the presence of vancomycin resistant enterococci (VRE) in animals in Europe and Australia. Humans acquire VRE via the food chain and amplify it through the medical use of other antibiotics similar in structure to vancomycin. VRE is now widespread in the developed world. In the United States, where avoparcin has never been used, VRE has spread primarily among hospitalized and acute care patients. There is no effective treatment, and some patients die from the infection.

The coexistence of VRE and MRSA in hospitals suggests that some populations of MRSA may soon become fully vancomycin-resistant. In fact, the emergence of MRSA that is not completely susceptible to vancomycin has been documented. A staphylococcus that is resistant to every weapon in the medical arsenal is a truly frightening creature.

Evolution Here and Now

Cases of antibiotic-resistant tuberculosis and pneumonia are increasing rapidly worldwide, and a number of other common bacteria are showing signs of increasing resistance. What can be done? Clearly, we need to control our use of antibiotics, both medically and agriculturally. Steps have already been taken to limit the routine use of some antibiotics in farm animals. Medically, antibiotics have been available for years, more or less on demand, even for treating viral infections such as colds and flu (which aren't affected by antibiotics). We need to use antibiotics much more selectively and only when and where we really need them. Most of us can get along fine using just soap and water for household chores, thereby not subjecting the bacteria around us to the selection pressure of constant exposure to antibacterials.

Remember, if you were never exposed to any infectious agents, you would have no resistance to them. (During the age of exploration, highly infectious diseases were often carried from Europe to new environments, where they were even more dangerous; a famous example was the importation of smallpox to the Americas. Native Americans, having never been exposed to this virus, had no resistance to it and died by the tens of thousands.) If bacteria aren't routinely exposed to antibiotics, resistant individuals will not be selected for.

8.16 Viruses

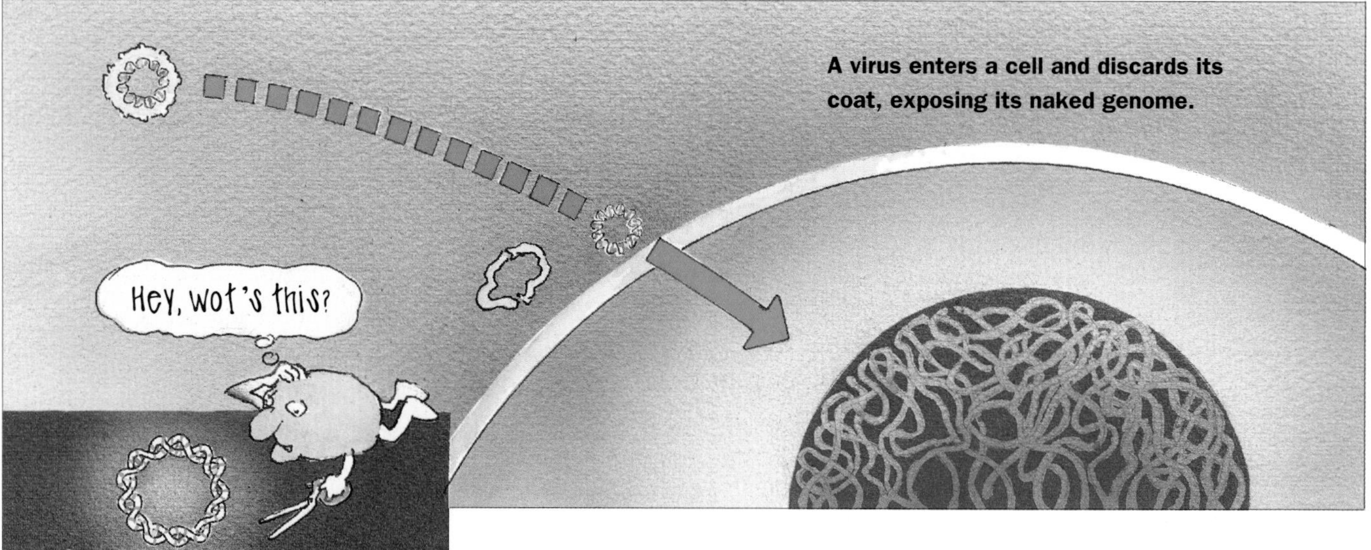

A virus enters a cell and discards its coat, exposing its naked genome.
Hey, wot's this?

It approaches a location on the unsuspecting host's genome...

...and gets spliced into it.

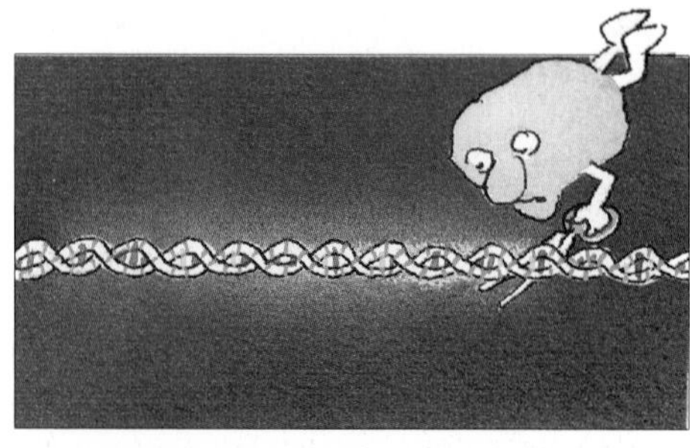

When it departs many cell generations later...

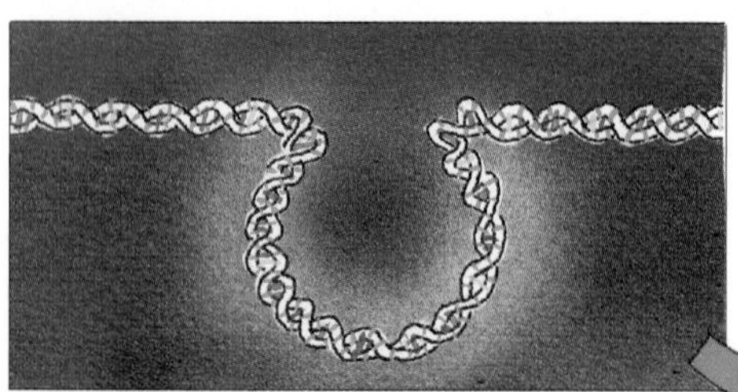

...it may take some of the host's genes along with it.

The Uninvited Guest

Certain independently reproducing pieces of genetic information have become obstreperous over the course of evolutionary history. These plasmids have evolved the ability to use the host cell's ATP and ribosome machinery to make protein coats for themselves — to become viruses. Shielded by their protective coats and making use of certain enzymes they produce, viruses make their escape from the host cell. From there, they invade other cells, take over their machinery, and make many more copies of themselves. Viruses can alter cells' function (as with the common cold) or destroy their function (as with AIDS, which is caused by HIV, the human immunodeficiency virus). Viruses may unobtrusively splice their genes into their victims' DNA, thereby subtly changing the infected cells' genetic character. And when they subsequently cut themselves out of the host's DNA, they may "kidnap" some of the host's genes. Then, as these viruses hop from one infected cell to another, they can transfer normal cell genes along with their own genetic material.

It appears, then, that viruses originally arose from cells and have interacted with cells throughout evolution — sometimes to the cells' detriment when they cause disease, sometimes in ways that produce evolutionary advantages. Viruses are a genetic shuttle between differing forms of life.

All this shuffling of genes inside cells and among cells means that life's information is constantly being reorganized. Simple mutations, gene transpositions (jumping genes), sexual recombination, and the existence of plasmids and viruses all contribute to enriching the ocean of variation fished by natural selection.

In this way, viruses carry new information from cell to cell.

The Bacterium's Nemesis

A bacteriophage (virus) injects its DNA into a bacterium.

The invading viral DNA orders the cell's machinery to make multiple copies of it...

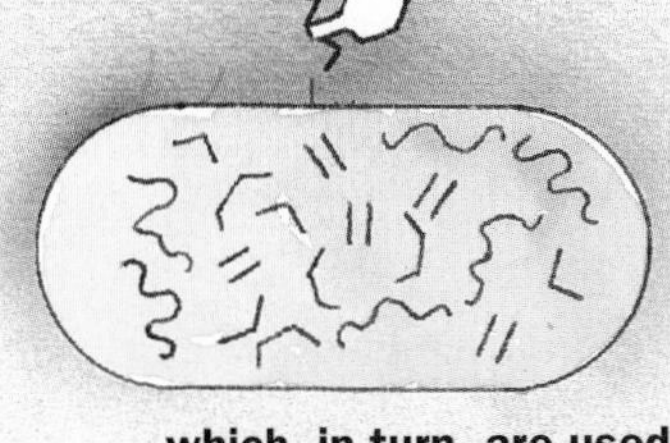

...which, in turn, are used to make multiple copies of the virus's proteins...

...which spontaneously assemble...

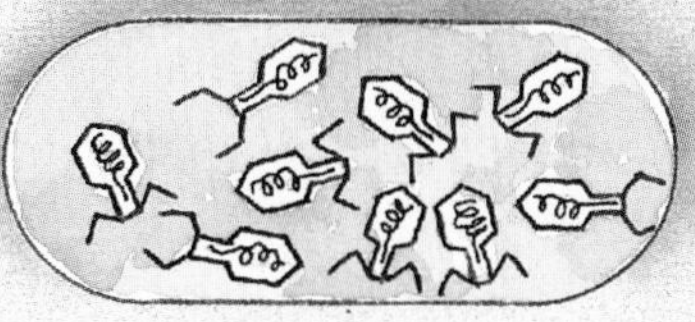

...into new viruses...

...which destroy the bacterium and escape.

Scientists have learned a lot about the relationship between viruses and their hosts by studying the behavior of a peculiar sort of virus called a bacteriophage (literally, "bacterium eater"). Phages (for short) are tiny DNA-filled syringes with protein coats. These marauders go on a take-over mission, attaching themselves to bacterial cells with spider-like "feet" and then injecting their DNA. Information in the phage's DNA prevents the bacterium from using its own protein-making machinery and diverts it to the construction of proteins for the phage. After about 20 minutes, the bacterium is chock full of 100 or so brand-new DNA-filled phages. In the ultimate insult, the phages instruct the bacterium to make an enzyme that breaks open the bacterium's wall. This kills the bacterium and releases the phages to go on to infect other cells!

Occasionally, after a phage has injected its DNA into a bacterium, nothing appears to happen; the bacterium goes right on growing. The phage's DNA, in this case, has spliced itself directly into the bacterium's DNA, where it lies dormant. Many bacterial generations later, the phage's DNA may emerge, subvert the bacterium's protein-making machinery to make new viruses, and then burst out of the bacterium to hunt for the next victim. Sometimes, when this happens, the phage carries along with it some of the bacterium's genes, transferring them to the next bacterium it attacks. Thus, many different species of bacteria seem to be connected by viruses in an immense gene pool in which information is constantly reshuffled.

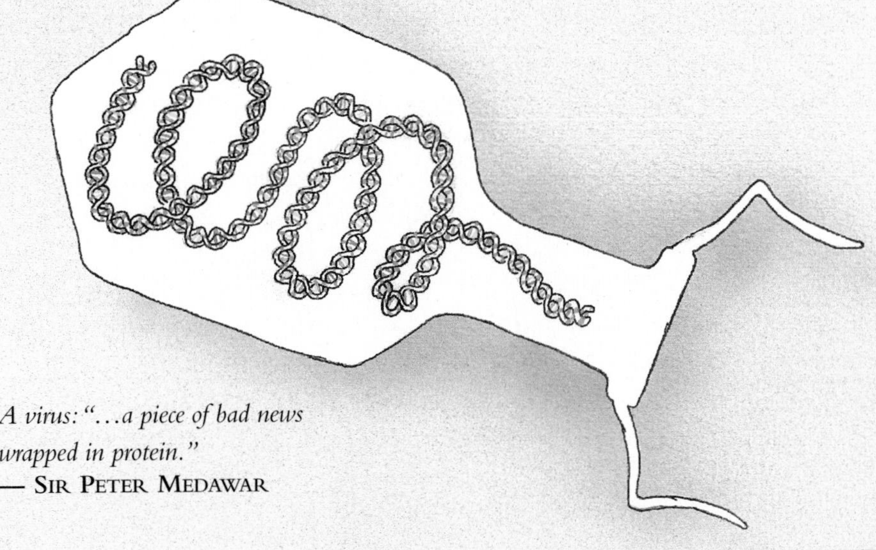

A virus: "...a piece of bad news wrapped in protein."
— Sir Peter Medawar

8.17 How New Species Arise

Necessity Is the Mother of Invention

Over life's history, the number of species — a species being a population of organisms that reproduces successfully only with its own members — has increased into the many millions. The wildebeest's trek across the Serengeti is a vivid illustration of how the basic mechanism of change and selection helps a species adapt and yet maintain its essential character. How does this mechanism work to bring new species into being?

We can think of the first form of life on Earth as a trunk from which new life forms have branched, and branched again, and again. Each pair of organisms, or species, has thus left behind it, forever, a common ancestor — just as a tree's branches spring from the trunk that produced them. As species mate and multiply, changing slowly or rapidly, depending on the demands, constraints, and opportunities of their environments, they branch farther out from the mother trunk (see page 345).

Remember that every species' potential for adapting to its environment — to change — hinges upon hidden capabilities residing in its pool of genes. This pool is stirred by sex, mutations, transpositions, and the other kinds of gene alternations we've discussed in previous pages. These produce changes in protein machinery that control a creature's ability to run or swim faster, to see better, to camouflage itself, to produce a useful digestive enzyme, etc.

Species begin to adapt and change when the environment presents new opportunities or dangers. In response, heretofore hidden skills of certain individuals within the population automatically come into play to help them get food, a mate, a home, or to avoid becoming someone else's meal. These new skills successfully reproduce in the population and eventually predominate. It is in this sense that the environment *selects*.

Thus, one species may branch into two, each adapted to two different food sources. They might even adapt to the same food if one eats by day and the other by night, or if they employ different food-gathering strategies (think of lions and cheetahs — teamwork vs. speed.)

Among several others, one important factor in the evolution of new species is geographic isolation. If some members of a species happen to get separated from a larger group, ending up in a very different environment — on an island or on the other side of a mountain range, glacier, or body of water — they diverge from the original population over the ensuing generations. If members of this now new species were to be reintroduced into the original population, their genetic pool would be so different that they would no longer be able to mate and bear offspring with them.

In this drive toward incredible profusion of form and function, we discern a dynamic pattern: The molecular machinery of cells, fueled by an inexhaustible flow of energy from the Sun — and abetted by chance changes in genes and selection of those changes by the environment — inexorably propels life toward greater complexity.

Beak Performance ▶

Isolated on different islands in the Galapagos, finches have, in the course of becoming food-gathering specialists, evolved into different species. All these finches resemble species from the South American mainland more than they resemble finches found in other parts of the world.

High-speed Natural Selection

Recently, scientists banded and observed some 20,000 birds over a period of twenty years in the Galapagos. They noted that within the flock there were finches with big beaks, which worked better for cracking tough spiky seeds; others with smaller beaks, better for tiny seeds. After a severe drought, the spiky-seed plants predominated. Predictably, the birds with bigger beaks enjoyed heartier meals and produced more well-fed offspring. Several years later, after a protracted wet season, the tiny-seed plants flourished, and the small-beak finches regained dominance.

These changes followed Darwinian principles — at a surprisingly rapid pace. It should be noted that the big beak and small beak variations occur within the *same* species. But it's not hard to imagine a situation in which a single finch population somehow gets separated into two groups — one inhabiting a "dry" island, the other a "wet" island. Over the generations, we'd expect the two groups could evolve into distinct "big-beak" and "small-beak" species. This is similar to the scenario Darwin encountered when he observed the finches in the Galapagos islands (see below).

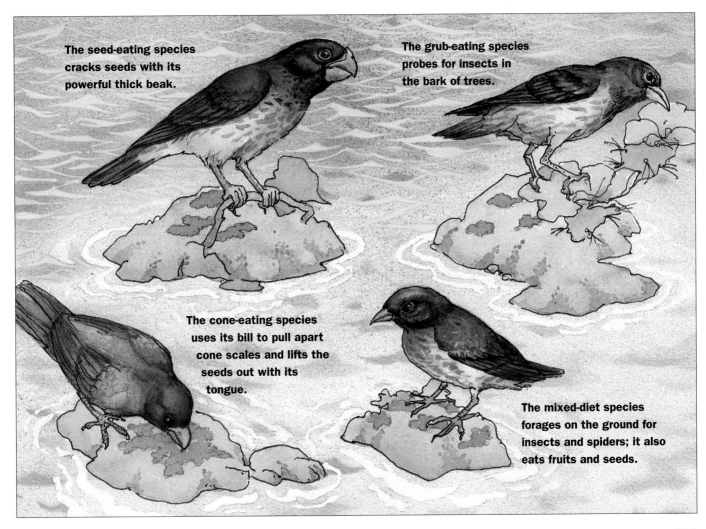

The seed-eating species cracks seeds with its powerful thick beak.

The grub-eating species probes for insects in the bark of trees.

The cone-eating species uses its bill to pull apart cone scales and lifts the seeds out with its tongue.

The mixed-diet species forages on the ground for insects and spiders; it also eats fruits and seeds.

The Way We Walk

When our early primate ancestors tentatively moved out of their densely forested environments onto flat, open savannahs some 1.8 million years ago, adaptations more suited to that new environment became increasingly successful. One of these was a skeletal structure suitable for efficient, upright walking and running.

It takes energy to sustain life, so successful organisms must be good at harvesting energy or efficient in their use of it, or both. Evolution tends to favor traits that allow more to get done with the same or even less energy.

The energy efficiency of the way we humans walk probably contributed to our success as a species. Our balanced upright position and gliding gait is more energy-efficient than the tilting motion of the ground-dwelling apes whose ancestors were related to ours. The illustration below shows that baboons, gorillas, and chimpanzees do not walk upright as we do. Comparatively more of their body weight (their center of gravity) lies in front of their walking legs. As a result, they have to exert a good deal of muscular energy to keep from falling forward as they walk, just as you would if you tried walking in a crouched position.

Besides differing from us in their distribution of body weight, apes differ in the way their leg muscles are attached to their skeleton. When an ape lifts its left or right leg, its body tends to tip to that side. To maintain its balance, it must bend its body

↓ = Primate centers of gravity

Notice that only in the human does the center of gravity parallel the vertical axis of the skeleton.

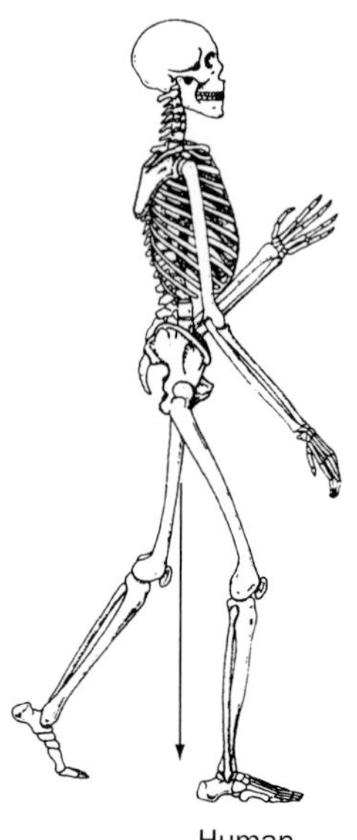

Human

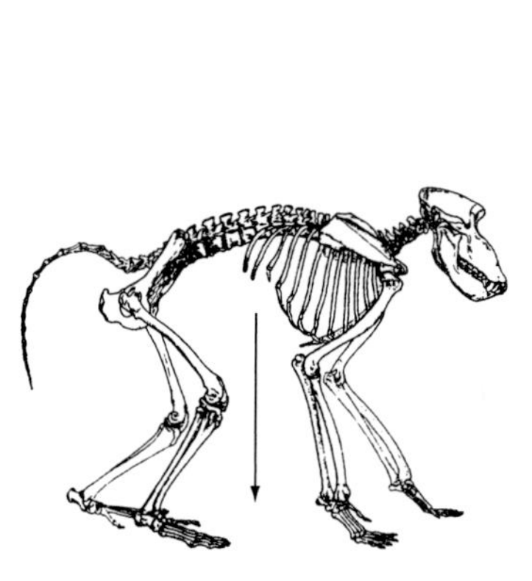

Baboon

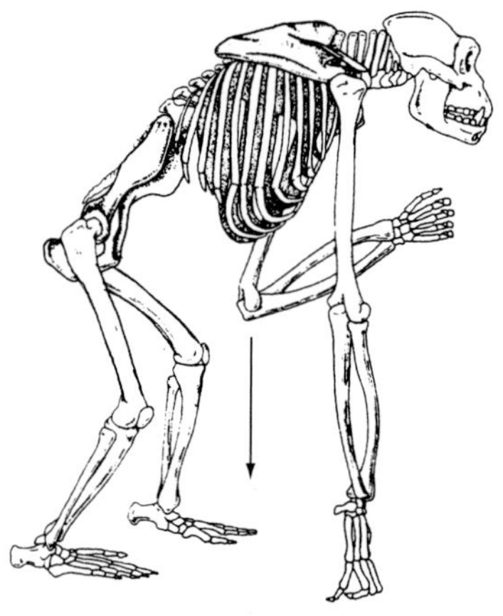

Gorilla

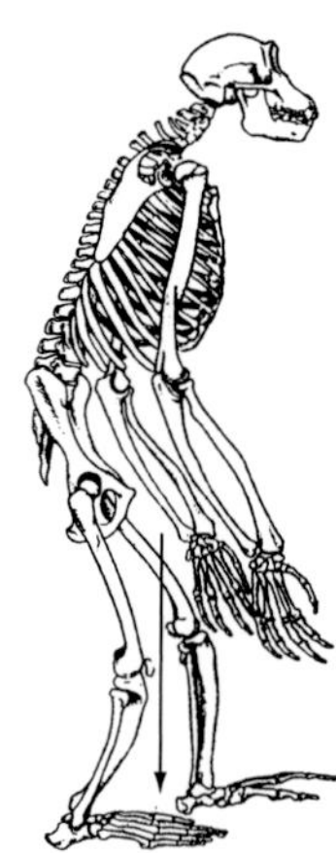

Chimpanzee

Working to walk

Every shift in the walking ape's center of gravity must be corrected for by contractions of counteracting muscles. Every contraction of a tiny actin-myosin complex requires ATP. The ape needs glucose and plenty of oxygen to keep its muscle cells respiring and making that ATP.

over the leg it has lifted. This bending causes the ape to tilt from side to side as it walks. The trunk muscles the ape has to use in order to compensate for the tilt make continual energy demands.

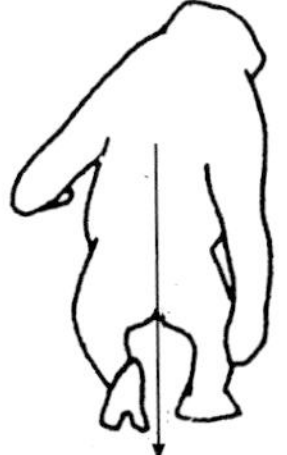

We can see another walking-related result of natural selection by comparing the chimp's foot to the human foot. In chimpanzees, the big toe is at an angle to the rest of the toes — good for grasping but not as good for walking. In humans, the big toe is parallel to the other toes — not much good for grasping but better for a balanced stride. The illustration also shows that the human foot is relatively narrower (and can act as a tripod to stabilize the distribution of weight). We can push off straight forward, and this enhances our ability to stride.

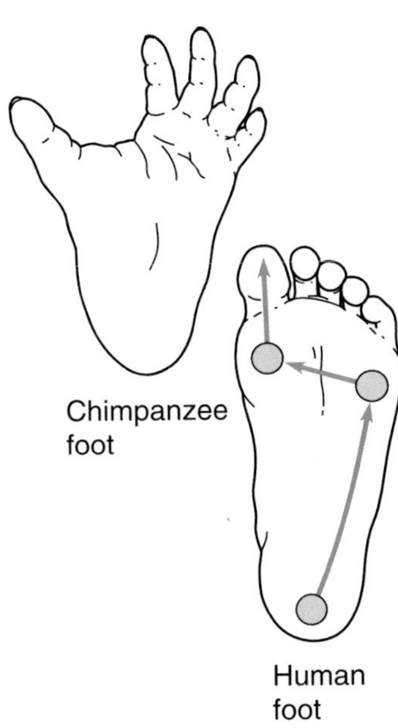

A tripod foot

While the chimp's foot is well adapted for grasping, the parallel placement of human toes enhances the ability to walk or run directly forward. The three circles indicate how balance is enhanced by a tripod for weight distribution. The arrows show how weight is dynamically distributed during "pushoff."

Here Today, Gone Tomorrow

Each species that exists on Earth today, or has ever existed, is the product of a unique set of genes — the "right stuff" for a given time and a particular ecological setting. Every species is a success story, at least in its own time and place. But time passes, and settings change. New competitors come into the picture. Climates vary. Some species become extinct; new species arise. That has been the history of life on our planet. Sometimes climates change catastrophically. Geological evidence shows that about 65 million years ago, when the dinosaurs were flourishing, a huge asteroid may have crashed into the Earth in what is now the Gulf of Mexico, sending a planet-wide shroud of dust and debris high into the atmosphere and causing enormous fires. The dust and smoke would have greatly reduced the amount of sunlight reaching the planet's surface. Climates may have grown cooler. Plant life could have been disrupted. If so, photosynthesis would have decreased, and the planet's food supply would have dwindled. During this time, called the Cretaceous extinction, many plant and animal species, including some of the largest, died out. It is thought that over half of existing species became extinct.

This enormous meteor crater in Arizona is probably far smaller than the one that would have been created by the asteroid thought to have precipitated the Cretaceous extinction.

Even before the dust settled, under those darkened skies long ago, life had probably already begun to demonstrate one of its fundamental attributes — the capacity to diversify. Specific adaptive variations in the organisms that survived were selected for, and over time, this variation and selection led to the development of new species. Maybe an animal with larger lungs, with slightly thicker fur or a quicker gait would result from the exchange of its parents' genes as sperm and egg came together. Many species had been obliterated, but the surviving gene pools still had the mechanism for making realignments, for "fleshing out" empty spaces and occupying new environments. Life re-diversified.

Today, biologists' estimates of how many species exist on Earth range from about 10 million to 100 million. Some entomologists say there may be 10 million species of insects alone!

Since its emergence, the human species has provided a new take on diversification, by doing so even without genetic change. The biological diversity represented by all other species has resulted from accidental changes in genes. Only *Homo sapiens* has diversified through its own inventiveness! Using the creative capacities of our large brains, we have been able to move into, or at least explore, virtually every environment occupied by other species. We may not be able to see as well as an eagle naturally, but we have invented devices that allow our eyes to vastly outperform those of eagles. We may not be able to breathe in water as fish do, but we can build and operate machines that take us to depths fish cannot penetrate. We have even gone where no other earthly species has gone before — into outer space. And we are the only species that can actually direct genetic change through genetic engineering.

Our very cleverness, though, may be a serious threat to other species, to the diversity that seems essential to a healthy biosphere. Researchers suggest that the Earth's overall extinction rate has increased by as much as a thousand times since humans came onto the scene. Some scientists think *Homo sapiens* may be as powerful an agent of extinction as the asteroid that presumably caused the demise of the dinosaurs and many other species millions of years ago. We are capable of diverting to our own use natural resources and spaces needed by other species. (In 1999, the Environment Ministry of Brazil announced that the rate at which its country's rain forests were being destroyed jumped 27 percent during the previous year.) We are polluting the water with our sewage and industrial wastes, and the air with gases that may cause global warming and threaten Earth's protective ozone layer.

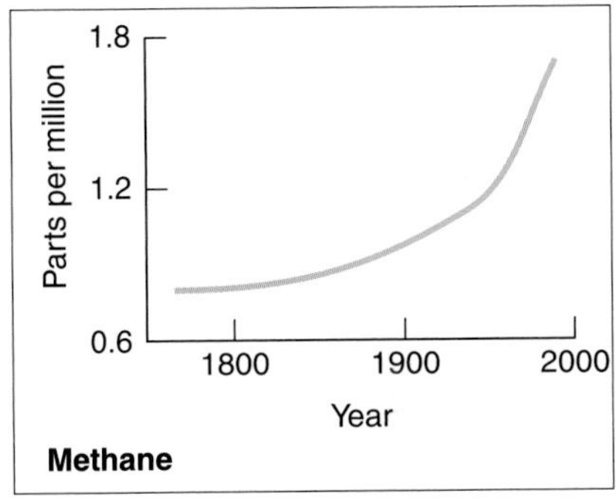

The graphs at right and below show how, in the last century and a half, the concentrations of potentially harmful gases in the atmosphere have climbed, largely as a result of human activities. Carbon dioxide, methane, and nitrogen oxides, the so-called greenhouse gases, may be contributing to global warming. Burning fossil fuels and rain forests releases huge amounts of carbon dioxide. Methane is a by-product of digestion in cattle and is also produced in great quantities by bacteria that inhabit rice paddies. As human agricultural activities have increased along with the growing world popluation, so has the production of methane. Nitrogen and sulfur oxides, produced by burning fossil fuels, are the source of the acid rain that has been so damaging to large tracts of vegetation and aquatic life.

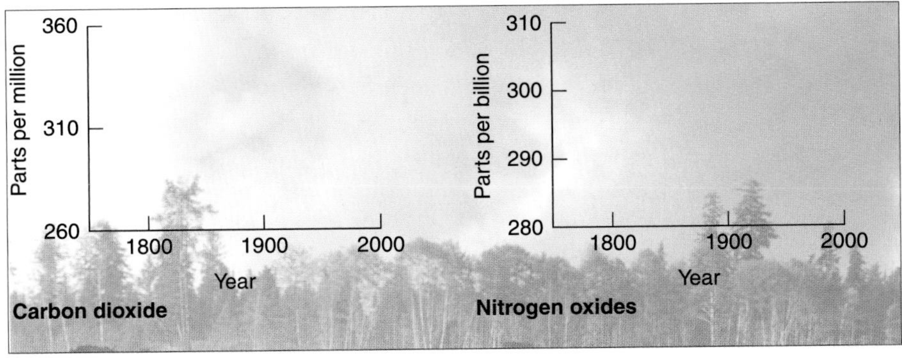

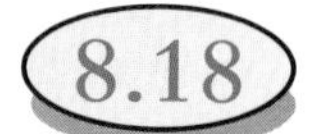

8.18 Co-evolution

Arms Races and Mutualism

Organisms don't *try* to evolve. But populations of organisms inevitably change because individuals are selected by changing environments. One of the most important features of an environment is the other organisms in it.

Evolutionary change in one creature will force changes in the creatures with which it closely interacts. If gazelles get faster, cheetahs will have to become either faster or smarter. If grass gets tougher, horses may evolve stronger teeth. And if humans develop antibiotics, bacteria may develop resistances to those drugs. These relationships can loosely be described as "arms races" that take place on an evolutionary time scale. Since each new development precipitates a counter-development, rarely does one "side" get to declare itself the winner. The process, however, can generate innovations on both sides. Thus, organisms end up in a dance of change that has the effect of keeping their relationships the same.

Digesting the Indigestible

Neither cows nor termites can independently digest cellulose — the tough chain of linked sugars that comprises the bulk of grass and wood. Fortunately, both species harbor a particular kind of bacterium in their guts that can do the job. Result: Everyone eats.

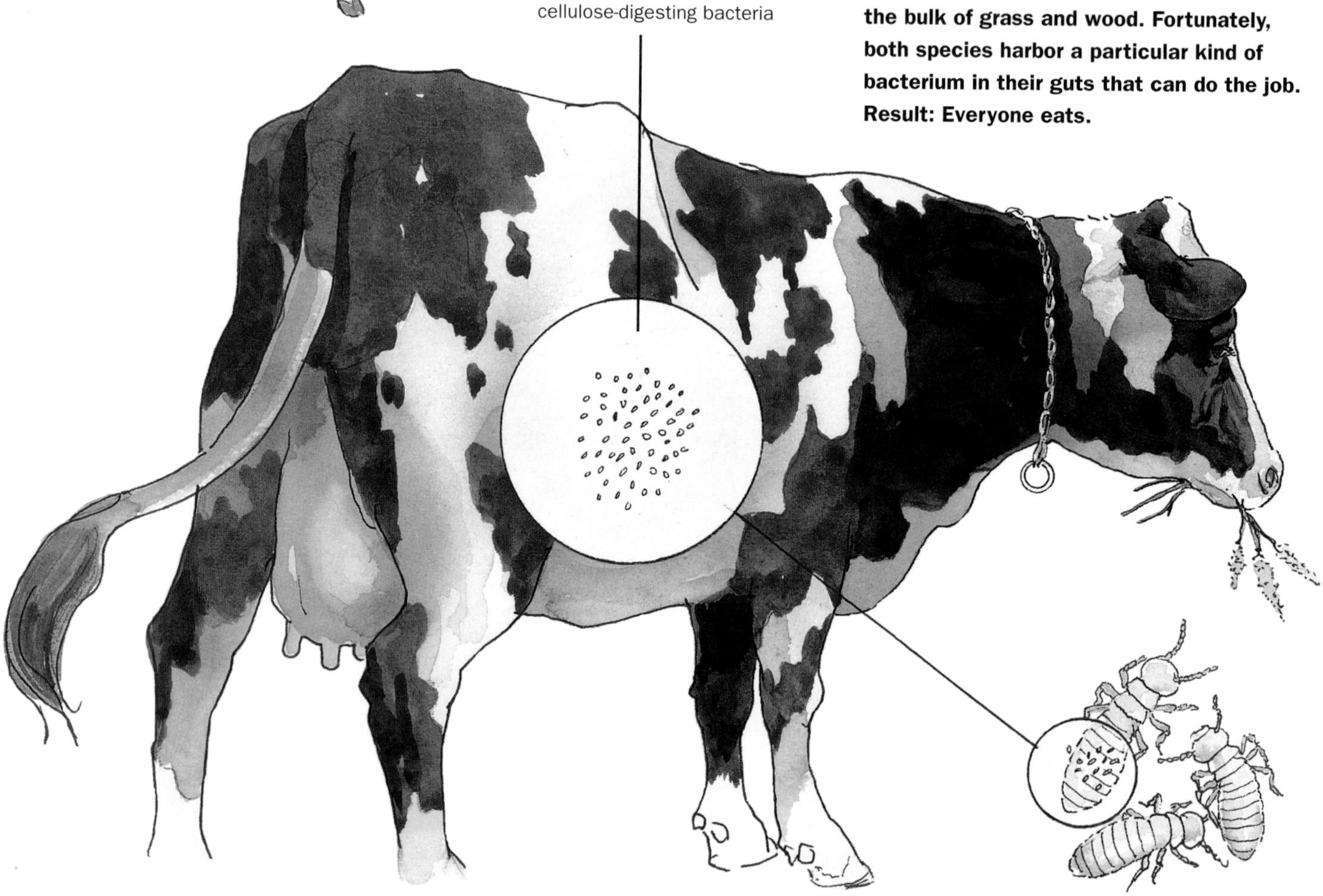

cellulose-digesting bacteria

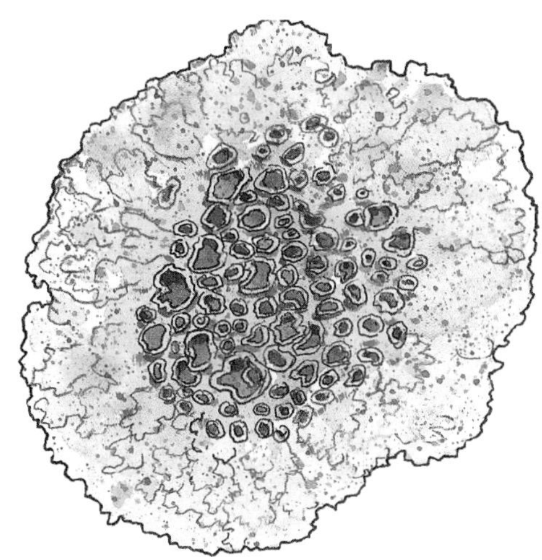

Two Organisms in One

Long ago, certain land-based fungi and water-dwelling photosynthesizing algae expanded their mutual horizon enormously by forming a permanent union — they banded together to become *lichens*. The algae provided energy through photosynthesis. The fungi enabled the algae to survive on small amounts of water without drying out. The combo can attach to rocks and other inhospitable surfaces anywhere from deserts to the Arctic — something neither could do alone.

Over time, arms races can mellow out. Enemies settle into cooperative relationships, combining their talents and pooling information. Such a relationship is called mutualism (see Chapter 2, *Patterns*).

A mutualistic leap occurred when bacteria living in the soil invaded the roots of legume plants (clover, alfalfa, and various members of the bean family). The bacteria stimulated the plants' roots to swell, creating nodules in which the invaders took up residence. The bacteria got sugar from the plants, and the plants got essential nitrogen from the bacteria. This was a particularly prized exchange for life in general because, while nitrogen gas is plentiful in air, plants can't use it in that ethereal form. The bacteria, however, can convert nitrogen into earth-bound packages of ammonia and nitrates, which plants can use to make their amino acids, nucleotides, etc. Without this arrangement, nitrogen could not flow through the living world to sustain multicellular life.

Such relationships demonstrate the power of combining information in "chunks." Genes from separate species collaborate to produce favorable evolutionary jumps that far exceed the incremental changes produced by hit-or-miss mutations.

Carnivores Perched on Herbivores

The oxpecker bird reaps a harvest of ticks and other parasites that live on the hide of rhinos and other large grazers. The birds get a free lunch; the rhinos get free pest control. As a bonus for the rhino, whenever a predator approaches, the birds swirl upward, loudly warning of danger.

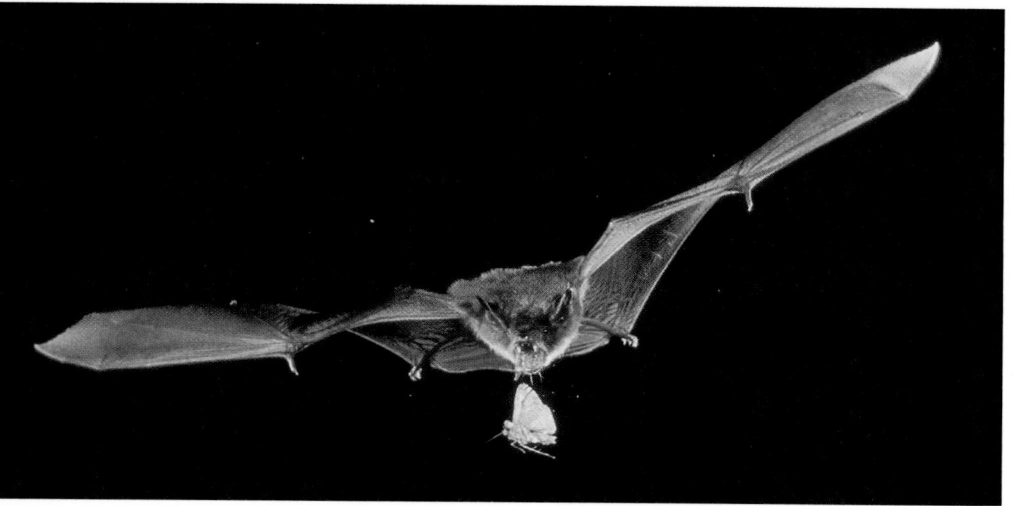

Responsive Evolution

The ability to locate prey or to sense predators in the dark is extemely important to nocturnal animals, which have evolved and specialized to function effectively in their almost lightless environment.

For example, some species of night-flying moths and bats have evolved in tandem as prey and predator. Acute hearing in some species of moths provides life-and-death information about the bats that are hunting them.

Bats navigate in their dark surroundings the way submarines do, using sonar. They emit high-pitched cries — sounds we cannot hear — and listen for the echoes. From the patterns of those echoes, they can tell what is around them, apparently in great detail. Some can catch night-flying moths in their mouths in mid-air. In response to this successful hunting strategy, evolution has shaped the community of nerve cells in some species of moths so as to make it less likely that they will become bat food. These moths have one ear (really just an eardrum) on either side of their body, under each wing, which allows them to hear bat cries and react to them fast enough to escape the mid-air attack. Time-lapse photos have shown such moths taking evasive action, diving suddenly, for example, in response to real or artificially generated bat cries.

Just four nerve cells in these moths detect where and how far away a bat is. Nerve cells in moths' ears can detect bat sounds, but how do the sounds provide usable information? They become bursts of electrical impulses. Since these impulses are the result of changes in the electric field across the nerve cell membranes, they can be detected by laboratory instruments sensitive to such activity. The impulses show up on a monitor as clusters of sharp spikes.

The readout below shows a pattern of nerve impulses (the tall spikes) generated in a moth's ear in response to the approach of a bat. When the bat is far away there are relatively few impulses in each cluster. As the bat approaches, the number of spikes per cluster increases. Thus, as the readout shows, the moth first hears the bat getting closer and then, at the right, moving away.

Locating the bat

Nerve impulses are the thin, sharp spikes on this readout. As the intensity of the sound increases, so does the number of impulses. When the intensity decreases, the number of impulses follows suit.

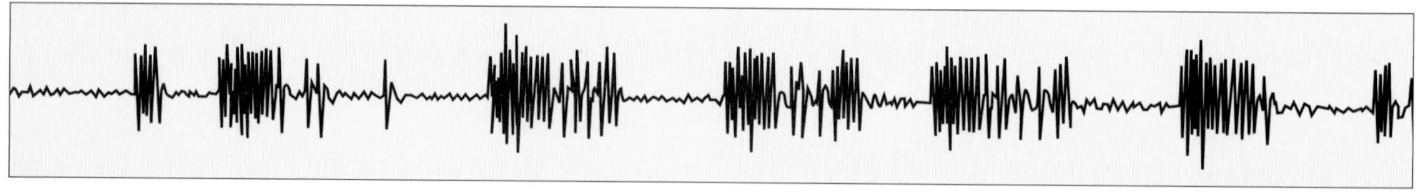

Bat approaching **Bat nearby** **Bat moving away**

It is clear that the moth can detect how far away the bat is, but can it also locate the bat in three-dimensional space? Yes, it can, because it has ears on both sides of its body. As a bat approaches a moth from the left, the moth's left ear perceives the bat cries slightly before the right ear does. The intensity of the sound in the left ear is also greater than the intensity in the right.

The readout below shows impulses from both of a moth's ears. At first, spikes appear in the left ear; then, a little less than a tenth of a second later, a single spike from the right ear appears. The bat is on the left, but it's not too close. The next set of sharp spikes from the left ear indicates closer bat cries. The bat is approaching. It is still on the left. Then, at the far right end of the readout, an equal number of clusters of spikes from the right ear appear at just about the same instant. The bat is directly over or under the moth but not, according to the number of impulses, dangerously close.

As the bat moves toward the moth from the left and hovers over it, the intensity of the sound the bat emits and the time it arrives at each of the moth's ears become equal, as reflected in the pattern of impulses.

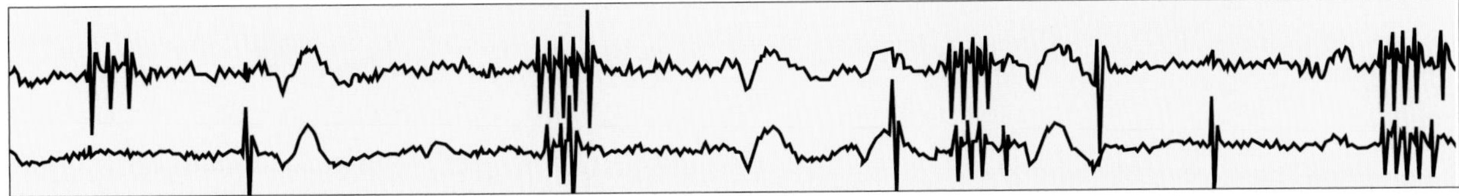

If the nerve cells in a moth's ears sent bursts of impulses like those below, the message immediately conveyed to its wing muscle cells would be "DIVE, DIVE!"

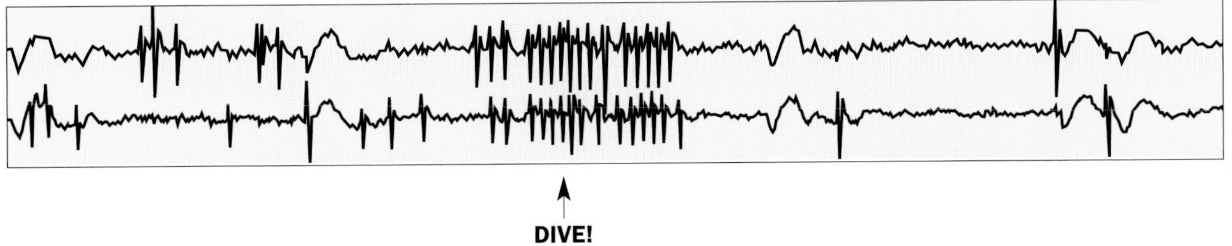

As these bats and moths have evolved in tandem, each species has selected out the less well-adapted individuals of the other, leaving the best adapted to survive and reproduce: the bat selects out the slow or less acutely perceptive moth (by eating it) and (less directly) the moth selects out the poorly navigating bat by escaping it — depriving it of nutrition.

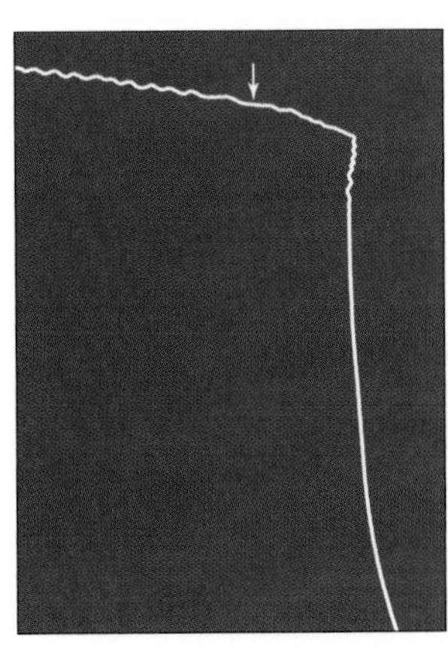

Dive!

A moth diving in response to a simulated "bat sound." The white arrow indicates the point in the moth's flight path at which the sound stimulus began.

Can Habits Be Inherited?

Lamarck vs. **Darwin**

Jean-Baptiste Lamarck (1744–1829) deserves credit for bringing about a shift in scientific thought about the evolution of life — from judgments based on theological absolutes to inquiries into connections and causes. He conceived the idea that species change over time, and that all organisms are related — essentially the first explicit theory of evolution.

Lamarck is best known, unfortunately, for his now-discredited theory of *inheritance of acquired characteristics* — that experiences of an organism could be passed on through inheritance — if an organism worked at something desirable, its children would inherit the fruits of that effort (see the illustration at right).

Darwin liked Lamarck's ideas about relatedness and change. They fit with his own theory that the accumulation of small changes could accomplish big results. And although Darwin had no explanation for why organisms varied (the science of genetics had not been born) and therefore couldn't rule out the possibility that acquired characteristics were inherited, he felt certain that evolution was not driven by the desires of organisms. Organisms change continuously. Those who coincidentally experience changes that better fit them to their environment have more offspring, and so their kind survives and flourishes.

The perspectives of Lamarck and Darwin differ fundamentally on the matter of purposeful design. Lamarck, while accepting change, couldn't drop the notion of a preordained plan behind evolution. Darwin saw natural selection as a powerful force, lacking a purpose, but creating the illusion of a planned goal.

Darwinism has decisively triumphed over Lamarckism. Evidence accumulated during the last fifty years has firmly established the generalization that information in living systems flows one way: from DNA to RNA to protein; there is no way the environment can influence the organism's proteins to change its DNA.

How Did Giraffes Acquire Long Necks? Two Theories on How Traits Develop

Lamarck's Theory

Back when there were plenty of leaves to eat, giraffes had short necks.

In time, the giraffes stripped the lower branches bare, so that the only available leaves were higher up.

Giraffes had to stretch their necks to reach the higher leaves.

The stretched-neck trait was passed on to their offspring, resulting in long-necked giraffes.

Darwin's Theory

Back when there were plenty of leaves to eat, most giraffes had short necks, but some did have longer necks.

In time, the giraffes stripped the lower branches bare so that the only available leaves were higher up.

The short-necked giraffes began to die off for lack of food; the long-necked giraffes survived and multiplied.

Eventually, only long-necked giraffes were left.

8.20 An Experiment in Evolution

Luria, S.E., and M. Delbruck. Mutations of Bacteria from Virus Sensitivity to Virus Resistance, *Genetics* 28: 491-511, 1943.

DOING *Science*

As late as the 1940s, scientists did not believe that bacteria, the most abundant and ancient forms of life on Earth, were governed by the rules of evolution. Bacteria multiplied and changed so fast that scientists thought their inheritance might be directly changed by their environment (in the Lamarckian mode). Nobel laureate biologist Salvador Luria, however, suspected that bacteria, like giraffes, obeyed Darwinian rules. In 1943, while watching slot machines being played at an alumni dance, Luria conceived an experiment that would conclusively settle the question.

Luria's Question:

Certain kinds of viruses kill bacteria. If bacteria are given food and grown in liquid in a tube for a day, the liquid becomes cloudy as their numbers swell to a billion or so. (Bacterial cells divide about every half-hour.) If you then add the bacteria-killing viruses to the tube, virtually all the bacteria are killed off within 20 minutes. But wait! A day later, the tube again has a billion bacteria in it — all immune to the virus. Is the immunity caused by the virus (the Lamarckian answer) or is an occasional bacterium already immune by chance — whether or not the virus is present — and then multiplies to produce a new immune population (the Darwinian answer)?

The Experiment:

Luria put an equal number of virus-sensitive bacteria into each of a hundred tubes and gave them food. The bacteria multiplied for a day. Next, he set out a hundred dishes, each covered with a gelatin-like material containing food and *bacteria-killing viruses.* He then spread the contents of each tube onto each of these dishes so that wherever a cell landed on the dish, it stayed put and began to multiply. A day later, any cell that was immune to the bacteria-killing virus would have multiplied into a clump of cells big enough to see.

The Reasoning:

Luria reasoned that if the bacteria *acquired* immunity — that is, if they somehow "learned" from contact with the virus how to avoid being killed by it — all the dishes would have about the same number of clumps because they'd all have been exposed to the same challenge. If, however, the immunity was caused by random mutations in the bacteria — which would occur whether or not the virus was present — the dishes would end up looking different. Many would have no clumps at all, some would have a few, and a rare one would have many clumps. Here is Luria's reasoning: Preserved or measurable mutations are rare events; they occur once in every 5 million or so cells. If a mutation that made the bacteria immune to the virus occurred soon after the cells were placed in one of the hundred tubes, that immune cell would have a long time to multiply and make many offspring; there'd be lots of clumps on the dish — a "jackpot." The *later* a mutation occurred, the fewer the clumps on the dish. Of course, many tubes would produce no mutant at all, so nothing would grow on those dishes.

The Result:

As Luria anticipated, there were widely differing numbers of bacterial clumps on the dishes and no clumps on most. This meant that the mutation that caused immunity was random and uninfluenced by the presence of the virus.

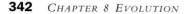

How Are Slot Machines Like Bacteria?

How Luria Got the Idea for His Experiment

A slot machine payoff is a very rare event.

But if you play lots of slot machines all night, the probability of payoff increases. Some machines will produce no payoffs, others a few payoffs, and a few big payoffs (of course, *you'll* probably still end up broke).

A mutation in a bacterial population that makes it immune to viral attack is also a rare event.

But if lots of separate bacterial populations multiply all night long, the probability of mutations occurring in some of them will increase.

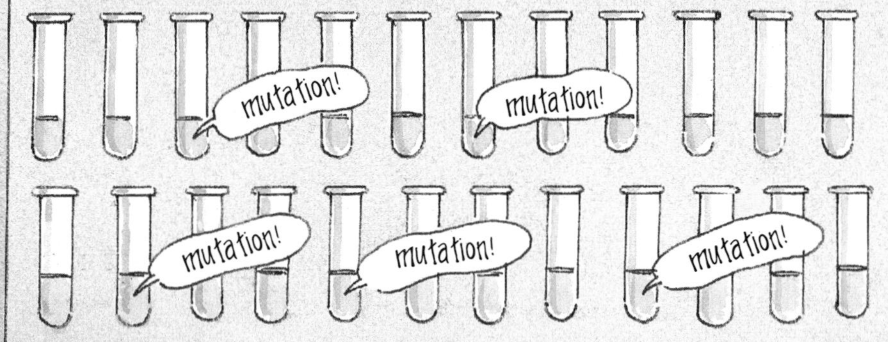

It occurred to Luria that it would matter very much *when* such a mutation occurred.

An early mutation would produce a huge number of off-spring — a jackpot — because the mutant cell's progeny would have all night to multiply. A mutation occurring later would have less time to produce off-spring; so there'd be fewer cells in the tube in the morning. This insight suggested an experiment that would conclusively test evolutionary theory (see previous page).

8.21 Evidence of Relatedness

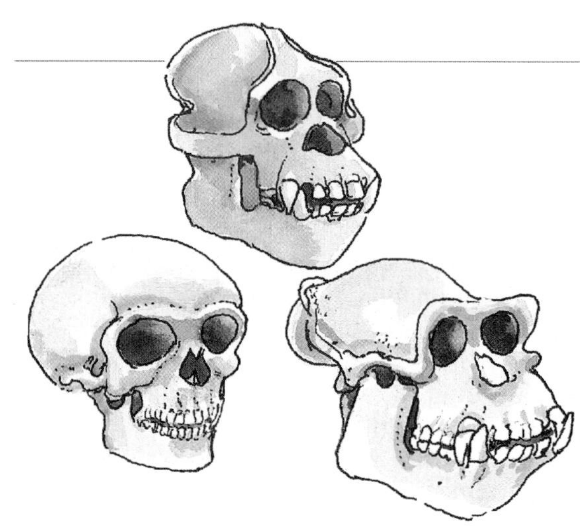

Comparative Anatomy

These skulls of a human, a gorilla, and an orangutan are obviously related. Do their anatomical features give any clues as to which is more closely related to which?

All in the Family

We know that all animals, plants, protists, fungi, and bacteria are related because they all use the same genetic code and much of the same molecular machinery for living. But how do scientists find out *how* closely any two species are related? That is, when did their common ancestor live? How long ago were they one and the same creature?

Over millions of years, genes accumulate mutations at an overall steady rate. The number of mutations that have accumulated in genes for the same function taken from members of two *different* species is a measure of the species' relatedness: The smaller the differences, the more closely related they are. Assuming that any two species have a common ancestor, the simplest way to connect them is to construct a genealogical tree.

Very old are the rocks
The pattern of life is not in their veins.
When the earth cooled, the great rains
came and the seas were filled.
Slowly the molecules enmeshed in
ordered asymmetry.
A billion years passed, aeons of
trial and error.
The life message took form, a spiral,
a helix, repeating itself endlessly,
Swathed in protein, nurtured by
enzymes, sheltered in membranes,
laved by salt water, armored with
lime.
Shells glisten by the ocean marge,
Surf boils, sea mews cry, and the great wind
soughs in the cypress.

Thomas H. Jukes, *Molecules and Evolution, 1966*

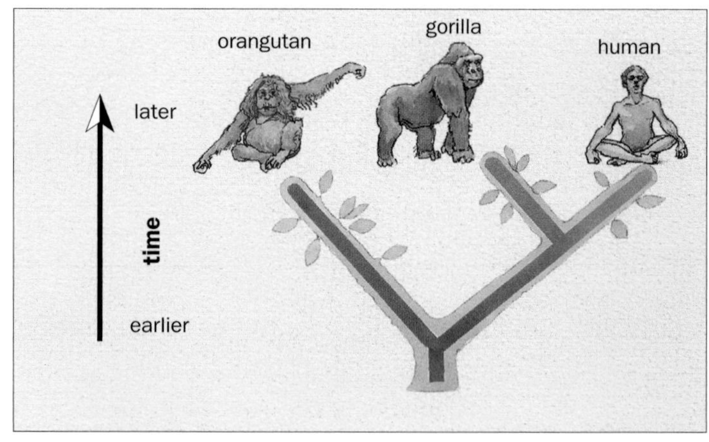

Genealogical Tree

This simple lineage is derived by determining the nucleotide sequence of a particular gene taken from each of three animals: an orangutan, a gorilla, and a human. Out of a sequence of 75 nucleotides in the gene, 12 differed between the human and the gorilla, and 20 between the human and the orangutan. Assuming that mutations have hit these genes randomly and at the same rate over time, this genealogical tree says that humans and gorillas are more closely related to each other than either of them is to the orangutan. In other words, humans and gorillas had a common ancestor that lived more recently than the common ancestor of all three.

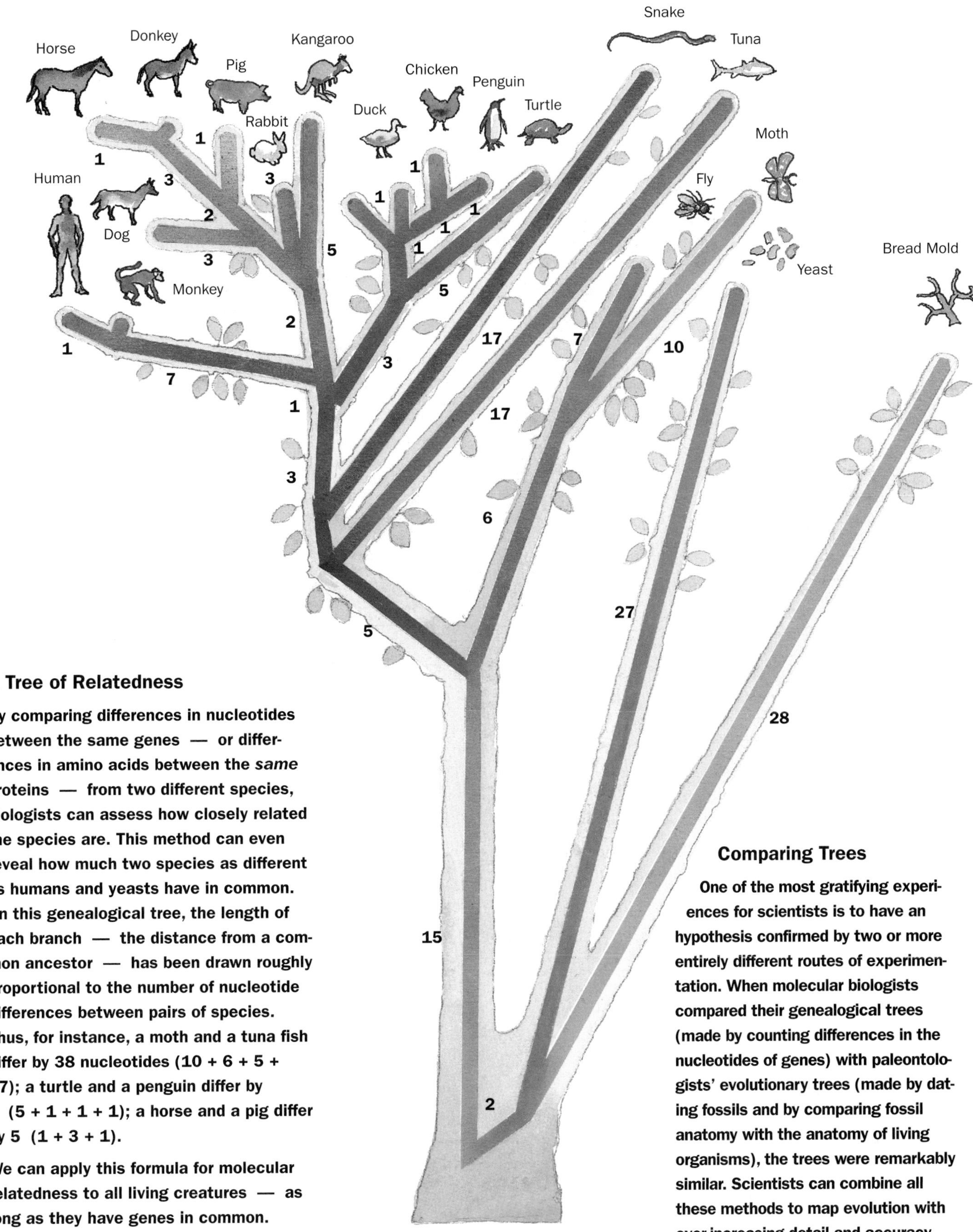

A Tree of Relatedness

By comparing differences in nucleotides between the same genes — or differences in amino acids between the *same* proteins — from two different species, biologists can assess how closely related the species are. This method can even reveal how much two species as different as humans and yeasts have in common. On this genealogical tree, the length of each branch — the distance from a common ancestor — has been drawn roughly proportional to the number of nucleotide differences between pairs of species. Thus, for instance, a moth and a tuna fish differ by 38 nucleotides (10 + 6 + 5 + 17); a turtle and a penguin differ by 8 (5 + 1 + 1 + 1); a horse and a pig differ by 5 (1 + 3 + 1).

We can apply this formula for molecular relatedness to all living creatures — as long as they have genes in common.

Comparing Trees

One of the most gratifying experiences for scientists is to have an hypothesis confirmed by two or more entirely different routes of experimentation. When molecular biologists compared their genealogical trees (made by counting differences in the nucleotides of genes) with paleontologists' evolutionary trees (made by dating fossils and by comparing fossil anatomy with the anatomy of living organisms), the trees were remarkably similar. Scientists can combine all these methods to map evolution with ever-increasing detail and accuracy.

Radiometric Dating

Fossils — the petrified remains of plants and animals — are most often found in sedimentary rocks: materials of the Earth's crust that have been laid down layer upon layer over millions of years. The *relative* age of fossils in these layers can be gauged by observing where they are in the sequence of layers: the lower the layer, the older the fossil. But their location tells us nothing about the *absolute* age of fossils.

Why do we find the bones of great fishes and oysters and corals and various other shells and sea snails on the high summits of mountains by the sea, just as we find them in low seas? And if you choose to say that it was the deluge which carried these shells away from the sea for hundreds of miles, this cannot have happened, since that deluge was caused by rain; because rain naturally forces the rivers to rush toward the sea with all the things they carry with them, and not to bear the dead things off the sea shores to the mountains.

Leonardo da Vinci, *Notebooks*, 1470–1480

Opening the book of time

Over thousands of years, the Colorado River has worn its way through layers of sedimentary rock to form the Grand Canyon. This photograph shows you the layers of sediment, which allow us to determine the relative ages of materials trapped in them.

To determine the age of the rocks in which fossils are found — and thereby discover the absolute age of the fossils — we measure the amounts of certain radioactive isotopes (forms of chemical elements that decay, or break down, over time, becoming other isotopes or elements) that occur naturally in the rocks. For example, the isotope uranium-235 breaks down to lead, and potassium-40 breaks down to argon. Each of these changes is accompanied by the release of energy in the form of radioactivity, which can be measured.

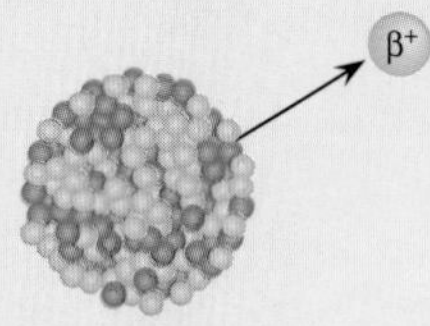

Step 1: Positron emission

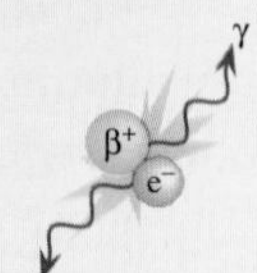

Step 2: Annihilation

A Decaying Atomic Nucleus

Here the nucleus of an atom decays (spontaneously converts a proton into a neutron and a subatomic particle — a ß particle). The ß particle emitted in this process collides with an electron. These particles annihilate, and their mass turns into energy — gamma radiation.

www.jbpub.com/connections

We measure the rate of an isotope's decay by its half-life: Half of a sample of uranium-235 becomes lead-207 in 0.7 billion years; half of a given amount of potassium-40 becomes argon in 1.3 billion years. So, for example, if we measure the amounts of uranium-235 and lead-207 in a rock, the ratio of the two amounts allows us to calculate the rock's age. Using the known half-lives of many isotopes, scientists have determined that the Earth's oldest rocks are approximately 3.9 billion years old. Other evidence, based on the ages of meteorites and moon rocks, push estimates of the time of Earth's formation back to 4.6 billion years ago.

Radiocarbon dating is an especially valuable tool for determining the age of once-living material from more recent times (i.e., alive within the last 50,000 years). Carbon-14 is a radioactive isotope of regular carbon (carbon-12). Its half-life is 5730 years. Carbon-14 is produced in small amounts in the Earth's atmosphere when nitrogen-14 is bombarded by cosmic radiation. Just like regular carbon, the radioactive isotope bonds with oxygen to form carbon dioxide (CO_2). Plants assimilate atmospheric carbon dioxide containing specific proportions of the two forms of carbon as they carry out photosynthesis, and animals ingest the same proportions when they eat plants. After death, intake of carbon stops, and the carbon-14 content of an organism decays steadily to nitrogen-14. So, for instance, if a sample of once-living material is found to have one-fourth as much carbon-14 as a living organism, we can conclude the sample is 11,460 years old (two half-lives of 5730 years).

Question.

The table below represents a cross section through a hypothetical hill that was blasted away to build a new exit ramp off an interstate. Multiple layers of rock from different periods of time were uncovered (from the most recent layer, labeled A, through the oldest layer, H), each with its own thickness, color, texture, etc. Many of the layers contained fossilized remains of organisms that existed during that particular geological period. Each different species of fossilized organism is represented by a number from 1 to 10, which indicates the layers in which the species was found.

a. Which species has been around the longest?
b. Which species was found in only one layer of rock?
c. Which species arose most recently?
d. Which species found in the oldest layer are most likely extinct today?
e. How does this evidence support Darwin's theory of evolution?

A (top layer)	1					6			9	10
B	1			4		6			9	10
C	1			4	5	6		8	9	10
D	1			4	5	6		8		
E	1			4	5	6	7	8		
F	1	2		4	5	6				
G	1	2		4	5	6	7	8		
H (bottom layer)	1	2	3							

Answer...

a. 1
b. 3
c. 9 and 10
d. 2 and 3
e. The species found in each layer change, showing that species can appear and disappear over time. Each period represented by a layer has certain characteristic species of fossils, some of which arose early and either remained somewhat unchanged over time or vanished. More recent periods have species that may be quite different from earlier ones. Each species was subject to natural selection pressure due to the conditions *in its particular era and region*. Less well-adapted species sometimes didn't last long if their reproductive fitness was negatively affected by their unfavorable traits.

Three Brains in One

Our brain has a "layered" architecture, with newer parts built on top of the earlier parts.

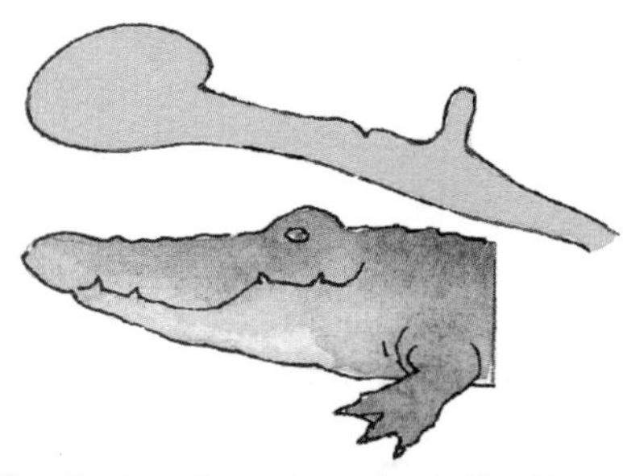

The first and most ancient, the R-complex (the R refers to reptiles), developed as an extension of the upper brain stem. This area controls respiratory and cardiovascular functions, and influences our territoriality, mating, and aggression. It's our basic "survival brain."

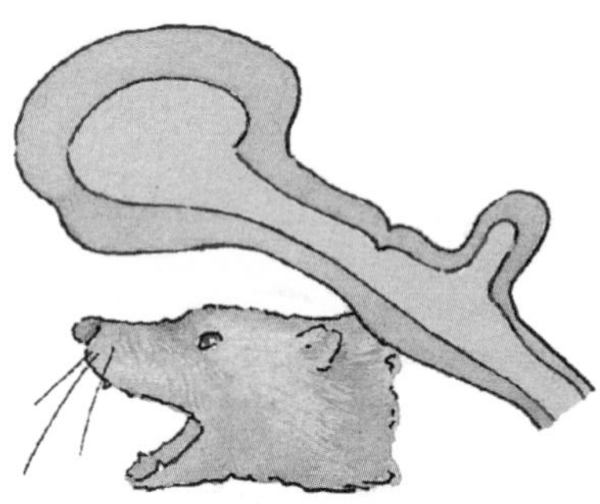

Above the R-complex lies the limbic system (evolved with the earliest mammals), which produces our emotional states — our "feeling brain."

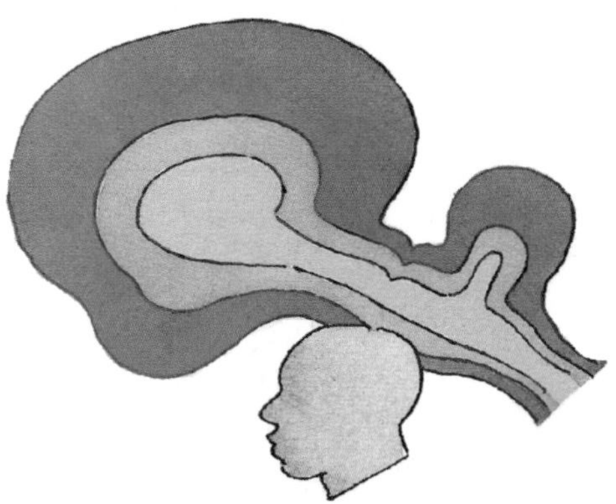

Our cerebral cortex is the thick, outer layer of our brain — our "thinking cap." With this new brain mass, we developed the traits that make us uniquely human.

8.22 The Evolution of Intelligence

Genes and Brains

No organ in the history of evolution has grown as fast as the human brain. From the time of our ape-like ancestors, some 5 million years ago, until modern humans appeared about 200,000 years ago, the proto-human brain has added a cubic inch to its total volume every 100,000 years. While adding on all this new mass, our brain's equipment for regulating our bodies and our instincts was left intact; new circuitry was installed right on top of the old (see the diagram at the left).

Humans at some point crossed a threshold; the information content of our brains surpassed that of our genes. There are about 3 billion nucleotides, or bits of information in each person's DNA. Each of us has a brain with some 10 *trillion* units of information, if we define a unit as a single connection between nerves conveying a digital (yes or no, on or off) message.

Our add-on-brain included adjustable hardware. It could modify its neural connections in response to experience — a feature essential to learning. Because of the powerful advantage learning confers, the better learners among our ancestors thrived and multiplied. Our earlier brain structures handled survival and reproduction; the add-on systems increasingly opened up the more abstract realms of reasoning, wonderment and creativity.

Language most probably evolved in stages, perhaps from making simple calls, then to representing objects (that is, naming things), and then to representing ideas. Our ability to use language to further ideas is a marvelous example of positive feedback. One idea led to another, building new connections in our brains — which, in turn, opened new doors of comprehension. It has been said that not only did we invent language, but language invented us.

The evolution of consciousness, like language, depended greatly upon cooperation. Brains of similar complexity interacted and lifted each other to higher levels of thinking. In this evolutionary phase, which probably occurred within the last 10,000 years, we gained both a sense of self and a sense of time. With "self" consciousness each individual is a "me" — a player in his or her own drama. A conscious person brings a measure of objectivity to an otherwise totally subjective life. With the dawn of consciousness, our species has evolved a mind that can observe itself.

Through the ability to imagine their past, present, and future, our ancestors could look both forward and backward in time. Agriculture, calendars, and a host of other cultural innovations flowed from this insight.

Our ability to look ahead and imagine the way life *could* be has given us a gift with far-reaching consequences: the gift of choice. We can make choices not only in our own lives but, to a degree unimagined by our ancestors, in service of the life of the entire biosphere.

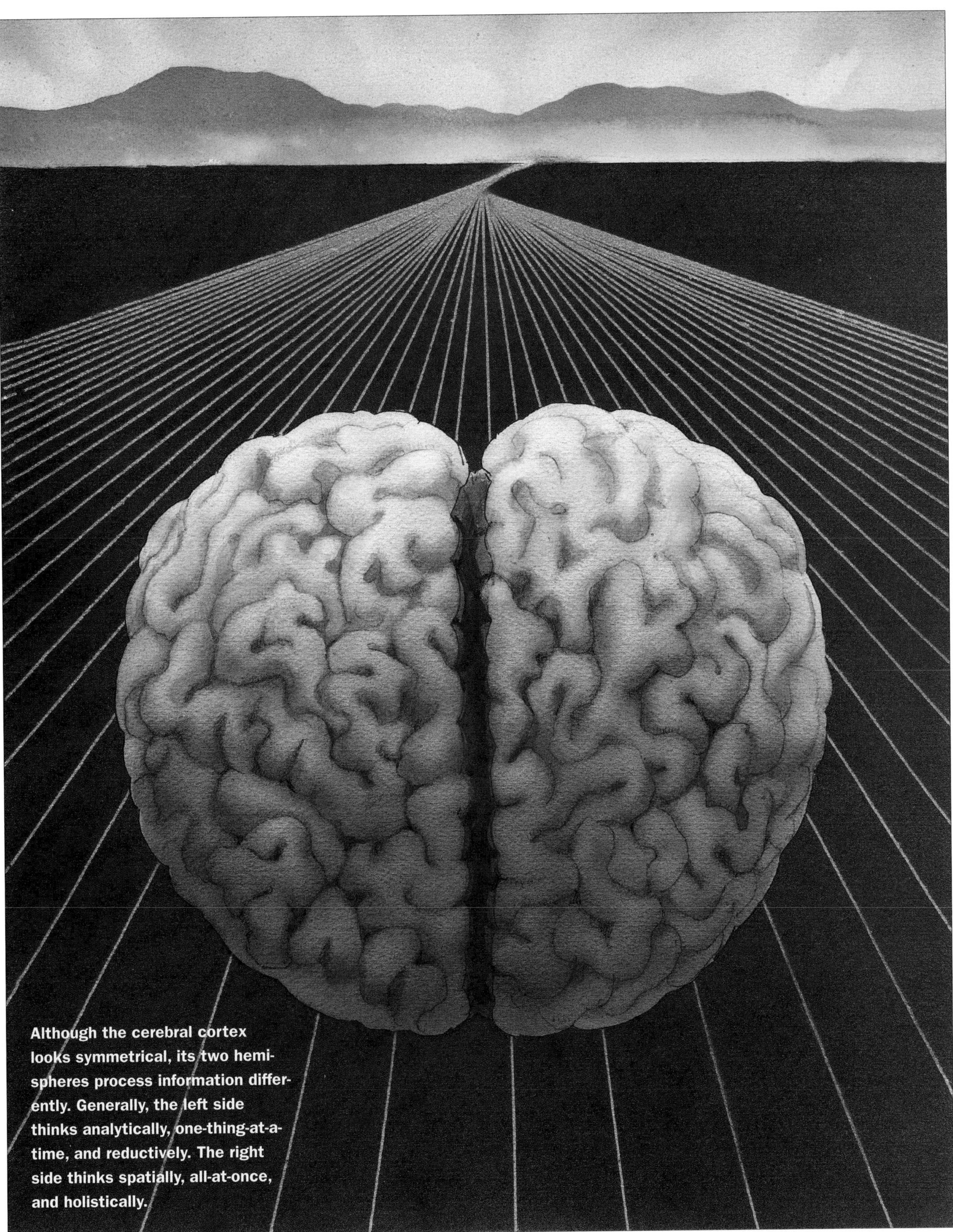

Although the cerebral cortex looks symmetrical, its two hemispheres process information differently. Generally, the left side thinks analytically, one-thing-at-a-time, and reductively. The right side thinks spatially, all-at-once, and holistically.

8.23 Cultural Evolution

Genes and Ideas

We come now to the furthest reach of evolution's information-building drive: the transmission of ideas through culture.

Ideas, like biological innovations, seem to follow the rules of evolution. From the random stream of thoughts, utterances, and writings, a few get selected and are reproduced, while the rest are weeded out. An idea born in one brain evolves as it passes through other brains. Those ideas that spread best will find more enduring lives in the world's libraries and CD collections.

Any discrete, memorable idea enters the competition: car headlights, microchips, Pinocchio, algebra, natural selection, and the TV jingle that you hate but can't get out of your head. The first order of business for an idea is to spread itself, whether or not it acts for the good of anything. The more it sticks in our collective minds, the more likely it is to survive.

With the rapid spread of ideas, cultural evolution has greatly sped up the rate of change on Earth. It has given us the tools to expand our range and lengthen our lives as well as to extract an ever-growing quantity of materials and energy from the bio-sphere. However, culture must fit within nature. By almost any measure, the natural environment we live in is under stress. Many organisms have been unable to keep pace with the changes we have produced — and extinction rates are rising. Evolution suggests to us that ideas that work for one environment may not work for another. In other words, the ideas that got us here may not be the ones to keep us here.

As some of our most "successful" ideas come to threaten the ecological balance of the natural world, we need to take another look at them and make choices that contribute to the well-being of the life system as a whole. One such choice is what biologist E. O. Wilson calls "biophilia," which he defines as our natural sense of connectedness to life. Wilson writes, "to explore and to affiliate with life is a deep and complicated process in mental development. To an extent still undervalued in philosophy and religion, our existence depends on this propensity, our spirit is woven from it, hope rises on its currents."

It may be that our growing appreciation of that truth will be the greatest legacy we can pass on to future generations.

As we deepen our imprint on the natural world, we increase our responsibility for it.

Illustration after M. C. Escher

Some of the Things You Learned About in Chapter 8

Questions About the Ideas in Chapter 8

1. What evidence suggests that all living things come from a single type of ancestor?
2. Critics of evolution argue that an organ like the human eye is far too complex, and dependent on the perfect interaction of its parts, to have evolved by natural selection. Comment.
3. If we speak of the evolution of resistance to antibiotics, what do we mean?
4. How is it that a process that leads to better and better adapted wildebeests can also lead to the creation of an organism quite different from a wildebeest?
5. What is a mutation?
6. What is one of the most important means by which new species arise?
7. Why is sex so important, given the fact that many successful organisms can reproduce asexually? Do bacteria have sex?
8. How, in principle, can you show that various living creatures are related to each other by examining the sequences of nucleotide units in their DNA or proteins?
9. Some people believe that the second law of thermodynamics, which states that the universe is inexorably moving from an ordered state to a more disordered one, from useful energy (sunlight) to less useful energy (heat), makes it impossible for the complex order of life to come into being without intelligent design. Are they right? Why or why not?

References and Great Reading

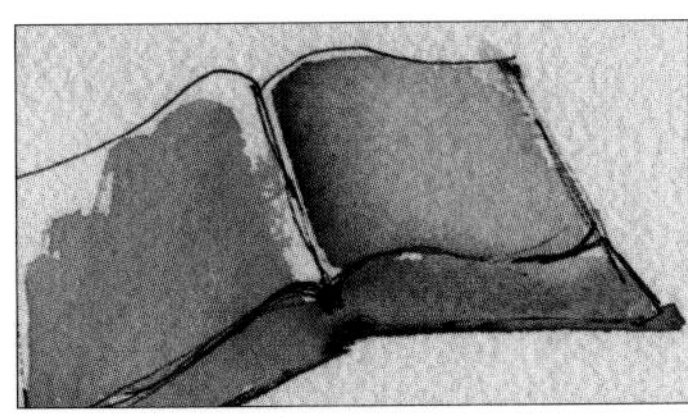

Ayala, F. 1978. The Mechanism of Evolution. *Scientific American.* (September) Volume 239.

Borror, Triplehorn, and Johnson. 1989. *An Introduction to the Study of Insects,* 6E. Philadelphia: Saunders.

Campbell, Neil. 1996. *Biology,* 4E. Benjamin Cummings.

Centers for Disease Control and Prevention. 1996. "Defining the public health impact of drug-resistant *Streptococcus pneumoniae:* report of a working group." *Morbidity and Mortality Weekly Report.* Vol. 45, No. RR-1.

Dillard, Annie. 1999. *A Pilgrim at Tinker Creek.* New York: HarperPerennial.

Da Vinci, Leonardo. *The Notebooks of Leonardo da Vinci, Vol. II,* ed. Jean Paul Richter, 1970. New York: Dover Publications, Inc.

Dawkins, Richard. 1987. *The Blind Watchmaker.* New York: W. W. Norton & Co.

Dawkins, Richard. 1989. *The Selfish Gene.* Oxford: Oxford University Press.

Dennett, Daniel C. 1991. *Consciousness Explained.* Boston: Little, Brown.

Eldridge, Niles. 2000. *The Pattern of Evolution.* New York: W. H. Freeman & Co.

Gershman, Kenneth. 1997. "Antibiotic Resistance and Judicious Antibiotic Use." *The Medical Reporter.* Vol. II, No. 11.

Gould, S. J. 1989. *Wonderful Life: The Burgess Shale and the Nature of Discovery.* New York: W. W. Norton.

Grant, P. and R. Grant. 1994. *The Beak of the Finch, A Story of Evolution in Our Time.* Alfred A. Knopf.

Jaynes, Julian. 1976. *The Origin of Consciousness in the Breakdown of the Bicameral Mind.* Boston: Houghton Mifflin Company.

Lloyd, James E. 1971. Bioluminescent communication in insects. *Annual Review of Entomology.* 16, 97-122.

Mlot, Christine. 1997. A Soil Story – Life inside a nematode: Creative chemistry and novel pest control. July 26, *Science News.* Vol. 152.

Roeder, K.D. 1963. *Nerve Cells and Insect Behavior.* Cambridge: Harvard University Press.

Stebbins, Ledyard. 1982. *Darwin to DNA: Molecules to Humanity.* New York: W. H. Freeman.

Strickberger, M.O. 2000. *Evolution,* 3E. Sudbury, MA: Jones and Bartlett Publishers.

Wilson, E.O. 1978. *On Human Nature.* Cambridge: Harvard University Press.

Glossary

A

absolute zero the temperature (-273 degrees Celsius) at which there is complete dispersal of energy. All motion stops. Nothing happens.

activation energy the minimum energy that atoms must have to overcome the repulsion of their electrons and allow a chemical reaction to occur between them.

active site the "docking site" for a substrate molecule on an enzyme.

actin a long, chainlike protein that, in association with another protein called *myosin*, provides the machinery for movement in muscle cells.

adaptive change a process that occurs when a species' environment presents new opportunities or dangers. The environment selects individuals with traits that are especially well-suited to it, while less well-suited individuals do not survive to reproduce.

adaptor a tRNA molecule that carries a triplet codon and its associated amino acid to a ribosome to make a lengthening amino acid chain.

ADP (adenosine diphosphate) the molecule that results when an ATP molecule has contributed energy from *one* of its high-energy phosphate bonds. "When life needs energy, it breaks off one of ATP's phosphates, like snapping a pop-bead off a necklace, which frees the energy inside the bond." In cells ADP, is continuously recycled into ATP.

aerobic/anaerobic an aerobic process is one that takes place in the presence of oxygen, as when sugar is burned in mitochondria (respiration). An anaerobic process takes place in the absence of oxygen; glycolysis is an example.

allosteric enzyme an enzyme that changes shape and function in response to a chemical signal. A chemical switch.

allostery (*Gr.* "other shape") the shape-changing capacity of proteins that allows them to regulate chemical processes.

amino acid activating enzyme a protein that energizes a specific amino acid with ATP and facilitates the chemical connection between the amino acid and it's correct transfer RNA.

amino acid any of 20 simple molecules, each of which has a similar "backbone" structure and a different "side group." The backbones of amino acids can link to one another to form a chain called a *protein*.

Archaea unicellular organisms that thrive in temperatures near the boiling point of water in oceanic steam vents and may be survivors of the earliest forms of life.

atom chemical unit consisting of negatively charged electrons orbiting around a positively charged nucleus. Only about 20 of the more than 100 kinds of known atoms are incorporated into living organisms.

ATP (adenosine triphosphate) the chemical "energy coin" of the cell. A molecule that contains accessible high-energy phosphate bonds and provides cellular energy by transferring energy from those bonds to create new bonds, making life's work possible.

autotroph (*Gr.* "self-feeder") a primary producer. An organism that captures the energy from sunlight and puts it into the chemical bonds of sugar in a process called *photosynthesis*.

B

bacteriophage ("bacterium eater") a virus that diverts a bacterium's protein-making machinery to the construction of its own protein, killing the bacterium and reproducing itself in the process. Phages may also incorporate some bacterial genes into their own DNA and thus transfer those genes from one bacterium to another.

bacterium (pl. bacteria) unicellular living organism that has no nucleus or membrane bound organelles and that, over time, originated all of life's essential chemical systems. A prokaryote.

base one of the four simple nitrogen-containing molecules (adenine, guanine, thymine and cytosine) that, joined with a sugar (deoxyribose) and a phosphate group, makes up one of DNA's nucleotide units.

blastula the hollow ball of approximately a hundred or so cells that forms after an egg is fertilized by a sperm and becomes the embryo and the placenta.

bond energy the amount of energy stored in the linkage between two atoms as measured by the amount of heat released when the bond is hydrolyzed (broken). More energy goes into making a bond than is actually stored in the bond.

C

Calvin Cycle "making sugar out of thin air" in the stroma of the chloroplast. A chemical cycle in which five enzymes, using energy supplied by ATP, transform carbon dioxide into sugar.

cancer cell a mutant cell whose signaling proteins have been damaged, allowing it to divide uncontrollably and migrate throughout an organism, disrupting normal functioning.

carbon an atom that plays a central role in life's molecules because it can form four separate bonds to other atoms. Carbon can thus form chain molecules of unlimited lengths and with many different side groups, allowing for great variety.

carnivore a predator or scavenger that eats other heterotrophs, or consumers.

cascading a biological positive feedback process in which events trigger other events in an ever-growing amplification, as in the growth of an embryo or in biological evolution.

catalyst a facilitator of chemical reactions. An enzyme is a biological catalyst—it allows chemical events to happen with less input of energy.

cell a small, specialized living unit that is self-sustaining and able to duplicate itself. Some kinds of cells can function alone; others aggregate into communities that make up the specialized parts of multicellular organisms.

cell death a kind of "pruning" of an initial overgrowth of cells in which, in response to a signal, specific cells "commit suicide" (often called *programmed cell death* or *apoptosis*). An important feature of the normal development of many organisms.

cellulose a chain of linked sugar molecules that makes up the bulk of grass and wood that cannot be digested by many multicellular organisms without the symbiotic help of bacteria.

cerebral cortex the most recently evolved part of the human brain—the thick outer layer that accounts for uniquely human attributes.

chance the random changes in the environment that challenge the capacities for survival of living things, and the random changes in information (DNA) that alter the traits of living things.

chemical bond a strong attraction between ions (ionic bond), a weak attraction between polar (oppositely charged) molecules (hydrogen bond), or a sharing of electrons (covalent bond) between or among atoms. Covalent bonds are storehouses of chemical energy, as evidenced by the release of heat when a bond is broken.

chemical signal any molecule that functions as a simple message and is perceived by a receptor on a living cell.

chemotaxis ("movement induced by chemicals") the motion of an organism toward or away from a chemical. Not a "random walk."

chlorophyll pigment molecule that absorbs and reflects specific wavelengths of sunlight and gives autotrophs their green color. Found in plant chloroplasts arrayed in clusters as solar antennae.

chloroplast bacteria-sized energy-producing organelles in plant cells that make sugar using carbon dioxide, water and the energy of sunlight and produce oxygen as a by-product. May have evolved from separate photosynthesizing cells, which were engulfed by other, larger cells.

chromosome a molecular structure consisting of double-stranded DNA spooled up on protein molecules residing in the nucleus of a cell. In organisms that reproduce sexually, chromosomes come in pairs, one from each parent. Each chromosome may contain a thousand or more genes.

cloning making identical copies of a gene or genome. May be used by genetic engineers to create organisms with identical genomes.

codon a sequence of three nucleotides (a triplet) that codes for a specific amino acid. A codon in messenger RNA is complementary to an anticodon in tRNA.

common ancestor the presumed life form from which two different existing species are descended. The degree to which the two species have diverged from the common ancestor can be assessed by counting the number of differences in nucleotides in a gene they have in common: the fewer the differences, the more closely related they are and the nearer to the common ancestor.

common body plan the similar developmental pattern of widely divergent organisms.

competition a situation in which two different organisms need the same resource, but only one will get it. Competition for available resources determines the success or failure of individuals, and thus their likelihood of producing offspring.

complementary nucleotides pairs of nucleotides that make weak bonds with each other. In DNA, guanine (G) bonds only with cytosine (C) and adenine (A) bonds only with thymine (T).

conjugation "primitive sex." An exchange of genetic information between single-celled organisms such as bacteria.

contact plan in cell differentiation, the process of induction by which one cell causes a neighboring cell to make proteins in the region nearest it. When the neighboring cell divides, two differing daughter cells are created. These cells, in turn, induce similar changes in neighboring cells.

covalent bonds strong bonds that involve a sharing of electrons between atoms; examples are those between amino acids in protein chains and nucleotides in DNA chains.

cristae the folds of the inner membrane of the mitochondrion through which enzymes channel electrons into the intermembrane sac during respiration.

cultural evolution the survival, innovation, and spread of ideas.

cytoplasm the fluid interior of a cell with its nucleus and other organelles. The "factory floor" where proteins are assembled.

D

daughter cells the two cells that form when a parent cell divides following replication of its DNA (see *mitosis*).

decomposer an organism that breaks down dead bodies and excretions of other organisms to simple molecules.

deoxyribose a sugar molecule that is one of two backbone components of DNA. Bonded to a phosphate in a nucleotide, it also bonds to the next nucleotide's phosphate and holds a sequence of nucleotides in order.

differentiation the process by which each new generation of cells becomes slightly different in structure and function as an organism develops. Cells often differentiate enough to become specialists at certain functions. They lose the capacity to create all of the proteins described in their DNA and then can no longer function alone.

diffusion the process by which a concentrated group of particles spontaneously spreads out to become less concentrated and evenly dispersed. An example of matter's tendency toward disorder and randomness.

DNA (deoxyribonucleic acid) long, double-stranded sequences of nucleotides that form the repository of a cell's information.

DNA duplication see *replication*

dominant trait one of a pair of contrasting versions (alleles) of a gene-determined characteristic (e.g. color, height, or shape) that is more likely than the other version to manifest itself in the organism's offspring.

double helix the characteristic form taken by the paired strands of a DNA molecule as a result of chemical forces at work within the molecule.

E

ecosystem a community of plants, animals, fungi, protests, and bacteria that interact with one another and their environment and depend on one another for survival. A macroscopic feedback loop.

electron tiny, negatively charged particle that orbits the nucleus of an atom.

electron acceptor a molecule that can carry electrons energized by sunlight and participate in the electron flow of photosynthesis and respiration.

electron flow during photosynthesis and respiration, the passage of energized electrons from one carrier molecule to another in the thylakoid, or inner mitochondrial membrane.

electron orbital (or shell) the space around an atom's nucleus that is occupied by its electrons. Overlapping orbitals holding shared electrons make a covalent bond between atoms.

embryo the early stage of development of a multicellular organism. Cells multiply, differentiate, and migrate during this stage. In humans, this stage of development lasts from the third to the eighth week after the egg cell is fertilized.

emergent pattern the complex pattern that is created when simple units follow basic organizational rules.

energy what moves matter and causes change, making it possible to do work. Energy has many forms — heat, electrical, nuclear, gravitational, chemical, radiant — and can be converted from one form to another.

energy flow (through life) sunlight to sugar to ATP to chemical bonds to heat.

entropy a measure of the disorder of a system. The second law of thermodynamics predicts a "downhill universe," in which energy inevitably disperses and ordered structures eventually become disordered. Thus, there is a directionality for all processes, from which our sense of time arises.

enzyme a protein molecule made of one or more long chains of amino acids, folded into a specific shape. Enzymes manipulate other molecules (as catalysts), regulate cellular production lines, "read" DNA's instructions, receive and react to chemical signals, and more.

epidermis the outer tissue layer of a multicellular organism.

equilibrium a state in which energy flows as readily backward as forward and no overall change takes place. Chemically, a state in which an equal number of reactions are converting reactants to products and products to reactants. Constantly adding reactants and removing products prevents equilibrium from being achieved.

eukaryote any cell that has a nucleus and other membrane-bound organelles. An organism made up of one or more of such cells.

evolution the process by which organisms change and diverge in form and function over long stretches of time as a result of the selective forces exerted by their environments.

F

factors Mendel's word for what controls the inheritance of traits in pea plants. Factors have since been found to be specific lengths of DNA called *genes.*

feedback a regulatory process that works either to increase or to decrease the difference between a current state of affairs and a desired or necessary state. A responsive loop that says "too much" or "not enough" and then makes a correction.

fermentation a chemical process by which sugar-converting enzymes make energy-rich molecules of ATP that can be used to fuel cells' activities. Fermentation does not use oxygen and is a common means of producing energy when no oxygen is available.

first law of thermodynamics energy can be gained or lost in chemical processes — shifted from one form to another — but it can't be created or destroyed.

food "chain" the several "consumer levels" of organisms through which energy percolates until it is finally dissipated as heat. Organisms that cannot convert the sun's energy to food are dependent on those that can.

fossil the mineralized remains of parts of once-living organisms found in rock layers. Dating the age of the rock in a layer allow us to learn the age of the fossils there.

fungus a heterotroph that secretes enzymes that decompose organic materials into their component molecules and then absorbs these molecules as nutrients.

G

gene a length of DNA in a chromosome that translates into a particular protein. A piece of genetic material that specifies one particular trait of an organism.

gene pool all the genes present in a population of organisms of the same species.

gene switch a regulatory protein that binds to a site on DNA adjacent to a specific gene. In response to a chemical signal, the switch can be activated to prevent the transcription of mRNA from the gene or deactivated to permit transcription.

genealogical tree a diagram of the ancestry and relatedness of different species of organisms that is compiled by studying homologies in organisms, by counting the differences in the nucleotides of genes, and by comparing the ages and anatomies of fossils.

genetic code the sixty-one triplet codons for the twenty amino acids.

genetics the study of the inheritance of traits, sexual recombination, and mutation.

genome all of the genes of a particular organism.

geographic isolation the physical separation of one group of organisms (or pool of genes) from a larger population. The smaller gene pool allows independent genetic change and selection over generations and can result in a new species. The most important single factor in the creation of new species.

gland a community of specialized cells that secretes signal molecules (hormones) that are carried throughout the body in blood to receptors within or on the surfaces of cells. Glands are located in various places in the body and secrete hormones that serve many different functions.

glycolysis cellular respiration; an energy-yielding process in cells in which each molecule of glucose gets broken into smaller pieces by a series of enzymes, generating two ATP molecules that supply the energy for the organism's activities. From the Greek words for "sweet" and "breaking down."

governor the agent that controls a feedback loop.

H

heat a form of energy generated by the random movement of molecules. The faster molecules move, the greater the heat.

helicase the "unzipper" protein that breaks the weak bonds between bases to separate the two strands of DNA so that each can be replicated.

herbivore an animal that consumes plant sugar directly. This group includes all the browsers and grazers, the seed- and fruit-eating birds, most insects, and the ocean-dwelling phytoplankton-eaters.

hereditary disease a disease that is the result of a particular protein's failing to perform its normal function because of a defect in a gene which can be passed on to the organism's offspring.

heredity the passing of information, and hence characteristics, from one generation to the next.

heterotroph (*Gr.* "other-feeder") an organism that cannot convert the sun's energy and is dependent for energy on those that can.

hormone a signal molecule, secreted from communities of cells called glands, that circulates in the blood and carries messages to other cells. The fall and rise of hormone levels in the blood also function as a negative feedback system to signal the release or restriction of more hormone output.

hox genes individual "master regulating" genes that, when activated by signal molecules during the development of an embryo, create proteins that turn on other genes that control the development of specific body segments. They are found in all animals, and the more complex the animal, the more hox genes it has.

hydrogen bond a weak chemical bond between hydrogen and another atom, usually oxygen or nitrogen, that is of critical importance in maintaining the shape of large molecules and also accounts for many of the unique properties of water.

hydrophilic (*Gr.* "water-liking") an electrically charged atom (ion) or group of atoms that is attracted to the opposite charge on a water molecule and forms a weak bond with it. Hydrophilic groups in proteins turn toward a watery environment, which tends to push lipophilic (fat-liking) groups toward the interior of the protein.

hydrogen the smallest atom, having one proton in its nucleus and one orbiting electron. Hydrogen is a necessary player in energy production in organisms. It loses its only electron to become a positively charged proton (hydrogen ion).

I

induction the process by which a cell signals its immediate neighbor to change by making one particular protein near that cell. When the neighbor cell divides, the two daughter cells will be different because they will have different amounts of the protein. Over many generations of cells, such induction results in an assortment of new and different cells.

information the instructions in DNA that direct the structure, function, and proliferation (replication) of self-organizing systems.

initiator protein in DNA replication, the protein that finds the place on the DNA strand to begin copying.

ion an atom with mismatched numbers of electrons and protons. It has lost one or more of its negatively charged electrons (becoming a positive ion) or has gained one or more electrons (becoming a negative ion).

J

jumping gene a gene that has changed its location on the DNA stand through transposition.

K

kinetic energy the energy of motion, in contrast to potential, or stored, energy.

L

lichen a symbiotic structure made up of a fungus and an alga or cyanobacterium. The alga or bacterium provides sugar from photosynthesis, the fungus provides water and structure. Together, the combination can survive in places where neither organism could survive alone.

ligase the "stitcher" protein that joins the short strands of replicated DNA into a long strand.

limbic system a part of the brain that evolved with the earliest mammals. It is the mediator of our emotional states. The "feeling brain."

lineage a genealogical tree that shows the relatedness of different species of organisms (their distance from a common ancestor). It is derived by determining the differences in a particular gene common to the several species in question.

lineage plan in cell differentiation, the process by which cells change from generation to generation because new protein ingredients are added to each successive generation and then are apportioned differently when the cell divides.

lipophilic (*Gr.* "fat liking") an uncharged molecule that does not interact with charged molecules such as water, and therefore tends to cluster with other lipophilic molecules away from watery fluids.

luciferin a molecule that is found, for example, in specialized cells in a firefly's tail and that emits light in a reaction with oxygen and ATP. The light-producing process is the exact opposite of photosynthesis.

M

messenger RNA (mRNA) a complementary copy of a length of DNA — a copy of a gene — that is translated into a protein on a ribosome.

meiosis gene recombination that occurs before sexual reproduction. Chromosome pairs double, transfer segments, separate into two cells and then into four eggs or sperm, each carrying a different combination of the mother's and father's genes.

metabolism the chemical reactions in cells by which materials are broken down to obtain energy to build living substance and drive vital processes.

mitochondria the cell's bacteria-sized sugar-burning organelles of respiration. They burn sugar in the presence of oxygen and produce ATP, water, carbon dioxide, and heat.

mitosis the process of doubling a cell's genes and dividing them into two new daughter cells.

molecule a chemical unit formed when two or more atoms collide and bond to each other. This bonding is not a random event - certain atoms have "the right fit" (bonding affinities with other atoms).

molecular communication the chemical relationships brought about by regulatory signal molecules and receptors that create a network of interconnections within cells, between cells, and among organs and tissues made up of those cells.

molecular relatedness the extent of differences in nucleotides between the same genes — or in amino acids between the same proteins — of two different species. Molecular relatedness reveals how much two different species have in common, and thus how far they are from an ancestor common to both.

morphogen (*Gr.* "maker of shape") a molecule that has an effect on all cells within an area as large as a square millimeter around it. As morphogen concentration varies, so does the effect on neighboring cells. Morphogens create positional concentration gradients that direct the development of limbs, sex organs, and the brain.

mutation a change in a gene. Though most mutations show no effect, they can produce new or altered traits. Rarely, such changes confer an advantage on the organism.

myosin a protein that, when energized by ATP, attaches to another protein called actin and changes shape. The shape change shortens the actin-myosin complex and is the mechanism for muscle contraction.

myxamoebae the individual cells of the slime mold *D. discoideum* that aggregate and differentiate to form a slug and fruiting body.

N

NAD+ (nicotinamide adenine dinucleotide) a molecule that acts as a carrier of "hot" hydrogens – activated hydrogen atoms. During respiration, hydrogens are extracted from fragments of sugar molecules and transferred to NAD+ (NAD+ + 2 electrons + 1 proton $\rightarrow$ NADH), then used to generate ATP.

NADP+ a molecule that acts as a carrier of "hot" hydrogens – activated hydrogen atoms essential for making sugar from carbon dioxide in photosynthesis (NADP+ = 2 electrons + one proton $\rightarrow$ NADPH). Thus, NADP+ is involved in the synthesis of sugar (anabolism), NAD+ in its breakdown (catabolism).

NADPH the NAD carrier molecule with energized electrons attached (NADP+ + 2 electrons + one proton).

natural selection the principal mechanism of evolution, which includes two processes that operate together: chance (random changes in the gene pool of a population) and selection (the non-random survival of what "works").

negative feedback a mechanism that acts as a damper on change. It reduces the difference between a current state of affairs and a desirable one.

nerve cell (neuron) a specialized cell with long extensions (axons and dendrites) that carry and transmit electrochemical messages. A rapid feedback facilitator.

nervous system an organized network of nerve cells that gathers information about the internal and external environment of a living organism and responds to that information.

nucleotides four similar three-part molecules (ATP, GTP, CTP and TTP), each made of a different base (adenine, guanine, cytosine and thymine) and identical sugar and phosphate groups. Each of the four nucleotides functions as an information unit in the DNA molecule.

nucleus the membrane-enclosed compartment in eukaryotic cell that houses the cell's information molecules. Also, the positively charged "core" of an atom, made up of positively charged protons and uncharged neutrons.

O

oxygen an atom with eight protons and eight electrons. Usually found as O_2 gas in the atmosphere, it reacts very readily with many other atoms, notably with hydrogen atoms in the final step of respiration. Oxygen molecules are a by-product of photosynthesis.

P

PCR (polymerase chain reaction) a method of amplifying DNA by rapidly multiplying selected gene sequences.

pheromone a small molecule that is released by an organism and fits receptor sites in a similar organism, triggering an alteration in its chemistry and/or behavior.

phosphate a small molecule, made of four oxygens and one phosphorus atom, that bonds to deoxyribose (in DNA) or ribose (in RNA) to form the backbone of a nucleotide chain. It serves many other purposes in biological systems.

phospholipid molecules fat molecules with water-liking heads and fat-liking tails. Phospholipid molecules are the most common component of cell membranes.

photon a packet of electromagnetic energy that radiates from the sun and can bump chlorophyll's electrons into higher-energy orbits.

photosynthesis the process in plants and some bacteria that converts the sun's energy, water, and carbon dioxide into glucose. Glucose in turn is used as an energy source for the photosynthetic organism or another one that eats it.

pigment a molecule that absorbs a particular wavelength of light to a greater extent than other wavelengths. Photosynthetic pigments transfer absorbed light energy to electrons.

pituitary gland a "master gland" in the brain that receives hormone signals from the hypothalamus and in turn sends out its own hormone signals that switch on hormone production in other glands. Rising blood levels of these secondary hormones act as a negative feedback signal that tells the pituitary to reduce its hormone output.

plasmid a circular segment of DNA that splits off from its host DNA and can replicate independently as a separate genetic unit.

polarity an unequal distribution of proteins within a cell that establishes a difference between one part of the cell and another. Polarity can lead to differentiation (two non-identical daughter cells) if a polarized parent cell divides.

polymerase the "builder" protein that assembles a complementary DNA strand along each of the separate, original strands by covalently bonding individual nucleotides to their complements.

positive feedback a mechanism that works as an amplifier of change by increasing the difference between the current state and the desirable state.

primary producers *autotrophs*; organisms (green plants and photosynthetic bacteria) that capture energy from sunlight and themselves are the energy source for herbivores, carnivores and decomposers.

products the molecules, atoms and energy present at the completion of a chemical reaction.

protein any of a wide variety of large "machinery" molecules made by linking combinations of the 20 different amino acids into long chains. The unique function of each protein is determined by the amino acids it contains and by their order in the chain, which in turn determine how it folds. Subtle shifts in their internal structures cause proteins to change shape and thereby do work. Proteins function as enzymes, material transporters, initiators of motion, structural elements, chemical regulators, chemical messengers and antibodies.

prokaryote a unicellular organism with no nucleus or membrane-bound organelles (a bacterium).

protist a single-celled organism with a nucleus and membrane-bound organelles.

proton the positively charged particle found in the nucleus of any atom. Each kind of atom has a unique number of protons in its nucleus. Also, a hydrogen atom stripped of its electron.

R

R-complex the first and earliest evolved "survival" layer of the brain. An extension to the upper brain stem first seen in reptiles, it influences territoriality, mating, and aggression in humans.

radiocarbon dating measuring the amount of radioactive carbon in the molecules of once-living tissues, as a way of determining the length of time since an organism died. Useful only for dating tissues less than 40,000 years old.

"random walk" motion undirected by any choice or goal.

reactants the atoms and molecules present at the start of a chemical reaction.

receptor a specialized protein that is embedded in a cell membrane and receives and transmits chemical messages. It is the basis for cellular communication.

recessive trait an inherited characteristic of an organism that is masked by a contrasting, dominant form of that characteristic.

recombination the mixing of segments of matching chromosomes prior to the making of sperm or egg. It is this exchange, and the subsequent union of the DNA in sperm and egg that provides variability in the genes of subsequent generations.

regulatory protein a protein that reversibly changes its function in response to a signal. Its surface has two working sites: one that recognizes a small signal molecule, and one that responds to the presence of the signal by changing its shape.

repair nuclease the "eraser" protein that finds poorly matching or defective nucleotides in a DNA strand and snips them out, replacing them with the correct complementary nucleotide.

replication the copying of all the DNA in a cell's nucleus. The DNA strands are separated and then complementary nucleotides are linked along each of the separated strands. The grand event preceding a cell's dividing into two.

repressor a gene regulator. A protein that binds to a site on DNA next to a gene and blocks the transcription of that gene, thus preventing the synthesis of the protein the gene prescribes.

respiration the cell's chemical process that burns sugar to release energy (create ATP). Respiration, unlike fermentation, requires oxygen.

ribose the sugar molecule that bonds to a phosphate group and forms the backbone of the nucleotide chain of an RNA molecule.

ribosome a cell organelle made of many RNA and protein molecules. In the cytoplasm, it performs as a machine that translates the information in messenger RNA into an amino acid chain — a protein.

RNA (ribonucleic acid) a close cousin to DNA but with uracil (U) replacing thymine (T) as one of these bases and with ribose instead of deoxyribose as the backbone sugar. There are three kinds of RNA: messenger (mRNA), transfer (tRNA), and ribosomal (rRNA).

RNA polymerase the enzyme that makes RNA by copying the sequence of bases on a DNA strand.

S

second law of thermodynamics the physical law that states the fact that energy eventually disperses, dissipates, scatters — that is, it is transformed from more usable forms such as photons and chemical bonds to a less usable form, namely heat (see *entropy*).

segmentation a type of body pattern that allows for many variations and was probably the result of a mutation that turned a one-part body into a two-part body.

selective breeding a way of purposefully and rapidly changing the traits of succeeding generations of an organism. An organism with the desired traits is selected and mated with another also having the desired traits. The most desirable offspring are then selected to reproduce, and this selection continues over several generations.

selection the retention in a population of advantageous traits that contribute to the organism's survival and production of offspring.

self-organizing system a system containing machinery and the information that allows it to build and maintain itself. A cell is one such system and you are another.

sexual reproduction the mingling of two sets of genes, one from the mother and one from the father. This process takes advantage of the differing attributes of two individuals to create a greater possibility of change in the offspring's genome.

side group group of atoms attached to the basic "backbone" structure of an amino acid and giving it its unique chemical character. Side groups can form weak bonds with each other, giving proteins their shape and flexibility.

signal molecule small molecule that activates a protein receptor on the surface of a cell or interacts with regulatory proteins within the cell.

"solar antenna" a group of chlorophyll pigment molecules on the surface of a thylakoid that absorbs solar energy and initiates the process of photosynthesis.

specialization the process by which cells acquire specific function and lose the capacity to create all the proteins prescribed by their DNA. During embryonic development, each new generation of cells becomes a little different from the generation before on the road to specialization. (see also *differentiation*)

species a population of organisms that mate only with others in the population and produce similar offspring.

spontaneous generation the obsolete idea that complex living organisms can arise repeatedly and without antecedent from non-living matter.

starch chains of sugar molecules created by plant cells as an energy-storage medium.

"straightener" a protein that binds to a single replicating strand of DNA and keeps it from tangling.

stroma the space in the chloroplast outside the "thylakoid where half-molecules of sugar are assembled using energy supplied by ATP.

substrate the specific molecule an enzyme recognizes and temporarily binds.

sugar life's universal food. A molecule produced from carbon dioxide, water, and sunlight by organisms capable of photosynthesis. Source of all of the energy and almost all of the building materials in cells.

superorganism a complex community, such as an ant or termite colony (or the internet for that matter), that shares information and can therefore perform functions impossible for individuals.

suppressor protein a protein that has a key role in preventing a cell from dividing.

symbiosis a cooperative relationship between organisms of different species that benefits both species.

T

thylakoid one of a series of flattened sacs stacked inside a plant chloroplast. The thylakoid membrane is the location for most of the chemical reactions that produce the energy to turn sunlight, carbon dioxide, and water into sugar.

thyroid gland a group of specialized cells that secrete hormones into the blood. The hormones bind to specific receptor molecules on some cell membranes and cause those cells' metabolism to speed up.

topoisomerase the "untwister." A protein that untwists the DNA helix, allowing helicase to separate the two strands to be replicated.

transcription the first stage of protein-making, in which a small section of a DNA strand (a gene) is copied onto a messenger RNA molecule.

transfer RNA (tRNA) the decoding unit that links information to its final protein product. Each tRNA molecule has a triplet nucleotide code (an anticodon) at one end and an amino acid acceptor site at the other. The anticodon is complementary to a codon on a particular mRNA molecule.

translation the last stage of protein-making, in which mRNA molecules attached to a ribosome are sequentially "read" by tRNAs. The resulting chain of amino acids (a protein) then folds up into its final shape.

transposition a change in the position of a gene on the DNA strand that can affect the proper functioning of nearby genes. Rarely, a transposition can lead to useful genetic innovation.

triggering event a small change in regulatory protein that alters the protein's time of action, how long it acts, its ability to bind, or some other function. Such small changes can introduce major evolutionary innovations.

triplet see *codon*

U

ultraviolet light light waves or photons more energetic than those of visible light which can damage nucleotides and thus threaten the integrity of DNA molecules.

V

virus an entity consisting of DNA and RNA in a protein coating, which can invade other cells and make copies of itself, often harming or destroying the function of the invaded cell or altering its genes. It probably evolved as a plasmid that acquired the ability to use a host cell's ATP and ribosomes to make a protein coat for itself.

W

weak bonds weak attractions between positive and negative charges on atoms that form at very close range and easily break and re-form. Such bonds between complementary bases in DNA allow the strands to separate readily.

X

X chromosome a sex-determining chromosome present in every cell.

Y

Y chromosome a sex-determining chromosome present in half of all sperm cells.

Notes

We have benefited from the study of many books, but note here our most valued general source: *Molecular Biology of the Cell* by Bruce Alberts, Dennis Bray, Julian Lewis, Martin Raff, Keith Roberts, James D. Watson, Garland Publishing, 1994. Another useful source was *A Guided Tour of the Living Cell* by Christian de Duve, Scientific American Library, 1984.

Chapter 2. Patterns

page 24 "Before a single plant or animal appeared..." A good discussion of the contributions made by our microbial ancestors can be found in Lynn Margulis and Dorion Sagan's *Microcosmos*, Summit Books, 1986.

page 49 "Wrinkly skin provides more surface area than smooth..." Elephants also gain more surface area through their oversize, blood-rich ears - another "creative mistake" of nature.

page 52 "...like a pair of Mickey Mouse ears..." This analogy came from M. Mitchell Waldrop's *Complexity*, Simon & Schuster, 1992.

page 58 "The steam engine with a governor..." Gregory Bateson effectively makes the case for life's self-corrective tendencies in *Mind and Nature*, Bantam Books, 1979.

page 67 "Life Maintains Itself by Turnover." Turnover also occurs in ecosystems through the birth and death of organisms. Individuals come and go but the overall characteristics of populations remain relatively stable.

page 74 "...from near boiling sulfur springs..." Recently scientists have focused their attention on single-celled microorganisms called thermophiles, which can thrive in deep ocean vents and hot springs at temperatures higher than the boiling point of water. Evidence suggests that organisms like these were the first living creatures on earth.

page 76 "Creatures are self-interested but not self-destructive..." Appreciation of the cooperative nature of life has been gaining ground in recent years. Lewis Thomas and Lynn Margulis, among others, have written extensively on the evolution of symbiotic arrangements. Another interesting treatment can be found in Robert Axelrod's *The Evolution of Cooperation*, Basic Books, 1994, in which the author uses game theory to show the effectiveness of cooperation as a strategy for survival.

page 77 "...originally acted as small predators..." The now widely accepted theory that mitochondria were once invading bacteria was championed by Lynn Margulis and later substantiated when it was found that mitochondria have their own DNA, different from that found in the cell nucleus.

Chapter 3. Energy

page 94 "This constant flow keeps our earth in an energized state..." Visible light, occupying just a tiny band on the electromagnetic spectrum, has just the right amount of energy to bounce electrons into higher orbits - the necessary first step in energy conversion. (Lower-frequency infrared light lacks the strength, while higher-frequency ultraviolet light carries so much energy it tends to break bonds and so disrupt the function of molecules.)

page 97 "How a Dog Shares His Fleas." This metaphor, taken from Heinz Pagels (*The Cosmic Code*, Bantam Books, 1983) also helps to make clear the statistical nature of the second law of thermodynamics. Individual fleas, like individual atoms or molecules, move about quite randomly. But given another, flealess dog, the probability is that the whole population of fleas will flow in one direction - from a more concentrated, or ordered state on one dog to a more dispersed or randomly distributed state on two dogs - until equilibrium us attained. Similarly, atoms and molecules will flow from a more concentrated to a more dispersed state. The probability of their flowing in the opposite direction is vanishingly small. Thus, the unidirectionality of events - time itself - arises from the statistical behavior of the atoms and molecules of matter.

page 105 "A fourth group, the 'decomposers'..." These organisms complete the cycle of materials by converting the substance of all other life into reusable forms in the soil, whereupon they are taken up once again by plants. All life on earth would quickly cease if the decomposers stopped work.

page 117 "Making Sugar Out of Thin Air." As mentioned in our note to page 74, there are many kinds of organisms (mostly bacteria) that can construct themselves without the aid of sunlight. Some of these, probably the most ancient, convert organic material (that is, short carbon chains such as sugar) derived from decaying life into ATP and thence their own substance by the process of glycolysis (see page 130). Others can use simple *inorganic* molecules from which they generate energetic electrons and ATP. They then use the electrons, hydrogen ions, and ATP to convert CO_2, much like the final steps of photosynthesis. An intriguing account of the habits of such organisms, many of which play important roles in the ecology of our planet, may be found in *The Outer Reaches of Life* by John Postgate, Cambridge University Press, 1994.

Chapter 4. Information

page 140 "The possibility that we could arise..." The 747 analogy comes from Fred Hoyle, quoted in *Origins: A Skeptic's Guide to the Creation of Life on Earth* by R. Shapiro (Summit Books, 1986).

page 152 "Chemical Units of Information." The four letters are the basis of a digital system. One advantage of digital systems is that information doesn't degrade even after making multiple copies of itself. If you copy a compact disk, then make a copy of the copy, etc., the hundredth copy would sound as true as the original. This would not be true of a phonograph record or a cassette tape.

page 174 "Copying Genes Into Messengers." In the later 1970s, scientists made the astonishing discovery that, unlike the simple state of affairs we have described that is characteristic of bacteria, genes in the larger, nucleated cells of higher organisms (eucaryotic cells) are interrupted by long sequences of nucleotides that do not code for a protein or a part of a protein. The sequences of nucleotides that do code for a protein are called exons, while the non-coding sequences are called introns. To make a complete protein, a long RNA copy is made of the stretch of DNA containing both exons and introns. Inside the nucleus, RNA-splicing enzymes cut out introns and link exons together to make a messenger RNA, which leaves the nucleus and is translated into protein on ribosomes in the cytoplasm.

It is now widely hypothesized that this split nature of genes is the most ancient and that modern bacteria, to grow more efficiently, lost their introns after their exons fully developed. It appears that, out of long, meaningless stretches of DNA, certain segments (exons) evolved that coded for useful *parts* of proteins (a special shape, a special affinity, etc.).

Mechanisms then arose (RNA splicing) for uniting useful parts into final functional proteins. The advantage of such a modular system is that parts can be combined in different ways to create a great variety of proteins with diverse functions.

Chapter 5. Machinery

Page 194 "Each adaptor recognizes a particular three-letter code." Earlier (pages 152,153) we compared nucleotides with letters and genes with paragraphs. Now we can extend the metaphor by comparing each three-letter code with a word. One word translates into one amino acid.

page 196 "From DNA to Protein - A Multistep Process." The story of the discovery, in the 1950s, of amino activation and of transfer RNA by Mahlon Hoagland, Paul Zamecnik, and their colleagues at the Massachusetts General Hospital is told in *Toward the Habit of Truth: A Life in Science* by Mahlon Hoagland (W.W. Norton, 1990).

page 198 "Assembling the Protein Chain." For clarity, we show one gene translated into one protein. In reality, many proteins, in their final working form, are multimers: two or more separate proteins that fit snugly together. Each protein is made from a messenger from a different gene, and these genes may not be contiguous in the DNA. But after ribosomes complete the assembly of each, they are brought together to form the final working unit.

Chapter 6. Feedback

page 209 "Signaling, Sensing, and Reacting." A general explanation of feedback principles by a pioneer in the field, Norbert Wiener, may be found in *The Human Use of Human Beings*, Avon Books, 1967.

"Feedback is a central feature of life." Because a feedback system treats its own state as information, some see it as similar to a mental process operating in all organisms and ecosystems - as well as in the minds of humans. Gregory Bateson advances this point of view in *Mind and Nature.*

page 214 We owe the discovery of allostery in biological systems to the great French microbiologist and Nobel laureate Jacques Monod and his colleague Jean-Pierre Changeux. Monod's book *Chance and Necessity* (Alfred A. Knopf, 1971) is rich in the scientific and philosophical background of many of the subjects discussed in this book.

page 220 "Controlling the Machinery that Makes the Machinery." Jacques Monod and Francois Jacob of the Pasteur Institute in Paris pioneered the exploration of regulating protein synthesis by repressors. They shared a Nobel Prize for this work in 1965.

page 236 "Ecology Loops." The cybernetic properties of the freshwater cycle are described in Barry Commoner's *The Closing Circle*, Bantam Books, 1971.

"...ecosystems operate not as single loops but as networks..." The scientist-inventor James Lovelock has offered numerous scenarios for the earth ecosystem operating as a single, large-scale network of feedback loops (*The Ages of Gaia*, W. W. Norton, 1990).

Chapter 7. Community

page 248 "Super-organisms?" Sources of this discussion include *Three Scientists and Their Gods* by Robert Wright (Times Books, 1988) and *The Insect Societies* by E. O. Wilson (Belknap Press, 1971).

pages 250-251 "Two-faced and Slimy." Much of this information comes from *Cells and Society* by John Tyler Bonner, Princeton University Press, 1955.

page 278 "A Chain of Command." Not all signaling originates within the embryo. In mammals, signals from the mother pass to the embryo by way of the placental bloodstream. In certain species, nurse cells adjacent to the egg signal the egg to begin its development. In honeybee populations, the queen determines the sex of bees by deciding which eggs to fertilize. Unfertilized eggs will become male; fertilized, female. In some cases, temperature acts as a signal: warm alligator eggs develop into males, cool ones into females.

Chapter 8. Evolution

page 288 "An Ancient Earth." In emphasizing the gradualness of geologic change, Hutton and his followers may have underestimated the role of past catastrophes in the evolution and extinction of life. Fossil records tell the story of several massive extinctions of species brought about by climate changes, impacts of asteroids, etc. After each one a great explosion of new life forms occurred, owing to the new possibilities opened up. In *Wonderful Life: The Burgess Shale and the Nature of History* (W. W. Norton, 1989), Stephen J. Gould gives a dramatic account of the explosive emergence of new life forms in the Cambrian period.

page 289 "Chance and selection are fundamental to any creative act." Often evolutionists find themselves in disagreement with those who believe that life is so complex and beautiful that it must have been designed. To many, the word "designed" means something like "planned in advance" — a misleading definition. Experienced designers, artists, and scientists know that design doesn't work that way. Creativity involves taking advantage of accidents, chance encounters, surprises, etc. In other words, design must access the random. If it did not, nothing new would happen.

page 292 "Self-replicating chains." The replication idea comes from Richard Dawkins in *The Selfish Gene*, Oxford University Press, 1989. This book and his later books, *The Blind Watchmaker* (W. W. Norton, 1986) and *River Out of Eden* (Basic Books, 1995) are especially fine statements of evolutionary evidence and theory.

page 298 "An Elephant-sized Mouse." This imaginary experiment comes from Ledyard Stebbins in *Darwin to DNA, Molecules to Humanity*, W. H. Freeman, 1982.

page 302 "Monkeys and Word Processors." This is a good place to state that metaphor can be stretched too far, or taken too literally. The key points we use the metaphor to illustrate are (1) that *chance* pecking produces, in small steps, useful or meaningful sequences of letters - i.e., information; (2) that the information *accumulates* by a selection process; and (3) that each level of complexity can set the stage for a higher level of complexity (a process that engineers call "bootstrapping"). Clearly, the metaphor breaks down when we recognize that monkeys are using computers, a human invention; and that we humans are dictating the goal, sonnets.

page 304 "...even the slightest advantages will take hold..." Richard Dawkins in *The Blind Watchmaker* (W. W. Norton, 1986) advances this idea most forcefully.

page 310 "The best information sets..." Labeling a gene "better" or "worse" usually depends on how well the protein it codes for works in its particular setting. It may not be equally adaptive in another, and no environment is static. Volcanoes, earthquakes, glaciers, asteroids, and continental drift can destroy the best-adapted creatures.

page 321 "An enzyme in the flower's cells makes pigment molecules." Assigning the flower's color to a single gene may be an oversimplification, but should convey the idea. Likewise, the white flower is meant to convey an absence of pigmentation, not a typical white flower which may be reflecting ultraviolet light invisible to us but not to the bee.

page 291 "Geographical Isolation." The creative possibilities inherent in small geographically isolated groups are analogous to those of such circles of artists as the French Impressionists in the late 1800s and the American Abstract Expressionists in the 1950s. Both groups were small, isolated from the mainstream museums and critics, and freely exchanging ideas within themselves. And both groups created major new movements (i.e., large changes) in a very short period of time.

page 330 "...toward greater variety and complexity." The idea that life inevitably evolves toward greater variety and complexity is a matter of some controversy among biologists. Fossil evidence suggests that a number of organisms have remained virtually unchanged for hundreds of millions of years (for example, the horseshoe crab). Nevertheless, a general drift toward the more complex, the more varied, and the more interactive seems undeniable.

We can also see a trend toward the more abstract. In a system that builds on top of its simpler foundations, as evolution does, the "add-ons" tend to operate at a more *abstract* level of logic — i.e., more indirectly. Regulatory genes (See *Genes as Switches*, page 264) offer a good example. These genes operate on other genes - the ones that produce the worker proteins — and must have evolved after them. Further evolution then produces a regulator that controls a whole set of regulators. Such a hierarchy of control seems to be a central feature of intelligence, and it is by such layering that progressive complexity evolves.

page 331 "High-speed evolution." The story of Peter and Rosemary Grant's exhaustive study of finches is told in the *Beak of the Finch, A Story of Evolution in Our Time*, Alfred A. Knopf, 1994.

page 336 "A key part of the environment for organisms is other organisms." It could equally be said that the main environment for an individual gene is other genes. Indeed, it is at this molecular level that we see the most fundamental cooperation.

page 345 "Tree of Relatedness." This phylogenetic tree is based on analyses of amino acid sequences in cytochrome c, a protein found in all the organisms shown. The study was done by Walter M. Fitch and Emanuel Margoliask and originally published in *Science* 155, 279-284, 1967. A modified version of the tree appears in *The Mechanism of Evolution* by Francisco Ayala, *Scientific American* 239, September 1978. Our version derives from the latter. As stated by Ayala, "The numbers of the branches are the minimum number of nucleotide substitutions in the DNA of the genes [of cytochrome c] that could have given rise to observed differences in amino acid sequences."

page 348 "Evolution of Intelligence." Excellent sources on the evolution of the brain are E. O. Wilson, *On Human Nature*, Harvard University Press, 1978, and *The Origin of Consciousness in the Breakdown of the Bicameral Mind*, Houghton Mifflin, 1976.

The idea that our brain evolved in three distinct stages was proposed by Paul D. Maclean in *Astride the Two Cultures*, edited by Harold Harris, Hutchinson, 1976. Our illustrations somewhat oversimplify in suggesting such marked divisions between reptiles, mammals, and humans. Contemporary species exhibit more subtle gradations and overlaps.

page 350 "Cultural Evolution." For a good discussion, see Daniel C. Dennet, *Consciousness Explained*, Little, Brown, 1991.

Answers (and Hints for Answering) the End-of-Chapter Questions

Chapter 1

1. There is no one correct answer - the idea behind any answer to this question is that, on the smallest scale, all living organisms are made up of the same "chemical ingredients" - molecules, chain molecules, and molecular structures, and that these units build on each other resulting in different levels of complexity.

2. To answer this one you have to enter a mindset that says "lets look at the similarities," rather than "lets see what's different."Then compare yourself to a mosquito, for instance, or a mosquito to a bird, and then list as many similarities as you can see or infer.

3. Over to you.

4. Hint - how would you go about finding out how a telephone, a remote phone, and a cell phone function?

5. Homologous patterns are variations on a common anatomical theme or pattern. They suggest that the creatures are more related than they appear to be and that the pattern may have existed longer than the creatures.

6. A chain molecule.

7. Carbon dioxide (CO_2), water (H_2O), and oxygen (O_2).

8. Sugar, nucleotides, and amino acids.

9. The dolls are an analogy for the levels of organization of living creatures. They show b) how a higher level (a larger doll) includes everything in the levels below it. They are NOT intended to imply that the lower levels are actually INSIDE the higher ones.

10. The differences are in the relative length, thickness, and angles of the bones. To answer these questions it's important to think about the environment and survival needs of each of these kinds of animals. The whale's forelimb is short and thick, which provides the strength and leverage necessary to angle it successfully through resistant water without breaking. The cat's long, narrow forelimb structure allows it the leverage and maneuverability to move a light body rapidly through a less resistant medium. To get what it needs to eat, the whale needs to be able to navigate accurately through a buoyant medium while the cat needs to move fast, climb, claw, and contend with gravity....

11. The possibilities are almost infinite. The unifying element is basically a raised platform on which to place one's posterior to keep it off the cold ground. Yet there are solid wooden kitchen chairs (which can double as ladders in a pinch) and soft, comfortable recliners for watching TV. Chairs may have a pedestal base made of stone or marble with no back, a low back, or a high back. The legs may be of turned wood - 3 for a milking stool, 4 for a "standard" dining chair. Leg lengths may vary depending on whether it is a child's chair or a barstool. Legs may be plain or elaborately carved, with varying proportions from thick to thin, curved to straight. Seats and backs can be upholstered in silk or cotton, or made out of bamboo sticks or aluminum bars or molded plastic so they can be left outside in the rain. And they are all still chairs, identifiable as such by their raised platform and function.

12. They separate based on density - gold particles are more dense than soil and sediment to the bottom of the gold pan easily and quickly. Nuclei, because of their size and density, are larger and more dense than smaller membrane pieces so will sediment faster (or at lower RPMs in the centrifuge) than other cell components.

Chapter 2

1. Information requires difference.

2. If information requires difference, it follows that the smallest number of units necessary for coding information is two. We know from computer language and Morse Code that two different units are sufficient to convey information.

3. Since water is both outside and inside a cell's membrane, and a single fatty layer has both a water-liking side and a water-shunning side, a cell membrane must be a double fatty layer. This way, the water-liking sides will face the water both inside and outside, and the water-shunning sides will face each other thereby avoiding water and creating an impermeable barrier to water.

4. Hard shells that cover the membranous wings, six legs, antennae, jointed legs, two eyes, bilateral symmetry.

5. It suggests that the beetle pattern is laid down in a common ancestor. The great variety came later in evolution.

6. It is likely that the better-camouflaged beetles that could eat what's to be found in a rose garden would survive longest. The first to go would be the very light-colored, very dark colored ones, and those with bright patterns. Three years from now the population of beetles would be far less diverse than the one you see on page 34. There would be very small beetles, green beetles, red beetles, beetles with dark spots on red or green backgrounds, and maybe some brightly colored beetles that taste VERY bad. (Other qualities, too, might also determine survival: whether they dwell on top of a leaf or on the underside; whether the predator birds are large or small; etc.)

7. They perform relatively simple tasks repetitively. Working together, they can perform complex operations.

8. They contain instructions for assembling proteins, the cell's robot machinery.

9. Errors in information may create new information - new instructions for making proteins. The new proteins consequently perform novel tasks that may confer novel properties on an organism. Accumulating novelties can result in novel forms of life.

10. Due to their polarity, i.e. the tendency of the oxygen to have a slight negative charge and the hydrogens to have a slight positive charge, water molecules interact with each other to form lattice-like structures. These lattices hold together and keep the water liquid at relatively high temperatures. CO_2 and CH_4 are not polar.

11. Self-correction is the organism's way of keeping life processes in control. When metabolic processes get too slow, the system takes note and speeds them up. When chemical products pile up, the system stops making them. Corrections are always such that a dynamic balance is maintained between too much and too little. A good example is the seemingly simple act of standing. When we stand still, nerve cells are constantly reporting the slightest tip one way or the other. Other nerves respond by signaling the appropriate muscles to bring us back to the "desired" upright position. Thus, we remain upright through the subtle and largely unconscious contractions of muscles.

12. Autopilot, thermostat, cruise control.

13. Oxygen, produced by photosynthesis in plants, is used by both plants and animals to burn sugar. The burning of sugar produces carbon dioxide, which is used by plants to construct themselves.
In another example, about 3% of our body's weight is nitrogen which is an essential component of RNA, DNA and protein. When we die, these molecules are broken down and become part of the soil. With the help of microorganisms in the soil, certain plants can use these broken down molecules to build the nitrogen back into their own molecules. These plants are then eaten by us and become, again, part of us. Other examples: chicken and egg; the flow of oxygen and carbon dioxide through the bloodstream; the migration of birds; the dying and regeneration of leaves on trees.

14. Turnover means the continual breakdown and replacement of components. Proteins, for example, are constantly being made in cells, perform their duties for a finite lifetime, then are broken up into their constituent amino acids again. At a higher level, most of the cells of plants and animals replace themselves regularly and their parts are reused.

15. Complex systems are unstable. They tend to break down randomly and therefore unpredictably. By using systems that regularly and controllably break down and rebuild components, life can maintain its stability - for a time.

16. Theoretically, any feature of an organism can continue to evolve to extreme limits. Legs can get longer, plumage gets fuller, branches grow longer. But the expansion of any variable, which might seem like a good thing, is likely to stress some other variable. Longer legs might be more fragile; too much plumage can hinder flight; lengthening branches requires a larger trunk or more roots. Generally what gets sacrificed is flexibility.

17. Opportunism.

18. A bacterium, one that is especially good at converting sugar into energy, invades another, larger bacterium. Over generations they begin to share products and functions. Over more generations their interdependence becomes such that the small bacterium evolves into a permanent and essential component of the larger cell - a mitochondrion.

19. Any two species closely linked to each other are likely to have co-evolved - like grazers and grass, predators and prey, parasite and hosts.

20. The elephant's big, blood-rich ears. The polar bear's fur. (Its fur is apparently transparent to UV light - allowing nearly 100% of it to reach the skin. But the fur allows no infrared out - i.e., its highly insulating.) We humans can only wear or remove clothes to help control our temperature. Getting wet is a reliable way for all three animals to cool down, since the process of evaporation takes up a lot of energy from the skin surface.

Chapter 3

1. The making and breaking of chemical bonds.

2. As energy "flows downhill" in this way, living systems can capture some of it in its concentrated, usable forms, put it to use, and then discard it in less usable, more dispersed states. In other words, life graphs the opportunity to create order from the universe's tendency to create disorder.

3. In a system in equilibrium there is no net accomplishment. Chemical reactions proceed at equal rates in either direction. This is the equivalent of death. When energy and material can enter and leave a system, it can be built and renewed, the equivalent of life.

4. A molecule rich in energy (ATP) that could transfer that energy to a reluctant molecule; and an enzyme (a protein molecule) to facilitate - catalyze - the transfer.

5. We humans, and all other animals, must eat other living things - ultimately plants - to exist. Photosynthesis uses the energy of sunlight to make chemical energy and living substance. Respiration breaks down living substance and generates chemical energy.

6. Photosynthesis produces glucose and oxygen. Respiration produces ATP, carbon dioxide, and water.

7. If you translate Franklin's "fluid fire" to mean solar energy and his "fluid air" to mean carbon dioxide, he has described the essence of photosynthesis long before that process was chemically defined.

8. Skunk cabbage in respiring loses heat every time a chemical reaction takes place (remember entropy). Unlike most organisms which produce more ATP from respiration and lose less heat, the skunk cabbage which starts to develop flowers while still covered with snow, can flower earlier than other plants because it has this extra heat-generating capacity.

9. No hint needed.

10. Hint - remember entropy?

11. Hint - review photosynthesis.

12. Kinetic energy (energy of motion) of the water is transferred to the mill wheel, making it rotate. This rotation can be transferred by a combination of gears and straps to the turning grindstone in a grain mill or to the turning saw blade of the lumber mill or to the straps, which run the weaving machinery in a fabric mill. The transfer is never 100% efficient. Some of the energy of the stream is lost as heat to the rotating grindstone or movement of gears. The mill machinery cannot be run backwards because the final product has less energy than the flowing water. Similarly, some of the energy of a beam (photon) of sunlight is captured and transferred to sugars during photosynthesis and from sugars to ATP during respiration in cells. The energy in ATP is then released to satisfy the energy needs of the cell such as synthesis of important molecular chains like DNA or proteins. However, some of the energy of the original sugar is also released as the heat, which is dispersed into the surroundings of the cell and can't be recaptured (run "backwards") to become sunlight.

Chapter 4

1. Difference.

2. They prescribe the structure of proteins.

3. A blueprint is a 2-dimensional representation of the object to be constructed. DNA is a digital - 1-dimensional - recipe, or set of instructions.

4. True and false - DNA cannot copy itself. A cadre of enzymes works in concert to accomplish this complex task.

5. You need egg to make chicken and chicken to make egg. So neither comes first. The egg is information, the chicken is machinery - a fundamental distinction in nature. The information in the egg needs help to start making the chicken - i.e., there must be some machinery, some chickenness, in the egg to get the process of chicken making started. On the other hand, the chicken contains all the information and machinery of chickenhood. Of course, if we follow the question "which came first" backward in time long enough, both the chicken and egg would gradually become something simpler, no longer recognizable as poultry! and making the question moot.

6. 1) Musical notation conveys information about the kind of music to be played (pitch, tempo, etc) using a staff of 5 lines, 4 spaces and various symbols placed on the staff to indicate pitch and to denote rhythmic values. 2) Coaches use X and O symbols with lots of arrows on a blackboard to describe the dynamic action of a football or basketball play. Differences in symbols' positions on the staff or on the blackboard in the coach's office indicate information to the players or musicians. 3) The signal flags used by ships to communicate.

7. If there are two complementary strands it leaves the way open for more information content being carried in the same molecule. It is theoretically possible to read and transcribe one strand in one direction to produce a gene product while the complementary sequence of the second strand read in the opposite direction has information for a completely different gene product. Also, if there is damage to one strand the other is there as a sort of "mirror image" for repair purposes.

8. Having separate tools means that each tool is designed to do one job optimally. Having separate proteins to unravel the helix, separate the strands, hold the strands open, read the bases, insert complementary bases into a new strand, and proofread the newly made sequence helps insure that the "fit" is correct at each step of the process. This system of checks and balances ensures fidelity of the new base sequence of DNA, which is so crucial for the life of the cell. This makes it worth the extra cost (a lot of cellular currency - ATP) to acquire the right tools.

9. The answer is not completely understood and is still debated, but one must think about the harsh atmosphere of the early Earth billions of years ago and how it is different from conditions on Earth today. Groups of chemicals (lots of water, methane, ammonia and other hot gases) in the anaerobic (no oxygen) atmosphere of early Earth were bombarded with ultraviolet light (a source of energy) and began to form molecular assemblies (early proteins and RNA) that were more "efficient" than the chemicals floating around them in the "soup." Eventually the molecules that are components of membranes developed and encircled these organic macromolecules, separating them from the disorganized materials around them. Some of these molecules acted as molecular "templates" for their own self-replication and could pass on their information content during formation of new molecules. Over billions of years, these organized and self-replicating molecules assembled, contained within a membrane, in ways to ensure efficient energy flow and could be considered the first living cells. Conditions on Earth today are much milder and very different (oxygen is present for one thing!). Although spontaneous alterations of chemicals by the environment might be theoretically possible today it is highly unlikely to occur now, so all present life must arise from living cells.

Chapter 5

1. As enzymes they catalyze chemical reactions; as muscle they facilitate movement; as key parts of the structure of cells and tissues they are "bricks and mortar"; as antibodies they defend against infection; and they serve as transporters of materials in and out of cells.

2. They must be activated (energized) by linkage to a portion of an ATP molecule. And they must then be attached, in this activated state, to a transfer RNA molecule.

3. DNA and the protein making machinery are geographically separated in the cell: DNA in the nucleus, machinery in the cytoplasm. Messenger RNA is the go-between. Messenger RNA is also, unlike DNA, expendable - it can be produced when needed, destroyed when not.

4. Using ATP, amino acids, ribosomes, mRNA and tRNA, the elements are 1) Amino acid activation; 2) transcription - copying genes into messenger RNA;
3) translation - linking activated amino acids to transfer RNAs so they may be aligned in order on messenger RNA on ribosomes prior to linking amino acid to one another.

5. Incubate ATP, activating enzymes, transfer RNAs, and ribosomes from bacteria with the appropriate messenger RNA from human cells. Then isolate the protein made and identify its human characteristics.

6. A triplet is a sequence of three nucleotides that codes for a specific amino acid.

7. a) would change the amino acid sequence of the protein chain. b) could possibly change the shape of the enzyme enough to interfere with its functioning. c) this could produce a tRNA that might carry the wrong codon/amino acid combination, thus changing the amino acid sequence of the protein change, and perhaps the shape of the protein.

8. Imagine opening and closing a clothespin with your fingers. This requires energy input (from your muscles!) to allow the ends to open (by absorbing the energy). Releasing the pin allows it to spring back into its original shape and attach the clothes to the clothesline. A protein does the same thing, absorbing energy from an outside source (usually ATP) then changing shape and springing back during its functioning. Changing shape quickly may be necessary for proteins, which need a rapid and reversible way to turn on and off (like a light switch). Regulatory proteins use this ability to recognize and work with their modulators, transport proteins to maneuver their loads across membranes and return , and signal or receptor proteins to communicate with other molecules. It is also one of the major mechanisms for activation or inhibition of enzyme activity in a cell.

9. There is some redundancy in the genetic code, i.e., more than one triplet sequence codes for some amino acids. Frequently the same amino acid will have a triplet that shares the first two letters of the code but varies ("wobbles") in the third letter. Example: CCU, CCC, CCA, and CCG all code for the amino acid called proline. Any of these triplets will allow proline to be inserted in that spot of the protein. So if a mutation changes CCC to CCG, it will make no difference to the structure of the completed protein. In addition, some amino acids are in relatively unused areas of the protein that are not critical for folding and function (analogous to a non load-bearing wall in a house), so if they mutate, there is still a chance that the protein will be functional even though slightly altered. And the third possibility is that a mutation may substitute an even "better" amino acid in the spot where the mutation occurs, thus increasing the function of that protein.

10. Multiple steps give you more control over any process. Each step uses a small amount of ATP rather than requiring a single large output of ATP in each of fewer steps. Each step provides multiple points where the process can be checked for fidelity, or even halted if necessary when the protein produced would be non-functional or in excess.

Chapter 6

1. Processes are never perfect; they inevitably deviate from course, run too fast or slow, and then must be corrected. This alternating deviation and correction produces the zig-zag.

2. In negative feedback, speeding up automatically leads to slowing down or vice versa, until a balance is reached. This equals control. Positive feedback means fast leads to faster - loss of control. You provide the examples.

3. Information that can control the rate or direction of a sequence of events feeds back to its beginning to alter that rate or direction.

4. Show parallelism between:
Airplane flight. A deviation in a compass bearing from the planned course induces a corrective course change.
A steam engine. Increased engine speed causes a corrective reduction of steam input and a

slowing rate.
Chemotaxis. Increased concentration of food molecules causes the bacterial motor to propel the bacterium toward even higher concentrations. When the bacterium finds itself in the highest concentration of food its motor reverses and keeps it there.
An owl getting its prey. A mouse's direction of movement causes the owl's eyes to signal its brain to notify its wings to alter course in response. In the four instances, a change in direction or rate is noted and induces a correction.
5. An enzyme catalyzes the first step in a sequence of chemical reactions leading to an essential end-product. If the end-product molecules can cause a shape change in that first-step enzyme, reducing its ability to do its job, the rate of the whole sequence of reactions slows. This results in less end-product, which in turn leads to faster action of the first-step enzyme. Voila: feedback control of an assembly line.
6. The genes that specify the structures of the proteins acting in the assembly line can be turned on and off by other proteins called repressors. When the end-products of assembly lines are not needed, repressors can grasp their genes and prevent them from being copied into messenger RNAs.

Chapter 7

1. No. All cells contain a full complement of DNA. Most of this DNA remains "silent" - unexpressed. Skin cells express only genes specifying skinness; brain cells, brainness, etc.
2. The colony is "smarter" than the ants. By cooperation and coordination among ants, the colony can accomplish tasks far more complex that any individual ant could.
3. Cells become different by changing their chemistry, their location, their shape, or their relationship to their neighbors. This is the beginning of pattern. You can't have a top or bottom, an outside or inside, a head or tail until you have differences between the participating cells.
4. One gene can code for a repressor - i.e., a protein that can bind to a gene coding for another protein. The repressor can be activated or inactivated by signal molecules - thus, it acts as a switch.
5. Some cells produce signal molecules that interact specifically with receptor proteins on the surfaces of other cells. A signal-activated receptor can, in turn, pass information to the interior of the receiving cell.
6. Early in life many brain cells (neurons) are produced to establish a vast array of connections. Later many of those connections prove to be weak or ineffective and the involved neurons self-destruct - sort of cellular suicide - called apoptosis.
7. If important constituents inside cells become, during early embryonic development, unevenly distributed, the two daughter cells of a division will be different. When these different cells again divide their contents, with similarly unevenly apportioned contents, their daughters will become even more different. So difference multiplies as cells multiply. This is lineage difference. In other cases, cells are influenced to change - e.g., to produce new proteins, by signals from their neighbors. This is difference induced by contact.
8. This is because of evolution's propensity to retain what works and discard what doesn't. Since early stages of embryonic development are the foundation for all subsequent stages, changes in the early stages are less likely to be tolerated - and so similarities tend to be preserved over time.
9. While it might seem inefficient to produce many cells to have only a few survive, it is critical to have the "choices" of all those potentially varied cells initially produced, so that the most appropriate ones for the task can survive in any given tissue or organism. And sometimes multiple cells are needed to accomplish a specific task, but when no longer necessary their continued growth and/or reproduction threaten the survival of the entire organism.
10. Where a cell "sits" or ends up after migration determines which other cells it can contact, both to receive information about gene expression (through direct contact, hormones or simple diffusion of small molecules) or to send information to other cells using the same mechanisms. These signals will determine the time and place for production of various components necessary in a differentiated cell or tissue.
11. 1) A cancer cell grows in an unregulated way, consuming resources such as food and space in the body of an animal and producing substances which damage other types of cells or induce their growth in the wrong time or place (e.g., blood vessel growth induced by solid tumor masses) and in general modifying the environment in the body so that normal cells cannot function. 2) When gypsy moths were introduced into the Northeastern United States, they destroyed trees by defoliation, consuming a resource that the rest of the community needed. Since trees serve many functions in a temperate climate forest (food and oxygen production by photosynthesis, shade for the forest floor under the canopy, homes for many animals, prevention of soil erosion, etc), the loss of these trees was a threat to the entire ecosystem. Likewise, rabbits introduced into Australian Outback quickly increased in numbers due to lack of natural predators, consuming the sparse vegetation, affecting competition among other species, and threatening the delicate ecosystem of this arid region.

Chapter 8

1. All creatures living today, from bacteria to humans, use DNA as information molecules, use the same 4 nucleotides to make their DNA (and RNA) and the same 20 amino acids to make their proteins, use the same genetic code, use the same methods of making proteins from DNA, and use similar mechanisms for generating energy and for many other processes. The probability that all these attributes evolved independently in separate organisms is vanishingly small.
2. Could the eye have arisen from no eye at all in a series of small steps each of which gave its possessor some incremental survival and reproductive advantage? The answer, inevitably, would seem to be yes: if there's enough time and the steps are small enough. Time is no problem given many millions of generations between us and our eyeless ancestors. It's common in life for nerve cells to connect with cells of other tissues. Suppose an eye began as a piece of transparent tissue on a body with a simple nerve connection to a primordial response circuit. Its possessor would benefit by sensing light and shadow and thereby avoiding potential danger. From there on incremental improvements, retained because they enhanced survival, would have led us to the eye.

The view that the eye either functions as a whole or not at all is at odds with our experience. In our own experience a small amount of vision is better than none. We're familiar with the value of partial vision in eyes with damaged corneas, lenses, and retinas. Furthermore, examination of the spectrum of modern animals reveals many different ways and degrees of seeing that seem to have evolved independently.
3. In a large population of bacteria mutations to antibiotic resistance occur spontaneously from time to time. The larger the population, the more frequent the event. In the absence of antibiotic the mutation is "silent" - it goes unnoticed. But in the presence of antibiotic the mutant survives and multiplies. This is a straightforward example of the chance appearance of a genetic trait that makes it possible for an organism to adapt to a hostile environment.
4. If a few wildebeests happened to get isolated in a separate and distinctly different environment, or if there were a sudden dramatic change in the population's habitual environment, different traits would be selected leading eventually to quite different animals adapted to their new setting.
5. Mutations are chance alterations in the sequence of nucleotides in DNA. There are many different kinds of such changes: deletion or addition of single nucleotides; substitution of one nucleotide for another; changes in or loss of longer stretches of nucleotides, etc. Such alterations may be caused by inevitable errors made by the replication machinery, as well as by radiation and chemicals in the environment impinging upon DNA.
6. New species arise when some members of a species get isolated by chance in a new environment: the other side of a mountain range or glacier, on an island or even on a continent that becomes separated from a larger land mass. The transplanted organisms thereafter evolve independently as they adapt, generation by generation, to their new habitat. New environments do not have to be geographical, however, they can be new chemical environments, as for a microorganism exposed to an antibiotic, or ecological, with any organism suddenly sharing its environment with a new type of predator or prey.
Major changes in environments produced by catastrophic events of many different kinds can wipe out many existing species and create new niches and opportunities for others.
7. Diploidy - i.e., having two copies of every gene, one from a mother and one from a father, is a kind of insurance policy. Damage to one copy leaves the other intact. Also, mixing genes from two parents can produce new genetic traits, conferring advantage. For example, mixing genetic information may contribute to resistance to disease. An invading organism may be foiled by the particular gene mix in a new generation.
8. Comparing the same stretches of DNA, or comparing the same proteins, taken from two organisms reveals differences in nucleotides or amino acids at certain locations. The fewer the differences, the more closely related the organisms.
9. Quite the contrary. Life takes advantage of the 2nd law to create order. The essence of life-building is creating chemical bonds. Life is an open system into which useful energy flows from the surrounding, is put to use, and is then dispersed as heat. Every time a bond is made, more energy goes into its formation than ends up in the bond. The excess energy is discarded as heat. Thus, the 2nd law is obeyed - useful energy is converted to less useful energy and the process - order is created.

Credits

Page 10 top By permission of the President and Council of the Royal Society. **Page 10 bottom** Used with permission of Jones and Bartlett Publishers, from *Living Images*, by Gene Shih and Richard Kessel. **Page 11 top** © Stanley Cohen/Science Photo Library/Photo Researchers **Page 11 bottom** Used with permission of Jones and Bartlett Publishers, from *Electron Microscopy*, 2E by John J. Bozzola and Lonnie D. Russell. **Page 12 Scanning electron micrographs** Used with permission of Jones and Bartlett Publishers, from *Electron Microscopy*, 2E by John J. Bozzola and Lonnie D. Russell. **Page 12 bacteria on pinpoint** © Tony Brain and David Parker/Science Photo Library/Photo Researchers. **Page 13** ©Rosalind Franklin/Cold Spring Harbor Laboratory Archives. **Page 15** Courtesy of Melissa Doppler. **Page 16** © Centre National de Recherches Iconographiques. **Page 17** © Molecules R Us. **Page 26 top** © John Cunningham/Visuals Unlimited. **Page 26 bottom** © Lee Simon/Science Photo Library/Photo Researchers. **Page 27 sponge** © Marc Rampulla/Peter Arnold. **Page 27 gecko** © Joe McDonald/Corbis. **Page 27 gecko foot** © Chris Mattison/Frank Lane Picture Agency/Corbis. **Page 30** © PhotoDisc. **Page 31** © S. Dalton/Animals Animals. **Page 34** © used with permission from W. H. Freeman from *Molecular Origin of Life* by Fox and Dose. **Page 35 top** NASA. **Page 35 bottom** U.S. Geological Survey. **Page 38** © Patrick Grace/Science Source/Photo Researchers. **Page 39 top left** © Kim Taylor/Bruce Coleman/PNI. **Page 39 top right** © Ralph Reinhold/Animals Animals. **Page 39 bottom** Used with permission from Art Siegel/University of Pennsylvania. **Page 46** © 1997 by Eric Sander. **Page 47** © John Doebley/Plimoth Plantation Agriculture. **Page 50** © Rockefeller University Press. **Page 51 top** © W. H. Freeman and Company. **Page 51 red blood cells** © Stanley Flegler/Visuals Unlimited. **Page 54** © George Bernard/Animals Animals. **Page 60 left** © Jim Mauseth. **Page 60 right** © Butch Gemin. **Page 64** © Charles Palek. **Page 65** © C. P. Hickman/Visuals Unlimited. **Page 68** © David M. Phillips/Visuals Unlimited. **Page 72** © Todd Barkman/University of Texas. **Page 78** © J. Frederick Grassle/Woods Hole Oceanographic Insitution. **Page 79 top** © Beth Davidow/Visuals Unlimited. **Page 79 bottom** NASA. **Page 82 top** Galileo Imaging Team/NASA. **Page 82 bottom** © IMP Team/JPL/NASA. **Page 107** National Museum of Art, Washington DC/Art Resource, NY. **Page 121** Used with permission of Jones and Bartlett Publishers, from *Living Images*, by Gene Shih and Richard Kessel. **Page 128** © E. R. Degginger. **Page 129 top** © Visuals Unlimited/Cabisco. **Page 129 bottom** © T. E. Adams/Visuals Unlimited. **Page 136 top** © L.L. Rue III/Earth Scenes. **Page 136 bottom** © S. Maslowski/Visuals Unlimited **Page 137** NASA. **Page 156** Used with permission of Jones and Bartlett Publishers, from *Essential Genetics* 2E, by Daniel L. Hartl and Elizabeth W. Jones. **Page 166** Used with permission of Jones and Bartlett Publishers, from *Essential Genetics* 2E, by Daniel L. Hartl and Elizabeth W. Jones. **Page 167 top** Courtesy Cellmark Diagnostics, Inc., Germantown, Maryland. **Page 167 bottom** © Mahlon Hoagland. **Page 172** Courtesy of Gerald Kingman. **Page 173** Used with permission of Jones and Bartlett Publishers, from *Essential Genetics* 2E, by Daniel L. Hartl and Elizabeth W. Jones. **Page 188** Used with permission from Springer-Verlag, from *Plant Cell Biology, Structure and Function*, by Brian S. Gunning and Martin W. Steer. **Page 189 top** © Linda Sims/Visuals Unlimited. **Page 189 bottom** Used with permission from Springer-Verlag, from *Plant Cell Biology, Structure and Function*, by Brian S. Gunning and Martin W. Steer. **Page 228 and 229** Courtesy of Richard Gomer. **Page 232** Courtesy of W. G. Eberhard. **Page 233** Courtesy of Chiappino, Nichols, and O'Conner. *Journal of Protozoology*, 31:228. 1984. **Page 238** © AP Photo/U.S. Department of Agriculture. **Page 239** National Museum of American Art, Washington DC/Art Resource, NY. **Page 240** Courtesy of USDA-ARS. **Page 247 left** © Hulton Deutsch/Corbis. **Page 247** right Used with permission of Jones and Bartlett Publishers, from *Living Images*, by Gene Shih and Richard Kessel. **Page 256** Used with permission from Springer-Verlag, from *Plant Cell Biology, Structure and Function*, by Brian S. Gunning and Martin W. Steer. **Page 257** Used with permission of Jones and Bartlett Publishers, from *Living Images*, by Gene Shih and Richard Kessel. **Page 273** Courtesy of James Mauseth, reprinted from *Botany, An Introduction to Plant Biology* 2E, by James Mauseth, with permission from Jones and Bartlett Publishers. **Page 276** © Richard G. Rawlins/Custom Stock Medical Photos. **Page 277** © Roslin Institute/PA Photos. **Page 284** Courtesy of Judith Austin. **Page 308** © John D. Cunningham/Visuals Unlimited. **Page 309** Courtesy of John M. Rensberger, reprinted with permission from Nature. **Page 318** © CNRI/SPL/Photo Researchers, Inc. **Page 319** © David P. Allison/Oak Ridge National Laboratory/Biological Photo Service. **Page 326** © D. M. Phillips/Visuals Unlimited. **Page 334** © Dell R. Foutz/Visuals Unlimited. **Page 338** © Merlin D. Tuttle/Bat Conservational International. **Page 346** Courtesy of Kathy Naylor.

Index

D

F

G

R

S